S0-BPJ-338

ELASTIC-PLASTIC FRACTURE

A symposium
sponsored by ASTM
Committee E-24 on
Fracture Testing of Metals
AMERICAN SOCIETY FOR
TESTING AND MATERIALS
Atlanta, Ga., 16-18 Nov. 1977

ASTM SPECIAL TECHNICAL PUBLICATION 668
J. D. Landes, Westinghouse R&D Center
J. A. Begley, The Ohio State University
G. A. Clarke, Westinghouse R&D Center
editors

04-668000-30

AMERICAN SOCIETY FOR TESTING AND MATERIALS
1916 Race Street, Philadelphia, Pa. 19103

Copyright © by AMERICAN SOCIETY FOR TESTING AND MATERIALS 1979
Library of Congress Catalog Card Number: 78-72514

NOTE

The Society is not responsible, as a body,
for the statements and opinions
advanced in this publication.

Printed in Tallahassee, Fla.
October 1972
Second Printing, July 1981
Baltimore, Md.

Ken Lynn

Dedication

It was with great sorrow and disbelief that we all learned of the sudden and untimely death of Ken Lynn during the summer of 1978. We have lost an imaginative and competent practitioner of the art of fracture mechanics who was able to cut through the many details of a problem and get to the essence of it to seek the practical solution. We have also lost a great friend who was intensely interested in the lives and achievements of his co-workers and contemporaries. It is with sincere appreciation for his fruitful technical life and his uplifting personal outreach that we dedicate this ASTM fracture mechanics symposium volume to his memory.

Ken grew up near Pittsburgh and in Florida; he received his B.S. in Mechanical Engineering in 1946 and M.S. in Engineering Mechanics in 1947 from Pennsylvania State University. His first employment was with the U.S. Steel Corporation, both in Kearny, New Jersey, and in Cleveland, Ohio, where he worked on brittle crack initiation and propagation in steels—a subject to which he would devote much of his efforts later in life. He was always proud of the fact that, while at U.S. Steel, he had established the strength of the cables which still support the original Delaware Memorial Bridge. In March of 1955, he joined the Lockheed Aircraft Corporation, and was employed at both the Burbank, California, and Marietta, Georgia, facilities. As a senior research engineer, he was in charge of structural materials research on the nuclear-powered bomber project as well as fatigue life prediction of aircraft wing structures. In August of

1957, he moved to the Rocketdyne Division of North American Rockwell Corporation in Canoga Park, California, where he began his serious development as a practitioner of fracture mechanics. Through a series of increasingly challenging assignments in experimental stress analysis and fracture mechanics evaluations, he became a lead consultant on structural problems and fracture mechanics for Rocketdyne hardware. A key responsibility of Ken's was for development of the fracture control plan for several critical Rocketdyne structures. It was at Rocketdyne that Ken became actively involved with ASTM, and with Committee E-24 in particular. He quickly recognized the consensus agreement value of the ASTM system and strongly promoted it. Ken's approach to ASTM was not to seek leadership, but rather to stay "down in the trenches" at the technical working level. He maintained this philosophy throughout his association with ASTM, especially in later years as he came to rely on ASTM E-24 more and more for consensus agreement. Ken next became intrigued by the technical challenges presented by the field of nuclear power generation. So, in January of 1971, he joined the Westinghouse PWR Division where he became deeply involved in applying advanced fracture mechanics techniques to the analysis of pressurized water components, mainly reactor pressure vessels. Because the nuclear industry was then in the process of upgrading safety analysis in terms of fracture mechanics, he eagerly helped promote the standardization of LEFM testing and analysis through ASTM. His Westinghouse experience led him to join the Atomic Energy Commission in August of 1972. At AEC he worked on applying fracture mechanics to thermal shock analysis problems and to flaw evaluation procedures which later were incorporated into the ASME Boiler and Pressure Vessel Code, Section XI. Recognizing greater opportunity for development and application of fracture mechanics, Ken joined the Division of Reactor Safety Research—now part of the Nuclear Regulatory Commission—whereupon he took over management of a series of research programs all directed at ensuring the safety of structures in the primary system of light water power reactors. Full of energy, Ken made many contributions to the understanding and application of fracture mechanics principles for the evaluation and solution of problems faced in primary system integrity. Included among these were thermal shock, crack arrest, crack growth rates, irradiation effects, and linear elastic and elastic-plastic analysis of vessels under overpressurization. With NRC, Ken undertook a front-line leadership of grounding technical advancements in fracture mechanics through ASTM Standards. His commitment to the ASTM E-24 Committee, and their efforts, was complete. He was especially looking forward to the ASTM standardization of test specimens and methods for both *crack arrest and for J-R curve testing of ductile steels, and personally assured that all work done under his direction was aimed at this goal.*

Because of his position as a program manager, Ken did not write many technical papers; he always felt that the individual researcher should take credit for the work, not himself. However, the technical literature today is filled with articles based on his understanding and direction of research and application in the field of fracture mechanics, and many acknowledgments and technical directions can be found in these papers. Because of his experience and competence in fracture mechanics, Ken was often asked to organize meetings and to chair some of the sessions. His summaries of the information presented and his conclusions and suggested directions were looked forward to, as we knew that if we did not understand what had happened, or what was truly significant, Ken usually did, and his evaluation would help to clarify the situation.

Ken was deeply devoted to his wife, Lois, and was thoroughly enjoying the experience of his two grandchildren, by his son David, who lives in Denver; and by his daughter, June Mesnik, who lives in Los Angeles. He was quite proud of his other daughter, Carol, and thoroughly enjoyed competing against his two younger sons, Gordon and Jerry, at golf or pool. In both his technical and personal life, Ken always strove for perfection and always challenged himself and his family to the same end. One of the true joys of his last few years was to be able to take Lois with him on several business trips to Europe, where they renewed many acquaintances they had made with Ken's contemporaries, who looked to him for technical leadership in fracture analysis of reactors, and also for good times after the job was done. At the time of his death, Ken was planning for several ASTM Meetings where crack arrest, fracture toughness, and crack growth rates were approaching, to his great satisfaction, true national and international standardization. We will no longer have the benefit of his contributions to his chosen discipline, and we will miss them. But most of all, we will miss Ken himself.

Foreword

The symposium on Elastic-Plastic Fracture was held in Atlanta, Georgia, 16-18 Nov. 1977. The symposium was sponsored by ASTM Committee E-24 on Fracture Testing of Metals. J. D. Landes, Westinghouse Research and Development Center, J. A. Begley, The Ohio State University, and G. A. Clarke, Westinghouse Research and Development Center, presided as symposium chairmen. They are also editors of this publication.

Related ASTM Publications

Cyclic Stress-Strain and Plastic Deformation Aspects of Fatigue Crack Growth, STP 637 (1977), $25.00, 04-637000-30

Use of Computers in the Fatigue Laboratory, STP 613 (1976), $20.00, 04-613000-30

Handbook of Fatigue Testing, STP 566 (1974), $17.25, 04-566000-30

References on Fatigue, 1965–1966, STP 9P (1968), $11.00, 04-0009160-30

A Note of Appreciation to Reviewers

This publication is made possible by the authors and, also, the unheralded efforts of the reviewers. This body of technical experts whose dedication, sacrifice of time and effort, and collective wisdom in reviewing the papers must be acknowledged. The quality level of ASTM publications is a direct function of their respected opinions. On behalf of ASTM we acknowledge with appreciation their contribution.

ASTM Committee on Publications

Editorial Staff

Jane B. Wheeler, *Managing Editor*
Helen M. Hoersch, *Associate Editor*
Ellen J. McGlinchey, *Senior Assistant Editor*
Helen Mahy, *Assistant Editor*

Contents

Introduction   1

ELASTIC-PLASTIC FRACTURE CRITERIA AND ANALYSIS

Instability of the Tearing Mode of Elastic-Plastic Crack Growth—
P. C. PARIS, H. TADA, Z. ZAHOOR, AND H. ERNST   5

The Theory of Stability Analysis of J-Controlled Crack Growth—
J. W. HUTCHINSON AND P. C. PARIS   37

Studies on Crack Initiation and Stable Crack Growth—C. F. SHIH, H. G. DELORENZI, AND W. R. ANDREWS   65

Elastic-Plastic Fracture Mechanics for Two-Dimensional Stable Crack Growth and Instability Problems—M. F. KANNINEN, E. F. RYBICKI, R. B. STONESIFER, D. BROEK, A. R. ROSENFIELD, C. W. MARSCHALL, AND G. T. HAHN   121

A Numerical Investigation of Plane Strain Stable Crack Growth Under Small-Scale Yielding Conditions—E. P. SORENSON   151

On Criteria for J-Dominance of Crack-Tip Fields in Large-Scale Yielding—R. M. MCMEEKING AND D. M. PARKS   175

A Finite-Element Analysis of Stable Crack Growth—I—M. NAKAGAKI, W. H. CHEN, AND S. N. ATLURI   195

A Comparison of Elastic-Plastic Fracture Parameters in Biaxial Stress States—K. J. MILLER AND A. P. KFOURI   214

Numerical Study of Initiation, Stable Crack Growth, and Maximum Load, with a Ductile Fracture Criterion Based on the Growth of Holes—Y. D'ESCATHA AND J. C. DEVAUX   229

EXPERIMENTAL TEST TECHNIQUES AND FRACTURE TOUGHNESS DATA

An Initial Experimental Investigation of the Tearing Instability Theory—P. C. PARIS, H. TADA, A. ZAHOOR, AND H. ERNST   251

Evaluation of Estimation Procedures Used in J-Integral Testing—
J. D. LANDES, H. WALKER, AND G. A. CLARKE   266

An Evaluation of Elastic-Plastic Methods Applied to Crack Growth Resistance Measurements—D. E. MCCABE AND J. D. LANDES   288

Elastic-Plastic Fracture Toughness Based on the COD and J-Contour Integral Concepts—M. G. DAWES   306

J-Integral Determinations and Analyses for Small Test Specimens and Their Usefulness for Estimating Fracture Toughness—J. ROYER, J. M. TISSOT, A. PELISSIER-TANON, P. LE POAC, AND D. MIANNAY   334

Effect of Size on the J Fracture Criterion—I. MILNE AND G. G. CHELL   358

Determination of Fracture Toughness with Linear-Elastic and
    Elastic-Plastic Methods—C. BERGER, H. P. KELLER, AND
    D. MUNZ   378
Minimum Specimen Size for the Application of Linear-Elastic
    Fracture Mechanics—D. MUNZ   406
Thickness and Side-Groove Effects on J- and δ-Resistance Curves for
    Steel at 93°C—W. R. ANDREWS AND C. F. SHIH   426
Computer Interactive J_{Ic} Testing of Naval Alloys—J. A. JOYCE AND
    J. P. GUDAS   451
Characterization of Plate Steel Quality Using Various Toughness
    Measurement Techniques—A. D. WILSON   469
Static and Dynamic Fibrous Initiation Toughness Results for Nine
    Pressure Vessel Materials—W. L. SERVER   493
Dynamic Fracture Toughness of ASME SA508 Class 2a Base and
    Heat-Affected-Zone Material—W. A. LOGSDON   515
Tensile and Fracture Behavior of a Nitrogen-Strengthened,
    Chromium-Nickel-Manganese Stainless Steel at Cryogenic
    Temperatures—R. L. TOBLER AND R. P. REED   537
Fracture Behavior of Stainless Steel—W. H. BAMFORD AND A. J. BUSH   553

APPLICATIONS OF ELASTIC-PLASTIC METHODOLOGY

A Procedure for Incorporating Thermal and Residual Stresses into
    the Concept of a Failure Assessment Diagram—G. G. CHELL   581
The COD Approach and Its Application to Welded Structures—
    J. D. HARRISON, M. G. DAWES, G. L. ARCHER, AND
    M. S. KAMATH   606
Fracture Mechanics Analysis of Pipeline Girthwelds—H. I. MCHENRY,
    R. T. READ, AND J. A. BEGLEY   632
An Elastic-Plastic R-Curve Description of Fracture in Zr-2.5Nb
    Pressure Tube Alloy—L. A. SIMPSON AND C. F. CLARKE   643
Correlation of Structural Steel Fractures Involving Massive
    Plasticity—B. D. MACDONALD   663
An Approximate Method of Elastic-Plastic Fracture Analysis for
    Nozzle Corner Cracks—J. G. MERKLE   674
Elastic-Plastic Fracture Mechanics Analyses of Notches—
    M. M. HAMMOUDA AND K. J. MILLER   703
Size Effects on the Fatigue Crack Growth Rate of Type 304 Stainless
    Steel—W. R. BROSE AND N. E. DOWLING   720
Use of a Compact-Type Strip Specimen for Fatigue Crack Growth
    Rate Testing in the High-Rate Regime—D. F. MOWBRAY   736

SUMMARY

Summary   755
Index   767

Introduction

Interest in elastic-plastic fracture mechanics grew as a natural extension of linear elastic fracture mechanics (LEFM) concepts when it became obvious that LEFM methods were not adequate to handle many problems in the design and reliability analysis of structural components. Some of the early elastic-plastic fracture parameters and crack-tip analyses were developed in the 1960's; in the early 1970's, however, work on elastic-plastic fracture characterization was greatly expanded. Many new parameters and methods of fracture prediction were introduced and interest in this topic became widespread.

This publication represents papers presented at the ASTM Committee E-24 sponsored Symposium on Elastic-Plastic Fracture held in Atlanta, Ga., in November 1977. The symposium was organized to provide a forum for presenting current work in this rapidly developing field. No single approach was taken in the papers presented; rather, a variety of parameters and methodologies was presented. For the most part the papers presented new approaches and new data; some of the papers presented summaries and applications of existing approaches.

The symposium was very successful in that a good cross section of workers presently engaged in elastic-plastic fracture studies was represented. Most of the methods presently being used were discussed. The work presented herein is a fairly accurate account of the present status of the elastic-plastic fracture field, which status is that work is progressing at a fairly rapid pace, new ideas are frequently introduced, different approaches are being attempted, and to date no single method has been adopted by all of the workers in this field.

The contents of this publication will be particularly useful to persons working in the elastic-plastic fracture field. This would include researchers involved in material property studies and structural analysis, designers, and persons concerned with safety and licensing. The contents of the book do not so much represent an end product in the development of elastic-plastic fracture; rather, they represent a step in the development of this field which should be followed by other important publications on the subject. Some of the papers may become dated as the technology advances and present techniques are discarded for new ones, while other papers may have more permanent value, marking the first introduction of a significant new concept.

Three major areas were covered in the symposium: fracture criteria and analysis; experimental evaluation and toughness testing; and applications of elastic-plastic methodology, including the application of elastic-plastic fracture concepts to fatigue crack growth analysis. The analysis papers dealt mainly with the assessment of new and existing criteria. The present emphasis is on extending fracture prediction based on an initiation criterion to include the characterization of stable crack growth and ductile instability in the fracture process. Some of the criteria are mainly empirical while others are based on the postulation of a fracture mechanism. Finite-element analysis remains the most popular method for evaluating crack-tip behavior in the elastic-plastic regime.

Fracture toughness test results were directed at determining properties of materials for specific applications and at evaluating present fracture criteria. Many of the materials evaluated were steels used in the nuclear industry. Of particular interest were pressure vessel steels tested under dynamic loading and stainless steels. Experimental evaluation of existing fracture criteria dealt with the evaluation of test specimen size, the evaluation of analysis methods, and the use of advanced testing methods such as computer-based data acquisition and reduction systems.

The application of elastic-plastic techniques to the evaluation of structural components is directed toward fracture problems in pressure vessels, pipelines, and other structural members. Specific areas of given structures were often considered, generally areas of high stress concentration such as nozzle corners and notches. Special note was given to the application of elastic-plastic techniques to fatigue crack growth studies, particularly in the high-strain low-cycle regime.

The variety of topics covered should be of interest to a large number of researchers working in the elastic-plastic area. This publication represents the first major collection of papers devoted solely to the topic of elastic-plastic fracture.

J. D. Landes
Westinghouse Electric Corp. Research and Development Center, Pittsburgh, Pa.; co-editor.

Elastic-Plastic Fracture Criteria
and Analysis

P. C. Paris,[1] *H. Tada,*[1] *A. Zahoor,*[1] *and H. Ernst*[1]

The Theory of Instability of the Tearing Mode of Elastic-Plastic Crack Growth

REFERENCE: Paris, P. C., Tada, H., Zahoor, A., and Ernst, H., "**The Theory of Instability of the Tearing Mode of Elastic-Plastic Crack Growth,**" *Elastic-Plastic Fracture, ASTM STP 668,* J. D. Landes, J. A. Begley, and G. A. Clarke, Eds., American Society for Testing and Materials, 1979, pp. 5-36.

ABSTRACT: This paper presents a new approach to the subject of crack instability based on the J-integral R-curve approach to characterizing a material's resistance to fracture. The results are presented in the chronological order of their development (including Appendices I and II).

First, a new nondimensional material parameter, T, the "tearing modulus," is defined. For fully plastic (nonhardening) conditions, instability relationships are developed for various configurations, including some common test piece configurations, the surface flaw, and microflaws. Appendix I generalizes these results for the fully plastic case and Appendix II treats confined yielding cases.

The results are presented for plane-strain crack-tip and slip field conditions, but may be modified for plane-stress slip fields in most cases by merely adjusting constants. Moreover, an accounted-for compliance of loading system is included in the analysis.

Finally, Appendix III is a compilation of tearing modulus, T, properties of materials from the literature for convenience in comparing the other results with experience.

KEY WORDS: crack instability, tearing instability, tearing modulus, elastic-plastic fracture, J-integral fracture mechanics, crack propagation

It has become common to characterize a material's static crack extension behavior under monotonically increasing deformation using a J-integral R-curve [1-4].[2] However, unlike the situation with linear-elastic (K-type) R-curves [5], crack instability phenomena have yet to be analyzed in terms of J-integral R-curves.

In low-strength steels, a common material for characterization via J-integral R-curves, there appears to be two distinct possibilities for instability. First is the so-called "cleavage" instability which is normally attributed to a local material instability on a microscopic scale (such as inclusion spac-

---

[1]Professor of mechanics, senior research associate, and graduate research assistants, respectively, Washington University, St. Louis, Mo.
[2]The italic numbers in brackets refer to the list of references appended to this paper.

ing). The second type of instability is associated with the global conditions in a test specimen or component and loading arrangement providing the driving force to cause continuous crack extension by a so-called "tearing" mechanism. Cleavage is associated with very flat fracture on crystalline planes whereas tearing is normally associated with dimpled rupture mechanisms on a microscopic level.

Moreover, in testing compact tension specimens to produce J-integral R-curves (for example, see Ref 2), cleavage is associated with a sudden instability where the crack jumps ahead, severing the test piece almost instantaneously. At low temperature this occurs prior to the beginning of tearing. At higher temperatures, just above transition, steady tearing commences first, followed after some amount of stable tearing by the sudden cleavage instability. At yet higher temperatures, much more extensive stable tearing occurs prior to cleavage, if the sudden cleavage occurs at all. (These patterns of behavior are more fully described in a later work [6].)

Indeed, the patterns of instability behavior versus stable tearing do not seem to be well understood, although attempts by Rice and co-workers [7,8] have provided some analysis of the mechanics of the situation. Herein a simple approach is taken to the mechanics of potential instability associated with the steady tearing portion of J-integral R-curves. The analysis is developed from simple examples of structural component (or test specimen) configurations with cracks, examining their instability possibilities individually, in order to draw more general conclusions about elastic-plastic cracking instability as contrasted to linear-elastic behavior. Finally, an attempt is made to model a more local cleavage-like instability for material in the fracture process zone just ahead of a crack tip.

The J-Integral R-Curve

It has been noted in many recent works that J may be interpreted as the intensity of the elastic-plastic deformation and stress field surrounding a crack tip (for example see Refs 3 or 4). As J is increased for a given crack situation, the response to increasing J is crack extension. Following Fig. 1, the crack extension first takes on the form of some minor lengthening due to flow and blunting of the tip until the conditions for tearing develop, whereupon increments of tearing extension, da, proceed approximately linearly with added increments of J, or dJ. As noted previously, sudden cleavage may occur either before or after the commencement of stable tearing, or not at all, depending on the temperature and material.

Here, at least initially, attention shall be directed to the stable tearing portion of the J-integral R-curve. It has been noted that plotting this R-curve, using J/σ_0 instead of J as the ordinate, results in a blunting and stable tearing behavior that is reasonably independent of temperature [6,9] for a given

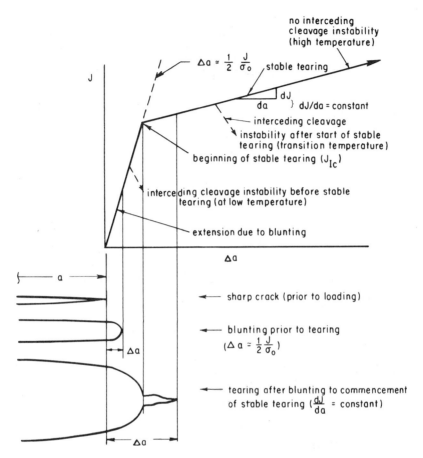

FIG. 1—*J-integral R-curve (with some diagrammatic details following Rice [4]).*

material and plane-strain conditions. This is simply schematically illustrated in Fig. 2.

Plane-strain conditions are, as usual [*1,3,4*], assumed to be present as long as the test specimen (or component) thickness and remaining ligament are large enough, by the criterion

$$\text{size} \geq 25 J/\sigma_0 \tag{1}$$

If this condition is met, then the J-integral R-curve is size independent [*2*] and is also reasonably independent of configuration with some reservations [*10*].

8 ELASTIC-PLASTIC FRACTURE

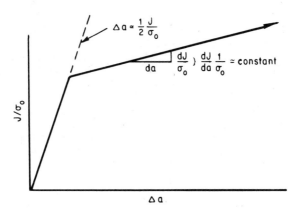

FIG. 2—*Temperature-independent plot of the blunting and stable tearing portion of the J-integral R-curves of a given material.*

Therefore, for present purposes, we may characterize a material's stable tearing properties by

$$\frac{dJ}{da} = constant \text{ (temperature dependent)}$$

or by (2)

$$\frac{dJ}{da} \times \frac{1}{\sigma_0} = constant \text{ (temperature independent)}$$

but most importantly, as shall be observed later, let

$$T = \frac{dJ}{da} \times \frac{E}{\sigma_0^2} = constant \qquad (3)$$

where T shall be termed the *tearing modulus* of the material. See Appendix III for the values of these material characteristics. Refer to the J-integral R-curves, plotted such as Figs. 1 or 2 from actual test data, to provide the values. Now it is acknowledged that the straight-line stable tearing portion of the R-curves in Figs. 1 and 2 involves perhaps some idealization, but that view is sufficient herein.

Moreover, although that portion of the R-curve is termed "stable tearing," it shall later be noted inherently as "stable" only for tests involving substantial bending loading imposed on a remaining ligament, such as deeply cracked compact tension or bend specimens. Indeed, the discussion now turns to other configurations where instability may be possible on the tearing portion of the R-curve; thus, it may simply be termed "tearing" or "steady tearing" rather than "stable tearing."

An Analysis of Possible Tearing Instabilities

Consider a center-cracked strip with crack length, $2a$, width, W, thickness, B, and length, L, as illustrated by Fig. 3. It is presumed that fully plastic behavior occurs, that is, the slip lines develop, prior to crack extension by the tearing behavior discussed previously. The appropriate slip lines are simply 45-deg slips from the crack tips to the edge, giving a limit load, P_L, by

$$P_L = \sigma_0(W - 2a)B \tag{4}$$

where for this case the flow stress, σ_f, on the remaining ligament is simply the flow stress for simple tension, σ_0.

The increased crack opening stretch, δ_T, at each crack can be viewed as contributed directly by the slips as they operate, increasing the length of the specimen, L and ΔL, by a like amount. Incrementally, that is

$$d(\Delta L_{\text{plastic}}) = d(\delta_T) \tag{5}$$

an increase in δ_T implies a corresponding increase in J by the usual relationship [3, 4, 7]

$$\delta_T = \alpha J/\sigma_0 \text{ or } d\delta_T = \alpha \frac{dJ}{\sigma_0} \tag{6}$$

where, for convenience, α will be taken as approximately 1 (α is more nearly 0.7 for plane strain).

Combining Eqs 5 and 6

$$d(\Delta L_{\text{plastic}}) = \frac{d(J)}{\sigma_0} \tag{7}$$

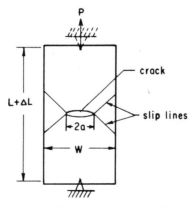

FIG. 3—*Fully-plastic center-cracked strip.*

Now the increment $d(J)$, increase in J, would imply a crack extension da from Eq 2 or Eq 3. As a consequence, from Eq 4 the limit load would be reduced by

$$dP_L = -2\sigma_0 daB \qquad (8)$$

and this load reduction would further imply an increment of elastic shortening of the length of the specimen[3]

$$d(\Delta L_{\text{elastic}}) = \frac{dP_L L}{AE} = \frac{-2\sigma_0 daL}{WE} \qquad (9)$$

Now, if the specimen were being tested in a rigid machine (fixed grips or end displacements) instability would ensue if the magnitude of elastic shortening exceeded the corresponding plastic lengthening required for crack extension. Equating Eqs 7 and 9 leads to the criterion of

$$T = \frac{dJ}{da} \times \frac{E}{\sigma_0^2} \leq \frac{2L}{W} \quad \text{(for instability of a center-cracked strip in tension)} \qquad (10)$$

for instability. The left-hand side of Eq 10 depends only on material properties, with dJ/da to be supplied from the tearing slope of the J-integral R-curve as implied in Eq 2 or Eq 3. The right-hand side of Eq 10 is a nondimensional configuration parameter. Before discussing the detailed implications of this instability criterion, Eq 10, further, consider first other configurations.

For instability of a fully plastic double-edge notched strip (see Fig. 4) the analysis proceeds identically except that the flow stress at the cracked section, σ_f, is elevated by the nature of the slip field (implied constraint) and is about three times the flow stress in a simple tension test, σ_0 [7,8,11]; thus since

$$\sigma_f \simeq 3\sigma_0 \text{ (double edge cracked strip)}$$

and for the rigid-plastic velocity field of a double-edge cracked strip

$$d(\Delta L_{\text{plastic}}) = \tfrac{1}{2} d\delta_t \qquad (11)$$

resubstituting as appropriate in the analysis preceding Eq 10 leads to

$$T = \frac{dJ}{da} \cdot \frac{E}{\sigma_0^2} \leq \frac{12L}{W} \quad \text{(for instability of a double-edge cracked strip in tension)} \qquad (12)$$

---
[3]Neglecting for the moment the elastic shortening due to the crack, which would be small here at any rate.

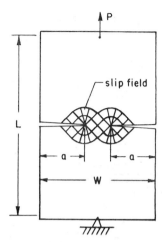

FIG. 4—*Double-edge cracked strip.*

It is noted that for the double-edge cracked strip the notches must be deep enough to induce the slip field shown in Fig. 4, if the analysis leading to Eqs 11 and 12 is to be correct. This simply requires $a/W \gg 1/3$.

Moreover, with both of the previous illustrations, center-cracked and double-edge cracked strips, tearing is likely to proceed in an unsymmetrical manner, causing bending and further undermining the stability of the situation. Thus in Eqs 10 and 12 the numerical coefficient on the right-hand side should be regarded as a lower limit for commencement of tearing instability. It is estimated that, due to bending, this numerical coefficient might be as much as a factor of 2 larger.

Approximate Analysis of Tearing Stability for a Deep Surface Flaw

A case of considerable interest in pressure vessels and other applications is that of a deep surface flaw where yielding ensues over the remaining ligament. An approximate analysis based on some methods first used by Irwin [12] is attempted here. Let the surface flaw be described as having a depth, a, and length, l, in a plate of thickness, t, subject to an applied stress, σ, as illustrated in Fig. 5. The stress, σ, is presumed to be below the yield stress, but at the remaining ligament, $t-a$, behind the crack it is presumed to induce flow, as noted by the shaded area in the Fig. 5 plan view. That is to say, a condition for the analysis is

$$\frac{\sigma_f(t-a)}{t} \leq \sigma < \sigma_0 \qquad (13)$$

where again, due to the nature of the slip field (see the side view in Fig. 5),

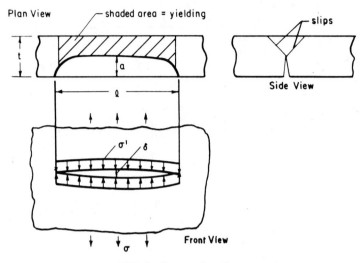

FIG. 5—*Deep surface flaw.*

the flow stress, σ_f, is just the yield strength in simple tension, σ_0. For a deep flaw this is not an unrealistic restriction.

Now the front view of Fig. 5 may be regarded as an elastic through-the-thickness crack with an average (through-the-thickness averaged) closure stress, σ', assisting in holding back the opening displacement, δ, at the center, where tearing instability shall be examined. Equilibrium gives

$$\sigma' t = \sigma_0 (t - a) \tag{14}$$

From elastic analysis, the opening displacement, δ, is

$$\delta = \frac{2l}{E}(\sigma - \sigma') \tag{15}$$

Combining Eqs 14 and 15 and noticing that for a deep crack ($a/t \geq \frac{1}{2}$) the displacement δ is a conservative estimate of the crack opening stretch

$$\delta_t \simeq \delta = \frac{2l}{E}\left[\sigma - \sigma_0\left(1 - \frac{a}{t}\right)\right] \tag{16}$$

Again, recalling the relationship of δ_t and J or

$$\delta_t \simeq \frac{J}{\sigma_0} \tag{17}$$

and examining for instability under a crack extension, da, from the corresponding, dJ, by combining Eqs 16 and 17, and differentiating, leads to[4]

$$T = \frac{dJ}{da} \times \frac{E}{\sigma_0^2} \leq \frac{2l}{t} \quad \text{(for tearing instability of a deep surface flaw)} \quad (18)$$

Note the similarity of this result, Eq 18, with the previous results, Eqs 10 and 12. A first observation is that

$$T = \frac{dJ}{da} \times \frac{E}{\sigma_0^2}$$

keeps reappearing as the left-hand member of the equation and may be interpreted as characterizing the material's resistance to tearing instability. On the other hand, the right-hand sides of these instability criteria are dependent on a nondimensional geometrical factor for the configuration plus a numerical factor which in part depends on the geometrical character of the flow field that develops (that is, the constraint).

Also very important and curious to note is that none of these instability criteria, Eqs 10, 12, and 18, contain the crack size, a. This is very unlike linear-elastic fracture mechanics instability criteria, where for the same geometrical configurations the crack size is strongly present! This rather striking difference will be further interpreted later, especially where potential explanations of fracture triggered by tiny flaws in a necked tensile bar or immediately ahead of a crack or notch are concerned.

The instability criterion for the surface flaw, Eq 18, has been formulated in a way that is conservative for assuring stability. First, the crack opening stretch, δ_t, was estimated, Eqs 16 and 17, only at the center of the crack border of the surface flaw; for deep flaws the estimate was conservative (too high) when considerations of bending due to the crack were made. Further, incompatibility of the slip fields at the ends of the crack assures a slight underestimate of the effective σ' in Eqs 14-16. Finally, the applied stress, σ, would be likely to diminish as crack growth occurs or, at most, remain constant. Thus, because of this nature, Eq 18 is put forward as a reasonable estimate for assuring against tearing instability for fully plastic surface flaws.

On the other hand, the instability criteria formulated for center-cracked and double-edge cracked strips, Eqs 10 and 12, assumed a rigid loading apparatus (fixed grips), and, unlike the surface flaw, they have a higher propensity for instability if the loading apparatus is flexible. This effect of the testing machine stiffness could be included as an additional term on the right-hand sides of Eqs 10 and 12; see Appendix I.

---

[4]An increment dl can simultaneously be considered but does not add significantly to results.

Tearing Instability in Bending

The 3-point bend specimen or its equivalent, the deeply cracked double-cantilever beam, shall now be considered; see Fig. 6. The remaining ligament, b, at the crack section is basically subject to pure bending and is also assumed to be small enough that all plasticity is confined to that region. This implies $b/W \leq 0.68$ for 3-point bending and for the double-cantilever beam.

The limit bending moment, M_L, is assumed to occur at the remaining ligament prior to crack extension, and from Green and Hundy [13] it is

$$M_L = P_L S = 0.35 \sigma_0 B b^2 \qquad (19)$$

With crack extension, da (or $-db$), the limit load, P_L, diminishes, differentiating Eq 19, by

$$dP_L = \frac{0.35 \sigma_0 B}{S} (-2b\,da) \qquad (20)$$

Again assuming a rigid loading apparatus during crack extension, simple beam theory shall be used for analysis of flexing of the elastic ends of the specimens. The reduction of beam bending due to diminishing load, dP_L, while the load-point displacements are fixed implies a further rotation, $d\theta$, to be absorbed at the plastic sections, where

$$d\theta = \frac{2d\Delta}{S} = \frac{2dP_L S^2}{3EI} \qquad (21)$$

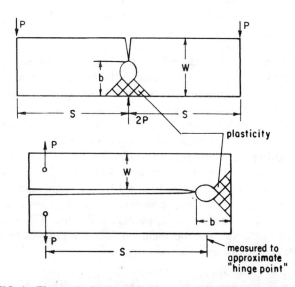

FIG. 6—*Three-point bend and deeply cracked double-cantilever specimens.*

where

$$I = \frac{BW^3}{12}$$

Now an increased rotation while at limit load further implies an increment of J, that is, dJ, by the usual Rice et al ([14] or [3]) pure bending analysis. While crack extension, da, implies a decrease

$$dJ = \frac{2}{Bb} M_L d\theta - \theta \sigma_0 da \qquad (22)$$

Simply combining Eqs 19–22 leads to

$$T = \frac{dJ}{da} \times \frac{E}{\sigma_0^2} \leq \frac{4b^2 S}{W^3} - \frac{\theta E}{\sigma_0} \text{ (for tearing instability in 3-point bending)} \qquad (23)$$

Now it is interesting here to notice that the remaining ligament size, b, comes into the instability criterion, Eq 23. Indeed, if instability occurs, b will then diminish with crack extension, and stability is regained. (The term $\theta E/\sigma_0$ is a small adjustment in the effect.) This is often observed in bending type (including deeply cracked compact tension) tests. Instability occurs and the crack runs rapidly but arrests before severing the ligament completely. This could occur with the tearing mechanism (not cleavage) in the following way.

Referring to Fig. 1, suppose that full plasticity at the ligament develops while on the initial blunting part of the R-curve. Deformation, that is, increasing J, continues without instability until the beginning of stable tearing is reached. At that point, the slope dJ/da suddenly changes to the value for "stable tearing," but if Eq 23 now predicts instability, sudden unstable crack growth should occur by tearing, until b becomes small enough to regain stability. However, if b were small enough in the first place, the situation would have remained stable throughout and the specimen would have been severed by slow stable tearing as loading (deformation) progressed.

Initial Note on Tearing Versus Cleavage Instability and Other Effects

Considerations that unstable tearing, implying enormous strain rates at the crack tip, might trigger cleavage in rate-sensitive materials are self-evident. But referring to the preceding paragraph, this possibility is shown to be ligament size, b, dependent in bending. The implications here are vast and it is seen especially that many past conclusions about the nature of cleavage may be in error. Indeed, sudden fracture might often be a result of a tearing instability, which rapidly changes to cleavage, so that cleavage might be an effect, not the cause of the instability. Indeed, cleavage fractures initially bordered by dimpled rupture are often observed [15].

Moreover, up to this point, the tearing portion of the R-curve in Fig. 1 has been regarded as a straight line. However, many experimentally determined R-curves appear to be curved concave downward as tearing progresses extensively; see Fig. 7 [16]. This implies that at least in some cases, dJ/da, the tearing slope, may diminish with crack extension, Δa. If so, then considering the instability criterion for bending, Eq 23, if dJ/da diminishes faster than

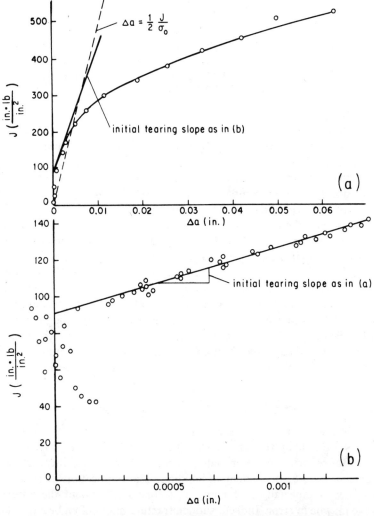

FIG. 7—*R-curve for 5083 aluminum alloy* [16] (a) *large scale* (b) *small scale.*

b^2, continuing instability would ensue in spite of the tendency for bend tests to regain stability. For such material behavior, advanced tearing instability considerations are appropriate.[5]

Approximate Tearing Instability Analysis of the Deeply Cracked Compact Specimen

In contrast to the previous analysis of 3-point bending, where the elastic bending of the beam provides the driving mechanism for instability, the deeply cracked compact tension specimen will now be considered. Indeed, if the specimen is deeply cracked, all of the significant deformations, both elastic and plastic, occur at the remaining ligament or very nearby. Referring to Fig. 8, the load-point displacement, Δ, is caused by rotation, θ, centered at about the middle of the remaining ligament or

$$\Delta = \theta(W - b/2) \tag{24}$$

Now the angle change θ is made up of its elastic, θ_{EL}, and plastic, θ_{PL}, parts, or Eq 24 becomes

$$\Delta = (\theta_{EL} + \theta_{PL})(W - b/2) \tag{25}$$

It is again assumed that the test machine and fixtures are rigid, or, examining for instability by presuming an increment of crack extension da (or $-db$), during that increment

$$d\Delta = 0 = \frac{\partial \Delta}{\partial \theta} d\theta + \frac{\partial \Delta}{\partial b} db$$

or (26)

$$0 = (d\theta_{EL} + d\theta_{PL})\left(W - \frac{b}{2}\right) + (\theta_{EL} + \theta_{PL})\left(-\frac{db}{2}\right)$$

Substituting from Eq 25 into Eq 26 and rearranging

$$d\theta_{EL} + d\theta_{PL} = \frac{\Delta}{2(W - b/2)^2} db \tag{27}$$

Assuming pure bending of the remaining ligament, the elastic deformation is [17]

---

[5]Indeed 5083 aluminum alloy is being used for liquefied natural gas storage tanks in large ships, where the hazards are enormous, but with minimal consideration of possible crack instabilities.

18   ELASTIC-PLASTIC FRACTURE

$$\theta_{EL} = \frac{16M}{EBb^2} \tag{28}$$

which is assumed to be occurring during limit load conditions, or the limit moment, M_L, is (as before)

$$M = M_L = 0.35\, \sigma_0 B b^2 \tag{29}$$

Combining Eqs 28 and 29

$$\theta_{EL} = \frac{16(0.35)\, \sigma_0}{E} \tag{30}$$

which is a constant unaffected by ligament changes, db, or

$$d\theta_{EL} = 0 \tag{31}$$

Now reexamining Eq 27, noting Eq 31, and that Δ and $(W - b/2)$ are positive, but that db is negative, the conclusion is

$$d\theta_{PL} = \text{negative} \tag{32}$$

This would imply a reduction in angle of plastic deformation or, consequently, a reduction in J. Indeed this shows that the situation is *always stable* when considering tearing instability of a very deeply cracked compact specimen in a rigid testing machine.

Tearing Instability Analysis of the Notched Round Bar

An instability analysis of the notched (cracked) round bar can proceed in the same manner for either internal or external notches; see Fig. 9. As

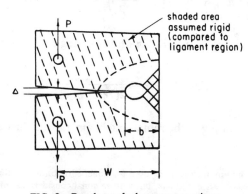

FIG. 8—*Deeply cracked compact specimen.*

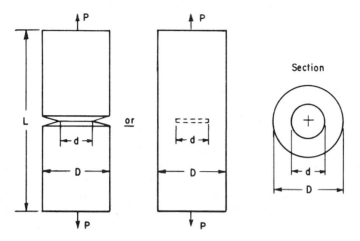

FIG. 9—*Notched round bar.*

before, a rigid test machine and fixtures are assumed. The analysis proceeds very much as with Eqs 4-10. Here the change in limit load with respect to crack size is

$$dP_L = \sigma_f \pi d\, da \tag{33}$$

or the elastic shortening is

$$d\Delta_{EL} = \frac{dP_L L}{\frac{\pi D^2}{4} E} \tag{34}$$

and, with no total shortening, the plastic lengthening is converted to crack opening stretch or

$$\beta d\delta_t = d\Delta_{PL} = -d\Delta_{EL} \tag{35}$$

As usual the change in δ_t implies an increment of J by

$$dJ = \sigma_0 d\delta_t \tag{36}$$

Combining and rearranging

$$T = \frac{dJ}{da} \times \frac{E}{\sigma_0^2} \leq \frac{4}{\beta}\left(\frac{\sigma_f}{\sigma_0}\right)\frac{Ld}{D^2} \quad \text{(for tearing instability of notched round bars)} \tag{37}$$

For *internal notches*

$$\frac{1}{\beta}\frac{\sigma_f}{\sigma_0} \simeq 1$$

whereas for *external notches*

$$\frac{1}{\beta}\frac{\sigma_f}{\sigma_0} > 6 \quad \left(\text{for } \frac{d}{D} < \frac{1}{\sqrt{3}}\right)$$

due to the constraint implied at the plastic section for each of these types of specimens. Of course a less than rigid test machine and fixtures, and bending and unsymmetrical cracking, all contribute to increasing the possibilities for instability in the notched round bar. Therefore, it is doubtful that, for external notches, the diminishing of d on the right-hand side of Eq 37 will in reality restore tearing stability under actual test conditions.

A Note on Instability of Microcracks at the Necked Section of a Tensile Bar

Consider Eq 37 or modifications of it for possible application to instability of tiny flaws at the necked section of a tensile bar at fracture. For tiny flaws, it is evident that the flow fields will be virtually undisturbed by the flaws while extreme deformation of the material takes place, especially at the neck. Initially, well before necking, the flow stress, σ_f, is by definition, σ_0, but as intense deformation occurs, it is evident in both Eq 33 and Eq 36 that it becomes the true stress at fracture, $\sigma_{t\,\text{fracture}}$. Meanwhile, the tearing resistance dJ/da might well diminish in the highly deformed material. Both of these considerations drive Eq 37 toward instability.

Consideration of the Possible Tearing Instability of Microflaws Ahead of a Crack Tip

In the plane-strain region ahead of a crack tip, many authors have depicted the slip field; for the most extensive analysis, see Rice [18], McClintock [19], or Kachanov [11], and, for effects of large deformation and hardening, see McMeeking [20]. For purposes of simplicity, the elementary slip field is shown here in Fig. 10. In Fig. 10 the stress conditions near the crack surface and ahead of the crack are depicted. The flow parameter, K, is $\sigma_0/\sqrt{3}$ or $\sigma_0/2$ for the Mises or Tresca criteria, so we see that the maximum normal stress ahead of the crack is about $3\sigma_0$. McMeeking [20] suggests that, with accounting for hardening and blunting, the stress ahead of the crack can easily reach $4\sigma_0$.

The situation for a microcrack ahead of the main crack is depicted in Fig. 11. Taking the view that the main crack's flow field dominates the sur-

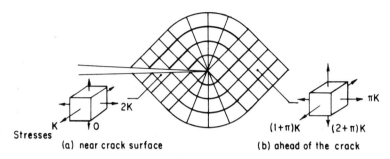

FIG. 10—*Elementary plane-strain slip field near the tip of a crack.*

rounding flow field leads to the stress, Fig. 11*b*. whereas above and below the microcrack the stresses are relaxed, Fig. 11*a*, for the shaded diamond-shaped region. (Note that this model for the relaxation of stresses above and below the microcrack has some objectionable features, but the discussion is continued as a model for its dimensional features.) Now computing the elastic strains over the height, *a*, for both fields of stress gives

$$\epsilon_y^{(a)} = [-3\mu]\frac{K}{E}$$

and (38)

$$\epsilon_y^{(b)} = [(2 + \pi) - (1 + 2\pi)\mu]\frac{K}{E}$$

which implies a reduction from (*b*) to (*a*) in elastic strains over the height, *a*,

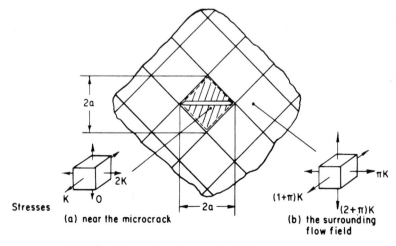

FIG. 11—*Microcrack within the flow field ahead of a crack.*

which would be turned into crack opening stretch, δ_t, at the tips of the microcrack, or

$$\delta_t = (\epsilon_y^{(b)} - \epsilon_y^{(a)})2a = [2 + \pi + 2\mu - 2\pi\mu]\frac{K}{E}2a$$

For $\mu \simeq 0.3$ and $K = \sigma_0/2$, this gives approximately

$$\delta_t = \frac{4\sigma_0 a}{E} \tag{39}$$

Again taking the usual relationship

$$\delta_t = \frac{J}{\sigma_0} \tag{40}$$

and combining Eqs 39 and 40 with an increment of crack extension, gives

$$T = \frac{dJ}{da} \times \frac{E}{\sigma_0^2} \leq 4 \quad \text{(for a model of tearing instability of a microflaw ahead of a crack)} \tag{41}$$

Again, as with the discussion of the tensile bar, undoubtedly deformation would increase the flow stress, σ_0, and reduce the tearing resistance, dJ/da, promoting instability. It is noted that the instability criterion resulting from this model does not contain the crack size, a, which may be a reasonable explanation for a local tearing instability from microflaws. The conclusions may be relevant even if the model is not numerically correct but is only dimensionally correct.

However, further exploration of this prospect should be pursued with a more realistic plasticity model such as those of Hutchinson and Shih [21,22].

Discussion

The preceding analysis has presented an approach to tearing instability criteria which is indeed proper from a basic mechanics viewpoint, hinged only on the concept that the J-integral R-curve is an appropriate representation of a material's tearing resistance.[6] In fact, the R-curve need not be either configuration or size independent or contain straight-line segments; it need only be appropriate for the particular situation for which tearing stability is being examined. However, the R-curve in reality is at least reasonably unvarying with size, configuration, etc., which adds to the value of this analysis, since simple laboratory tests for the R-curve characteristic may be used to predict behavior of other components, etc.

The methods of plasticity, that is, slip field analysis, for displacement rela-

---

[6]An alternative crack opening stretch, δ_T, approach can be used which is equivalent but not any further enlightening.

tionships and limit loads used herein are only approximations but very reasonable tools from at least a dimensional viewpoint. Therefore, the instability relationships presented are to be regarded as at least dimensionally correct but numerically approximate.

The weakest assumptions of the analysis are that the plasticity solutions are formed for constant geometry conditions and then subjected to crack length changes, that is, differentiated with respect to crack length. This causes "nonradial loading" of elements near the crack tip and other discrepancies. However, if these analyses are only reasonable approximations, increment by increment, as crack length changes are occurring, then they are dimensionally correct and only numerically approximate. Moreover, since local crack-tip "error" will occur both in developing the R-curve and in its companion application, a strong tendency for compensating error effects will occur within crack-tip fields. This compensation is indeed relied upon in currently widely accepted elastic R-curve analysis, when considering disturbances due to the crack-tip plastic zone in elastic analysis. It is desirable to be equally optimistic here. Thus if the instability criteria presented herein are not corrected for large crack length changes, they are at least good approximations for small increments of crack extension and, in any event, are regarded as dimensionally correct.

The material's resistance to tearing instability is clearly identified here as the tearing modulus, T, where $T = dJ/da \times E/\sigma_0^2$ depends only on the slope of the J-integral R-curve and other well-known properties, the flow stress, σ_0, in simple tension, and the elastic modulus, E. Appendix III has, for comparative purposes, a brief table of these properties for some rotor steels and a cast steel. It is noted that some materials have a tearing modulus, T, of over 100 (nondimensional), implying a very high degree of stability against tearing mechanisms for all crack configurations. However, other materials have a tearing modulus, T, below 10, which virtually ensures tearing instability in some configurations (such as double-edge cracked, see Eq 12) as soon as J_{Ic} and limit load are reached. Therefore, tearing instability is a reality to be dealt with in such materials, and the size and geometry effects for tearing instability of various crack configurations become of practical interest for such materials.

The situation where tearing instability may trigger cleavage in rate-sensitive materials is a reality more than an important possibility and bears further study. Past observations have almost always led to conclusions that cleavage instability was a cause any time fractures were rapid and displayed a high amount of cleavage on the fracture surface. Clearly different cause, tearing instability, is possible. Indeed, many perplexing cracking instability situations, unexplained previously, seem perfectly logical with a little study of the possibilities for tearing instability. The possibilities seem endless for reconsidering everything from K_{Ic} test behavior observed, the effects on material property changes (in T) for irradiation-damaged nuclear materials, and cutting and machining problems, to cracking under forming and other large scale plastic flow problems. It would be too much to consider all the immediate possibilities here.

Continuation of the detailed development of other aspects of tearing instability criteria also seems relevant. In Appendix I an attempt has been made to generalize the preceding analysis of the tearing instability criteria for various configurations. The generalization is developed for any situation where the elastic components of deformation are linear, where limit load occurs, etc. The tearing modulus, T, again reappears as the key material parameter. It is shown that testing machine stiffness can easily be taken into account. The effect of changes in geometrical aspects of instability due to deformation itself appears, in the term in Eq 51 involving J as a measure of deformation. Further studies should account for such effects as work hardening, elastic nonlinearity, and geometric nonlinearity (large deformation), which seem a bit out of place in this first discussion of the concept of tearing instability.

Finally, it is evident that to date no systematic experimental programs have been performed to explore tearing instability; this is perhaps the first thing that should be done. Previous J-integral fracture testing provides many examples of tearing instability (though not interpreted as such at the time) which should be reexamined. But most of these previous tests were on bending configurations, that is, those of the most natural stability; thus the more unstable types of test configurations should be employed, and a wide variety (extremes) of a material's tearing moduli, T, should be included. Again, the possibilities seem endless, and it is evident that judiciously chosen critical experimentation is needed.

Acknowledgments

This work was made possible through a contract from the U.S. Nuclear Regulatory Commission (NRC) with Brown University, Providence, R.I., during the summer of 1976 and a later contract from NRC with Washington University (St. Louis, Mo.). In addition to the financial support, the continued encouragement of the regulatory staff, and especially of Messrs. W. Hazelton and R. Gamble, is gratefully acknowledged.

The special efforts and encouragement of Professor J. R. Rice (upon whose work and methods [4,7,18] much of this current work is based) during the first author's visit to Brown University (1974–1976) are due special acknowledgment and thanks. The continued assistance of this work by Professor Rice and Professor J. W. Hutchinson of Harvard University, as consultants to the current NRC contract at Washington University, is also gratefully acknowledged.

APPENDIX I

A General Analysis of Tearing Instability, Including the Effect of Testing Machine Stiffness for the Fully Plastic Case

The preceding descriptions of tearing instability criteria for various crack configurations are all based on relaxation of "global" contributions to reduction in elastic displacements which causes increases in plastic displacements which drive the crack ahead. (The only exception is the analysis of the microcrack in the flow field ahead of a large crack.) For a general analysis, consider the arbitrary configuration shown in Fig. 12.

Now, during an examination of stability, the displacement of the loading train remains constant; thus

$$\Delta_P = \Delta_{EL} + \Delta_{PL} + \Delta_M = \text{constant} \quad (42)$$

where Δ_{EL} and Δ_{PL} are the elastic and plastic components of the specimen displacements and Δ_M is the testing machine displacement. During crack extension then

$$d\Delta_{EL} + d\Delta_{PL} + d\Delta_M = 0 \quad (43)$$

Elastic displacements are (normally) linearly proportional to load and have the form:

$$\Delta_{EL} = \frac{P}{BE} \times f\left(\frac{a}{W}, \frac{B}{W}, \frac{L}{W}, \text{etc.}\right) \quad (44)$$

where $f( )$ is a nondimensional function of specimen dimensions. The plastic displacement, Δ_{PL}, presuming we are at limit load for the cracked section, will have a linear relationship to the crack opening stretch, δ_t, or

$$\Delta_{PL} = \delta_t \times g\left(\frac{a}{W}, \frac{B}{W}, \frac{L}{W}, \text{etc.}\right) \quad (45)$$

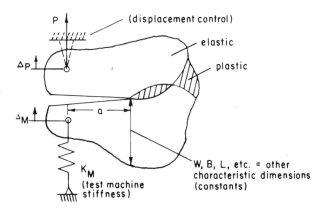

FIG. 12—*An arbitrary configuration.*

ELASTIC-PLASTIC FRACTURE

and finally the testing machine will be regarded as a linear spring or

$$\Delta_M = \frac{P}{K_M} \tag{46}$$

where K_M is the spring constant. The load, P, is assumed to be at limit load, P_L, which will depend linearly on the flow stress, σ_0, and thickness, B; therefore

$$P = P_L = \sigma_0 B W h \left(\frac{a}{W}, \frac{B}{W}, \frac{L}{W}, \text{etc.} \right) \tag{47}$$

and the usual relationship between J and δ_t is assumed with a constant, α, adjustable approximately to suit the configuration, or

$$\delta_t = \alpha \frac{J}{\sigma_0} \tag{48}$$

Now during an increment of crack extension, da, the variables in the foregoing expressions which change are, a, P, δ_t, and J. These cause changes in the displacements; therefore, writing the differentials of Eqs 44–48 gives

$$d\Delta_{EL} = \frac{dP}{BE} \times f(\ ) + \frac{P}{BE} \frac{\partial f(\ )}{\partial a} da \tag{44a}$$

$$d\Delta_{PL} = d\delta_t g(\ ) + \delta_t \frac{\partial g(\ )}{\partial a} da \tag{45a}$$

$$d\Delta_M = \frac{dP}{K_M} \tag{46a}$$

$$dP = dP_L = \sigma_0 B W \frac{\partial h(\ )}{\partial a} da \tag{47a}$$

and

$$d\delta_t = \alpha \frac{dJ}{\sigma_0} \tag{48a}$$

Now substituting Eqs 47, 48, 47a, and 48a in the right-hand sides of Eqs 44a, 45a, and 46a, so that the remaining differentials there are da and dJ, gives

$$d\Delta_{EL} = \frac{\sigma_0 W}{E} \frac{\partial}{\partial a} [h(\ ) \times f(\ )] da \tag{44b}$$

$$d\Delta_{PL} = \alpha \frac{dJ}{\sigma_0} \times g(\ ) + \frac{\alpha J}{\sigma_f} \cdot \frac{\partial g(\ )}{\partial a} da \tag{45b}$$

$$d\Delta_M = \frac{\sigma_0 B W}{K_M} \frac{\partial h(\ )}{\partial a} da \tag{46b}$$

Finally, substituting Eqs 44b–46b into Eq 43 and rearranging gives

$$\frac{dJ}{da}\frac{E}{\sigma_0^2} = \frac{W}{\alpha g(\ )}\left\{\frac{\partial}{\partial a}[h(\ ) \times f(\ )] + \frac{EB}{K_M}\frac{\partial h(\ )}{\partial a}\right.$$
$$\left. + \frac{\alpha JE}{W\sigma_0^2}\frac{\partial g(\ )}{\partial a}\right\} \quad (49)$$

But Eq 49 can be further simplified by noting that (with signs chosen for usual behavior)

$$-\frac{W}{g(\ )}\frac{\partial}{\partial a}[h(\ ) \times f(\ )] = p\left(\frac{a}{W}, \frac{B}{W}, \frac{L}{W}, \text{etc.}\right)$$

$$-\frac{W}{g(\ )}\frac{\partial h(\ )}{\partial a} = q\left(\frac{a}{W}, \frac{B}{W}, \frac{L}{W}, \text{etc.}\right)$$

and

$$\frac{W}{g(\ )}\frac{\partial g(\ )}{\partial a} = r\left(\frac{a}{W}, \frac{B}{W}, \frac{L}{W}, \text{etc.}\right) \quad (50)$$

Therefore, substituting Eq 50 into Eq 49, the simplification leads to

$$T = \frac{dJ}{da}\frac{E}{\sigma_0^2} \leq \frac{1}{\alpha}\left\{p(\ ) + \frac{EB}{K_M}q(\ ) - \frac{JE}{W\sigma_0^2}r(\ )\right\} \quad (51)$$

as the general form for the tearing instability criterion for any specimen under fully plastic conditions.

All of the tearing instability criteria given previously herein fit this general form, Eq 51. In the previous criteria a rigid testing machine was assumed, that is, $K_M = \infty$, so the term with $q(\ )$ did not enter. Moreover, considering Eqs 45 and 50 for most configurations, $g(\ )$ did not contain crack size; thus $r(\ )$ was zero. (The exception is the deeply cracked compact specimen, where $h(\ ) \times f(\ )$, which did not contain crack size; thus $p(\ )$ was zero, and $r(\ )$ turned out positive, thus assuring an always stable situation).

Moreover, even though the microcrack within the flow field ahead of the crack does not fit the physical model here as depicted by Fig. 12, the criterion which resulted still fits the form of Eq 51.

Finally, all of these tearing instability criteria have $T = (dJ/da) \times (E/\sigma_0^2)$ as a key material parameter in resisting tearing instability. Also note that the term with $r(\ )$ sometimes enters as a partly geometrical term assisting resistance to tearing instability.

APPENDIX II

Tearing Instability of the General Small-Scale Yielding Case[7]

All of the preceding discussion addresses tearing instability which occurs following development of full plasticity on the remaining ligament at the cracked section. Here a

---
[7] Attention to this topic was drawn at the suggestion of James R. Rice of Brown University.

general analysis of tearing instability is done where the remaining ligament is predominately elastic; that is to say, only small-scale yielding at the crack tip is present. The effects of elasticity of the loading apparatus, for example, testing machine stiffness, are also considered.

The analysis begins by following the notation and arrangement as shown in Figure 13. For an arbitrary configuration, such as Fig. 13, the form of the crack-tip stress intensity factor, K, is always

$$K = \frac{P\sqrt{a}}{BW} Y \left(\frac{a}{W}, \frac{L}{W}, \frac{B}{W}, \text{etc.}\right) \qquad (52)$$

Assuming small-scale yielding or $J = \mathcal{G}$, then

$$J = \mathcal{G} = \frac{K^2}{E} = \frac{P^2 a}{EB^2 W^2} Y^2 \left(\frac{a}{W}\right) \qquad (53)$$

Taking differentials associated with a crack length change, da, results in

$$dJ = \frac{2Pa}{EB^2 W^2} Y^2 \left(\frac{a}{W}\right) dP + \frac{P^2}{EB^2 W^2} H \left(\frac{a}{W}\right) da \qquad (54)$$

where[8]

$$H \left(\frac{a}{W}\right) = Y^2 \left(\frac{a}{W}\right) + 2 \left(\frac{a}{W}\right) Y \left(\frac{a}{W}\right) Y' \left(\frac{a}{W}\right)$$

Now during the increment of crack extension, the total displacement, Δ_P, will be constant. Again referring to Fig. 13, this may be written

$$\Delta_P = \Delta_{\text{specimen}} + \Delta_M = \text{constant}$$

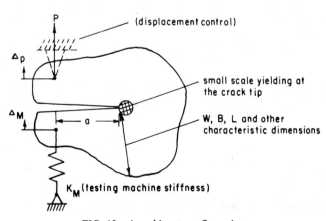

FIG. 13—*An arbitrary configuration.*

[8] The function $Y(a/W, L/W, B/W,$ etc.), which is a nondimensional function of configuration dimensions, will be denoted $Y(a/W)$ for simplicity in the analysis to follow. Similar simplifications will be taken here for other functions, such as $f(\ )$.

or again writing differentials

$$d\Delta_P = d\Delta_{sp} + d\Delta_M = 0 \tag{55}$$

For an elastic small-scale yielding situation, the specimen displacement Δ_{sp} has the form

$$\Delta_{sp} = \frac{P}{EB} f\left(\frac{a}{W}\right) \tag{56}$$

and the testing machine behavior is expressed by

$$\Delta_M = \frac{P}{K_M} \tag{57}$$

Differentiating and substituting Eqs 56 and 57 into Eq 55 gives

$$d\Delta_P = \frac{dP}{EB} f\left(\frac{a}{W}\right) + \frac{P}{EBW} f'\left(\frac{a}{W}\right) da + \frac{dP}{K_M} = 0 \tag{58}$$

or rearranging Eq 58 gives

$$dP = \frac{-\dfrac{P}{W} \dfrac{f'\left(\dfrac{a}{W}\right)}{f\left(\dfrac{a}{W}\right)}}{1 + \dfrac{EB}{K_M f\left(\dfrac{a}{W}\right)}} da \tag{59}$$

Equation 59 gives the load change, dP, associated with a crack length change, da, when examining for instability. It may therefore be substituted in Eq 54, which results in

$$dJ = \frac{-\dfrac{2}{E} \dfrac{P^2}{B^2 W^2} \dfrac{a}{W} Y^2\left(\dfrac{a}{W}\right) \dfrac{f'\left(\dfrac{a}{W}\right)}{f\left(\dfrac{a}{W}\right)}}{1 + \dfrac{EB}{K_M f\left(\dfrac{a}{W}\right)}} da + \frac{P^2}{EB^2 W^2} H\left(\frac{a}{W}\right) da \tag{60}$$

Now let

$$G\left(\frac{a}{W}\right) = \frac{a}{W} Y^2\left(\frac{a}{W}\right) \frac{f'\left(\dfrac{a}{W}\right)}{f\left(\dfrac{a}{W}\right)}$$

and recall

$$H\left(\frac{a}{W}\right) = Y^2\left(\frac{a}{W}\right) + \frac{2a}{W} Y\left(\frac{a}{W}\right) Y'\left(\frac{a}{W}\right)$$

where $Y(a/W)$ and $f(a/W)$ are configuration factors in the K and Δ_{sp} formulas. Finally, Eq 60 can be rearranged as the instability criterion,

$$T = \frac{dJ}{da}\frac{E}{\sigma_0^2} \leq \left\{\frac{-2G\left(\frac{a}{W}\right)}{1 + \frac{EB}{K_M f\left(\frac{a}{W}\right)}} + H\left(\frac{a}{W}\right)\right\} \left(\frac{\sigma}{\sigma_0}\right)^2 \quad (61)$$

where in addition it is noted that:

$$\sigma = \frac{P}{BW}$$

In Eq 61 it is noted that the $H(a/W)$ and the stress $(\sigma/\sigma_0)^2$ combine to drive the situation toward tearing instability and that $H(a/W)$ depends only on the form for the stress-intensity factor solution, Eq 52. The other term with $G(a/W)$ involves the relative stiffness of the testing machine (or loading arrangement); note that if $K_M = 0$, this term is zero, and that as loading arrangement stiffness, K_M, is added, the negative sign implies adding to the stability of the situation. It is easy to apply the criterion.

Plastic Zone Correction

The instability criteria for small-scale yielding may be reformulated using the usual plastic zone correction in the analysis. Adding the plastic zone correction to the crack length gives

$$a_{\text{eff}} = a + r_y \quad (62)$$

where

$$r_y = \frac{1}{\gamma\pi}\left(\frac{K}{\sigma_0}\right)^2 = \frac{JE}{\gamma\pi\sigma_0^2}$$

Differentiating Eq 62 with respect to crack size gives

$$da_{\text{eff}} = \left[1 + \left(\frac{dJ}{da}\frac{E}{\sigma_0^2}\right)\frac{1}{\gamma\pi}\right] da \quad (63)$$

For correcting the instability criterion, a and da in Eq 60 should be replaced by a_{eff} and da_{eff} from Eqs 62 and 63, which gives

$$T = \frac{dJ}{da}\frac{E}{\sigma_0^2} \leq \frac{\{\ \}\left(\frac{\sigma}{\sigma_0}\right)^2}{1 - \frac{1}{\gamma\pi}\{\ \}\left(\frac{\sigma}{\sigma_0}\right)^2} \quad (64)$$

where

$$\{\ \} = \left\{ \frac{-2G\left(\frac{a_{\text{eff}}}{W}\right)}{1 + \frac{EB}{K_M f\left(\frac{a_{\text{eff}}}{W}\right)}} + H\left(\frac{a_{\text{eff}}}{W}\right) \right\}$$

However, the second term in the denominator of the right-hand side of Eq 64 is normally small compared to one, so that within it a may be used instead of a_{eff} with no appreciable error. Nevertheless, a_{eff} would be retained in the numerator. Thus Eq 64 represents the plastic zone-corrected instability criterion in the usual linear elastic fracture mechanics spirit.

Dugdale Strip-Yield Zone Correction

Although the Dugdale strip-yield zone methodology is most appropriate for "plane stress" conditions, the tearing instability criterion can be appropriately formulated for the mixed mode and here may be restricted to plane stress if necessary. The analysis proceeds as follows.

The form of crack opening stretch, δ_T, solutions from strip-yield models is

$$\delta_T = \frac{\sigma_0 a}{E} f\left(\frac{\sigma}{\sigma_0}, \frac{a}{W}, \frac{B}{W}, \frac{L}{W}, \text{etc.}\right) \tag{65}$$

As in earlier analysis

$$\delta_T = \alpha \frac{J}{\sigma_0}$$

Differentiating and rearranging gives

$$T = \frac{dJ}{da} \frac{E}{\sigma_0^2} \leq \frac{1}{\alpha} f\left(\frac{\sigma}{\sigma_0}, \frac{a}{W}, \frac{B}{W}, \frac{L}{W}, \text{etc.}\right)$$
$$+ \frac{a}{\alpha} \frac{d}{da} f\left(\frac{\sigma}{\sigma_0}, \frac{a}{W}, \frac{B}{W}, \frac{L}{W}, \text{etc.}\right) \tag{66}$$

From here, further details follow in the manner of Eq 54 and subsequent development. But the form is already clear with Eq 66 and is not developed further herein.

APPENDIX III

Table of Some Material Properties[9] (Including "Tearing Modulus," T).

Material	Condition[a] and Reference	Specimen Type	Temperature (°F)	$\sigma_0 = \dfrac{\sigma_y + \sigma_{ult}}{2}$ (ksi)	J_{Ic} $\left(\dfrac{\text{in.-lb}}{\text{in.}^2}\right)$	$\dfrac{dJ}{da}$ $\left(\dfrac{\text{lb}}{\text{in.}^2}\right)$	$T = \dfrac{dJ}{da} \times \dfrac{E}{\sigma_0^2}$ (—)
ASTM-A469 (GE) rotor steel (Ni-Mo-V)	[7]	1-in. CT	75	95.9	(760)	...	...
			100	95.3	1260	3.8×10^4	123.0
			125	94.5	...	3.0×10^4	101.0
			150	93.7	1340	3.8×10^4	130.0
			175	93.0	1050	3.6×10^4	124.0
			250	(88.0 estimated)	...	3.4×10^4	131.0
			300	80.0	720	2.8×10^4	131.0
ASTM-A470 rotor steel (Cr-Mo-V)	[1]	1-in. CT	300	90.0	470	6.9×10^3	25.5
			500	90.0	413	7.0×10^3	25.8
			800	85.9	493	1.2×10^4	48.6
ASTM-A471 rotor steel (Ni-Cr-Mo-V)	[2]	1-in. CT	50	146.1	860	1.45×10^4	20.4
			150	134.0	594	3.25×10^3	5.4
			250	128.5	670	6.54×10^3	11.9
			250	138.0	575	2.03×10^4	31.9
			300	125.0	500	1.47×10^4	27.9
			500	115.0	325	9.58×10^3	21.6
			800	104.0	325	8.75×10^3	24.1
AISI-403 (12Cr-stainless) rotor steel	[3]	1-in. CT	75	109.0	572	7.1×10^3	17.9
			200	103.0	425	7.9×10^3	22.3
			300	96.3	460	7.3×10^3	23.6
			500	94.5	417	5.83×10^3	19.6
			800	86.2	372	4.08×10^3	16.5
ASTM-A217	[4]	1-in. CT	−50	79.5	905	2.5×10^4	119.0

Material		Ref.	Specimen	Temp.				
(2¼Cr-1Mo) cast steel				75	71.9	860	1.89×10^4	109.0
				300	67.0	666	3.33×10^4	221.0
				500	64.5	433	2.40×10^4	172.0
				800	62.8	333	2.33×10^4	176.0
ASTM-A453 (A286) (Discaloy) stainless steel		[2]	1-in. CT	−452	167.0	815	3.87×10^4	41.5
				75	135.0	692	2.0×10^4	32.8
				400	119.0	600	1.2×10^4	25.6
				800	112.0	517	9.3×10^3	22.4
ASTM-A453 stainless steel (gas tungsten arc welds)		[4]	1-in. CT	−452	107.8	1666.7	3.54×10^4	91.4
ASTM-A540 (AISI 4340) steel		[2]	CT	−100	(153.5 estimated)	600	3.7×10^4	47.1
				50	146.4	800	2.15×10^4	30.1
				75	145.5	652	2.10×10^4	29.8
				250	143.0	730	1.75×10^4	25.7
				300	142.5	720	1.0×10^4	14.8
				550	133.0	725	1.45×10^4	24.6
Inco LEA (low expansion alloy) solution-treated and aged		[4]	CT	−452	187.2	194	2.53×10^3	2.16
				75	137.2	263	1.07×10^3	1.71
6061-T651 aluminum		[3]	1-in. CT	75	43.25	98.4	4.92×10^2	2.79
3105 stainless steel (shielded metal arc welds)	[5] VIM-VAR/ST	[4]	CT	−452	152.9	336.5	1.03×10^4	13.27
	DO-/ST		CT	75	120.4	320	4.64×10^3	10.2
	DO-/STDA		CT	−452	127.0	240	1.72×10^3	3.6
Inconel-X750	AAM-VAR/STDA		CT	−452	139.0	138	5.26×10^3	0.9
	VIM-/STDA			−452	184.0	135	5.09×10^3	50.4
	HIP ····			−452	213.7	500	2.06×10^4	15.1
	HIP-/STDA			−452	172.6	465	4.57×10^4	51.4
AISI-3105 stainless steel	[6] STQ			−452	185.3	233	1.89×10^4	18.4
	sensitized			−452	154.0	1600	1.2×10^2	0.152
				−452	152.0	675	7.5×10^4	97.1

[9] The J-R-curves used in preparing this table were not corrected for influences of crack growth on J values.

Appendix III (continued)

Material	Condition[a] and Reference	Specimen Type	Temperature (°F)	$\sigma_0 = \dfrac{\sigma_y + \sigma_{ult}}{2}$ (ksi)	J_{Ic} $\left(\dfrac{\text{in.-lb}}{\text{in.}^2}\right)$	$\dfrac{dJ}{da}$ $\left(\dfrac{\text{lb}}{\text{in.}^2}\right)$	$T = \dfrac{dJ}{da} \times \dfrac{E}{\sigma_0^2}$ (—)
Kromarc-58 stainless steel	[6] CW(L-t)	CT	−452	227.3	452	3.0×10^3	1.74
	CW(T-L)		−452	216.0	388	1.0×10^4	6.43
	STQ(T-L)		−452	175.0	1250	2.0×10^3	1.96
	[4] GTAW	CT	−452	181.4	615.5	9.3×10^3	8.48
	CW/GTAW	½ in. thick		199.9	725	9.17×10^3	6.89
	GTAW/CN			214.2	216.7	2.71×10^3	1.77
	GTAW/CW/AN			199.0	429	1.95×10^4	14.8
	GTAW/CW/AN			175.7	812	3.62×10^4	35.19
Iconel-706	[4] VIM-EFR/STDA	CT	−452	211.5	530	1.21×10^4	8.13
	VIM-VAR/STDA		−452	207.0	660	1.94×10^4	13.6
			75	166.0	357	1.88×10^4	20.45
Pyromet-538	[4] GATW		−452	209.1	166	5.19×10^2	0.35
(21Cr-6Ni-9Mn) stainless steel			−452	137.3	862.3	3.27×10^4	51.69

[a]Denotes the following material conditions
VIM-VAR = vacuum induction melted followed by vacuum arc remelt.
AAM-VAR = air arc melted followed by vacuum arc remelt.
VIM = vacuum induction melted.
GTAW = gas tungsten-arc welds.
SMAW = shielded metal-arc welds.
and heat treatments:
ST = solution treated.
STDA = solution treated and double aged.
CW = cold worked, 30 percent reduction in thickness.
STQ = 2000°F, 1-h water quenched.

References (for Appendix III only)

[1] Logsdon, W. A. and Begley, J. A., *Engineering Fracture Mechanics*, Vol. 9, 1977, pp. 461-470.
[2] Logsdon, W. A. in *Mechanics of Crack Growth, ASTM STP 590*, American Society for Testing and Materials, 1976, pp. 43-61.
[3] Begley, J. A., Logsdon, W. A., and Landes, J. D. in *Flaw Growth and Fracture, ASTM STP 631*, American Society for Testing and Materials, 1977, pp. 112-120.
[4] Wells, J. M., Logsdon, W. A., Kossowsky, R., and Daniel, M. R., "Structural Materials for Cryogenic Applications," Research Report 76-9D9-CRYMT-R1, Westinghouse Research Laboratories, Pittsburgh, Pa., 1976.
[5] Logsdon, W. A., *Advances in Cryogenic Engineering*, Vol. 22, 1977, pp. 47-58.
[6] Logsdon, W. A., Wells, J. M., and Kossowsky, R. in *Proceedings*, Second International Conference on Mechanical Behavior of Materials, Boston, Mass., Aug. 1976, pp. 1283-1289.
[7] Clarke, G. A., Andrews, W. R., Schmidt, D. W., and Paris, P. C. in *Mechanics of Crack Growth, ASTM STP 590*, American Society for Testing and Materials, 1976, pp. 27-43.

References

[1] Begley, J. A. and Landes, J. D. in *Fracture Analysis, ASTM STP 560*, American Society for Testing and Materials, 1974, pp. 170-186.
[2] Clarke, G. A., Andrews, W. R., Schmidt, D. W., and Paris, P. C. in *Mechanics of Crack Growth, ASTM STP 590*, American Society for Testing and Materials, 1976, pp. 27-42.
[3] Paris, P. C. in *Flaw Growth and Fracture, ASTM STP 631*, American Society for Testing and Materials, 1976, pp. 3-27.
[4] Rice, J. R., "Elastic Plastic Fracture Mechanics," *The Mechanics of Fracture*, F. Erdogan, Ed., American Society of Mechanical Engineers, 1976.
[5] *Fracture Toughness Evaluation by R-Curve Methods, ASTM STP 527* (papers on linear-elastic R-curve analysis), American Society for Testing and Materials, 1974.
[6] Paris, P. C. and Clarke, G. A., "Observations of Variation in Fracture Characteristics with Temperature Using a J-Integral Approach," submitted to the Symposium on Elastic-Plastic Fracture, Atlanta, Ga., American Society for Testing and Materials, 1977.
[7] Rice, J. R., "Elastic-Plastic Models for Stable Crack Growth," *Mechanics and Mechanisms of Crack Growth*, British Steel Corp., 1973.
[8] Ritchie, F. O., Knott, J. F., and Rice, J. R., *Journal of the Mechanics and Physics of Solids*, Vol. 21, 1973, pp. 395-410.
[9] Private communication on fracture test results, Westinghouse Research Laboratories, Fracture Mechanics Group, E. T. Wessel, Manager, Pittsburgh, Pa., 1976-1977.
[10] Begley, J. A. and Landes, J. D., *International Journal of Fracture Mechanics*, Vol. 12, No. 5, Oct. 1976, pp. 764-766.
[11] Kachanov, L. M. in *Foundations of the Theory of Plasticity*, H. A. Lauwener and W. M. Koiter, Eds., North Holland Publishing Co., 1971.
[12] Irwin, G. R., private communication on ligament flow (δ), 1969.
[13] Green, A. P. and Hundy, B. B., *Journal of the Mechanics and Physics of Solids*, Vol. 4, 1956, pp. 128-144.
[14] Rice, J. R., Paris, P. C., and Merkle, J. G. in *Progress in Flaw Growth and Fracture Toughness Testing, ASTM STP 536*, American Society for Testing and Materials, 1973, pp. 231-245.
[15] Private communication of observations by E. T. Wessel, Manager, the Fracture Mechanics Group, Westinghouse Research Laboratories, Pittsburgh, Pa., 1976-1977.
[16] Argy, G., Paris, P. C., and Shaw, F. in *Properties of Materials for Liquefied Natural Gas Tankage, ASTM STP 579*, American Society for Testing and Materials, 1975, pp. 96-137.
[17] Tada, H., Paris, P. C., and Irwin, G. R., *The Stress Analysis of Cracks Handbook*, Del Research Corp., 226 Woodbourne Dr., St. Louis, Mo., 1973.

[18] Rice, J. R., "Mathematical Aspects of Fracture," *Fracture*, Academic Press, New York, Vol. 2, 1968.
[19] McClintock, F. A., "Plasticity Aspects of Fracture," *Fracture*, Academic Press, New York, Vol. 3, 1968.
[20] McMeeking, R. M., "Large Plastic Deformation and Initiation of Fracture at the Tip of a Crack in Plane Strain," Brown University report, Providence, R. I., Dec. 1976.
[21] Shih, C. F. in *Mechanics of Crack Growth, ASTM STP 590*, American Society for Testing and Materials, 1976, pp. 3-26.
[22] Shih, C. F. and Hutchinson, J. W., "Fully Plastic Solutions and Large-Scale Yielding Estimates for Plane Stress Crack Problems," Harvard University, Rep-deap.-S-14, Cambridge, Mass., July 1975.
[23] Goldman, N. L. and Hutchinson, J. W., *International Journal of Solids and Structures*, Vol. 11, 1975, pp. 575-591.

J. W. Hutchinson[1] *and P. C. Paris*[2]

Stability Analysis of *J*-Controlled Crack Growth

REFERENCE: Hutchinson, J. W. and Paris, P. C., "**Stability Analysis of *J*-Controlled Crack Growth**," *Elastic-Plastic Fracture, ASTM STP 668*, J. D. Landes, J. A. Begley, and G. A. Clarke, Eds., American Society for Testing and Materials, 1979, pp. 37–64.

ABSTRACT: The theoretical basis for use of the J-integral in crack growth analysis is discussed and conditions for *J*-controlled growth are obtained. Calculations related to the stability of crack growth are carried out for several deeply cracked specimen configurations. Relatively simple formulas are obtained which, in certain cases, permit an assessment of stability using data from a single load-displacement record. Numerical results for a bend specimen and for a center-cracked specimen illustrate the influence of strain-hardening and system compliance on stability.

KEY WORDS: crack propagation, fracture (materials), plastic deformation, stable crack growth

This paper builds upon the report of Paris et al [1][3] which promulgates an approach to the stability analysis of crack growth based on the concept of a J-integral resistance curve. We start by presenting a theoretical justification for use of the J-integral of the deformation theory of plasticity in the analysis of crack growth. Restrictions on such use are discussed in detail with particular emphasis on application in the large-scale yielding range.

When applicable, the approach of Ref *1* and the present paper is the natural extension of Irwin's resistance curve analysis (for example, see Ref *2*) for small-scale yielding based on the elastic stress intensity factor *K*. In a sense this approach is less fundamental, and less ambitious, than studies based on flow (that is, incremental) theories of plasticity which attempt to identify and calculate a single near-tip parameter governing the initiation and continuation of crack growth. Studies along such lines [*3,4*] have attempted to discuss the source of stable crack growth but they have not cleared the way for much progress in its analysis. In part, this is because there is not yet agreement on a suitable near-tip growth criterion; but it is

---

[1] Professor of applied mechanics, Harvard University, Cambridge, Mass. 02138.
[2] Professor of mechanics, Washington University, St. Louis, Mo. 63130.
[3] The italic numbers in brackets refer to the list of references appended to this paper.

also due to the difficulties of carrying out crack growth calculations using a flow theory of plasticity. Deformation theory has distinct computational advantages leading in some instances to closed-form solutions which would otherwise be unobtainable. Illustrations of this point will be found within the present paper. Moreover, the conditions for J-controlled growth are derived to ensure essentially identical results from deformation theory and flow theory on the topics herein.

Following the discussion of the applicability of J to analyzing crack growth, we discuss the stability of crack growth with emphasis on the role of the compliance of the entire system under given prescribed loading conditions. An analysis of a deeply cracked bend specimen is carried out. Relatively simple formulas are obtained for assessing stability. Numerical results for a bending specimen in plane-stress conditions are presented to illustrate the influence of strain hardening and compliance on stability. A possible scheme for measuring the resistance curve experimentally is mentioned. The final sections of the paper deal with the analyses of deeply cracked edge and center-notched specimens.

Applicability of J-Integral to Analysis of Crack Growth

Crack growth invariably involves some elastic unloading and distinctly nonproportional plastic deformation near the crack tip. The J-integral [5] is based on the deformation theory of plasticity which inadequately models both of these aspects of plastic behavior. At just a glance it would appear that use of J must be restricted to the analysis of stationary cracks. In the following, a rationale is given for use of J to analyze crack growth and stability under conditions which will be called J-controlled growth. The argument relies on the fact that many metals sustain only very small amounts of crack growth relative to other dimensions for overall deformations well beyond initiation of growth. If J-controlled growth is to exist, it is essential (and by implication sufficient) that nearly proportional plastic deformation occurs everywhere but in a small neighborhood of the crack tip. This results from the fact that when nearly proportional deformation occurs, the differences between a deformation theory of plasticity and the corresponding flow (incremental) theory become essentially negligible [6], and both theories are surely physically realistic.

In small-scale yielding it is widely accepted that either the elastic stress intensity factor K or J ($J = K^2/\text{modulus}$) can be used in a resistance curve analysis of stable crack growth, at least for amounts of growth which are small compared to all other relevant geometric lengths. Under limiting conditions of plane stress or plane strain the increase in crack length Δa has a unique functional relationship to K or J which is otherwise configuration-independent. In small-scale yielding the existence of such a relationship rests on the fact that K or J do uniquely measure the intensity of the fields surrounding the immediate vicinity of the crack tip. In this paper we shall be concerned mainly with growth under large-scale yielding, including fully plastic situations, where extra conditions must be met for J to be meaningful and for the relation of Δa to J to be configuration-independent. When these conditions are met, a J-resistance curve analysis is the appropriate generalization of small-scale yielding resistance curve analysis.

Figure 1 displays a schematic sketch of J versus Δa for a typical intermediate-strength steel under nominally plane-strain conditions as obtained by large-scale yielding testing techniques [7]. Leaving aside for the moment the issue of the validity of experimentally measured J-values used to generate such data, the main feature of importance to our argument in the following is the relatively large increase in J above the initiation value J_{Ic} (that is, an increase which can be as much as several times J_{Ic}) needed to produce an increase in crack length of, say, only several millimetres. Emphasis in this discussion is on this range of small growth with its attendant large increases in J. We look for conditions under which it can be expected that the dominant singularity crack-tip fields of deformation theory, with amplitude J, continue to be relevant in the presence of small amounts of growth as depicted in Fig. 2.

Consider a material with a strain hardening index n such that the plastic strain is proportional to the stress to the nth power well into the plastic range. The strain field of the dominant singularity at the crack tip according to deformation theory is [8, 9]

$$\epsilon_{ij} = k_n J^{n/(n+1)} r^{-n/(n+1)} \tilde{\epsilon}_{ij}(\theta) \tag{1}$$

where r, θ = planar-polar coordinates centered at the tip. The θ variation, $\tilde{\epsilon}_{ij}$, depends on n and on whether plane strain or plane stress pertains; k_n is a dimensional constant not needed in the present discussion. Let R denote the characteristic radius of the region controlled by Eq 1 in the deformation theory solution. In small-scale yielding, R will be some fraction of the plastic

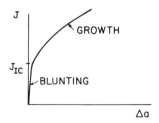

FIG. 1—*Material J-resistance curve for small amounts of crack growth.*

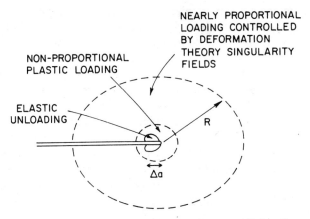

FIG. 2—*Schematic of crack tip conditions for J-controlled crack growth.*

zone size,[4] while in a fully yielded specimen R will be some fraction of the uncracked ligament. Since the wake of elastic unloading and the region of distinctly nonproportional plastic loading will be of the order Δa in length, one condition for J-controlled growth is

$$\Delta a \ll R \tag{2}$$

Next we examine the strain increments determined from deformation theory under a *simultaneous* increase in J and crack length. The crack lies along the x-axis and is assumed to advance by an amount da in the x-direction. For deformation theory, the strain field, Eq 1, continues to hold in the presence of growth with r and θ centered at the current tip location. From Eq 1, the strain increments are

$$d\epsilon_{ij} = \frac{n}{n+1} k_n J^{-1/(n+1)} \, dJ r^{-n/(n+1)} \tilde{\epsilon}_{ij}(\theta)$$
$$- k_n J^{n/(n+1)} \, da \, \frac{\partial}{\partial x} [r^{-n/(n+1)} \tilde{\epsilon}_{ij}(\theta)] \tag{3}$$

Using

$$\frac{\partial}{\partial x} = \cos\theta \, \frac{\partial}{\partial r} - \frac{\sin\theta}{r} \, \frac{\partial}{\partial \theta}$$

---

[4] Adjusting Eq 1 to represent alternatively the elastic field, R would be substantially larger than the plastic zone size for small-scale yielding.

Eq 3 becomes

$$d\epsilon_{ij} = k_n J^{n/(n+1)} r^{-n/(n+1)} \left\{ \frac{n}{n+1} \frac{dJ}{J} \tilde{\epsilon}_{ij} + \frac{da}{r} \tilde{\beta}_{ij} \right\} \quad (4)$$

where

$$\tilde{\beta}_{ij}(\theta) = \frac{n}{n+1} \cos\theta \, \tilde{\epsilon}_{ij} + \sin\theta \, \frac{\partial}{\partial \theta} \tilde{\epsilon}_{ij}$$

The first term in the braces in Eq 4 corresponds to a proportional loading increment (that is, $d\epsilon_{ij} \sim \epsilon_{ij}$), while the second term is not proportional. Since $\tilde{\epsilon}_{ij}$ and $\tilde{\beta}_{ij}$ are of comparable magnitude, the first term in the braces will overwhelm the second term if

$$\frac{da}{r} \ll \frac{dJ}{J} \quad (5)$$

That is, predominantly proportional loading will occur in the dominant singularity region where Eq 5 holds.

Define a material-based length quantity D as

$$\frac{1}{D} = \frac{dJ}{da} \frac{1}{J} \quad (6)$$

where for records such as those in Fig. 1, just beyond initiation, D can be interpreted approximately as the crack advance associated with a doubling of J above J_{Ic}. Equation 5 can be restated as

$$D \ll r \quad (7)$$

If

$$D \ll R \quad (8)$$

then there exists an annular region

$$D \ll r < R \quad (9)$$

in which plastic loading is predominantly proportional *and* the singularity field, Eqs 1 or 4, is dominant. In other words, if Eq 8 is satisfied, it can be expected that there will be little difference between the strain fields predicted by flow theory and deformation theory for $r \gg D$ and, most importantly, J uniquely measures, or physically controls, the fields in the Eq 9 region surrounding the tip.

To summarize, we note that the foregoing argument is somewhat analogous to that made for the relevance of the elastic analysis for K in the presence of small-scale yielding. Here, however, the argument has dealt with two factors: (1) small-scale growth requiring Eq 2 and (2) applicability of deformation theory and J requiring Eq 8. The foregoing argument relies on the existence of some strain hardening since the θ-variation, $\tilde{\epsilon}_{ij}$, of the strain field of the dominant singularity is not unique for an ideally plastic material but will, in general, depend on the overall geometry. In effect, R diminishes to zero with vanishing strain hardening. This same point is at issue in the use of J in analyzing initiation [10] and has been addressed experimentally by Landes and Begley [11] using inherently different specimen types. In making the foregoing argument, we have also tacitly assumed that, if predominantly proportional loading occurs throughout most of the dominant singularity region, it will occur outside (that is, $r > R$) this region as well. This can be expected for the same reasons which apply to the configuration with a stationary crack under monotonic loading.

For a specimen or configuration which has fully yielded, R will be some fraction of the relevant uncracked ligament, b (or other characteristic distance from the crack tip to a boundary or loading point if smaller than b). Introduce a nondimensional parameter as defined by

$$\omega = \frac{b}{J} \frac{dJ}{da} \tag{10}$$

Thus, for a fully yielded specimen, the condition for J-controlled growth, Eq 8, can be restated

$$\omega \gg 1 \tag{11}$$

together with Eq 2.

This Eq 11 requirement for proportionality of strain in the singularity field is different than an earlier J-integral test specimen size requirement ([7] and discussion to Ref *11*). The earlier requirement may be stated as

$$\frac{b\sigma_0}{J} \gg 1 \tag{12}$$

where σ_0 is flow stress. This requirement, Eq 12, may be interpreted as keeping the crack opening displacement small compared to the ligament dimension, b, and is regarded as applying both to the initiation and growth phases of cracking. Indeed, perhaps both requirements must be met to maintain a proper singularity field during crack growth; however, the later one, Eq 12, will not be discussed further here.

One consequence of J-controlled growth, which follows immediately from the foregoing arguments, is that J will be approximately independent of the integration path when calculated using the standard line integral definition in conjunction with a flow (incremental) theory of plasticity as long as the points on the path satisfy (7). Shih et al [12] have found less than 5 percent variation in J over a wide range of paths for as much as a 5 percent increase in crack length in their finite-element analysis of a fully yielded, plane-strain compact tension specimen of A533-B steel. Using their values for b, J_{Ic}, and the initial slope dJ/da following initiation, we find that $\omega \simeq 40$ for their specimen. Shih et al [12] also found that their computed values of J were in good agreement with J-values obtained using the experimentally measured load-deflection curve and a deformation theory formula for a deeply cracked specimen. These two facts strongly suggest that J-controlled growth is in evidence in their specimens. An important question which remains to be answered is, What is the *smallest* value of ω which will guarantee J-controlled growth? To a certain extent the answer will depend on specimen configuration and on strain hardening. Closely related is the need for a systematic study of the amount of growth allowable under J-controlled conditions.

An additional consequence of J-controlled growth is that the material resistance curve of J versus Δa obtained under large-scale yielding conditions must coincide with that obtained under small-scale yielding conditions, assuming that the same plane-strain or plane-stress conditions prevail in both instances. [As discussed earlier, a relation of J (or K) versus Δa in small-scale yielding can be meaningful independent of the condition of Eq 8.]

Stability of J-Controlled Growth

Elaborating further in the development of Paris et al [1], we derive some general expressions related to growth and stability based on a J-resistance curve analysis.

Consider a two-dimensional specimen (in plane strain or plane stress) with a straight crack of length a and of a material characterized by deformation theory. As depicted in Fig. 3, let P be the generalized load acting on the specimen and let Δ be the load point displacement of the specimen through which P works. The specimen is loaded in series with a linear spring with compliance C_M (which can, if desired, be identified with the testing machine compliance) such that the total load point displacement is

$$\Delta_T = C_M P + \Delta \qquad (13)$$

Let J have the usual deformation theory definition [5], equivalently as the path-independent line integral, or as

$$J = \int_0^P \left(\frac{\partial \Delta}{\partial a}\right)_P dP = -\int_0^\Delta \left(\frac{\partial P}{\partial a}\right)_\Delta d\Delta \qquad (14)$$

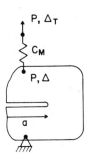

FIG. 3—*Typical specimen geometry.*

It will be important to draw a distinction between applied values of J and the values of J on the material resistance curve such as that in Fig. 1. For this purpose, values of J falling on the material resistance curve will be denoted by J_{mat} and will be regarded as a function solely of the increase in crack length Δa. For given material properties and overall specimen geometry, the "applied" J in Eq 14 can be regarded as a function of P and current crack length $a = a_0 + \Delta a$, where a_0 is the initial crack length. At any P and current length, a, equilibrium based on the resistance curve data requires

$$J(a, P) = J_{mat}(\Delta a) \qquad (15)$$

Stability will be considered with the total load point displacement Δ_T held fixed. (Then, note that $C_M = 0$ corresponds to a rigid test machine while $C_M = \infty$ corresponds to a dead-load machine.) Stability of the equilibrium state, Eq 15, will be ensured if

$$\left(\frac{\partial J}{\partial a}\right)_{\Delta_T} < \frac{dJ_{mat}}{da} \qquad (16)$$

The Eq 15 state is assumed to be unstable if

$$\left(\frac{\partial J}{\partial a}\right)_{\Delta_T} > \frac{dJ_{mat}}{da} \qquad (17)$$

and the onset of instability is associated with equality in Eq 16 or 17. Following Ref *1* we introduce nondimensional quantities

$$T = \frac{E}{\sigma_0^2}\left(\frac{\partial J}{\partial a}\right)_{\Delta_T} \quad \text{and} \quad T_{mat} = \frac{E}{\sigma_0^2}\frac{dJ_{mat}}{da} \qquad (18)$$

where E is Young's modulus and σ_0 is an appropriate flow stress. In terms of these quantities, Eqs 16 and 17 become

$$T < T_{mat} \text{ (stability)} \quad (19)$$

$$T > T_{mat} \text{ (instability)} \quad (20)$$

A general expression for $(\partial J/\partial a)_{\Delta_T}$, to be used in later sections, will now be derived by regarding J and Δ as functions of a and P. For arbitrary da and dP

$$dJ = \left(\frac{\partial J}{\partial a}\right)_P da + \left(\frac{\partial J}{\partial P}\right)_a dP \quad (21)$$

But with Δ_T held fixed, from Eq 13

$$d\Delta_T = C_M dP + \left(\frac{\partial \Delta}{\partial a}\right)_P da + \left(\frac{\partial \Delta}{\partial P}\right)_a dP = 0$$

and thus

$$dP = -da \left(\frac{\partial \Delta}{\partial P}\right)_P \left[C_M + \left(\frac{\partial \Delta}{\partial P}\right)_a\right]^{-1} \quad (22)$$

Combining Eqs 21 and 22 gives

$$\left(\frac{\partial J}{\partial a}\right)_{\Delta_T} = \left(\frac{\partial J}{\partial a}\right)_P - \left(\frac{\partial J}{\partial P}\right)_a \left(\frac{\partial \Delta}{\partial a}\right)_P \left[C_M + \left(\frac{\partial \Delta}{\partial P}\right)_a\right]^{-1} \quad (23)$$

If C_M is identified with the compliance of the testing machine, then specialization to the two limiting cases noted in the foregoing can be made. With $C_M = 0$, Eq 23 applies to a rigid test machine. If $C_M \to \infty$, the test machine applies a dead load and Eq 23 reduces to

$$\left(\frac{\partial J}{\partial a}\right)_{\Delta_T} = \left(\frac{\partial J}{\partial a}\right)_P \quad (24)$$

Analysis of Deeply Cracked Bend Specimen

In the spirit of Rice, Paris, and Merkle [13], a formula will be derived for $(\partial J/\partial a)_{\Delta_T}$ for a deeply cracked bend specimen which, up to initiation, involves quantities that can be taken from a single experimental test record. First, however, we derive a more general result for J than that given in Ref 13, which allows for determination of J under changing crack length. In the configuration of Fig. 4a, M is the applied moment per unit thickness, a is the current crack length, and a_0 is the initial crack length. The specimen

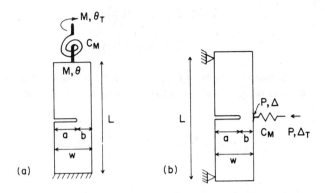

FIG. 4—*Bend specimen (a) and three-point bend specimen (b).*

load-point rotation θ at any given M can be decomposed into two parts according to

$$\theta = \theta_{nc} + \theta_c \tag{25}$$

where θ_{nc}, by definition, is the rotation of the uncracked specimen under the same M, and θ_c is the remainder (that is, the contribution due to the presence of the crack). It is assumed that the current ligament length $b = W - a$ is sufficiently small compared with W such that θ_c at a given M depends only on b and not on L or W. From dimensional considerations it then follows that θ_c must be a function of the combination M/b^2, that is

$$\theta_c = f(M/b^2) \text{ or } M = b^2 F(\theta_c) \tag{26}$$

In Ref *13* it is noted that Eq 26 implies

$$\left(\frac{\partial \theta}{\partial a}\right)_M = \left(\frac{\partial \theta_c}{\partial a}\right)_M = \frac{2M}{b}\left(\frac{\partial \theta_c}{\partial M}\right)_a \tag{27}$$

Using Eq 27 in the definition of J from the first form of Eq 14, that is

$$J = \int_0^M \left(\frac{\partial \theta}{\partial a}\right)_M dM \tag{28}$$

gives

$$J = \frac{2}{b}\int_0^{\theta_c} M d\theta_c = 2b \int_0^{\theta_c} F(\theta_c) \, d\theta_c \tag{29}$$

However, it is noted here that using the second form of Eq 14 with the second form of Eq 26 leads to this same result (Eq 29) more directly. The definition of J in Eq 28 and the derivation leading to Eq 29 require a to be held fixed in the integration. Nevertheless, by virtue of its deformation theory definition, $J(a, \theta_c)$ is independent of the history, giving rise to the current values of a and θ_c. (Equivalently, J can be regarded as a function of a and M, but for present purposes θ_c is a more convenient independent variable.) For arbitrary increments in a and θ_c it follows from Eq 29 that

$$dJ = 2bd\theta_c F(\theta_c) - 2da \int_0^{\theta_c} F(\theta) \, d\theta_c \qquad (30)$$

$$= \frac{2M}{b} d\theta_c - \frac{J}{b} da$$

Since dJ is a perfect differential, the general expression for J from Eq 30, that is

$$J = 2 \int_0^{\theta_c} \frac{M}{b} d\theta_c - \int_{a_0}^{a} \frac{J}{b} da \qquad (31)$$

holds for any history of a and θ_c leading to the current values and is necessarily independent of the history. With no change in crack length, Eq 31 reduces to Eq 29. In the presence of growth, the second term represents a correction to Eq 29 which should be but is not currently used to determine J from experimental data [7,11]. For small amounts of growth, this correction is usually small but not necessarily negligible. In addition, b should take on its variable value in the first term in Eq 31.

As in the general discussion in the previous section, let C_M be the compliance of a linear spring in series with the bend specimen. The total load point rotation is given by

$$\theta_T = C_M M + \theta = C_M M + \theta_{nc} + \theta_c \qquad (32)$$

To obtain an expression for $(\partial J/\partial a)_{\theta_T}$, it is first necessary to evaluate $(\partial J/\partial a)_M$. From Eq 28

$$\left(\frac{\partial J}{\partial a}\right)_M = \int_0^M \left(\frac{\partial^2 \theta}{\partial a^2}\right)_M dM \qquad (33)$$

Using Eq 26, it is readily verified that

$$\left(\frac{\partial^2 \theta}{\partial a^2}\right)_M = \left(\frac{\partial^2 \theta_c}{\partial a^2}\right)_M = \frac{6M}{b^2}\left(\frac{\partial \theta_c}{\partial M}\right)_a + \frac{4M^2}{b^2}\left(\frac{\partial^2 \theta_c}{\partial M^2}\right)_a \qquad (34)$$

from which it follows that

48 ELASTIC-PLASTIC FRACTURE

$$\left(\frac{\partial J}{\partial a}\right)_M = 6 \int_0^{\theta_c} \frac{M}{b^2} d\theta_c + 4 \int_0^{\theta_c} \frac{M^2}{b^2} \left(\frac{\partial^2 \theta_c}{\partial M^2}\right)_a dM \quad (35)$$

Here again it is important to note that $(\partial J/\partial a)_M$ depends only on the current values of a and θ_c or M. In Eq 35 it is understood that the integrals are to be evaluated with a held fixed at the current value. The first term in Eq 35 is $3J/b$. The second term can be integrated by parts and combined with the first to give

$$\left(\frac{\partial J}{\partial a}\right)_M = -\frac{J}{b} + \frac{4M^2}{b^2}\left(\frac{\partial \theta_c}{\partial M}\right)_a \quad (36)$$

For the bend specimen, the general expression, Eq 23, translates to

$$\left(\frac{\partial J}{\partial a}\right)_{\theta_T} = \left(\frac{\partial J}{\partial a}\right)_M - \left(\frac{\partial J}{\partial M}\right)_a \left(\frac{\partial \theta_c}{\partial a}\right)_M \left[C_M + \left(\frac{\partial \theta}{\partial M}\right)_a\right]^{-1} \quad (37)$$

This expression can be substantially simplified without approximation using Eqs 27 and 36

$$\left(\frac{\partial J}{\partial M}\right)_a = \frac{2}{b} M \left(\frac{\partial \theta_c}{\partial M}\right)_a \quad (38)$$

and

$$\left(\frac{\partial \theta}{\partial M}\right)_a = \frac{d\theta_{nc}}{dM} + \left(\frac{\partial \theta_c}{\partial M}\right)_a \equiv C_{nc} + \left(\frac{\partial \theta_c}{\partial M}\right)_a \quad (39)$$

Here C_{nc} is the compliance of the uncracked specimen. In general, C_{nc} may be a function of M, but for a deeply cracked specimen it will usually be the elastic compliance since M will seldom exceed initial yield of the uncracked specimen. The resulting exact reduction of Eq 37 is

$$\left(\frac{\partial J}{\partial a}\right)_{\theta_T} = -\frac{J}{b} + \frac{4M^2}{b^2}\left[\frac{C}{1 + C\left(\frac{\partial M}{\partial \theta_c}\right)_a}\right] \quad (40)$$

where C is the combined compliance

$$C = C_M + C_{nc} \quad (41)$$

In a rigid machine, $C_M = 0$ and Eq 40 applies with $C = C_{nc}$. In a dead-load machine, $C_M \to \infty$ and Eq 40 reduces to the expression for $(\partial J/\partial a)_M$ in Eq 36. For a fully yielded, elastic-perfectly plastic specimen, M is the limit moment which is independent of θ_c for fixed a, so that Eq 40 becomes very simply just

$$\left(\frac{\partial J}{\partial a}\right)_{\theta_T} = -\frac{J}{b} + \frac{4M^2 C}{b^2} \qquad (42)$$

The analysis of the three-point bend specimen of Fig. 4b is essentially identical to that carried out in the foregoing. Now Δ_T is the total load point displacement through which P works and Δ_C is the displacement of the specimen due to the presence of the crack defined similarly to Eq 25. With C as the combined compliance, now as

$$C = C_M + \frac{d\Delta_{nc}}{dF} \equiv C_M + C_{nc}$$

one finds

$$\left(\frac{\partial J}{\partial a}\right)_{\Delta_T} = -\frac{J}{b} + \frac{4P^2}{b^2}\left[\frac{C}{1 + C\left(\frac{\partial P}{\partial \Delta_C}\right)_a}\right] \qquad (43)$$

Both Eqs 40 and 43 are exact for deeply cracked specimens. It is now possible to make contact with the simpler approximate analysis of a fully yielded, elastic-perfectly plastic three-point bend specimen given by Paris et al [1]. At the limit state, $M = PL/4 = A\sigma_0 b^2$, where σ_0 is the tensile yield stress and $A \simeq 0.35$ for plane strain and $A \simeq 0.27$ for plane stress. In a rigid testing machine, $C = C_{nc} = L^3/(4EW^3)$. Under these circumstances, Eq 43 immediately leads to

$$T = \frac{E}{\sigma_0^2}\left(\frac{\partial J}{\partial a}\right)_{\Delta_T} = 16A^2 \frac{b^2 L}{W^3} - \frac{EJ}{\sigma_0^2 b} \qquad (44)$$

The first term in Eq 44 is the same as that obtained in Ref 1, while the second term has been approximated in the analysis of Ref 1.

Prior to the initiation of crack growth, Eqs 40 and 43 involve quantities which can all be obtained from a single test record. Furthermore, all quantities on the right-hand side of Eqs 40 and 43 are continuous across the initiation point and, thus, so are $(\partial J/\partial a)_{\theta_T}$ and $(\partial J/\partial a)_{\Delta_T}$. It follows that stability of crack growth initiation can be assessed using Eqs 40 or 43 from quantities obtained directly from the experimental test record just prior to initiation. For a fully yielded specimen with little strain hardening, it may be possible in some instances to neglect $C(\partial M/\partial \theta_c)_a$ in Eq 40 (or the analogous term in Eq 43 and thereby use Eq 42). When this term is not negligible, it will be necessary to estimate $(\partial M/\partial \theta_c)_a$ in order to assess stability using Eq 40 beyond the initiation of crack growth.

Numerical Results for a Deeply Cracked Bend Specimen in Plane Stress

The estimation procedure of Shih and Hutchinson [*14*] will be used to relate J and θ_c to M for the deeply cracked, plane stress bend specimen of Fig. 4a. These relations are then sufficient to calculate $(\partial J/\partial a)_{\theta_T}$ using Eq 40. The material is assumed to be governed by a Ramberg-Osgood stress-strain curve in uniaxial tension, that is

$$\epsilon/\epsilon_0 = \sigma/\sigma_0 + \alpha(\sigma/\sigma_0)^n \tag{45}$$

where σ_0 is the effective yield stress and $\epsilon_0 = \sigma_0/E$. The estimation procedure uses the linear elastic solution and a fully plastic power-law solution, which is given in Ref *14*, to interpolate over the entire range from small-scale yielding to fully yielded conditions.

The results of the procedure as applied to the deeply cracked plane stress bend specimen are

$$\frac{EJ}{\sigma_0^2 b} = 1.135\,\psi^3 \left(\frac{M}{M_O}\right)^2 + \alpha h_1(n)\left(\frac{M}{M_O}\right)^{n+1} \tag{46}$$

$$\frac{\theta_c}{\epsilon_0} = 4.238\,\psi^2 \left(\frac{M}{M_O}\right) + \alpha h_3(n)\left(\frac{M}{M_O}\right)^n \tag{47}$$

where $M_O = A\sigma_0 b^2$ and $A = 0.27$. The plasticity adjustment for the effective crack length in the low M range is incorporated through the ψ factor [*14*], which for the deeply-cracked specimen is given by

$$\frac{1}{\psi} = \frac{b_{\text{eff}}}{b} = 1 - 0.1806\left(\frac{n-1}{n+1}\right)\left(\frac{M}{M_O}\right)^2, \quad M \leq M_O \tag{48}$$

$$= 1 - 0.1806\left(\frac{n-1}{n+1}\right), \quad M \geq M_O \tag{49}$$

The numerical coefficients in the first terms in Eqs 46 and 47 and in Eq 48 are from the appropriate deeply cracked limits for the linear elastic problems which can be found in Ref *15*. For $h_1(n)$ and $h_3(n)$ we have used the numerical values presented in Table 2 of Ref *14* which were computed for $b/W = 1/2$. (These are listed again in Table 1 of the present paper.) As discussed in Ref *14*, these values are expected to be very close to the deeply-cracked limit for $b/W \to 0$ when $n \geq 3$.

TABLE 1—*Numerical values for $h_1(n)$ and $h_3(n)$, taken from Ref 14.*

	$n = 2$	$n = 3$	$n = 5$	$n = 10$
Bend				
$h_1(n)$	0.957	0.851	0.717	0.551
$h_3(n)$	2.36	2.03	1.59	1.12
Center-cracked				
$h_1(n)$	...	1.09	0.906	0.717
$h_3(n)$	...	1.93	1.35	0.88

From Eqs 47 and 48

$$\frac{M_0}{\epsilon_0}\left(\frac{\partial \theta_c}{\partial M}\right)_a = 4.238\,\psi^2 + 3.062\,\psi^3 \left(\frac{n-1}{n+1}\right)\left(\frac{M}{M_0}\right)^2 \quad$$
$$+ \alpha n h_3(n)\left(\frac{M}{M_0}\right)^{n-1} \quad (50)$$

for $M \leq M_0$. For $M > M_0$ we take

$$\frac{M_0}{\epsilon_0}\left(\frac{\partial \theta_c}{\partial M}\right)_a = 4.238\,\psi^2 + 3.062\,\psi^3 \left(\frac{n-1}{n+1}\right)$$
$$+ \alpha n h_3(n)\left(\frac{M}{M_0}\right)^{n-1} \quad (51)$$

The second term in Eq 51 would be absent if we had used Eq 49. It is included to ensure continuity of $(\partial \theta_c/\partial M)_a$ at $M = M_0$. Both this quantity and $(\partial J/\partial M)_a$ are discontinuous across $M = M_0$ according to the estimation procedures of Ref *14*, leading to Eqs 46-49, where in fact these two quantities should be continuous. In this paper, where we are primarily interested in displaying the trends due to strain hardening, the effect of including or omitting the second term in Eq 51 makes relatively little difference. For future work, however, the estimation procedure will need to be improved upon in this regard.

Equation 40 can be written in nondimensional form as

$$T = \frac{E}{\sigma_0^2}\left(\frac{\partial J}{\partial a}\right)_{\theta_T} = -\frac{EJ}{\sigma_0^2 b}$$
$$+ 4A^2 \left(\frac{M}{M_0}\right)^2 \overline{C}\left[1 + A\overline{C}\frac{\epsilon_0}{M_0}\left(\frac{\partial M}{\partial \theta_c}\right)_a\right]^{-1} \quad (52)$$

where $\overline{C} = Eb^2 C = Eb^2(C_M + C_{nc})$ is the nondimensional combined com-

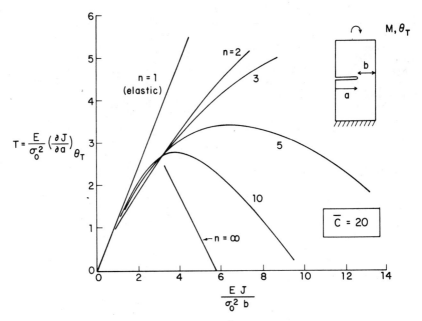

FIG. 5—*Numerical results for T versus $EJ/(\sigma_0^2 b)$ for a deeply cracked plane-stress bend specimen for various values of the strain hardening index n. Nondimensional combined compliance is $\bar{C} = 20$, corresponding to a typically dimensioned specimen in a rigid test machine (see text).*

pliance. Values of T and $EJ/(\sigma_0^2 b)$ for values of M/M_O are generated using Eqs 46-52. We have taken $\alpha = 3/7$ for $n > 1$ and for $n = 1$ we have used the linear elastic formulas. In Figs. 5 and 6 we have cross-plotted T as a function of $EJ/(\sigma_0^2 b)$ for various values of the hardening index n. The curve in Fig. 5 labeled $n = \infty$ is obtained from the fully yielded result, Eq 42, for the elastic-perfectly plastic specimen, that is

$$T = -\frac{EJ}{\sigma_0^2 b} + 4A^2\bar{C} \qquad (53)$$

Fully yielded conditions set in when $M > M_O$ and this occurs when $EJ/(\sigma_0^2 b)$ is a bit larger than 2 for essentially all the higher n-values.

For the bend specimen of Fig. 4a

$$Eb^2 C_{nc} = 12Lb^2/W^3 \qquad (54)$$

The results in Fig. 5 are for a combined nondimensional compliance $\bar{C} = 20$. In a rigid test machine, this corresponds to a specimen with $Lb^2/W^3 = 5/3$, which is representative of a typical test specimen. For $n \geq 5$, it can be seen

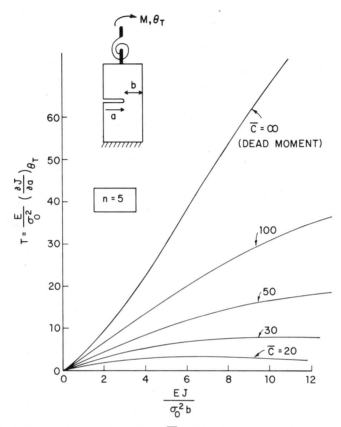

FIG. 6—*Influence of combined compliance $\bar{C}$ on T for plane-stress bend specimen for strain hardening index* $n = 5$.

from Fig. 5 that T will always be less than about 4 in plane stress. Fully plastic power-law solutions for bend specimens are not yet available for plane strain. However, judging from Eqs 52 and 53, we can perhaps expect T-values to be larger in plane strain by a factor of as much as 2, due to the presence of A^2. Thus it is reasonable to expect that T will always be less than some number around 10 in plane strain when $\bar{C} = 20$ and when $n \geq 5$. In contrast, T_{mat} spans the range $1 < T_{mat} < 200$ for a wide range of steels under nominally plane-strain conditions just following initiation (Ref *1*, Appendix II). Thus, only those specimens of steels in the low end of the range of T_{mat} will display unstable crack growth upon crack growth initiation in a bend specimen rigid test machine (compare Eqs 18 and 19).

The effect of increasing the combined compliance $\bar{C}$ is seen in Fig. 6 for $n = 5$. The lowest curve is from Fig. 5. The uppermost curve is for $\bar{C} = \infty$, corresponding to a specimen loaded under dead (constant) moment. Paris et al ([*1*], Part II) varied the combined compliance in their test program by including an extra bar in series with a three-point bend specimen. In this way they were able to achieve a tenfold increase in compliance with the associated large increase in T.

The parameter ω in Eq 10 related to the validity of J-controlled growth in the fully yielded range can be expressed in the revealing form

$$\omega = \left(\frac{\sigma_0^2 b}{EJ}\right) T \tag{55}$$

involving only the ordinate and abscissa of Figs. 5 and 6. For the results of Fig. 5 for $\overline{C} = 20$, ω does not exceed unity. Consequently, as a result of Eq 11, it is unlikely that the conditions for J-controlled growth are satisfied for an increment of crack growth with θ_T fixed under these low compliance conditions. This should be no cause for concern if $T_{\text{mat}} \gg T$ since then crack growth is almost certainly highly stable anyway. The parameter T in Eq 55 increases with increasing compliance. At the compliance associated with the onset of instability

$$\omega = \left(\frac{\sigma_0^2 b}{EJ}\right) T_{\text{mat}} \tag{56}$$

In the plane-strain tests of Paris et al ([1], Part II), $T_{\text{mat}} \simeq 36$ and $\omega \simeq 15$, using Eq 56 with $J = J_{\text{Ic}}$.

Two-Specimen Method for Determining (dJ_{mat}/da) and $(\partial M/\partial \theta_c)_a$ in J-Controlled Growth

Here we note a method which, in principle, can be used to measure (dJ_{mat}/da) without recourse to direct measurement of crack length changes. It will permit an experimental determination of $(\partial M/\partial \theta_c)_a$ once growth starts.

Consider two deeply cracked bend specimens of identical material, denoted by A and B, with differing initial ligament lengths $b_A{}^0$ and $b_B{}^0$. Equation 26 applies to both specimens, that is

$$M_A = b_A^2 F(\theta_c) \text{ and } M_B = b_B^2 F(\theta_c) \tag{57}$$

where b_A and b_B are the current ligaments. The $F(\theta_c)$ in each of Eqs 57 are the same since $F(\theta_c)$ depends only on material properties. Thus at the same value of θ_c, from Eq 29

$$J_A/J_B = b_A/b_B \tag{58}$$

Consequently, initiation, that is, $J = J_{\text{Ic}}$, will occur at a larger value of θ_c for

Specimen A than for Specimen B, if $b_A^0 < b_B^0$. With $b_A^0 < b_B^0$, Specimen A can be used to measure $F(\theta_c)$ from

$$F(\theta_c) = M_A/(b_A^0)^2 \qquad (59)$$

in the range of θ_c prior to initiation in A, as depicted in Fig. 7.

In Fig. 7 we have also depicted the curve of $M_B/(b_B^0)^2$ versus θ_c, where (b_B^0) is the initial ligament of B. Prior to initiation in B, the two curves in Fig. 7 must coincide, assuming both specimens have the same material and same degree of plane-strain constraint. From the second equation in Eq 57

$$dM_B = 2b_B \, db_B F + b_B^2 F' \, d\theta_c$$

where $F' = dF/d\theta_c$. With $da = -db_B$ the foregoing equation can be rearranged to give

$$da = \frac{b_B}{2} \left[\frac{F'(\theta_c)}{F(\theta_c)} d\theta_c - \frac{dM_B}{M_B} \right] \qquad (60)$$

This equation provides a relation for indirectly obtaining increments in crack length in Specimen B in terms of the measured relation between M_B and θ_c and from $F(\theta_c)$ determined by using A. The associated change in J_B, that is, $dJ_B = dJ_{mat}$, is given by Eq 30. Combining Eqs 60 and 30 gives

$$\frac{dJ_{mat}}{da} = 4F(\theta_c) \left[\frac{F'(\theta_c)}{F(\theta_c)} - \frac{1}{M_B} \left(\frac{dM_B}{d\theta_c} \right) \right]^{-1} - \frac{J_B}{b_B} \qquad (61)$$

These results, Eqs 60 and 61, may be used to assess crack length changes, da, and dJ_{mat}/da up to values of θ_c where initiation occurs in Specimen A, or

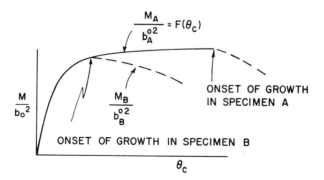

FIG. 7—*Curves of normalized moment as a function of θ_c for two deeply cracked bend specimens with differing initial uncracked ligaments ($b_A^0 < b_B^0$).*

$J_A = J_{Ic}$. This limit can be assessed from the onset of growth in Specimen B, where $J_B = J_{Ic}$, and using Eqs 29 and 58.

The practicality of using Eqs 60 and 61 together with experimental records must await further work.[5] One desirable feature of these relations is that the resistance curve data, dJ_{mat}/da versus Δa, are generated without having to specify a precise definition of initiation.

Using Eq 57 we also note that

$$\left(\frac{\partial M_B}{\partial \theta_c}\right)_a = b_B^2 F'(\theta_c) \qquad (62)$$

Thus $F'(\theta_c)$ obtained from Specimen A also provides the one term in the general expression, Eq 40, for $(\partial J/\partial a)_{\theta_T}$ which cannot be obtained from Specimen B itself. For small amounts of growth, the replacement of b_B by its initial value in Eqs 60–62 will introduce little error.

Analysis of Deeply Cracked Center-Notched and Edge-Notched Specimens Under Tension

In this section, expressions are obtained for $(\partial J/\partial a)_{\Delta T}$ for deeply cracked center and edge-notched specimens in plane strain or plane stress and for the deeply cracked round bar. First the two-dimensional plane specimens in Fig. 8a and b will be considered. In each case we now write the load-point displacement of the specimen Δ as the sum of the elastic part Δ_e and the plastic part Δ_P according to

$$\Delta = \Delta_e + \Delta_P \qquad (63)$$

For $b/w \ll 1$, dimensional analysis implies the general functional dependence

$$\Delta_P = bf(P/b) \qquad (64)$$

where P is the load per unit thickness carried by each ligament. Let

$$J_e = \int_0^P \left(\frac{\partial \Delta_e}{\partial a}\right)_P dP \qquad (65)$$

denote the value of J for an elastic specimen at P. Then

$$J = \int_0^P \left(\frac{\partial \Delta}{\partial a}\right)_P dP = J_e + \int_0^P \left(\frac{\partial \Delta_P}{\partial a}\right)_P dP \qquad (66)$$

---

[5] Although these methods were used in Ref 1, Part II, for data reduction, their full application and limitations to practical testing procedures are yet to be explored.

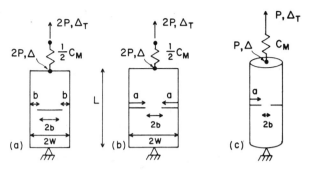

FIG. 8—*Center-cracked specimen (a), edge-cracked specimen (b), and edge-cracked round bar specimen (c).*

Using Eq 64 and following the development in Ref *13*, one can show that

$$J = J_e + b^{-1} \left\{ 2 \int_0^{\Delta_P} P d\Delta_P - P\Delta_P \right\} \tag{67}$$

Next using Eqs 64 and 66 we find

$$\left(\frac{\partial J}{\partial a}\right)_P = \left(\frac{\partial J_e}{\partial a}\right)_P + \int_0^P \frac{P^2}{b^2} \left(\frac{\partial^2 \Delta_P}{\partial P^2}\right)_a dP \tag{68}$$

But for deeply cracked specimens

$$J_e = kP^2/(2bE) \tag{69}$$

where

$$k = (1 - \nu^2)8\pi/(\pi^2 - 4) \text{ (plane strain, center-notched)}$$
$$= 8\pi/(\pi^2 - 4) \text{ (plane stress, center-notched)}$$
$$= (1 - \nu^2)8/\pi \text{ (plane strain, edge-notched)}$$
$$= 8/\pi \text{ (plane stress, edge-notched)}$$

and where ν is Poisson's ratio.

So, $(\partial J_e/\partial a)_P = J_e/b$. Integrating the second term in Eq 68 by parts and using Eq 66 gives

$$\left(\frac{\partial J}{\partial a}\right)_P = -\frac{J}{b} + \frac{2J_e}{b} - \frac{P\Delta_P}{b^2} + \frac{P^2}{b^2}\left(\frac{\partial \Delta_P}{\partial P}\right)_a \tag{70}$$

With 1/2 C_M as the compliance of a linear spring in series with the specimen, the total load point displacement is

$$\Delta_T = C_M P + \Delta = C_M P + \Delta_e + \Delta_P \tag{71}$$

58 ELASTIC-PLASTIC FRACTURE

To reduce the general expression, Eq 23, for $(\partial J/\partial a)_{\Delta_T}$, we note the following relations

$$\left(\frac{\partial \Delta_P}{\partial a}\right)_P = -\frac{\Delta_P}{b} + \frac{P}{b}\left(\frac{\partial \Delta_P}{\partial P}\right)_a$$

$$\left(\frac{\partial \Delta_e}{\partial a}\right)_P = \frac{kP}{Eb} = \frac{2J_e}{P}$$

Let $C_e = (\partial \Delta_e/\partial P)_a$ be the compliance of the cracked elastic half-specimen and denote by C the combined elastic compliance

$$C = C_e + C_M \tag{72}$$

Then Eq 23 can be reduced by algebraic manipulation without approximation to

$$\left(\frac{\partial J}{\partial a}\right)_{\Delta_T} = -\frac{J}{b} + \left[\frac{2J_e}{b} - \frac{P\Delta_P}{b^2} + \frac{P^2}{b^2}\left(\frac{\partial \Delta_P}{\partial P}\right)_a\right] \times$$

$$\left[C - \frac{k}{E} + \frac{\Delta_P}{P}\right] \times \left[C + \left(\frac{\partial \Delta_P}{\partial P}\right)_a\right]^{-1} \tag{73}$$

In a dead-load machine, $C = \infty$ and Eq 73 reduces to $(\partial J/\partial a)_P$. For a fully yielded, elastic-perfectly plastic specimen, P is at the limit load and $(\partial P/\partial \Delta_P)_a = 0$ so that Eq 73 becomes

$$\left(\frac{\partial J}{\partial a}\right)_{\Delta_T} = -\frac{J}{b} + \frac{P^2}{b^2}\left(C - \frac{k}{E} + \frac{\Delta_P}{P}\right) \tag{74}$$

The same comments made in connection with the analogous formulas for the bend specimen apply here; namely, Eq 73 allows determination of $(\partial J/\partial a)_{\Delta_T}$ from a single experimental record prior to initiation. Furthermore, $(\partial J/\partial a)_{\Delta_T}$ is continuous across the point of initiation. Beyond initiation, $(\partial P/\partial \Delta_P)_a$ must be estimated or perhaps neglected, as in Eq 74, if the specimen is fully yielded and strain hardening is not significant.

For the deeply cracked edge-notched round bar of Fig. 8c we write, copying the procedure for the bend specimen

$$\Delta_T = C_M P + \Delta = C_M P + \Delta_{nc} + \Delta_c \tag{75}$$

Now, with b as the radius of the circular ligament

$$\Delta_c = bf(P/b^2) \tag{76}$$

where P is the total load. Omitting details, we find

$$J = \frac{1}{2\pi b^2}\left[3\int_0^{\Delta_c} P\,d\Delta_c - P\Delta_c\right] \tag{77}$$

$$\left(\frac{\partial J}{\partial a}\right)_P = -\frac{J}{b} - \frac{P\Delta_c}{\pi b^3} + \frac{2}{\pi b^3}P^2\left(\frac{\partial \Delta_c}{\partial P}\right)_a \tag{78}$$

$$\left(\frac{\partial J}{\partial a}\right)_{\Delta_T} = -\frac{J}{b} - \frac{P\Delta_c}{\pi b^3} + \frac{1}{2\pi b^3}\left[4P^2C + 4P\Delta_c\right.$$

$$\left. - \Delta_c^2\left(\frac{\partial P}{\partial \Delta_c}\right)_a\right] \times \left[1 + C\left(\frac{\partial P}{\partial \Delta_c}\right)_a\right]^{-1} \tag{79}$$

where

$$C = \frac{d\Delta_{nc}}{dP} + C_M \tag{80}$$

For a fully yielded, elastic-perfectly plastic specimen, Eq 79 reduces to

$$\left(\frac{\partial J}{\partial a}\right)_{\Delta_T} = -\frac{J}{b} + \frac{P\Delta_c}{\pi b^3} + \frac{2P^2C}{\pi b^3} \tag{81}$$

Numerical Results for a Deeply Cracked Center-Notched Specimen in Plane Stress

We again employ the estimation procedure of Ref *14* together with the general formula, Eq 73, to generate numerical results for T for the deeply cracked center-notched specimen of Fig. 8a. The Ramberg-Osgood stress-strain curve, Eq 45, is used. Of the quantities needed in Eq 73, J_e has been given previously while J and Δ_P are

$$\frac{EJ}{\sigma_0^2 b} = \frac{k}{2}\psi\left(\frac{P}{P_0}\right)^2 + \alpha h_1(n)\left(\frac{P}{P_0}\right)^{n+1} \tag{82}$$

$$\frac{\Delta_P}{\epsilon_0 b} = k\ln\psi\left(\frac{P}{P_0}\right) + \alpha h_3(n)\left(\frac{P}{P_0}\right)^n \tag{83}$$

where $P_0 = \sigma_0 b$ and

$$\frac{1}{\psi} = \frac{b_{\text{eff}}}{b} = 1 - \frac{k}{4\pi}\left(\frac{n-1}{n+1}\right)\left(\frac{P}{P_0}\right)^2, \quad P \leq P_0$$

$$= 1 - \frac{k}{4\pi}\left(\frac{n-1}{n+1}\right), \quad P \geq P_0 \tag{84}$$

60   ELASTIC-PLASTIC FRACTURE

The linear elastic solution for the limit of a deeply cracked specimen from Ref *15* has been used to arrive at the first terms in Eq 82 and ψ; the first term in Eq 83 follows from the same limiting solution plus the definition of Δ_P in Eq 63. From Eqs 83 and 84

$$\frac{P_0}{\epsilon_0 b}\left(\frac{\partial \Delta_P}{\partial P}\right)_a = k \ln \psi + 2k(\psi - 1) + \alpha n h_3(n) \left(\frac{P}{P_0}\right)^{n-1} \qquad (85)$$

To ensure continuity at P_0, this same expression is used for $P \geq P_0$. Values of h_1 and h_3 for the power-law solution are listed in Table 1. These values are converted from Table 1 of Ref *14* using $h_1 = (a/w)g_1$ and

$$h_3 = (a/w)g_3/(1 - a/w) \text{ for } a/w = 3/4$$

Equation 73 can be expressed in nondimensional form for T, involving only the foregoing quantities and the nondimensional combined compliance

$$\bar{C} = EC = E(C_e + C_M) \qquad (86)$$

Curves of T as a function of $EJ/(\sigma_0^2 b)$ are shown in Figs. 9 and 10. For the deeply cracked specimen

$$EC_e = L/W + k \ln (W/b) \qquad (87)$$

so the value $\bar{C} = 10$ in Fig. 9 can be regarded as being fairly representative of a test specimen with typical dimensions in a rigid test machine. In contrast to the bend specimen of Fig. 5 in the same range of $EJ/(\sigma_0^2 b)$, a decrease in strain hardening increases T, that is, decreases stability. In both cases the effect of strain hardening is quite significant. Only for the bend specimen does T decrease at sufficiently large increasing J; this is most evident from the formula for the perfectly plastic case, Eq 42. The effect of increasing the combined compliance $\bar{C}$ is seen in Fig. 10 for $n = 5$. Here the trends are very similar to those for the bend specimen.

Discussion

Two open questions are the maximum allowable amount of crack growth and the minimum admissible value of the nondimensional parameter ω to ensure J-controlled growth to a reasonable approximation. For a fully yielded specimen with ligament b, the most relevant material-based estimate of ω just following initiation is Eq 56, that is

$$\omega = \frac{b}{J_{Ic}}\frac{dJ}{da} = \left(\frac{\sigma_0^2 b}{EJ_{Ic}}\right) T_{mat}$$

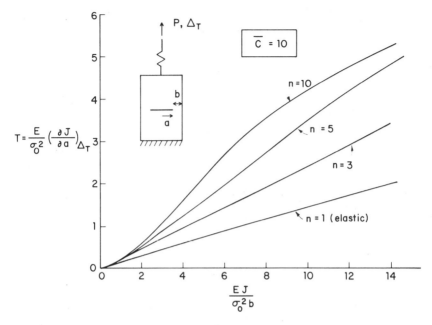

FIG. 9—*Curves of T as a function of $EJ/(\sigma_0^2 b)$ for various levels of strain hardening index n for a deeply center-cracked plane-stress tension specimen. Nondimensional combined compliance is $\bar{C} = 10$, corresponding to a typically dimensioned specimen in a rigid test machine (see text).*

The range of this parameter for the steels listed by Paris et al ([1] Appendix II) is roughly $0.1 < \omega < 100$ with most entries satisfying $\omega > 10$. Thus it does seem likely that there will be an important class of metals whose properties are such that a limited amount of growth can be analyzed, both for equilibrium and stability, using the deformation theory J. However, Landes and Begley ([16] and private communications) have noted that the J-integral resistance curves for compact or bend specimens have a different slope, dJ_{mat}/da, than those for center-cracked specimens. On the other hand, their curves were plotted with J-values which were uncorrected for effects of crack length changes, such as is illustrated by the second term in Eq 31. Moreover, in their tests ω was not evaluated and their ligaments sizes, b, did not quite meet the other requirements, as illustrated by Eq 12. In addition, they report unsymmetrical crack extension for the two crack tips in their center-crack specimens. Nevertheless, this perplexing point cannot be dismissed; thus further exploration is warranted until a reasonable explanation is found.

It should be most interesting to compare results from deformation theory calculations and flow theory calculations for precisely the same prescribed growth conditions. In this way it should be possible to learn more about the

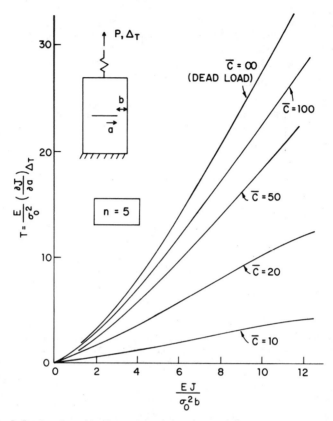

FIG. 10—*Influence of combined compliance on T for deeply center-cracked plane-stress tension specimen for strain-hardening index n = 5.*

influence of strain hardening and configuration on the minimum permissible value of ω. It is worth bearing in mind that the simplest flow theories of plasticity, based on a smooth yield surface, have limitations which may also be important in the analysis of crack growth. In particular, certain of the incremental moduli tend to be overestimated by the simplest flow theories, leading in some problems to unrealistically high resistance to plastic deformation. In this connection, we also note that our argument requiring $\omega \gg 1$ may be relaxed somewhat by appealing to the total loading concept for justifying deformation theory as discussed by Budiansky [6].

In this paper we have emphasized the analysis of deeply cracked configuration because relatively simple formulas for T can be obtained which, in some instances, are directly applicable to test specimens. Work is underway [17] using estimation procedures based on the power-law solutions to predict T for arbitrary crack lengths. At the moment, however, power-law solutions are available only for a very few configurations and, in plane strain, only for the center-cracked strip [18].

Finally, within this work, suggestions have been made for methods of analysis of load-displacement records which permit establishing a material's J-integral R-curve without direct measurements of crack length changes. Though a special case of this approach was used with success in earlier tests [1], the method is unexplored, but holds great promise for simplifying testing. Indeed, those methods can be extended to eliminate the requirement for deeply cracked specimens, but that will be a topic for subsequent discussions. Furthermore, since those methods determine dJ_{mat}/da from load-displacement relations as influenced by crack growth, they are a most natural way to assess material parameters affecting stability. That is true because stability itself depends directly on the influence of crack growth on load-displacement behavior, as is observed throughout the analysis herein.

Acknowledgments

The first author (J. W. H.) acknowledges support of this work by the National Science Foundation, Grant No. ENG76-04019, and in part by the Division of Applied Sciences of Harvard University. The second author (P. C. P.) acknowledges the support of this work by the United States Nuclear Regulatory Commission, Contract No. NRC-03-77-029 with Washington University. The work was also significantly assisted by many stimulating discussions with several co-workers, including J. R. Rice, H. Tada, A. Zahoor, H. Ernst, C. F. Shih, and R. Gamble.

References

[1] Paris, P., Tada, H., Zahoor, A., and Ernst, H., "A Treatment of the Subject of Tearing Instability," U. S. Nuclear Regulatory Commission Report NUREG-0311, Aug. 1977. (See also papers in this publication by these authors.)
[2] *Fracture Toughness Evaluation by R-Curve Methods, ASTM STP 527*, American Society for Testing and Materials, 1974.
[3] McClintock, F. and Irwin, G. R. in *Fracture Toughness Testing and Its Applications, ASTM STP 381*, American Society for Testing and Materials, 1965, pp. 84-113.
[4] Rice, J. R. in *Mechanics and Mechanisms of Crack Growth*, British Steel Corp., 1973.
[5] Rice, J. R. in *Fracture*, Vol. 2, Academic Press, New York, 1968.
[6] Budiansky, B., *Journal of Applied Mechanics*, Vol. 26, 1959, pp. 259-264.
[7] Begley, J. A. and Landes, J. D. in *Fracture Analysis, ASTM STP 560*, American Society for Testing and Materials, 1974, pp. 170-186.
[8] Hutchinson, J. W., *Journal of the Mechanics and Physics of Solids*, Vol. 16, 1968, pp. 13-31.
[9] Rice, J. R. and Rosengren, G. F., *Journal of the Mechanics and Physics of Solids*, Vol. 16, 1968, pp. 1-12.
[10] McClintock, F. A. in *Fracture*, Vol. 3, Academic Press, New York, 1971.
[11] Landes, J. D. and Begley, J. A. in *Fracture Toughness, ASTM STP 514*, American Society for Testing and Materials, 1972, pp. 1-20 and 24-39.
[12] Shih, C. F., de Lorenzi, H. G., Andrews, W. R., Van Stone, R. H., and Wilkinson, J. P. D., "Methodology for Plastic Fracture," Fourth Quarterly Report by General Electric Co. to Electric Power Research Institute, 6 June 1977.
[13] Rice, J., Paris, P., and Merkle, J. in *Progress in Flaw Growth and Fracture Toughness Testing, ASTM STP 536*, American Society for Testing and Materials, 1973, pp. 231-245.

[14] Shih, C. F. and Hutchinson, J. W., *Journal of Engineering Materials and Technology*, Vol. 98, 1976, pp. 289–295.
[15] Tada, H., Paris, P. C., and Irwin, G. R., *The Stress Analysis of Cracks Handbook*, Del Research Corp., 226 Woodbourne Drive, St. Louis, Mo., 1973.
[16] Begley, J. A. and Landes, J. D., *International Journal of Fracture Mechanics*, Vol. 12, No. 5, Oct. 1976.
[17] Zahoor, A., work in progress.
[18] Goldman, N. L. and Hutchinson, J. W., *International Journal of Solids and Structures*, Vol. 2, 1975.

C. F. Shih,[1] H. G. deLorenzi,[1] and W. R. Andrews[1]

Studies on Crack Initiation and Stable Crack Growth

REFERENCE: Shih, C. F., deLorenzi, H. G., and Andrews, W. R., "**Studies on Crack Initiation and Stable Crack Growth,**" *Elastic-Plastic Fracture, ASTM STP 668*, J. D. Landes, J. A. Begley, and G. A. Clarke, Eds., American Society for Testing and Materials, 1979, pp. 65–120.

ABSTRACT: Experimental results are presented which suggest that parameters based on the J-integral and the crack opening tip displacement δ are viable characterizations of crack initiation and stable crack growth. Observations based on some theoretical studies and finite-element investigations of the extending crack revealed that J and δ when appropriately employed do indeed characterize the near-field deformation. In particular, the analytical and experimental studies show that crack initiation is characterizable by the critical value of J or δ, and stable crack growth is characterizable in terms of the J or δ resistance curves. The crack opening angle, $d\delta/da$, appears to be relatively constant over a significant range of crack growth. Thus, appropriate measures of the material toughness associated with initiation are J_{Ic} and δ_{Ic}, and measures of material toughness associated with stable crack growth are given by the dimensionless parameters $T_J [= (E/\sigma_o^2)(dJ/da)]$ and $T_\delta [= (E/\sigma_o)(d\delta/da)]$. The two-parameter characterization of fracture behavior by J_{Ic} and T_J or δ_{Ic} and T_δ is analogous to the characterization of deformation behavior by the yield stress and strain hardening exponent.

KEY WORDS: fracture, cracks, crack initiation, crack growth, crack opening displacement, crack opening angle, J-integral, resistance curve, plastic deformation, elastic properties, plastic properties, fracture toughness, tearing modulus, crack propagation

The progress of ductile fracture from an existing sharp-tipped flaw may be separated into four regimes: the blunting of the initially sharp crack tip, initial crack growth, stable crack growth, and unstable crack propagation. An illustration of these regimes is given in Fig. 1. For low-toughness materials, there is relatively little crack-tip blunting, and essentially no stable growth when fracture proceeds under plane-strain conditions. For high-toughness materials on the upper shelf, there is significant crack-tip

---

[1] Mechanical engineers, Corporate Research and Development, and metallurgical engineer, Materials and Processes Laboratory, respectively, General Electric Co., Schenectady, N.Y.

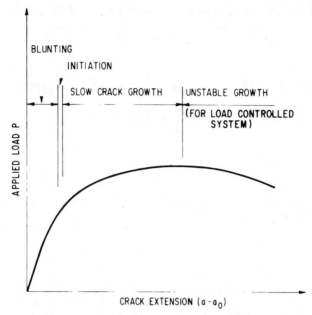

FIG. 1—*Stages of crack extension, showing crack tip blunting, initiation, and growth.*

blunting and substantial stable crack growth. More importantly, the nominal 'resistance' of the material to further crack extension increases with increasing crack growth. In this case, the conventional assessment of the margin of safety of flawed structures based on the onset of crack extension is conservative, since the nominal or measured 'toughness' of the material may increase significantly with relatively small crack extension.

Stable crack growth can occur under small-scale yielding conditions when the plastic zone size is small compared to crack length, and under large-scale yielding conditions when the plastic zone extends across the remaining ligament [1-4].[2] Under small-scale yielding conditions, the characterization of material resistance to crack growth by the resistance curve procedure based on the stress intensity factor K has received considerable attention [4]. Extensive investigations on the characterization of crack growth by J-resistance curves and on approximate methods for predicting fracture instability are reported in Ref 5.

In the fracture of ductile materials, the central portion of the crack typically advances in a thumbnail fashion, and eventually the outer edges fail in shear. The stress and deformation state ahead of the crack that extends in this manner is rather complex, and it is unlikely that the crack

---

[2] The italic numbers in brackets refer to the list of references appended to this paper.

extension can be characterized by a single parameter valid for the changing modes of fracture. This may be argued from the completely different strain energy density fields associated with plane-strain and plane-stress conditions at the crack tip. However, some theoretical and experimental results suggest that a single-parameter characterization may apply if crack extension occurs by a single mode of fracture, that is, when it is either strictly flat or shear fracture under either plane-strain or plane-stress conditions [1-5,16,29]. The critical value of the parameter will of course be different for the two different modes of fracture.

In this investigation, we seek characterizing parameters for macroscopically flat fracture (initiation and growth) under large-scale yielding conditions. Throughout the investigation, a close coupling was maintained between testing and analysis. The basic material employed in the experimental program is A533B steels. Most tests were carried out at upper-shelf temperatures, where the mode of fracture is dimpled rupture.

Potential Fracture Criteria

In this section we discuss the background for the potential fracture criteria that have been studied during our investigations. In particular, the J-integral and the crack opening displacement (COD) are discussed in detail, because, as is demonstrated in this paper, they seem to have considerable potential for predicting the behavior of growing cracks.

The works of Hutchinson [6] and Rice and Rosengren [7] revealed that, for stationary cracks, the stresses and strains in the vicinity of the crack tip under both small-scale yielding and fully plastic conditions may be represented by

$$\sigma_{ij} = \sigma_o \left(\frac{EJ}{\sigma_o^2 I_n r}\right)^{1/(n+1)} \tilde{\sigma}_{ij}(\theta, n)$$

$$\epsilon_{ij} = \frac{\sigma_o}{E}\left(\frac{EJ}{\sigma_o^2 I_n r}\right)^{n/(n+1)} \tilde{\epsilon}_{ij}(\theta, n)$$

(1)

where

J = J-integral defined by Rice [8],
σ_o = yield stress,
E = elastic modulus,
r = radial distance from crack tip,
$\tilde{\sigma}_{ij}, \tilde{\epsilon}_{ij}$ = known dimensionless functions of the circumferential position θ and the hardening exponent n, and
I_n = constant which is a function only of n [6,9].

In Eq 1, the J-integral is the amplitude of the stress and strain singularity.[3]

For an ideally plastic material ($n = \infty$), the strain fields exhibit a $1/r$ singularity. In this case the foregoing expressions may also be rewritten in terms of the crack tip opening displacement δ_t by exploiting the relationship [8,10]

$$J = \alpha \, \sigma_o \, \delta_t \qquad (2)$$

where α is a parameter of the order unity. Thus

$$\sigma_{ij} = \sigma_o \, \tilde{\sigma}_{ij} \, (\theta, n = \infty)$$

and

$$\epsilon_{ij} = \frac{\alpha \, \delta_t}{I_n \, r} \, \tilde{\epsilon}_{ij} \, (\theta, n = \infty) \qquad (3)$$

The functions $\tilde{\sigma}_{ij}(\theta, n = \infty)$ and $\tilde{\epsilon}_{ij}(\theta, n = \infty)$ are the stress and strain variations associated with the Prandtl field [6-10], which represents the nonhardening limit of Eq 1. Expressions for strain hardening materials where δ appears as the amplitude of the singular fields have been discussed by Tracey [11]. Equations 1 and 3 are known as the Hutchinson-Rice-Rosengren (HRR) singularities or fields, and J and δ are the HRR field parameters. When the HRR field encompasses the fracture process zone (and there is evidence that it is indeed the case for both small- and large-scale yielding conditions at the onset of crack growth for certain crack configurations), the HRR parameters, J and δ, are natural candidates for characterizing fracture.

In crack-growth situations, the near-tip field is far more complex than in the stationary case. To date there is no complete description of the stress and strain fields ahead of an extending crack. Some features have emerged from studies due to Rice [12], Chitaley and McClintock [13], and Amazigo and Hutchinson [14]. In general these studies revealed a milder singularity—an $\ln(1/r)$ strain singularity for elastic-perfectly plastic materials. The studies by Rice, based on a J_2 flow theory of plasticity for an ideally plastic material ($n = \infty$), showed that the incremental strains in the immediate vicinity of the crack are related to an increase of the crack

---

[3] The J-integral characterizes the crack tip field in the same spirit that the elastic stress-intensity factor characterizes the elastic singular field under small-scale yielding conditions [33]. Expressions similar to Eq 1 may also be written for mixed-mode situations (cracks subjected to combined tensile and shear loadings). The crack-tip fields are now characterized by two parameters, the J-integral, which is again the amplitude of the singularity, and a parameter M^p which is a measure of the relative strength of the tensile and shear stress directly ahead of the crack tip [9].

opening displacement $d\delta$, and the increment of the crack extension da, through the relationship [12]

$$d\epsilon_{ij} = \frac{d\delta}{r} f_{ij}(\theta) + \frac{\sigma_0 \, da}{E \, r} \ln \frac{R(\theta)}{r} g_{ij}(\theta) \qquad (4)$$

In terms of the rate of change of the strain field with crack growth, we can rewrite Eq 4 in the form

$$\frac{d\epsilon_{ij}}{da} = \frac{1}{r}\frac{d\delta}{da} f_{ij}(\theta) + \frac{\sigma_0}{E}\frac{1}{r} \ln \frac{R(\theta)}{r} g_{ij}(\theta) \qquad (5)$$

Here, $R(\theta)$ is a measure of the distance to the elastic-plastic boundary and g_{ij} is a dimensionless function of order unity. The first term of Eq 4 represents the additional strain due to crack-tip blunting if the crack did not advance during load/displacement increments, and f_{ij} is related to the stationary crack expression (Eq 3) by the relationship $f_{ij} = \alpha \, \tilde{\epsilon}_{ij}(\theta)/I_n$, and is of order unity. The second term in Eq 4 represents the additional plastic strain induced by the incompatible elastic strain increments caused by the advance of the stress field through the material.[4]

For ductile metals (A533B steels are an example), subsequent results will show that the first term in Eqs 4 and 5 will be dominant terms over a significant interval (compared to δ_t) except right at the crack tip, where the $(1/r) \ln (R/r)$ singularity will dominate. In other words, the strains at the crack tip are uniquely characterized by the crack-tip opening angle, $d\delta/da$,[5] if

$$\frac{d\delta}{da} \gg \frac{\sigma_0}{E} \ln \left(\frac{R(\theta)}{r} \right) \qquad (6)$$

For r varying between $0.1 R$ and R, Eq 6 expresses the condition that the crack opening angle must be large compared with the yield strain σ_0/E.

An expression for the incremental strains during crack growth has recently been derived by Hutchinson and Paris [15] on the basis of J_2 deformation theory of plasticity. For an ideally plastic material, this expression reduces to

$$d\epsilon_{ij} = \frac{1}{\alpha \sigma_0} \frac{dJ}{r} f_{ij}(\theta) + \frac{1}{\alpha \sigma_0} J \frac{da}{r^2} \beta_{ij}(\theta) \qquad (7)$$

---

[4] For a perfectly plastic or low hardening material, the stress field in the immediate vicinity of the crack tip is essentially determined by the yield condition and the stress equilibrium equations. The stress field moving with the crack tip therefore generates an elastic strain increment which is not derivable from a set of displacement increments; the incompatible elastic strain increment in turn induces additional plastic strain at the crack tip.

[5] For a Cartesian coordinate system fixed in space, where x is parallel to the crack plane, $d\delta/da \approx -\partial\delta/\partial x$. Thus the crack opening angle reflects the actual slopes of the crack faces.

where $\beta_{ij}(\theta)$ is a dimensionless quantity of order unity. Rewriting Eq 7 in terms of the rate of change of strain with respect to crack extension, we get

$$\frac{d\epsilon_{ij}}{da} = \left(\frac{1}{\alpha\sigma_o}\right)\frac{1}{r}\frac{dJ}{da}f_{ij}(\theta) +$$

$$\left(\frac{1}{\alpha\sigma_o}\right)\frac{J}{r^2}\beta_{ij}(\theta) \qquad (8)$$

Hutchinson and Paris argued that J uniquely characterizes the near field if the first term in Eq 7 is the dominant term, that is, if[6]

$$\frac{dJ}{da} \gg \frac{J}{r} \qquad (9)$$

We note that while Eqs 5 and 8 are derived from distinctly different approaches (that is, J_2 flow theory and J_2 deformation theory, respectively), they have a similar structure. Their first terms represent proportional increments in the strain fields due to the increase in size or strength of the HRR singularity, while the second terms represent the nonproportional strain increments due to the advance of the HRR field with the extending crack. Therefore, if the HRR field increases in size more rapidly than it advances, then the crack opening angle, $d\delta/da$, and dJ/da describe the crack-tip environment for an extending crack. When the fracture process zone is enclosed in the region dominated by $d\delta/da$ or dJ/da, Eqs 5 and 8 coupled with the respective conditions (Eqs 6 and 9), provide the basis for a COD-based or a J-based resistance approach for stable crack growth.

Other fracture parameters were also examined in this investigation. They include the work density $\overline{W}$ in a process zone of size l [16] and the energy separation rate [17,18]. These parameters were found to be rather dependent on the finite-element mesh size and the crack-tip mesh configuration used in the analysis, and involve the introduction of an additional length parameter. Details of these parameters are given in Refs 16-18. In this discussion, we have focused our attention on the J-based and COD-based approaches.

Strategy

In order to ascertain which parameters are viable fracture criteria, a philosophy was followed whereby the results of tests were analyzed with

---

[6] The characterization of the singular strain fields by the J-integral assumes some amount of strain hardening since the θ-variation of the strains is not unique for perfectly plastic material [34]. We employ the perfectly plastic idealization to simplify the structure of Eqs 7 and 8 and for comparison with Eqs 4 and 5.

detailed finite-element computations. The tests were done on compact specimens, and center-cracked panels with varying remaining ligaments. Finite-element calculations in plane strain were carried out using the ADINA code [19] suitably modified to allow crack-tip blunting and growth and the computation of appropriate fracture parameters [20]. The analytical investigations were carried out in two phases and these are discussed in the following.

Phase I—Initial Filter

The objective of this first analytical phase is to observe the behavior of potential fracture criteria when crack growth in the finite-element model is forced to follow the measured relationship between load-line displacement (LLD) and crack extension. From experimental measurements, whether they be made by heat tinting, by the unloading compliance technique, or by other procedures, it is possible to determine the crack extension, $a - a_o$, as a function of the measured load line displacement (LLD). This information is the basic input to the finite-element crack-growth simulation calculations. In the finite-element model, the rate of crack growth is prescribed to follow the experimentally measured LLD $- (a - a_o)$ relationship. Thus at an applied displacement of, say, LLD$_j$, the crack is extended by an amount $a_j - a_o$ (determined from the experimental data) as illustrated in Fig. 2. The process is repeated for the entire finite-element crack growth calculations. The various potential fracture parameters, J and dJ/da at near-field and remote contours, COD and crack opening angle, energy separation rates, work density over a process zone, etc., are computed during the crack-growth simulation. The calculations are repeated for different test configurations. Here again, the experimentally measured LLD $- (a - a_o)$ relationship is employed to control the crack-growth rate in the corresponding finite-element model. In this phase of study, the compact specimen and the center-cracked panel which exhibit different deformation fields in the fully plastic state are employed; these are shown in Fig. 3. The parameters which appear to be viable are retained for analysis in the subsequent evaluation phase. The viability of the criteria is determined by a number of requirements that are discussed in the following.

Certain checks are carried out during the filter phase of the study. The calculated load-deflection relationship is compared with the experimental load-deflection record. The degree of agreement is a direct measure of the capability of the finite-element model and the crack extension technique for modeling the complete range of crack extension.

Phase II—Evaluation

In the evaluation phase, the analytical process is reversed. The selected

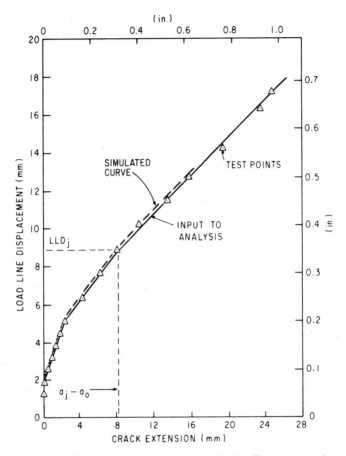

FIG. 2—*Load-line displacement versus crack extension for 4T compact specimen T52.*

viable parameters and their critical values (or their resistance curve values), which are obtained in Phase I, are now employed in finite-element crack-growth calculations for a variety of cracked configurations. In these calculations, the crack growth is governed by the fracture parameter itself. The results and, in particular, the LLD-crack extension relationship and the LLD-load relationship, are compared with the experimental measurements for the corresponding crack configurations. Based on these evaluations, final fracture criteria can be selected.

Requirements for Viable Fracture Parameters

In the two phases of the investigation, the fracture parameters are

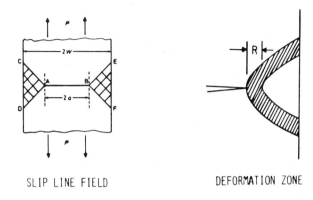
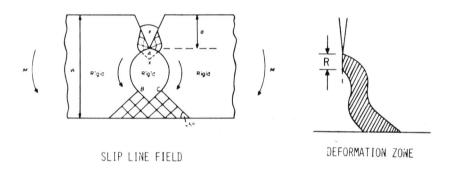

FIG. 3—*Global deformation fields under fully plastic conditions.*

assessed on the basis of several requirements which a viable fracture criterion should satisfy:

1. The parameter must be a measure of the stress and deformation state in the vicinity of the crack tip.
2. The critical value of the parameter must be independent of initial crack length and specimen geometry.
3. The parameter must be employable in instability analyses.
4. The parameter should be applicable to three-dimensional crack geometries, for example, to an elliptic surface flaw.
5. The parameter should be generalizable to mixed-mode fracture, though the critical value will depend on the relative ratio of shear to flat fracture.

6. The parameter should remain constant during crack extension. This is an attractive but not an essential feature. In an R-curve approach, the resistance parameter will increase with crack extension.

From the computational viewpoint, the viable parameter should possess the following additional properties:

7. The parameter must be relatively insensitive to finite-element modeling, to mesh size, process zone size, and load/displacement increment size.

8. The parameter should be computable within reasonable computational cost.

From the experimental viewpoint the viable parameter should have an additional feature:

9. The parameter should be obtainable from global measurements remote from the crack, but, if this is not possible, it must be obtainable from local measurements near the crack tip.

These requirements are the basis of the subsequent discussions on experimental and analytical results.

Finite Element Modeling of Blunting and Crack Growth

From a micromechanics viewpoint, ductile fracture involves three distinct processes—void nucleation, void growth, and void sheet coalescence [21]. Attempts to quantify these processes using rather simple models are discussed in Refs 22-24. However, the development of a quantitative fracture methodology based on microstructural aspects of dimpled rupture is a long way off. The alternative quantification of fracture at a macroscopic scale is adopted in this investigation. At this level, fracture occurs by crack-tip blunting followed by crack advance. This process has been observed using three separate techniques—rubber infiltration of a crack specimen, metallographic study of interrupted crack propagation tests, and macrofractograph of fractured specimens [16]. The crack-tip profiles observed in A533B steels tested at the upper shelf are shown in Figs. 4-6. These figures reveal a macroscopically flat but a microscopically tortuous crack path.

Our effort is focused on the identification of parameters that characterize the stress and strain state in a region (with an extent of several times the crack-tip opening displacement) that encompasses the fracture process zone. At this scale, a schematic of the crack tip profile is shown in Fig. 7. A finite-element model of the crack tip region using 8-noded isoparametric elements is shown in Fig. 8; in this figure, only the corner nodes are indicated. The remaining body is also modeled by 8-noded elements. In our two-dimensional studies, the original sharp crack is modeled with degenerate elements in which the two corner nodes and the mid-side node are initially collapsed to a common point—the crack tip—giving a $1/r$

FIG. 4—*Optical micrograph showing a blunted crack in an interrupted A533B compact specimen tested at 93°C (200°F).*

strain singularity as discussed by Barsoum [25]. As the load is increased, the crack-tip blunting is modeled by the separating of the nodes at the common point, as shown in Fig. 8b. Crack extension is modeled by shifting the current crack tip node as discussed by Shih et al [26] and is illustrated in Fig. 8c.[7] The adjacent mid-side nodes are also shifted so that they remain midway between the corner nodes. This shifting continues at each load or displacement increment as long as a controlling parameter for growth (for example, crack opening angle) is at a critical value; shifting ceases when its value is less than this critical value. When the ratio of the remaining element ligament to the overall element size reaches a critical ratio, the crack tip node is released together with the corresponding mid-side node. The shifting process is repeated with the nodes of the next element.

---

[7] As the crack extends, the element ahead of the crack is distorted, which causes the integration points (Gauss points) to be "dragged" along by the extending tip. Our studies revealed that if the crack tip extends by small increments (compared to the element size), the shift in the Gauss points at each increment is small, and the stresses associated with the Gauss points could be "dragged" along since the error incurred is small compared to the stress redistribution due to crack growth and additional external loading. Typically, the crack extends through an element in about 30 increments. For larger increments, the stresses associated with the new location of the Gauss points are obtained by linear interpolation.

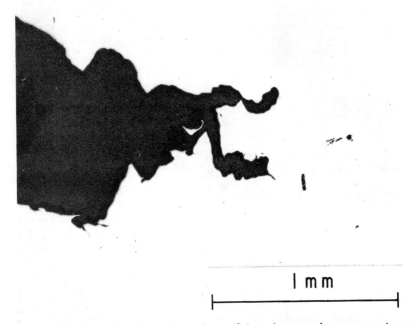

FIG. 5—*Optical micrograph showing crack growth in an interrupted compact specimen of A533B at 93°C (200°F).*

Eight-noded isoparametric elements are employed for several reasons. Nagtegaal, Parks, and Rice [27] have shown that conventional 4-noded isoparametric elements are not appropriate for analyses in the fully plastic range. Numerical studies by deLorenzi and Shih [28] showed that the 8-noded isoparametric element is ideally suited for fully plastic analyses. Furthermore, the latter element allows the modeling of crack tip blunting by the rather simple technique discussed in the foregoing, and does not require the degree of mesh refinement associated with constant-strain elements. Accurate procedures for evaluating the J-integral by contour integration and further details on crack-tip studies in the elastic and fully plastic range are discussed in Refs *16* and *20*.

Fracture Investigations

The basic material employed in our investigation is A533B steels, which are employed in the fabrication of pressure vessels. Because of the high toughness of A533B in the upper-shelf region, compact specimens of the size 1T to 4T invariably exhibit the formation of shear lips. The goal of this investigation is to identify parameters that will characterize flat fracture. Consequently, compact specimens of varying thickness and side

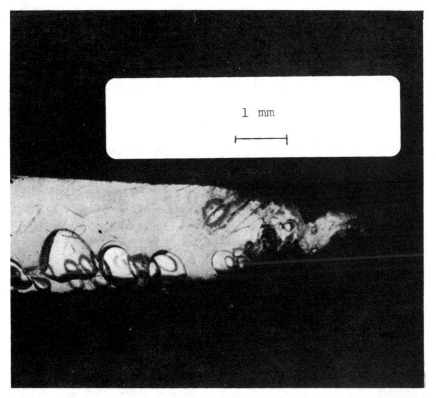

FIG. 6—*Silicone rubber replica of crack profile in 4T Specimen T71 at 93°C (200°F)*.

groove depth were tested at 93°C (200°F). A description of these tests, showing the effect of thickness and side grooving on fracture toughness measurements, has been given by Andrews and Shih [29]. The experiments showed that side grooves 12.5 percent or deeper promoted flat fracture surfaces with straight crack fronts as illustrated in Fig. 9. Some lateral contraction occurred for 12.5 percent grooves, but 25 and 50 percent grooves produced no appreciable lateral contraction. Three-dimensional elastic analysis of grooved and smooth compact specimens were carried out by Shih et al [30]. The analyses revealed that while 25 percent side grooves were sufficient to promote a uniform plane-strain condition across the crack front, its effect on the stress-intensity factor and compliance is minimal. In addition, the linear crack front allows the crack extension to be measured without ambiguity. Consequently, crack configurations with 25 percent side grooves were employed in the experimental program and the corresponding finite-element calculations assumed plane-strain conditions.

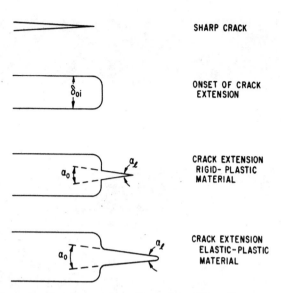

FIG. 7—*Schematic of crack extension illustrating COD and two definitions of angle between separated surfaces.*

Experimental Data for Compact Specimens

Comprehensive crack growth experiments have been carried out on 4T compact specimens of A533B steel at 93 and 260°C (200 and 500°F). Details of material properties, specimen geometries, remaining ligaments, and experimental procedures are given in Refs *16* and *29*. Two different heats were used: one for the thickness and side groove study (Material 1), and one for the comprehensive crack growth experiments (Material 2). Measurements were made on the load-line displacement, load, crack opening displacement, J-integral, and the unloading compliance using the techniques described in Refs *16* and *29*. Test results for the J-integral and COD as a function of crack extension are summarized in Figs. 10 through 15. The thickness and side groove studies reported in Figs. 10 and 11 show that the side grooving procedure did not affect the fracture behavior of the plate tested at 93°C (200°F) as long as the crack extension is measured at the midsection, where presumably plane-strain conditions prevail. Based on these results, comprehensive crack growth experiments were performed on 4T specimens with side grooves either 12.5 or 25 percent deep, since the crack extension can be measured unambiguously by the compliance technique. The heat-tint technique was also employed to check some of the measurements. The variations of J and COD (measured at the original crack tip) with crack extension are relatively independent of initial crack length for a/W ranging from 0.5 to 0.85. These are shown in Figs. 12 and

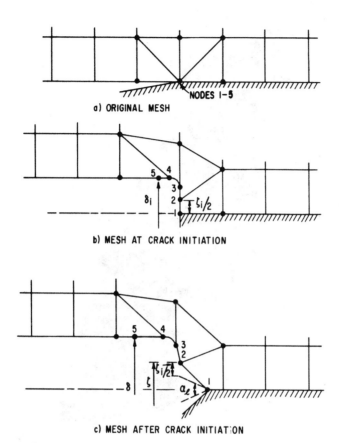

FIG. 8—*Finite-element model for crack tip blunting and growth.*

13. Similar studies were also conducted at higher temperatures—260°C (500°F)—and the test results are given in Figs. 14 and 15. The good agreement between the test results at different temperatures and for specimens with widely different initial crack lengths suggests that the J-integral or the COD possesses features suited for characterizing crack initiation and growth. Further discussion of the experimental investigation is given in Refs *16* and *29*.

Analytical Studies of Compact Specimens

Extensive analytical studies were carried out for all the compact specimens tested. The following discussion focuses on studies based on Tests T52 and T61. These 4T specimens have remaining ligaments ($W - a_o$) of 85.98 and 40.46 mm (3.385 and 1.593 in.), respectively. All the other specimens had ligaments between these two limits and, in general, values

80 ELASTIC-PLASTIC FRACTURE

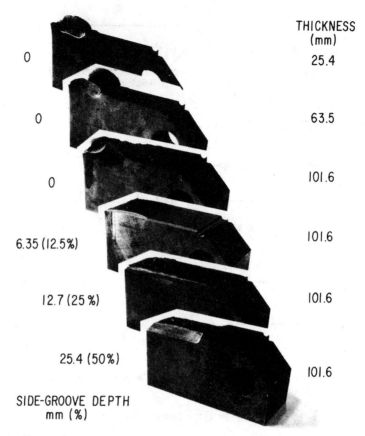

FIG. 9—*Fractured compact specimens of A533B steel tested at 93°C (200°F)—various thickness and side grooves.*

of the fracture parameters calculated from these specimens fall between the range of results reported for T52 and T61.

The finite-element mesh employed for Specimen T52 is shown in Fig. 16; that employed for T61 had a finer mesh near the crack tip. The size of the elements at the crack tip ranges between 2.5 and 5.0 mm (0.1 and 0.2 in.); this is about 5 to 10 times the blunted crack-tip diameter δ at initiation. The stress-strain curve employed for all the calculations is given in Fig. 17.

In the initial filter phase, the crack extension in the finite-element model is prescribed to follow the load-line displacement versus crack-extension data obtained from experiments. For Test T52, the basic input to the finite-element calculations and the relationship reproduced by the finite-element simulation are in good agreement; they are shown in Fig. 2. Two sets of calculations were performed, one based on J_2 flow theory and the other employing an incremental form of J_2 deformation theory of plasticity.

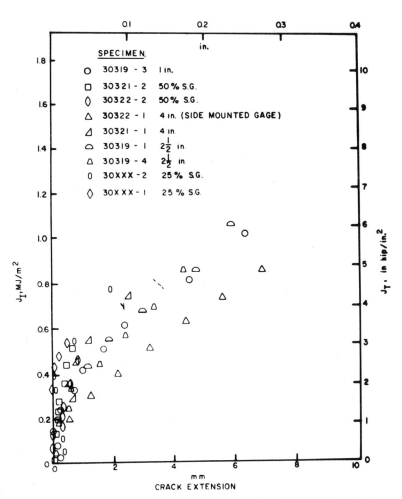

FIG. 10—*J-resistance curves for A533B Material 1 tested at 93°C (200°F)—4T-plan compact specimens.*

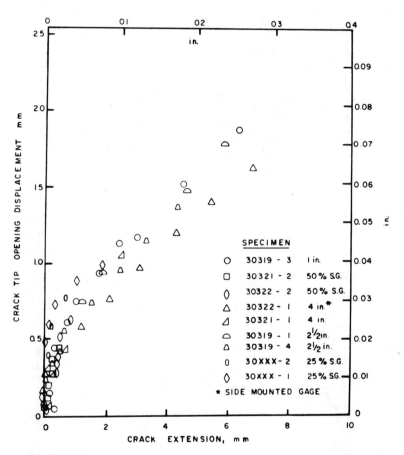

FIG. 11—*COD-resistance curves for A533B Material 1 tested at 93°C (200°F)—4T-plan compact specimens.*

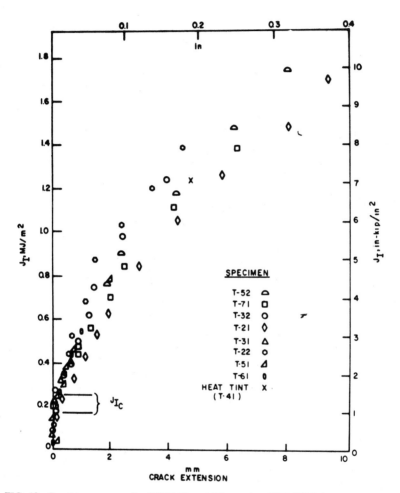

FIG. 12—*J-resistance curves for A533B Material 2 tested at 93°C (200°F)—4T side-grooved compact specimens.*

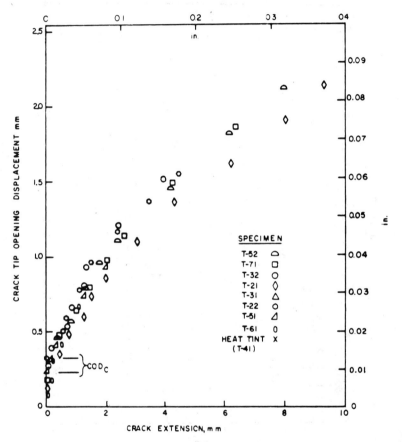

FIG. 13—*COD-resistance for A533B Material 2 tested at 93°C (200°F)—4T side-grooved compact specimens.*

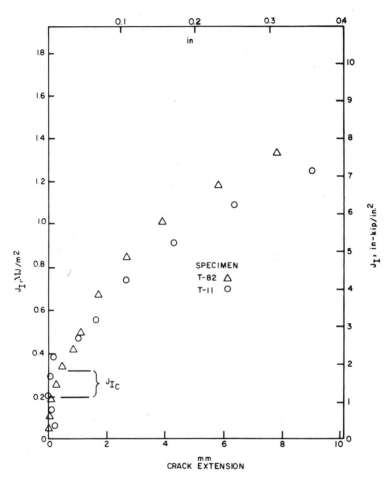

FIG. 14—*J-resistance curves for A533B Material 2 tested at 260°C (500°F)—4T side-grooved compact specimens.*

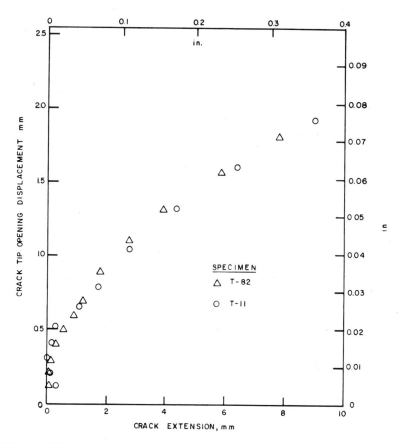

FIG. 15—*COD-resistance curves for A533B Material 2 tested at 260°C (500°F)—4T side-grooved compact specimens.*

The calculated load versus load-line deflection based on the two theories of plasticity are almost identical, and are also in excellent agreement with the experimental data. They are shown in Fig. 18; here the total load was computed on basis of the net thickness, which is 76.2 mm (3.0 in.). The good agreement strongly suggests that test specimen net thickness is the appropriate value to employ in calculations based on the plane-strain assumption.

The foregoing calculations were repeated for the deeply cracked configuration based on Test T61. The experimentally determined relationship which governed the crack growth simulation is shown in Fig. 19. The calculated applied load versus load-line displacement and the experimental results are compared in Fig. 20; the calculated curve is slightly lower than the experimental results, but the trend is in complete agreement.

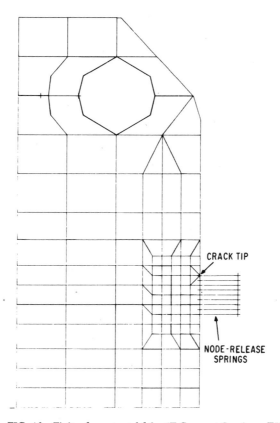

FIG. 16—*Finite-element model for 4T Compact Specimen T52.*

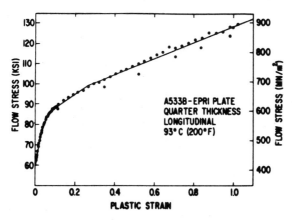

FIG. 17—*Stress-strain curve for A533B plate (Material 2) at 93°C (200°F).*

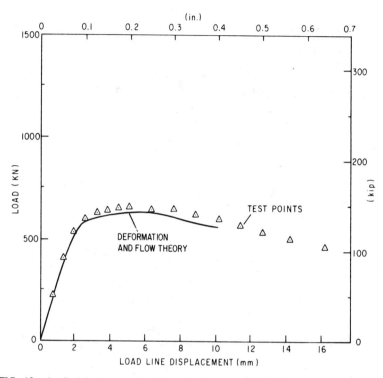

FIG. 18—*Applied load versus load-line displacement for 4T Compact Specimen T52, 25 percent side-grooved; W − a_o = 86 mm (3.385 in.).*

Crack Opening Displacement and Angle Criteria—The computed relationship between the crack tip opening displacement δ_o defined at the original crack tip (given by the nodal displacement for Node 5 in Fig. 8) and the experimental measurements is shown in Fig. 21 for the T52 configuration. The calculated curves are slightly lower than the experimental values but follow the complete trend of the test data. Interestingly, the COD obtained from a deformation theory calculation is slightly larger than that from a flow theory and is in better agreement with the experimental results. This is in accord with the expectation that the deformation theory of plasticity or nonlinear elasticity will in fact lead to a larger COD [*12*]. The calculated and measured COD for Test T61 is shown in Fig. 22 and similarly good agreement is noted.

Two definitions of the crack opening angle (COA) have been employed in our work. The definition of the average COA, α_o, is based on the crack extension $a − a_o$ measured from the original crack tip and COD, δ_o, measured at the original crack tip. The local COA α_l is based on the opening

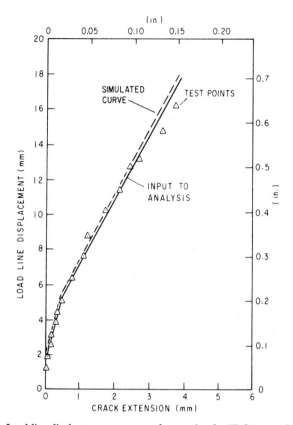

FIG. 19—*Load-line displacement versus crack extension for 4T Compact Specimen T61.*

displacement δ_l measured at a fixed distance, Δa, behind the current crack tip. Thus, as shown in Fig. 8

$$\alpha_o = (\delta_o - \delta_{oi})/(a - a_o) \quad \alpha_l = \delta_l/\Delta a \qquad (10)[8]$$

where δ_{oi} denote the opening displacement at initiation.

It was noted in the Potential Fracture Criteria section that the plastic strain increments for an extending crack have a $(1/r) \ln (1/r)$ singularity which gives rise to a crack opening profile that has the form $r \ln (1/r)$ [12]. Thus while the opening displacement vanishes at the tip, the opening profile has a vertical tangent at the crack tip corresponding to an opening angle of π radians. The angle cannot be defined unambiguously near the

---

[8] The local and average COA are computational measures, at slightly different scales, of the angle between the fractured surfaces behind the advancing crack tip.

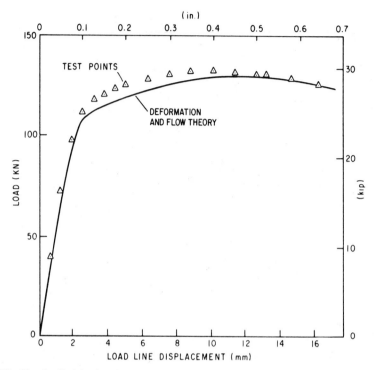

FIG. 20—*Applied load versus load-line displacement for 4T Compact Specimen T61, 25 percent side-grooved; W − a_o = 40 mm (1.593 in.).*

crack tip; however, at a small but finite distance away from the tip, a meaningful definition is possible as our subsequent discussion shows.

The calculated COA for Test T52 defined by Eq 10 from both deformation and flow theory analyses is shown in Fig. 23. The COA varies considerably during the initial stages of crack extension but appears to approach a constant value with further growth. As expected, the angle computed from the deformation analysis is slightly larger. The crack-tip element for the T52 configuration has lengths of about 5 mm (0.2 in.), which is about 10 times the COD at initiation. Perhaps this level of representation is not fine enough to capture an adequate description of the near-tip deformation. Thus, based on these calculations, the COA's approach a value of about 0.21 rad after 3 mm (0.12 in.) of crack extension. With a finer mesh, the angles would presumably approach a constant after smaller extensions. This expectation is in fact borne out in the calculations for T61; here the crack tip element has size of the order of 2.5 mm (0.1 in.). In this case the angles approach about 0.23 rad after 1.5 mm (0.06 in.) of crack extension as shown in Fig. 24. The computed angles for the T61 specimen are slightly larger than the corresponding angles for the T52

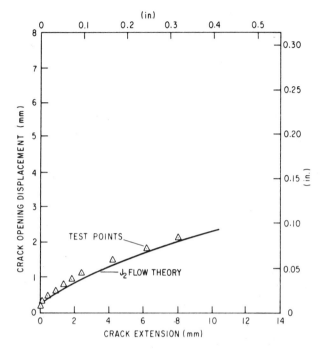

FIG. 21—*Crack opening displacement versus crack extension for Specimen T52.*

specimen, and are consistent with the slight mesh sensitivity associated with the COA. Calculations for T61 with larger near-tip elements gave angles that are in closer agreement with the values for T52. On the basis of these results, the COA appears to be a viable toughness parameter for crack growth; higher values of the angle correspond to higher resistance to crack growth.

J-Integral and dJ/da *Criteria*—Crack-growth calculations based on a J_2 flow theory, for the T52 configuration, showed that the J-integral computed from a remote contour, J_{ff}, and that computed using the Merkle-Corten expression [31], J_{MC}, are in excellent agreement for significant intervals of crack extension. Subsequent calculations employing a J_2 deformation theory gave values of J that are essentially identical to those values obtained on the basis of J_2 flow theory. The variation of J with load-line displacement for deformation and flow theories and test measurements is given in Fig. 25; the experimental J is determined from the measured load-displacement record using the Merkle-Corten expression based on net thickness. Crack-growth studies for the T61 configuration confirmed the foregoing observations; the results are shown in Fig. 26.

These observations suggest that the highly nonproportional deformation due to crack growth and to the elastic unloading at the wake of the ad-

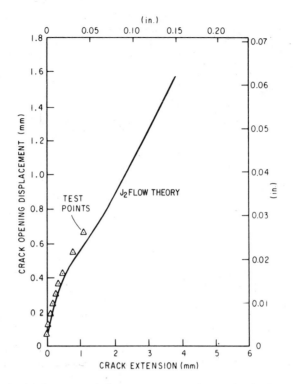

FIG. 22—*Crack opening displacement versus crack extension for Specimen T61.*

vancing crack is rather localized (of the order of the crack extension), and does not appear to appreciably influence the region at distances greater than several times the blunted tip diameter away from the crack tip. Finite-strain studies, based on a J_2 flow theory by McMeeking [32], showed that the J-integral is path dependent at distances less than 5 δ_o from the tip. Based on these observations, it will be convenient to distinguish two regions in the vicinity of the crack tip. At distances less than 5 δ_o, the deformation is highly nonproportional and will not be characterizable by the HRR field; this will be defined as the crack-tip field. Beyond this tip region, the deformation is characterizable by the HRR singularity if certain conditions are met [15] (also discussed earlier in the Potential Fracture Criteria section); this region will be defined as the near field. The size of the near field will in general depend on specimen geometry, material strain hardening, the plastic zone size, and the amount of crack growth. The region at large distances from the crack is called the far field. An illustration of these regions is given in Fig. 27.

To distinguish these different regions, a typical mesh configuration in the vicinity of the crack is shown in Fig. 28. The characteristic element

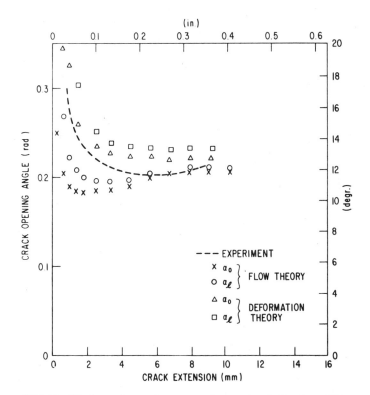

FIG. 23—*Crack opening angle versus crack extension for Specimen T52.*

size, l, is typically about 5 δ_o. Thus the J-integral evaluated in the tip field ($< 5\ \delta_o$) is termed J_{tip} and that evaluated in the near field ($> 5\ \delta_o$) is termed J_{nf}; J_{ff} is evaluated along contours remote from the crack tip.

The variation of the J's, evaluated along the different contours, with crack extension is shown in Figs. 29 and 30 for the T52 and T61 configurations, respectively. J_{tip} deviates from path independence almost immediately and the ratio J_{tip}/J_{ff} approaches zero after some crack growth. However, J_{nf} is in good agreement with the J's evaluated along remote contours and the J computed from the Merkle-Corten expression for crack extensions up to 6 percent of the remaining ligament. The path independence of the near-field J-integral suggests that the deformation in the near field is essentially proportional and that the J-integral would be an appropriate parameter for characterizing the near-tip deformation during crack growth.

To obtain a direct comparison between the predictions of deformation and flow theories of plasticity, the foregoing calculations were repeated with an incremental form of J_2 deformation theory. The far-field J's are essentially identical with those summarized in Figs. 29 and 30 for flow theory. The near-field J's are in good agreement up to about 6 percent of

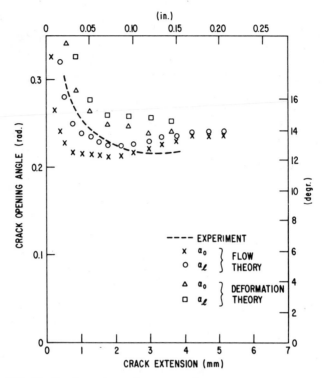

FIG. 24—*Crack opening angle versus crack extension for Specimen T61.*

crack growth. Beyond this range, the J's from deformation theory calculations remain, for practical purposes, path independent. The slight deviation from path independence is probably due to finite-element computational and discretization errors. A direct comparison of the predictions of deformation and flow theory for Specimen T52 is given in Fig. 31. The near-field and far-field J's are evaluated along contours that advance at the same rate as the crack tip. While the J's from deformation theory show slight path dependence, J_{nf} (flow theory) deviated from J_{ff} and J_{MC} by about 10 percent at 5-mm (0.2 in.) crack extension. At 8-mm (0.32 in.) crack extension the deviation of J_{nf} (flow theory) from the remote J's exceeds 20 percent. The former and latter correspond to crack extension of 6 and 10 percent, respectively, of the remaining ligament.

The foregoing calculations demonstrate that the predictions of deformation and flow theories of plasticity are in agreement for a limited range of crack growth. Therefore, the near-field environment during initiation and some amount of growth is characterizable by the J-integral. One of the requirements for a J-controlled growth concerns the slope of the J resistance curve. The variation of dJ/da with crack extension for the T52 and T61

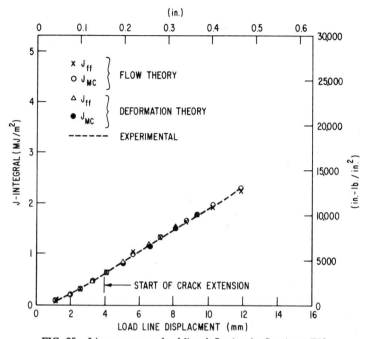

FIG. 25—*J-integral versus load-line deflection for Specimen T52.*

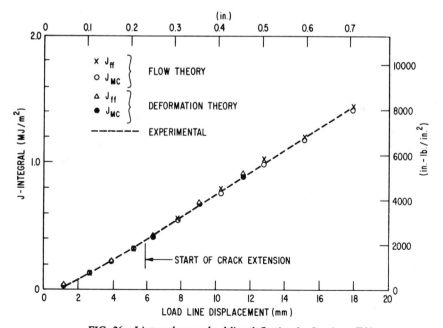

FIG. 26—*J-integral versus load-line deflection for Specimen T61.*

96 ELASTIC-PLASTIC FRACTURE

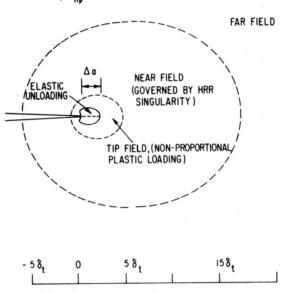

FIG. 27—*Schematic of the fields surrounding a growing crack.*

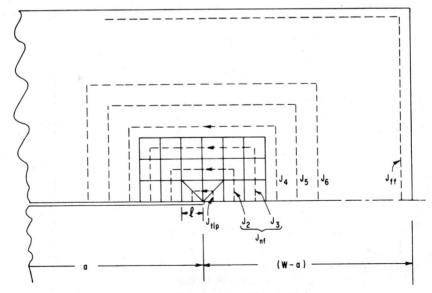

FIG. 28—*Paths for J-integral evaluations.*

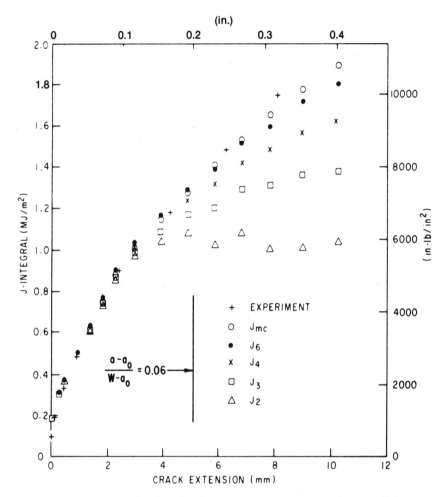

FIG. 29—*J-integral evaluated at different contours for Compact Specimen T52.*

configurations is shown in Figs. 32 and 33. Although the two configurations have very different crack lengths, the trends of the curves are very similar. In particular, $(dJ/da)_{nf}$ rapidly drops to zero after about 6 percent crack extension while $(dJ/da)_{ff}$ appears to level off. The results also suggest that the dJ/da measured from experiments [5] is a meaningful tearing modulus parameter only for a restricted range of crack growth. Although the measured dJ/da may remain finite or constant after this range of growth, $(dJ/da)_{nf}$ rapidly declines to zero and this violates the requirements of a *J*-controlled growth [15].

Sensitivity of Mesh and Step Size—In a series of calculations, the mesh-size and the crack propagation step-size were varied for the T52 crack

98 ELASTIC-PLASTIC FRACTURE

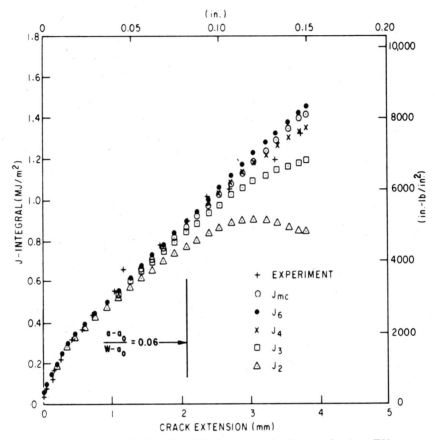

FIG. 30—*J-integral evaluated at different contours for Compact Specimen T61.*

configuration. The coarse mesh typically has dimensions that are twice as large as the regular mesh while the fine mesh is typically twice as fine as the regular mesh; the large step-size is about twice as large as the regular step-size. The load-deflection relationship is only slightly sensitive to the mesh variation and the step-size variation; this is shown in Fig. 34. The COA measured at the original crack tip and consequently the average COA, α_o, are also relatively insensitive to mesh and step-size variation; however, the local COA, α_l, is moderately sensitive. The sensitivity of these COD-based parameters is summarized in Figs. 35 and 36. For the same mesh and step-size variations, the relationship between J (evaluated distances of the order of 10 δ_o) and crack extension is shown in Fig. 37. It is apparent that the J-integral is only slightly affected by mesh and step-size variation. Thus from a finite-element modeling viewpoint, the J-integral

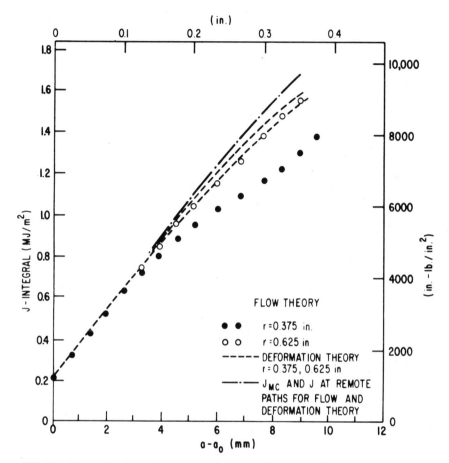

FIG. 31—*J-integral evaluated for contours advancing with crack tip, Compact Specimen T52.*

and the COD (or average COA) defined at the original crack tip are attractive fracture parameters.

The Evaluation Phase—In the next series of calculations, the COD-based criterion was employed to govern the crack extension for Configuration T61. The crack initiates at a critical value of the COD, δ_{oi}, and its propagation is determined by the critical value of the average angle α_o. From the initial filter phase, the upper and lower bounds of the COD at initiation are 0.508 and 0.417 mm (0.02 and 0.0164 in.), respectively. The angles range from 0.21 to 0.33 rad. Three crack-growth calculations were carried out using these limiting values to control the rate of crack extension. An intermediate value of δ_{oi} = 0.508 mm (0.02 in.) and α_o = 0.21 rad was also used. The prescribed relationship between COD (measured at the original crack tip) and the crack extension is illustrated in Fig. 38. The

100   ELASTIC-PLASTIC FRACTURE

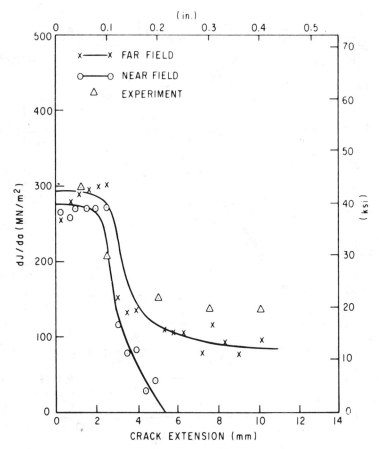

FIG. 32—*dJ/da for Compact Specimen T52*, a/W = 0.577.

calculated load-displacement relationship and the experimental measurements are compared in Fig. 39. Curves I and II reach maximum load at 12.7 mm (0.5 in.) of crack extension, and follow the trend of the experimental results while Curve III continues to rise. This is not unexpected since the average angle, α_o, for the third calculation was deliberately fixed at a value much higher than the values typically encountered in the initial filter phase.

In another series of crack growth analyses, the crack extension in Specimen T52 was governed by the *J*-resistance curve. The *J* (far field) versus crack extension $(a - a_o)$ obtained in the initial filter was employed to control the rate of crack extension. The governing curve is given by J_{MC} in Fig. 29. A comparison of the calculated load-displacement relationship and the test measurements is given in Fig. 40; the agreement is remarkably

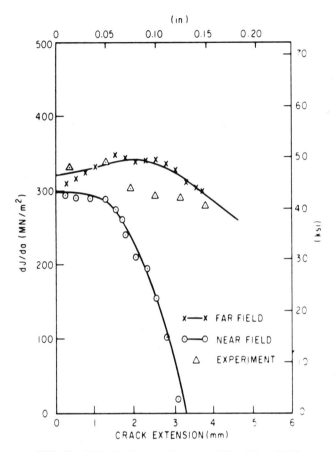

FIG. 33—dJ/da *for Compact Specimen T61, a/W = 0.801.*

good. These computations were repeated using a COD resistance criterion; the same general agreement with experiment was observed, as shown in Fig. 40.

Analytical and Experimental Investigations for Center-Cracked Panels

Experiments were carried out for center-cracked panel geometries, with some interesting results. It was found that for A533B panels with short remaining ligament [about 25 mm (1 in.)], the failure mode in the fully plastic state is apt to be due to slip along shear bands and subsequent fracture along 45-deg lines as opposed to crack growth ahead of the crack.

Side-grooved and smooth center-cracked panels made from A533B steel (Material 2) were tested at 93°C (200°F). The specimens have a total width, $2W$, of 150 mm (6 in.) and a thickness of 50 mm (2 in.). Some

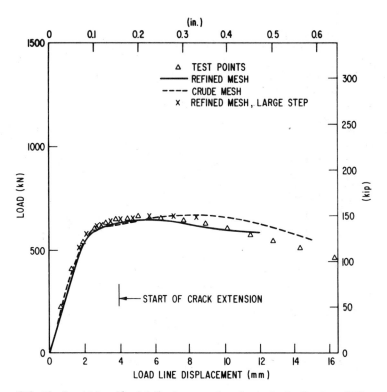

FIG. 34—*Sensitivity of load-deflection to mesh and step size for Specimen T52.*

specimens were side-grooved to 25 percent, and the total remaining ligaments, $2(W - a)$, ranged from 25 to 50 mm (1 to 2 in.). In all the tests, intense shear deformation developed along approximately 45-deg bands, emanating from the original crack tip as shown in Fig. 41. Examination of the fracture surfaces and of silicone rubber castings of the crack tips revealed that initiation of cracking occurred after the crack tip had opened into a three-cornered profile as illustrated by the insert in Fig. 41. Extension at mid-thickness occurred at a COA in excess of the COA observed in compact specimens. At the surface of smooth-faced specimens, separation had occurred in the direction of the shear bands. Apparently the shear deformation controlled both crack initiation and extension. The value of the J-integral as determined from the load-displacement record is approximately 1.8 MJ/m^2 (10 000 in.-lb/in.^{2}) at the point shown in the figure, where the experiment was terminated. Finite-element calculations for the identical configuration indeed showed that the deformation is concentrated on narrow shear bands emanating from the crack tip at 45 deg to the crack plane; these deformations are shown in Fig. 42. An examination of the stress fields and the crack-tip profile revealed that crack-tip fields do not

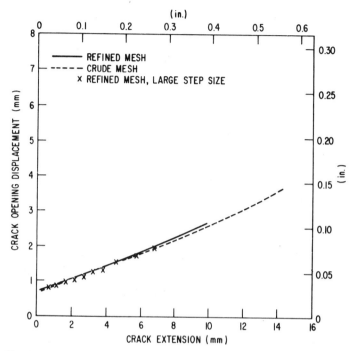

FIG. 35—*Sensitivity of COD measured at original crack tip to mesh and step size in Specimen T52.*

approach the HRR fields, but appear to be strongly influenced by the global shear deformation. It is clear that the minimum size requirements for fracture were not met in the center-cracked panel and this is discussed further in the next section.

Size Requirements for Fracture Toughness Testing in the Fully Plastic Range

The success of a characterizing parameter approach for fracture based on J or δ requires that the Hutchinson-Ride-Rosengren (HRR) singular field govern over the fracture process zone. In the case of small-scale yielding, the HRR field is embedded in the elastic singular K-field [33], and consequently J and K are related by [8]

$$J = K^2/E' \qquad (11)$$

where $E' = E$ for plane-stress and $E' = E/(1 - \nu^2)$ for plane-strain conditions. For small-scale yielding, the invariance of the critical stress intensity factor, K_{Ic}, to crack length and specimen geometry strongly

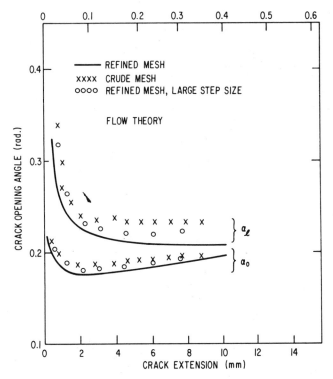

FIG. 36—*Sensitivity of average and local crack opening angles to mesh and step size in Specimen T52.*

suggests that the HRR field governs the deformation state in the fracture process zone. In this instance, K_{Ic} is a valid fracture toughness parameter if the plastic zone size, r_P, is small compared to crack length, thickness, and remaining ligament. Since the crack-tip fields can also be represented in terms of J and δ via Eqs 1 and 3, the K_{Ic} criterion would also imply a J_{Ic} or δ_{Ic} fracture criterion. It also follows from Eq 11 that

$$J_{Ic} = K_{Ic}^2/E' \qquad (12)$$

In small-scale yielding, there is no intrinsic advantage in characterizing toughness in terms of J_{Ic} or δ_{Ic} since K_{Ic} is an adequate toughness parameter, and elastic solutions to numerous crack configurations are available in terms of K. However, if toughness characterization beyond the regime of small-scale yielding is considered, as is the case in this study, then characterization in terms of J_{Ic} or δ_{Ic} becomes advantageous as the following discussion will show.

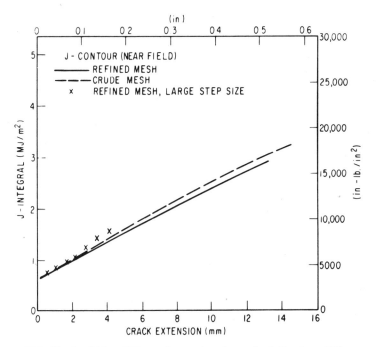

FIG. 37—*Sensitivity of J-integral to mesh and step size in Specimen T52.*

In the fully plastic range, the HRR singularity is embedded in a plastic zone that extends throughout the remaining ligament. This situation is illustrated in Fig. 43. There is no elastic K-field in this case, and any toughness characterization must now be made in terms of HRR field parameters and the J-integral or δ_t are therefore natural candidates. Now some important points regarding the size of the HRR fields must be made.

In small-scale yielding, the size of the plastic zone r_P, and therefore of the HRR field r_o, is governed by the ratio of the applied load to the load corresponding to net section yield and is otherwise relatively independent of the specimen geometry and the strain-hardening properties. The size requirements [ASTM Test for Plane-Strain Fracture Toughness of Metallic Materials (E 399-74)] for a valid K_{Ic} test can therefore be stated in terms of the ratio of $(K_{Ic}/\sigma_o)^2$. In contrast, our experimental results suggest that the size of the HRR field beyond the regime of small-scale plasticity is influenced by specimen geometry and strain-hardening properties. This has also been alluded to by McClintock [34] and Rice [35]. Thus the specimen size requirements for a valid toughness test in the fully plastic range will differ for different material properties and crack geometry.

Under general yield or fully plastic conditions, the only characteristic length is the COD. The size requirements may be stated in terms of this

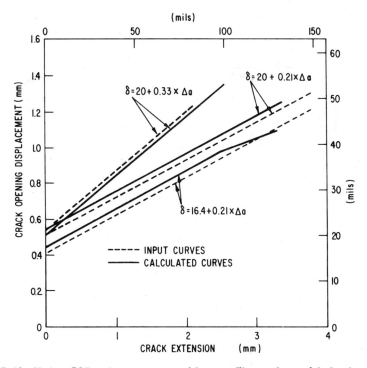

FIG. 38—*Various COD-resistance curves used for controlling crack growth in Specimen T61.*

characteristic length δ_t. For the HRR parameters (namely, J or δ_t) to characterize the near-tip field, the crack length, remaining ligament, and thickness must be large compared to the crack tip opening displacement. Just how large the three dimensions must be compared to the crack opening displacement δ_t appears to depend on the material strain-hardening properties and specimen geometry. In the fully plastic center-cracked panel, the deformation is concentrated along slip planes at 45 deg from the crack plane. For this configuration, the global deformation will dominate the deformation state over a considerable interval in the vicinity of the crack tip. The situation with the single-edge crack subjected to bending is quite different. Here the deformation in the fully plastic state closely resembles the Green and Hundy slipline field solution [*36*]. In the vicinity of the crack tip, the imposed deformation, due to the Green and Hundy field, is very similar to the HRR field. Thus one expects that the HRR field will govern over relatively larger intervals in the bend specimen, compared with the center-cracked panel at similar levels of loading. The global deformation fields for the two geometries are illustrated in Fig. 3.

To explore these issues more precisely, finite-element calculations were carried out for the two geometries with a/W ratio of 0.5 and a ligament of

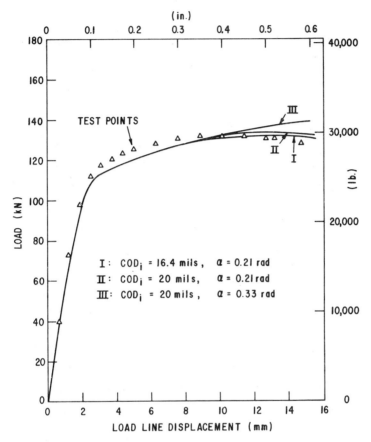

FIG. 39—*Comparison of load-deflection relationships generated by crack-growth simulation based on COD-resistance curve for Specimen T61.*

26 mm (1 in.). Calculations based on elastic-perfectly plastic idealization attained limit loads that are 2 to 3 percent higher than the theoretical limit loads for the center-cracked panel and the single-edge cracked bend bar; the details are given in Fig. 44. The variation of the tensile stress σ_{yy}, or the 'crack opening stress' across the remaining ligament under fully plastic conditions at the same level of J is shown in Fig. 45. The Prandtl field, which represents the limit of the HRR field for elastic-perfectly plastic materials, requires the tensile stress to reach a maximum of 2.97 σ_o ahead of the crack. The HRR or Prandtl field is attained in the bend bar. For the center cracked panel, however, the stress at a distance of 0.02 ($W - a_o$) ahead of the crack is only approximately twice the yield stress. For the latter configuration, the Prandtl field is attained over a rather small interval. The introduction of strain hardening affects the stress field some-

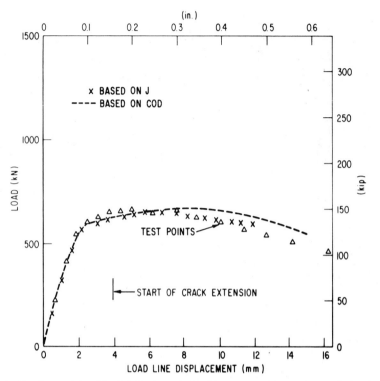

FIG. 40—*Comparison of load-deflection relationships generated by crack-growth simulations based on J and COD-resistance curves for Specimen T52.*

what. For a power-law curve corresponding to $n = 10$, the variation of the tensile stress for the two geometries at approximately about the same level of applied J is also shown in Fig. 45. It is clear that, between the two configurations, the HRR field associated with the bend bar dominates over a comparatively larger interval. The calculations are repeated for the compact specimen and the center-cracked panel, both with a/W ratios of 0.7 and a remaining ligament of 19 mm (0.75 in.), using the stress-strain curve for A533B steels. The stress fields are shown in Fig. 46.

These calculations for stationary cracks show that, when the uncracked ligament is subjected primarily to bending, the HRR field dominates over a comparatively large interval. If the loading in the ligament is primarily tensile, the HRR field is attained over comparatively small intervals and the triaxial stress state is relatively low. Thus the minimum size requirements for fracture toughness testing to evaluate J_{Ic} or δ_{Ic} will be significantly larger for the center-cracked panel. We may also infer from the stationary crack solutions that the size requirements for obtaining J or COD resistance curves will vary with crack configuration and material properties.

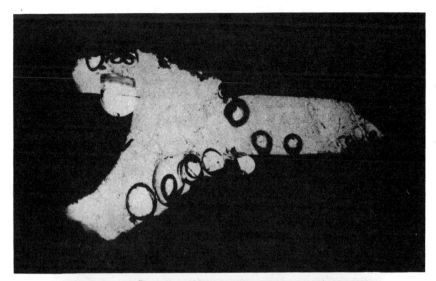

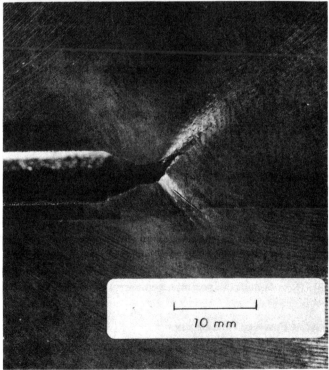

FIG. 41—*View of deformed crack tip in center-cracked panel specimen, showing shear bands emanating from crack tip at 45 deg; $W - a_o = 19.6$ mm (0.784 in.). Top photo is profile of silicon rubber casting of crack tip.*

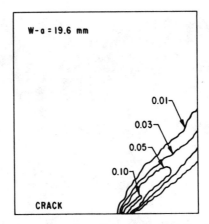

FIG. 42—*Effective strain contours from finite-element solution for center-cracked panel shown in Fig. 41 at* $J = 1.75$ *MJ/m^2* (*10 000 in.-lb/in.^{2}*).

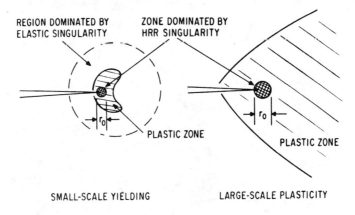

FIG. 43—*Crack tip fields.*

It may be noted that in the early stages of these same calculations the HRR field was attained in all the configurations under conditions of small-scale plasticity.[9] These results are consistent with that absolute size requirement (independent of specimen geometry) placed on valid K_{Ic} tests.

Assessment of Fracture Parameters

The requirements that viable fracture criteria must satisfy have been discussed already in the Strategy section. A comparison of the contending

---

[9] In the center-cracked panel, the crack opening stress σ_{yy} gradually 'unloads' due to the stress redistribution caused by the intense formation of the intense shear band as the limit load is approached.

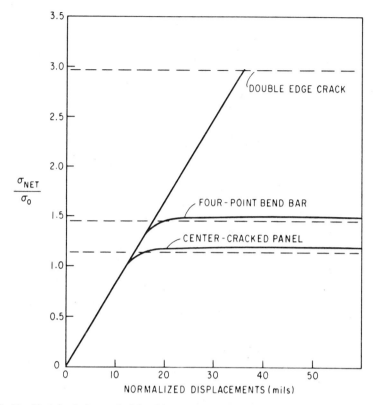

FIG. 44—*Limit loads for cracked bend bar, center-cracked panel, and double-edge cracked panel.*

fracture parameters on the basis of these requirements is summarized in Table 1. Most of the parameters satisfy the first six requirements. This is not unexpected since all these parameters, when appropriately defined and employed, characterize the crack near-field. However, in terms of the last four requirements, the trend is clear. From the computation and measurement point of view, the viable candidates are the J-integral and the COD and COA.

The results from our analytical and experimental investigations support the conclusions reached by Begley and Landes [1] that the onset of plane-strain flat fracture under small- or large-scale yielding conditions is characterizable by J_{Ic} or equivalently by δ_{Ic}. Thus our experimental and analytical investigations gave values of J_{Ic} ranging from 0.175 to 0.263 MJ/m^2 (1000 to 1500 in.-lb/in.^{2}) and δ_{Ic} ranging from 0.25 to 0.40 mm (0.010 to 0.016 in.) for A533B steels at 93°C (200°F). These values are consistent with K_{Ic} toughness numbers reported elsewhere [37]. In addition, our studies

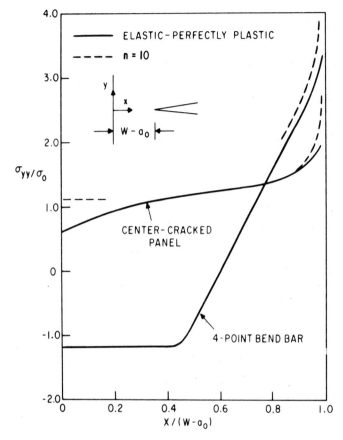

FIG. 45—*Variation of tensile stress across remaining ligament for 4-point bend bar and center-cracked panel for elastic-perfectly plastic and strain-hardening materials.*

showed that some amount of stable crack growth is characterizable by J or COD resistance curves of the COA [16,29]. Paris et al [5] proposed to characterize resistance to crack growth by the tearing modulus $[T = (E/\sigma_o^2) \cdot (dJ/da)]$. In subsequent paragraphs we discuss the requirements for a J or COD characterization of crack growth and possible limitations.

In arguments based on considerations involving a deformation theory of plasticity, Hutchinson and Paris [15] identified the requirements for J-controlled crack growth. The first requires that crack extension, Δa, be small compared to the characteristic radius, R, of the region controlled by the HRR singularity. In addition, dJ/da must be large compared to J/r (see the Potential Fracture Criteria section for details). The latter requirement may be stated in terms of a material-based length quantity D defined by

$$\frac{1}{D} = \frac{dJ}{da}\frac{1}{J} \tag{13}$$

Thus the two requirements for a J-controlled growth are

$$\Delta a \ll R \tag{13a}$$

and

$$D \ll r < R \tag{13b}$$

The deformation field ahead of an extending crack was derived by Rice [*10*] using a J_2 flow theory of plasticity. The study revealed that the near field of an extending crack may be characterized by the local COA (see the Potential Fracture Criteria section for more details). On the basis of Rice's work, Shih et al [*16*] suggested the use of COA as a stable growth

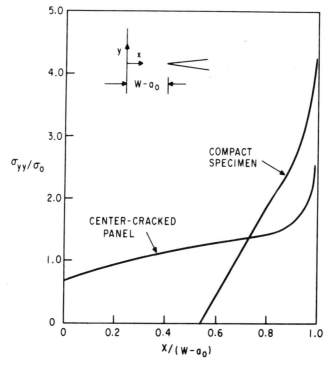

FIG. 46—*Variation of tensile stress across remaining ligament for compact specimen and center-cracked panel for A533B steels.*

TABLE 1—*A comparison of some fracture parameters.*

	J-Integral, J	Crack Opening Displacement, COD	Local Crack Opening Angle	Work Done in a Process Zone	Generalized Energy Release Rate
Measure of near-tip field	yes	yes	yes	yes	yes
Constant during crack extension	no[a]	no[a]	yes	yes	yes
Independent of specimen geometry	mildly dependent	mildly dependent	mildly dependent	insufficient data	mildly dependent
Applicability to instability analyses	yes[b]	yes	yes	yes	yes
Generalizable to mixed-mode fracture	probable	uncertain	uncertain	probable	probable
3D crack geometries	uncertain	probable	probable	probable	uncertain
Sensitivity to mesh, process zone, and increment size	no	mildly	moderately	moderately	yes
Computation cost (estimated ratio)	1	1	1	1	3
Direct local measurement	no	yes	yes (difficult)	no	no
Direct global measurements	yes	no	no	no	no

[a] J and COD are employed as resistance parameters.
[b] For limited crack extension.

parameter. The requirement for a COA-controlled growth is, as shown in the Potential Fracture Criteria section

$$\frac{d\delta}{da} \gg \frac{\sigma_o}{E} \ln\left(\frac{R}{r}\right) \tag{14}$$

The crack opening profile has a vertical tangent at the crack tip (corresponding to a COA of π rad at the tip) and thus the angle cannot be defined in any meaningful manner close to the current crack tip. If the requirement given by Eq 14 is satisfied, however, then the crack profile exhibits a well-defined angle at a small but finite distance away from the crack tip.

The requirements for COA-controlled growth can be restated in terms of a tearing modulus T_δ based on the COA

$$T_\delta = \frac{d\delta}{da} \frac{E}{\sigma_o} \gg 1 \tag{15}$$

In other words, the crack opening angle, $d\delta/da$, must be large compared to the yield stress divided by the elastic modulus (or the yield strain). For A533B steels on the upper shelf, direct measurements by rubber infiltration [16] and finite-element crack growth calculations reported in this paper showed COA's of the order of 0.2 to 0.3 rad. This is significantly larger than σ_o/E, which is about 0.002.

From Eq 13b, the requirement for a J-controlled growth can be restated as

$$T_J = \frac{dJ}{da} \frac{E}{\sigma_o^2} \gg 1 \tag{16}$$

By exploiting deformation theory for crack growth, namely

$$\frac{dJ}{da} = \alpha \sigma_o \frac{d\delta_t}{da} \tag{17}$$

it may be argued that the inequalities given by Eqs 15 and 16 are equivalent when the near field is governed by the HRR field. For A533B steels at the upper shelf, T_J ranges from 100 to 300 and T_δ ranges from 100 to 150.[10] Thus T_J or T_δ can be viewed as toughness parameters for stable crack growth; T_δ is more appealing because it is a more fundamental quantity and is relatively more constant for a given material. How large T_J or T_δ must be for a J-controlled or COD-controlled growth is yet to be explored. This will be the subject of further investigations.

[10]T_J generally falls between 20 and 150 for a wide variety of materials reported in Ref 5.

An additional limitation placed on the J resistance approach is expressed by Eq 13a. Our finite-element investigations based on actual experimental data suggest that, for A533B steels in the upper shelf, the J resistance approach will be valid for crack growth up to 6 percent of the remaining ligament when the mode of loading is primarily bending. The amount of crack extension where the J resistance approach is valid is expected to depend on specimen geometry and material properties, and in particular on the strain hardening. A similar restriction on the COA's does not appear to be necessary.

It is instructive to compare the J resistance and the COD resistance approaches with the onset of growth and stable growth. In terms of the J-integral, there are basically two quantities, J_{nf} and J_{ff}. J_{nf} is evaluated in the region dominated by the HRR singularity while J_{ff} is evaluated along a remote contour. The derivatives of these quantities are $(dJ/da)_{nf}$ and $(dJ/da)_{ff}$. Our analytical studies show that J_{ff} is the quantity measured in experimental investigations [1-3,5,16,29]. Similarly, the dJ/da referred to in Ref 5 is in this context $(dJ/da)_{ff}$. There are also two possible definitions of the COD. One definition is based on the opening displacement at the original crack tip, δ_o, while the other is based on the opening displacement at a fixed distance behind the current crack tip δ_l. The average crack opening angle, α_o, is determined from the first definition while the local crack opening angle, α_l, is determined from the latter.

A typical variation of J, dJ/da, δ, and α with crack extension is illustrated in Fig. 47. These figures show that while J_{ff} continues to increase, J_{nf} begins to level off after some crack extension. Thus $(dJ/da)_{nf}$ falls to zero and violates one of the requirements for J-controlled growth expressed by Eq 16. On the other hand, both the crack opening angles, α_o and α_l, after an initial transient remain constant for a considerable range of crack growth. This satisfies the requirement for COA-controlled growth specified by Eq 15. It also appears that beyond the initial stage of crack growth, a constant critical value of the average or local COA may be employed to characterize stable growth. This is not unexpected since the COA is derived from more fundamental considerations.

We also note that, while the local crack opening angle α_l is the fundamental measure of the crack growth fields, it is not an easily measured quantity. However, the average crack opening angle α_o is easily obtained from linear variable differential transformer (LVDT) measurements over the entire range of crack extension [29]. Our calculations show that both angles are essentially identical for crack extension up to 10 times the blunted tip opening, δ_{oi}, at initiation. Therefore, even though the average angle α_o is not a fundamental measure of the crack field, it is an attractive quantity from a practical viewpoint. Similarly $(dJ/da)_{ff}$ is a parameter readily determined from experimental measurements, while $(dJ/da)_{nf}$ can be obtained only through theoretical and numerical crack-growth studies.

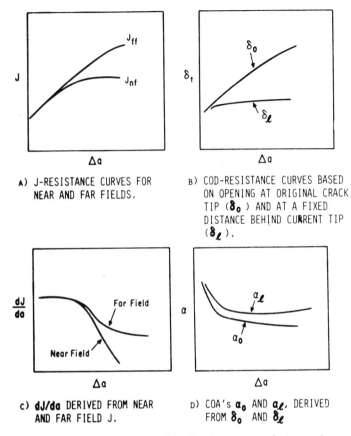

FIG. 47—*Typical behavior of J and COD-based parameters during crack growth.*

a) J-RESISTANCE CURVES FOR NEAR AND FAR FIELDS.

b) COD-RESISTANCE CURVES BASED ON OPENING AT ORIGINAL CRACK TIP (δ_0) AND AT A FIXED DISTANCE BEHIND CURRENT TIP (δ_ℓ).

c) dJ/da DERIVED FROM NEAR AND FAR FIELD J.

d) COA's a_0 AND a_ℓ, DERIVED FROM δ_0 AND δ_ℓ.

This study shows that it is appropriate to characterize the toughness properties of ductile metals in terms of initiation and growth. The material toughness associated with initiation is characterizable by J_{Ic} or δ_{Ic}, while the material toughness associated with crack growth is characterizable by the dimensionless parameters $[T_J = (E/\sigma_o^2)(dJ/da)]$ and $[T_\delta = (E/\sigma_o) \cdot (d\delta/da)]$. The two-parameter characterization of fracture toughness properties by J_{Ic} and T_J or δ_{Ic} and T_δ is analogous to the characterization of material deformation properties by the yield stress and strain-hardening exponent. In fact the ambiguities inherent in the definition of the yield stress are present in the definition of J_{Ic} or δ_{Ic}. Thus the range of variation in J_{Ic} and δ_{Ic} for any given material is significantly larger than the variation in T_J and T_δ [16,29].

Conclusions

This experimental and analytical investigation was directed toward the identification of viable criteria for the characterization of flat fracture under essentially plane-strain conditions in the large-scale yielding range. The following summarizes our studies:

1. Macroscopically flat fracture surfaces with a straight leading edge can be produced by employing side grooves on test specimens. Side grooves 25 percent of the specimen thickness are recommended, since they promote an essentially uniform plane-strain constraint along the crack front while producing minimal effect on specimen compliance and stress-intensity factor.

2. The experimentally determined J and COD (measured at the original crack tip) resistance curves appears to be independent of specimen size and initial crack length when plane-strain flat fracture occurs and if certain minimal size requirements are met.

3. Analytical investigations also reveal that J and COD resistance curve can be employed to characterize crack initiation and growth. The slope of the J-resistance curve (dJ/da) appears to be constant for a relatively short interval of crack extension, while both the local and average crack opening angle remain constant over the entire range of crack extension explored in our experimental and analytical investigations. The J-based criteria appear to be valid for limited amounts of crack growth. For A533B steel on the upper shelf, this amount is about 6 percent of the original remaining ligament for test specimens subjected to bending. The range of validity will depend on the strain-hardening exponent and specimen geometry. The COD-based criteria appear to be valid for larger amounts of crack growth.

4. The tearing modulus $[T_J = (E/\sigma_o^2)(dJ/da)]$ proposed by Paris and co-workers as a measure of material toughness during stable growth is constant over relatively short intervals of growth. Our investigations suggest that a tearing modulus based on the COA $[T_\delta = (E/\sigma_o)(d\delta/da)]$ is an attractive alternative. The latter modulus is measurable directly and appears to be constant over the entire range of stable growth. Fracture toughness associated with crack initiation is measured by J_{Ic} or δ_{Ic}, while material resistance associated with crack growth is measured by T_J or T_δ. The two-parameter characterization of fracture properties by J_{Ic} and T_J or δ_{Ic} and T_δ is analogous to the characterization of material deformation properties by the yield stress and the strain-hardening exponent.

5. Certain size requirements must be met for fracture toughness testing in the fully plastic range. These requirements are analogous to the size requirements for valid K_{Ic} testing in linear elastic fracture mechanics. For the J-based or COD-based parameters to govern over size scales that encompass the fracture process zone, the remaining ligament, crack length, and specimen thickness should be large compared to the crack tip

opening displacement. The precise magnitudes of these quantities with respect to the COD appear to depend on material strain hardening and specimen geometry.

Acknowledgments

The authors wish to acknowledge helpful discussions with J. W. Hutchinson of Harvard University and J. R. Rice of Brown University. We are grateful for the assistance rendered by J. P. D. Wilkinson, R. H. VanStone, M. D. German, S. Yukawa, and D. F. Mowbray of the General Electric Co. Some of the analyses presented were carried out by R. H. Dean of Harvard University and Figs. 4 and 5 were kindly provided by R. H. VanStone. Discussions with G. T. Hahn, M. F. Kanninen, and E. F. Rybicki of Battelle Columbus Laboratories who are engaged in a similar program are gratefully acknowledged. This work was sponsored by the Electric Power Research Institute, Palo Alto, Calif. and we wish to thank R. E. Smith and T. U. Marston for their encouragement.

References

[1] Begley, J. A. and Landes, J. D. in *Fracture Toughness, ASTM STP 514*, American Society for Testing and Materials, 1972, pp. 1-23 and pp. 24-39.
[2] Clarke, G. A., Andrews, W. R., Paris, P. C., and Schmidt, D. W. in *Mechanics of Crack Growth, ASTM STP 590*, American Society for Testing and Materials, 1976, pp. 27-42.
[3] Griffis, C. A. and Yoder, G. R., *Transactions*, American Society of Mechanical Engineers, *Journal of Engineering Materials and Technology*, Vol. 98, 1976, pp. 152-158.
[4] *Fracture Toughness Evaluation by R-Curve Methods, ASTM STP 527*, (papers on linear-elastic R-curve analysis), American Society for Testing and Materials, 1973.
[5] Paris, P. C., Tada, H., Zahoor, A., and Ernst, H., this publication, pp. 5-36.
[6] Hutchinson, J. W., *Journal of the Mechanics and Physics of Solids*, Vol. 16, 1968, pp. 13-31 and pp. 337-347.
[7] Rice, J. R. and Rosengren, G. F., *Journal of the Mechanics and Physics of Solids*, Vol. 16, 1968, pp. 1-12.
[8] Rice, J. R., *Journal of Applied Mechanics*, Vol. 35, 1968, pp. 379-386.
[9] Shih, C. F. in *Fracture Analysis, ASTM STP 560*, American Society for Testing and Materials, 1974, pp. 187-210.
[10] Rice, J. R. and Tracey, D. M. in *Numerical and Computer Methods in Structural Mechanics*, S. J. Fenves et al, Eds., Academic Press, New York, 1973, pp. 585-623.
[11] Tracey, D. M., *Transactions*, American Society of Mechanical Engineers, *Journal of Engineering Materials and Technology*, Vol. 98, 1976, pp. 146-151.
[12] Rice, J. R. in *Mechanics and Mechanisms of Crack Growth (Proceedings*, Conference at Cambridge, England, April 1973), M. J. May, Ed., British Steel Corporation Physical Metallurgy Centre Publication, 1975, pp. 14-39.
[13] Chitaley, A. D. and McClintock, F. A., *Journal of the Mechanics and Physics of Solids*, Vol. 19, 1971, pp. 147-163.
[14] Amazigo, J. C. and Hutchinson, J. W., *Journal of the Mechanics and Physics of Solids*, Vol. 25, 1977, pp. 81-97.
[15] Hutchinson, J. W. and Paris, P. C., this publication, pp. 37-64.
[16] Shih, C. F. et al, "Methodology for Plastic Fracture," 1st through 5th Quarterly Reports, General Electric Co., Schenectady, N. Y., Sept. 1976-Dec. 1977.

[17] Hahn, G. T. et al, "Methodology for Plastic Fracture," 1st through 5th Quarterly Reports, Battelle Columbus Laboratories, Columbus, Ohio, Sept. 1976-Dec. 1977.
[18] Kfouri, A. P. and Rice, J. R., "Elastic/Plastic Separation Energy Rate for Crack Advance in Finite Growth Steps" in *Fracture 1977 (Proceedings,* 4th International Congress on Fracture, Waterloo, Ont., Canada, 1977), D. M. R. Taplin, Ed., Vol. 1, 1977.
[19] Bathe, K. J., "ADINA—A Finite Element Program for Automatic Incremental Nonlinear Analysis," Report No. 82448-1, Massachusetts Institute of Technology, Cambridge, Mass., 1975.
[20] deLorenzi, H. G. and Shih, C. F., "Application of ADINA to Elastic-Plastic Fracture Problems" in *Proceedings,* ADINA User's Conference, Report No. 82448-6, Cambridge, Mass., 1977.
[21] Cox, T. B. and Low, J. R., Jr., *Metallurgical Transactions,* Vol. 5, 1974, pp. 1457-1470.
[22] Argon, A. S. and Im, J., *Metallurgical Transactions,* Vol. 6A, 1975, pp. 839-851.
[23] Rice, J. R. and Tracey, D. M., *Journal of the Mechanics and Physics of Solids,* Vol. 17, 1969, pp. 201-217.
[24] McClintock, F. A., *Journal of Applied Mechanics,* Vol. 35, 1968, pp. 363-371.
[25] Barsoum, R. S., *International Journal for Numerical Methods in Engineering,* Vol. 11, 1977, pp. 85-98.
[26] Shih, C. F., deLorenzi, H. G., and German, M. D., *International Journal of Fracture Mechanics,* Vol. 12, 1976, pp. 647-651.
[27] Nagtegaal, J. C., Parks, D. M., and Rice, J. R., *Computer Methods in Applied Mechanics and Engineering,* Vol. 4, 1974, pp. 153-177.
[28] deLorenzi, H. G. and Shih, C. F., *International Journal of Fracture Mechanics,* Vol. 13, 1977, pp. 507-511.
[29] Andrews, W. R. and Shih, C. F., this publication, pp. 426-450.
[30] Shih, C. F., deLorenzi, H. G., and Andrews, W. R., *International Journal of Fracture,* Vol. 13, 1977, pp. 544-548.
[31] Merkle, J. G. and Corten, H. T., *Transactions,* American Society of Mechanical Engineers, *Journal of Pressure Vessel Technology,* 1974, pp. 286-292.
[32] McMeeking, R. M. in *Flaw Growth and Fracture, ASTM STP 631* American Society for Testing and Materials, 1977, pp. 28-41.
[33] Irwin, G. R., *Journal of Applied Mechanics,* Vol. 24, 1957, pp. 361-364.
[34] McClintock, F. A. in *Fracture: An Advanced Treatise,* H. Leibowitz, Ed., Vol. 3, Academic Press, New York, 1971, pp. 47-225.
[35] Rice, J. R. in *The Mechanics of Fracture,* F. Erdogan, Ed., Applied Mechanics Division, American Society of Mechanical Engineers, Vol. 19, 1976, pp. 23-53.
[36] Green, A. P. and Hundy, B. B., *Journal of the Mechanics and Physics of Solids,* Vol. 4, 1956, pp. 128-144.
[37] Shabbits, W. O., Pryle, W. H., and Wessel, E. T., "Heavy Section Fracture Toughness Properties of A533 Grade B Class 1 Steel Plate and Submerged Arc Weldment," WCAP-7414, Westinghouse Electric Corp., Pressurized Water Reactor Systems Division, Pittsburgh, Pa., Dec. 1969 (also available as HSSTP-TR-6).

M. F. Kanninen,[1] *E. F. Rybicki,*[1] *R. B. Stonesifer,*[1] *D. Broek,*[1] *A. R. Rosenfield,*[1] *C. W. Marschall,*[1] *and G. T. Hahn*[1]

Elastic-Plastic Fracture Mechanics for Two-Dimensional Stable Crack Growth and Instability Problems

REFERENCE: Kanninen, M. F., Rybicki, E. F., Stonesifer, R. B., Broek, D., Rosenfield, A. R., Marschall, C. W., and Hahn, G. T., "**Elastic-Plastic Fracture Mechanics for Two-Dimensional Stable Crack Growth and Instability Problems,**" *Elastic-Plastic Fracture, ASTM STP 668*, J. D. Landes, J. A. Begley, and G. A. Clarke, Eds., American Society for Testing and Materials, 1979, pp. 121–150.

ABSTRACT: An elastic-plastic fracture mechanics methodology for treating two-dimensional stable crack growth and instability problems is described. The paper draws on "generation-phase" analyses in which the experimentally observed applied-load (or displacement) stable crack growth behavior is reproduced in a finite-element model. In these calculations a number of candidate stable crack growth parameters are calculated for the material tested. The quality of the predictions that can be made with these parameters is tested with "application-phase" analyses. Here, the finite-element model is used to predict stable crack growth and instability for a different geometry, with a previously evaluated parameter serving as the criterion for stable growth. These analyses are applied to and compared with measurements of crack growth and instability in center-cracked panels and compact tension specimens of the 2219-T87 aluminum alloy and the A533-B grade of steel.

The work shows that the crack growth parameters $(COA)_c$, J_c, dJ_c/da, and the linear elastic fracture mechanics (LEFM)-R, which sample large portions of the elastic-plastic strain field, vary monotonically with stable crack extension. However, the parameters $(CTOA)_c$, $\mathcal{R}$, $\mathcal{G}_o$, and F_c, which reflect the state of the crack tip process zone, are essentially independent of the amount of stable growth when the mode of fracture does not change. Useful, stable growth criteria can therefore be evaluated from the crack tip state at the onset of crack extension and do not have to be continuously measured during stable crack growth. The possibility of making accurate predictions for the extent of stable crack growth and the load level at instability is demonstrated using only the value of J_c at the onset of crack extension.

KEY WORDS: plastic fracture mechanics, finite-element models, stable crack growth, crack instability, J-integral, J-integral derivative, generalized energy release rate, process zone, computational process zone energy, crack opening angle, node force, linear elastic fracture mechanics resistance curve, center-cracked panel, compact tension specimen, aluminums, steels, crack propagation

[1] The authors are members of the staff, Battelle Columbus Laboratories, Columbus, Ohio.

ELASTIC-PLASTIC FRACTURE

Nomenclature

- a Crack length
- a_0 Initial crack length
- a_c Crack length at crack growth instability
- Δa Crack growth increment
- b Plate thickness
- COA Crack opening angle evaluated from crack opening at position of initial crack tip
- $(COA)_c$ Critical value of COA for stable crack growth
- COD Crack opening displacement
- $CTOA$ Crack opening angle evaluated from slope of the crack faces at the crack tip
- $(CTOA)_c$ Critical value of $CTOA$ for stable crack growth
- E Elastic modulus
- F Crack tip node force in finite-element model of crack growth process
- F_c Critical value of F for stable crack growth
- G Energy release rate based on LEFM concepts
- $\mathcal{G}$ Generalized energy release rate based on computational process zone concept
- $\mathcal{G}_0$ Work of separating crack faces per unit area of crack growth
- $\mathcal{G}_{0c}$ Critical value of $\mathcal{G}_0$ for stable crack growth
- $\mathcal{G}_z$ Energy change in computational process zone per unit of crack growth
- $\mathcal{G}_{zc}$ Critical value of $\mathcal{G}_z$ for stable crack growth
- J Value of the J-integral evaluated on a contour remote from the crack tip
- J_c Critical value of J for stable crack growth
- J_{Ic}, J_{ci} Critical values of J for initiation of crack growth under plane strain and plane stress, respectively
- dJ/da Rate of change of J with crack growth
- dJ/da_c Rate of change of J_c with crack growth
- K_{Ic} Fracture toughness for initiation of crack growth
- LEFM   Linear elastic fracture mechanics
- P_1 Scaling parameter $= (K_{Ic}/\sigma_Y)^2/b$
- R Critical value of G for stable crack growth
- $\mathcal{R}$ Critical value of $\mathcal{G}$ for stable crack growth
- T Surface traction
- u Displacement
- w Plate width
- W Strain energy density
- V Volume of the computational process zone
- ϵ Strain

Γ   Contour for evaluation of J
σ   Stress
σ_Y Yield stress
σ_o Applied stress at initiation of stable crack growth
σ_c Applied stress at crack growth instability
N Number of load increments used in finite-element solution procedure during a crack growth step
P Applied load
ΔP Increment of applied load

Existing linear elastic fracture mechanics (LEFM) and J-integral analyses are well suited for safety assessments of high-strength/low-toughness materials. These analyses apply only to the onset of crack growth, which is usually tantamount to crack instability and structural failure in that class of materials. However, this is not the case for the lower-strength, higher-toughness grades when crack instability may be preceded by extensive stable crack growth under rising load. Here, a substantial margin of safety may exist even when the onset of crack growth is imminent. Attempts at accurate calculation of this margin of safety by extending LEFM or J-analyses into the stable growth regime are precluded by their inability to treat the inelastic effects arising from large-scale plasticity and material unloading [1].[2] Consequently, improved methods are needed.

This paper describes research leading to a plastic fracture mechanics methodology designed to treat two-dimensional large-scale yielding and stable crack growth problems. While the intended applications are the steels and failure modes encountered in nuclear pressure vessels, the approach has greater generality. The research draws on elastic-plastic finite-element analyses. Although some closed-form solutions exist [2,3] the virtual impossibility of obtaining closed-form treatments for the conditions of interest necessitate numerical analyses. A special "toughness-scaled" aluminum alloy was used to avoid the cost of full-scale experiments on reactor pressure vessel steel (A533-B). This alloy approximates the flaw size/plastic zone size/structural size relations of the steel vessel wall in thinner, more manageable test pieces. The experiments employed both center-cracked panels and compact tension specimens and are compared with results for the A533-B steel generated by a concurrent and closely related program at the General Electric Co. [4-6].

A key element of the research is the "generation-phase" analysis procedure in which the experimentally observed applied-load (or displacement) stable crack growth behavior is reproduced in a finite-element model. In these calculations, each of a number of candidate stable growth parameters is calculated for the material tested. The quality of the predictions that can

---

[2]The italic numbers in brackets refer to the list of references appended to this paper.

be made with these parameters is tested with "application-phase" analyses. Here, the finite-element model is used to predict stable crack growth and instability for a different geometry, with a previously evaluated parameter serving as the criterion for stable growth. The feasibility, economy, and accuracy of the application-phase calculations offer a basis for appraising the various candidate criteria.

One finding of this work is that parameters truly reflecting the state of the crack tip process zone are not a function of the extent of stable crack growth when the mode of fracture (full shear or flat) remains fixed. The possibility exists, therefore, that useful, stable growth parameters can be evaluated from the state of the crack tip at the onset of crack extension. Consequently, they will not have to be measured separately. The possibility of making accurate predictions of stable crack growth and crack instability directly from J_{ci} in this way is demonstrated.

Background Discussion

The fracture criteria examined in this program include the J-integral, its rate of change during crack growth dJ/da, the crack tip opening angle ($CTOA$), and the average crack opening angle (COA). In addition, two new candidates are examined. One is the generalized energy release rate $\mathcal{G}$. This corresponds to the energy flowing into a computational process zone surrounding the tip of the extending crack per unit area of crack extension. The other candidate is the crack tip force F which acts at the crack tip nodes in a finite-element model during the stable crack growth process.

The J-integral—more specifically, its derivative with crack length dJ/da—has been proposed by Paris et al [7] as a geometry-independent material parameter for a limited amount of stable crack growth. The crack opening angle has also been proposed [8,9]. It should be recognized, however, that there are two distinct definitions of the COA that have been used. De Koning [8] uses the angle ($CTOA$) that reflects the actual slopes of the crack faces at the crack tip. Green and Knott [9] use an average value (COA) based on the COD at the original crack tip position. These are appealing because of their readily grasped physical significance and the opportunity offered for direct measurement. Garwood and Turner [10] have apparently been successful in extending infiltration techniques to the tip of a stably growing crack. The average COA can be obtained with a displacement gage mounted near the original crack tip and a measurement of the crack length.

Energy release rate concepts have been examined by a number of investigators [11-17]. The inherent difficulty caused by the dependence on the computational model, originally pointed out by Rice [18], has not yet been completely resolved. But, as argued by Kfouri and Rice [19], for example, this can be circumvented by appealing to micromechanical considerations.

A possibly more satisfactory alternative approach is described in the next section of this paper.

Analytical Approach

Finite-Element Analysis

The finite-element program being used in this study utilizes constant-strain triangular elements and quadrilateral elements that are composed of four triangular elements. The use of these simple elements allows crack growth to be accommodated and permits various candidate fracture parameters to be calculated readily. The finite-element program satisfies two further important requirements. These are the ability to model strain hardening plasticity and elastic unloading. The use of more sophisticated elements could possibly increase the accuracy/cost ratio for the types of analyses presented here, but the use of higher-order elements would also greatly complicate the manner in which the crack growth is simulated.

Several methods for modeling crack extension are in the literature. These are based on uncoupling nodal points ahead of the crack tip by relaxing the forces holding them together. Two methods for releasing nodes are common. Kobayashi et al [20], de Koning [6], and Light et al [21] first apply forces to the nodes that are equal to, but opposite in direction to, those holding the nodes together; then, these are generally relaxed. Andersson [22,23] and Newman and Armen [24], in contrast, reduce the stiffness associated with coupling the nodes together. While both of these approaches are conceptually similar, they are procedurally different. In the work reported here, the first technique was adopted.

These two methods for modeling crack extension can be used when the nodal spacing along the path of crack growth is small relative to the total crack extension. If one uses higher-order elements, however, the nodal spacing will generally be increased and could possibly become comparable to the total crack extension. In this case, a more sophisticated scheme for modeling the crack extension is required. One such scheme is to shift the crack tip node along the crack growth path. This method is currently being used by Shih [6].

Computational Procedure

The finite-element method is utilized in the fracture analysis procedure in two conceptually different ways. In the first, the finite-element analysis is used to further analyze data from fracture experiments on simple geometries. In this role, experimentally measured load (or displacement) versus crack growth records are used as input to the analysis. The outputs of the analysis are the candidate fracture criteria and their dependence on crack extension. This mode of analysis is called a generation-phase analysis, that is, *generation* of fracture parameters from experimental measurements. In the second mode of analysis, the role of the finite-element analysis is reversed. The input is a selected fracture criterion and its dependence on crack growth. The output is the extent of stable crack growth, and the load level giving crack growth instability. This type of analysis is called an application-phase analysis, that is, *application* of a fracture criterion to determine the structure's response.

In the generation-phase analysis, the relation between the applied load and the crack length is known beforehand. Therefore, it is possible to increase the external load simultaneously with the release of the crack tip nodes. This approach results in a piecewise linear approximation to the experimental load versus crack length curve. Another approach is to maintain a constant load during the crack-tip-node release increments. However, this approach results in a stepwise approximation to the experimental curve and, therefore, is not as representative of the physical situation for finite crack growth increments.

In an application-phase analysis, where the load/crack growth relation is the desired output, one must either iterate to determine the increase in load for the given increment in crack growth or be satisfied with a stepwise approximation. Since the iteration procedure would be as much as three times as expensive as the stepwise approach in the application mode, the latter choice is more attractive. However, if the less expensive option is used only for the application phase, consistency between the generation phase and the application phase will be lost. Since the computed fracture parameters could be sensitive to differences in the manner of load application for the finite increments of crack growth considered here, this should be avoided. For analyses presented here, the generation and application phases have been conducted in a consistent manner using the piecewise linear approach and iteration in the applications phase. The one exception to this is the application-phase analysis which used the critical node tip force, F_c.

The application-phase analysis using F_c was conducted using a stepwise application of load and therefore did not require the additional expense involved in iteration. Since F_c is determined during the application analysis, no generation-phase analysis is required. Therefore, the question of consistency does not exist. As will be made more clear in discussing the applications-phase calculations, this is a very attractive feature, unique (at least at present) to this particular fracture criterion.

Computation of Candidate Fracture Criteria

The parameters evaluated in this program can all be calculated without recourse to a special crack tip element. The J-integral calculation is accomplished by evaluating an integral over a closed path containing the crack tip. The expression is

$$J = \int_\Gamma \left\{ W dy + \overline{T} \cdot \frac{\partial \overline{u}}{\partial x} ds \right\} \quad (1)$$

where Γ denotes a counterclockwise path surrounding the crack tip, W is the strain energy density given by

$$W = \int \sigma_{ij} d\epsilon_{ij} \quad (2)$$

and $\overline{T}$ and $\overline{u}$ are the surface traction and displacement vectors, respectively. Note that a path remote from the crack tip is used.

The crack opening angle values are obtained in an obvious way from the node point displacements. The average COA is obtained from the displacement at the initial crack tip position while the crack tip value ($CTOA$) is obtained from the displacement of the nodes nearest the crack tip. The parameter F, of course, is obtained directly. Thus, only the generalized energy release rate needs further elaboration. This is as follows.

The generalized energy-release rate is intended as a direct extension of the basic energy-balance concept that has proven itself in LEFM. Two generalizations are included in this approach. First, a small region surrounding the crack tip is identified which contains the three-dimensional heterogeneous processes that must be excluded from continuum-mechanics considerations. Second, a direct computation is made of the plastic-energy dissipation rate for the material outside the excluded region. Two key assumptions are then required. The first is that the energy dissipation rate in the excluded region can be taken as a material property, independent of crack length and other dimensions of the body containing the crack. The second assumption is that $\mathcal{G}$, the energy flow rate to the excluded region, is unaffected by the details of the deformation occurring within it. Thus, the computation can be made entirely by two-dimensional continuum mechanics techniques.

The basic parameter involved in the approach is the critical computational process zone energy-dissipation rate $\mathcal{R}$: the energy dissipation accompanying ductile crack extension from processes such as hole growth and coalescence that occur within a small region surrounding the crack tip.[3] The approach will be successful if $\mathcal{R}$-values can be found that are independent of the structural geometry and of the crack length. This hinges on a third key assumption—that the geometry-dependent portion of the energy dissipation rate accompanying crack extension can be accurately

---

[3]The word "rate" is used here, and throughout this paper, just as in conventional LEFM, to mean per unit area of crack growth.

calculated by applying a continuum plasticity formulation to the material outside the computational process zone. Then, the geometry dependence can be separated from the material dependence. Taken together, these assumptions will lead to an approach that has two primary virtues: (1) recognition is given to the micromechanical nature of the crack growth process, and (2) at the same time, computations can be carried out conveniently, using continuum-mechanics techniques, for example, the finite-element method.

Computationally, the generalized energy-release rate is the sum of two terms. That is, $\mathcal{G} = \mathcal{G}_o + \mathcal{G}_Z$ where $\mathcal{G}_o$ is related to the work done in separating the crack faces according to

$$\mathcal{G}_o = \frac{1}{b \, \Delta a} \int_{\xi=0}^{\xi=\Delta a} \overline{T}_i(\xi) \, \overline{u}_i(\Delta a - \xi) d\xi \tag{3}$$

and $\mathcal{G}_Z$ is related to the change in the energy contained in the computational process zone, from prior to crack extension—State 1—to following crack extension—State 2—according to

$$\mathcal{G}_Z = \frac{1}{b \, \Delta a} \int_V \left\{ \int_{(1)}^{(2)} \sigma_{ij} d\epsilon_{ij} \right\} dV \tag{4}$$

where

b = plate thickness,
Δa = increment of crack growth,
$\overline{T}_i$ = tractions holding the crack tip closed,
$\overline{u}_i$ = crack-opening displacements behind the crack tip, and
V = volume of the computational process zone.

Crack growth then proceeds such that

$$\mathcal{R} = \mathcal{G} \equiv \mathcal{G}_o + \mathcal{G}_Z \tag{5}$$

in this approach. Crack instability (fracture) will then occur when $\mathcal{G} > \mathcal{R}$ for the prescribed loads or displacements at some crack length.

It should be emphasized that this analysis scheme is *not* based on what might be termed a "recoverable energy" criterion. Although the approach is based on the idea of an energy balance which does include recoverable elastic energy, there are fundamental differences which circumvent the pitfalls of a technique based solely on this idea. In particular, as Rice [16] has shown, for a material that saturates at large plastic strain (for example, an elastic-perfectly plastic material), the elastic energy rate supplied to the crack tip is exactly equal to the plastic energy dissipation rate. Hence, in this case the crack-driving force is identically zero for all load levels. The feature of the approach described here that eliminates this fundamental objection is the use of a crack tip computational process (CP) zone.

While the size and shape of the CP-zone must be somewhat arbitrary, there are some physical considerations that guide the choices. Most importantly, for a through crack in a thin plate, the 45 deg through-thickness mode of plastic relaxation will dominate. In order to cope with this three-dimensional effect within the bounds of a two-dimensional analysis, in plane-stress conditions, the dimensions of the CP-zone were taken to be roughly equal to the plate thickness. Specifically, for plate thicknesses of 6.35 mm, calculations of $\mathcal{G}_z$ are made using a computational process zone having a height of 10 mm in the direction normal to the crack plane, and a length of 4.5 mm in the direction along the crack plane. In thicker plates where plane-strain conditions are approached, a smaller CP-zone height probably could be used but with attendant higher computation costs. The effect of the CP-zone size has not yet been thoroughly explored. However, it is clear that $\mathcal{R}$ will depend on the size of the zone, so that these dimensions must be known in order to apply the result.

Experimental Verification

Toughness-Scaled Materials

The finite-element analyses are based upon and verified by systematic measurements of load extension curves, COD, *COA*, stable crack growth, and instability. The verification task is greatly simplified (1) by limiting both the finite-element models and the experiments to essentially two-dimensional events, and (2) by reducing the scale of the experiments relative to actual vessels. This was done by using 6.35-mm-thick panels of 2219-T87 aluminum, a material that matches the flaw size/plastic zone size/structural size relations of the full-scale vessel.

Toughness scaling, accomplished by preserving the relation between the plane-strain plastic zone size and the plate thickness of a nuclear pressure vessel, is expressed by the scaling parameter $P_1 = (1/b)(K_{Ic}/\sigma_Y)^2$. The scaling parameter has a value $P_1 = 1.42$ for a 200-mm-thick plate of A533-B steel (K_{Ic} = 220 MPa·m$^{1/2}$, σ_Y = 413 MPa); see Refs *4-6*. The same value of the scaling parameter is obtained with 2219-T87 aluminum K_{Ic} = 36 MPa·m$^{1/2}$, σ_Y = 379 MPa) with a panel thickness b = 6.35 mm. The aluminum plates can thus be regarded as 1/32-scale models of much larger steel plates.

The cracks produced in the aluminum panels actually extended by the full shear mode. Thus, these experiments model full-scale (200 mm-thick) steel plate loaded in the same fashion that also fail with a full-shear mode. Failures with some amount of shear are expected since the thickness requirement for a flat plane-strain fracture is b = 710 mm for the A533-B steel.

Verification for the flat (ductile) plane-strain mode of crack extension, and for the toughness scaling concept in general, was obtained by analyzing measurements on an A533-B steel test piece performed by Shih et al [5,6]. The measurements included load, load point displacement, and crack opening displacement as a function of stable crack growth for a 4T compact tension specimen with 25 percent side grooves tested at 93 °C. More information on this experiment (No. T52) and on the test material is given in Refs 5 and 6.

A piecewise-linear approximation of the stress-strain curve derived from specimens of the 2219-T87 aluminum panels is shown in Fig. 1. The material exhibited essentially no anisotropy. A similar representation of the stress strain curve for the A533-B steel, reported by Shih et al [6], is also given in Fig. 1.

Experimental Details

Experiments were performed on center-cracked panels as well as on compact tension specimens. The center-cracked panels were 6.35-mm-thick 2219-T87 aluminum, either 305 mm wide by 1016 mm long with three different initial crack lengths, $2a_0 = 25.4$, 102, and 204 mm, and 152 mm wide by 813 mm long with $2a_0 = 102$ mm. The experiments were performed in a closed-loop electrohydraulic testing machine of 2.3-MN dynamic capacity (3 MN static). The specimens had a central 6.24-mm-diameter hole with a 2-mm-wide milled slit at both sides of the hole. The slits were fatigue cracked at a cyclic load equal to one-third or less

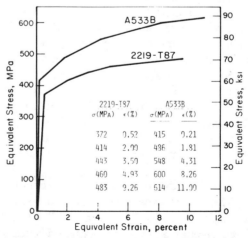

FIG. 1—*Piecewise linear stress-strain curves used for analysis of 2219-T87 and A533-B steel fracture specimens.*

of the expected failure load. Antibuckling guides were applied. These consisted of four steel angle sections, two by two bolted together outside the specimen. The specimens were pulled to fracture in times varying between 60 and 120 s.

Three clip gages were used, one mounted in the central hole, one at the end of the slit, and one at the end of the fatigue crack. They were each supported by 2 spring pins 1.6 mm in diameter mounted in holes of this size spaced 12.5 mm apart and located symmetrically with respect to the notch. In addition, the compliances of the specimens were determined through measurement of the displacement of the specimen end by means of a linear variable differential transformer (LVDT) induction coil. The three clip gages and the LVDT were recorded as a function of the load on four separate X-Y recorders. A digital load recorder was placed in front of the specimen. Moving pictures at 20 frames per second were made of each experiment to obtain a record of crack size versus load.

The compact tension specimens complied with ASTM proportions, except for the thickness. They were 127-mm-wide, 122-mm-high, and 6.35-mm-thick 2219-T87 aluminum. The tests were performed in an Instron testing machine. Cover plates were applied as antibuckling guides with a porthole to observe crack growth by means of an optical microscope. Simultaneouly, crack growth measurements were made ultrasonically, using a probe at the edge of the specimen opposite the crack.

The most relevant data from the center-cracked panel tests are given in Table 1. Actual load-crack extension curves for the three panels are reproduced in Fig. 2. Some results of the tests on compact tension specimens are presented in a later section where they are compared with analytical results.

The average COA was obtained from the experimental data in the following way. A clip gage was mounted over the tip of the original fatigue crack. The displacement at this point was determined at the onset of slow stable crack growth. Any further displacements were assumed to be entirely due to crack growth. Thus the additional displacement was divided by the instantaneous amount of slow crack growth to give COA.

Computational Results

This section describes the results of several generation-phase and application-phase analyses for center-cracked panels and compact tension specimens. Figures 3 and 4 show typical finite-element grids used for the two geometries. The grid for the center-cracked specimens represents one quadrant of the panel. Typical models contain approximately 325 elements and 700 degrees of freedom. The compact tension specimen shown is a 2T specimen with an initial crack length of 40 mm. It contains approximately the same number of elements and degrees of freedom as the center-cracked

132 ELASTIC-PLASTIC FRACTURE

TABLE 1—*Test results for 2219-T87 aluminum center-cracked panels.*

Specimen No.	Plate No.	σ_Y MPa	σ_Y ksi	$2a_0$ Intended mm	$2a_0$ Intended in.	$2a_0$ Actual Surface mm	$2a_0$ Actual Surface in.	$2a_0$ Actual Center mm	$2a_0$ Actual Center in.	$2a_0$ Actual Avg mm	$2a_0$ Actual Avg in.
2Cm 2.01[a]	(2C)1	372	53.9	25.4	1	25.5	1.00	27.2	1.07	26.3	1.04
2Cm 2.04[a]	(2C)3	383	55.5	101.6	4	102	4.02	104	4.09	103	4.06
2Cm 2.08[a]	(2C)2	373	54.0	203.2	8	204	8.03	205.5	8.09	204.7	8.06
2Cs 2.04[b]	2E1	372	53.9	101.6	4	100	3.94	101	3.98	101.5	4.0

Specimen No.	$2a_c$ mm	$2a_c$ in.	$2a_c$ Tongue mm	$2a_c$ Tongue in.	Fracture Left	Fracture Right	σ_0 MPa	σ_0 ksi	σ_c MPa	σ_c ksi	σ_0/σ_Y	σ_c/σ_Y
2Cm 2.01[a]	53.5	2.11	42	1.65	SS[c]	DS[d]	297	43.0	322	46.7	0.80	0.87
2Cm 2.04[a]	130	5.12	[f]	[f]	SS[c]	SS[c]	149	21.6	214	31.0	0.39	0.56
2Cm 2.08[a]	227	8.94	209	8.23	SS[c]	DS[d]	81	11.8	102	14.8	0.22	0.27
2Cs 2.04[b]	108	4.25	114	4.49	SS[c]	SS[c]	90	13.0	104	15.1	0.24	0.28

[a] Overall dimensions: 305 mm wide by 1016 mm long.
[b] Overall dimensions: 152 mm wide by 813 mm long.
[c] Single shear.
[d] Double shear.
[e] Double shear with steps.
[f] No tongues, smooth conversion to shear.

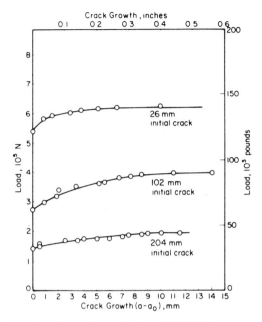

FIG. 2—*Experimental crack growth data for three 2219-T87 aluminum center-crack panels ($2w = 305$ mm).*

panels. In all cases, the stress-strain curve was represented as being piece wise linear. Incremental plasticity relations based on the Prandtl-Reuss equations were used in all calculations. The aluminum center-cracked panels and compact tension specimens were all 6.35 mm thick and therefore were modeled under the assumption of plane stress. The A533-B specimen, however, was nominally 102 mm thick (76 mm at root of side groove) and therefore was modeled under the assumption of plane strain.

The hardening rule used for the analyses assumed that the yield strength of the material upon reloading is unchanged from initial yield. Since there is little reloading in the analyses due to the simple geometries and loading, the type of hardening rule is expected to be of secondary importance.

The program simulates a crack growth step by releasing the nodal force at the pair of nodes representing the crack tip. The nodal forces are released incrementally. For an initial semicrack length a_o, the external load P is increased by an amount ΔP in each of N steps such that $N\Delta P = P_2 - P_1$. Because it is possible only to match the curve at discrete points, a piecewise linear representation of the experimental curve is actually used. During each load step, the crack-tip nodal constraint force F is relaxed and the free crack surface extended one element ahead. This stepwise process is continued until the onset of unstable crack growth is reached. Usually, five relaxation increments were employed to release the nodal

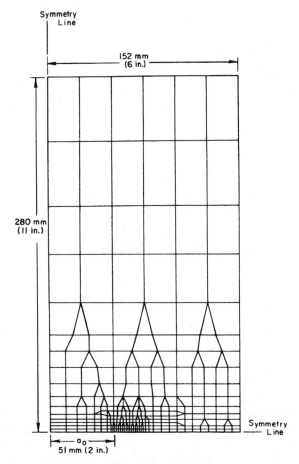

FIG. 3—*Finite-element model of a center-crack panel.*

force at the crack tip; that is, $N = 5$. For the 1.5-mm crack growth increment used in the majority of the analysis, approximately 50 incremental solutions were required for a stable growth of 15 mm.

Generation-Phase Computations

Generation-phase computations were made for three aluminum center-cracked panels, an aluminum compact tension specimen, and a steel compact tension specimen. Computational results for the different fracture parameters during stable crack growth are presented in Fig. 6. The quantities that reflect the toughness of the material in the locale of the crack tip, G_{∞}, $\mathcal{R}$, $\mathcal{G}_{Zc}$, $(CTOA)_c$, and F_c, are relatively invariant during stable crack

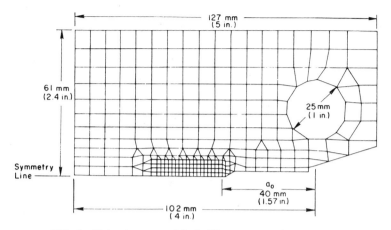

FIG. 4—*Finite-element model of a 2T compact tension specimen.*

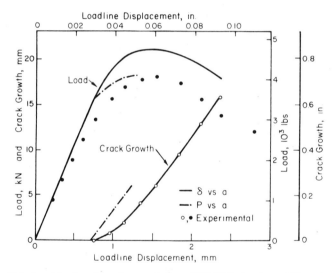

FIG. 5—*Calculated load and crack growth for a 2219-T87 aluminum 2T specimen (w = 102 mm, a_o = 40 mm, b = 6.35 mm).*

growth. The parameter dJ_c/da, which might be expected to be valid only for limited amounts of stable crack growth, shows some systematic crack growth dependence. Of these quantities, F_c and the $(CTOA)_c$ appear to be most nearly constant. All of the local quantities, and also dJ_c/da, reflect a loss in crack growth resistance in the first 2 mm of crack extension, but are then constant. In contrast, J_c, the LEFM-R,[4] and $(COA)_c$, which can re-

---

[4] These R-values do not contain a plastic zone size adjustment. The adjusted R values could be close to the J_{c_i}-values.

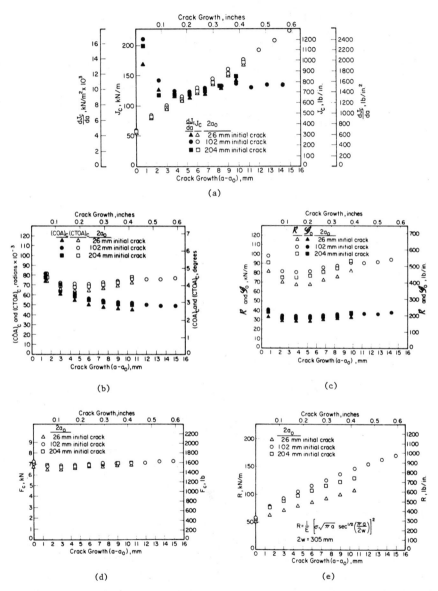

FIG. 6—*Calculated values of candidate fracture criteria for three 2219-T87 aluminum center-crack panels (2w = 305 mm, b = 6.35 mm).*

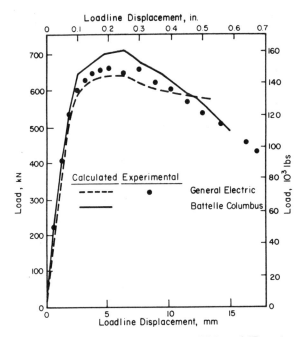

FIG. 7—*Load-displacement curves calculated for an A533-B steel 4T specimen with 25 percent side grooves* ($w = 203$ mm, $a_o = 163$ mm, $b = 76$ mm).

flect changes in the character of the strain field remote from the crack tip, vary strongly but systematically. With the exception of the LEFM-R, all of the parameters were relatively insensitive to the initial flaw size, with F_c and J_c showing the least dependence.

Figures 5 and 7 contain the load line displacement versus crack length measurements that serve as inputs for the generation-phase calculations for the aluminum and steel compact specimens. These figures also compare the calculated load-displacement curves with the measurements. The latter results illustrate that the finite-element model for the aluminum compact specimen is 7 to 15 percent stiffer than the actual test piece. For this reason the calculations for the aluminum test piece were repeated using load versus crack length as the input. This established a lower bound for the different fracture parameters.

The results of the generation-phase analysis of the aluminum compact specimen are presented in Figs 8 and 9, where they are compared with the values for the center-cracked panels.[5] The results for the A533-B steel compact tension specimen are given in Fig. 10. Qualitatively the results for the steel and the aluminum compact specimen are very similar despite the

---

[5] As reflected by the different J_{ci}-values for the compact tension specimen and the center-cracked panel (see Fig. 8a), there were some metallurgical differences in the materials used.

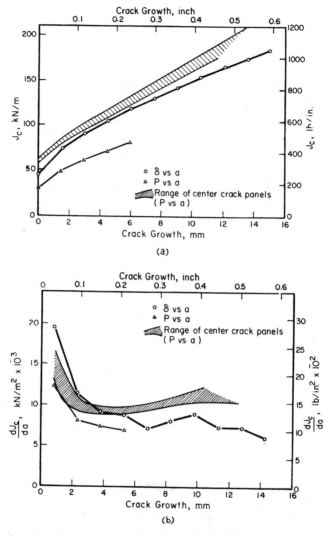

FIG. 8—*Calculated values of* J_c *and* dJ_c/da *for a 2219-T87 aluminum 2T compact tension specimen* (w = 102 mm, a_o = 40 mm, b = 6.35 mm).

fact that both the modes of fracture (ductile shear for the 2219-T87, ductile flat for the A533-B) and the absolute values of the toughness parameters differ greatly. The parameters that reflect the deformation in the crack tip process zone, $(CTOA)_c$, $\mathcal{R}$, and F_c, are essentially independent of the extent of stable crack growth. In contrast, the parameters that sample larger portions of the elastic and plastic strain field, $(COA)_c$, J_c, and dJ_c/da, vary monotonically with stable crack extension. The variation of $(COA)_c$ is confirmed by actual measurements as shown in Fig 9c. Within the precision of the analyses, comparison of the 2219-T87 center-cracked panel and compact specimen results indicates that $\Re$, $(CTOA)_c$ and possibly the $(COA)_c$ are independent of geometry. The quantities J_c and dJ_c/da show some geometry dependence after about 4 mm of stable crack growth (6 percent of the remaining ligament).

Figure 11 summarize results of generation-phase analyses of a 2219-T87 center-cracked panel ($2w = 305$ mm, $2a_o = 102$ mm) for three different mesh sizes in the crack growth region. The calculations indicate that J_c, $(COA)_c$, and $\Re$ are not sensitive to mesh size. The $(CTOA)_c$ shows some dependence. The quantities $\mathcal{G}_{oc}$ and F_c are strongly mesh size dependent, approaching zero as the mesh size approaches zero, as reported by Rice [18].

Application-Phase Analyses

Application-phase analyses are shown in Figs. 12 and 13. These were carried out on two center-cracked panels: 2Cm2.04, a load-controlled experiment, and 2Cs2.04, a displacement-controlled experiment (see Table 1). The first analysis employed $J = J_c$ as the initiation criterion and $\mathcal{G} = \Re$ as the stable crack growth criterion; the second employed $J = J_{ci}$ for initiation and $F = F_c$ for stable growth. However, an even more important distinction is that the first calculation relied on two separately measured toughness values (J_{ci} and $\Re$) which were obtained by averaging the results of three previously performed generation-phase analyses. The second calculation relied on a single toughness value as input: the value of J_{ci} at the onset of crack extension to characterize both initiation and stable growth. This was made possible by virtue of the fact that F_c is essentially constant during stable crack growth (see Figs. 5, 9, and 10). Its value is determined by the finite-element at the load level producing $J = J_{ci}$ (that is, F_c was calculated from J_{ci} at the onset of growth). The results of the two application-phase computations presented in Figs. 12 and 13 show that the model predicted, with quite good accuracy, the load versus stable crack growth behavior of the test pieces, including the maximum load and corresponding crack length, and the instability condition.

Discussion of Results

The present findings illuminate the basic cause of stable growth in elastic-plastic materials. In the cases analyzed here, crack stability cannot be attributed to an increase with crack growth of the toughness of the material in the process zone. This is supported by the constancy of the CP-zone energy and the $CTOA$ with the unchanging fracture mode in the 2219-T87 and A533-B test pieces. The constancy of $\Re$ coupled with increasing load

140 ELASTIC-PLASTIC FRACTURE

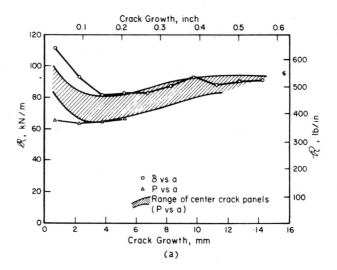

(a)

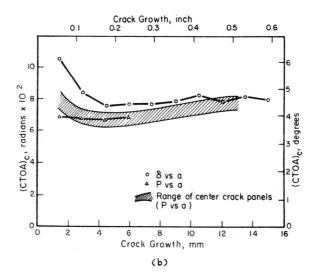

(b)

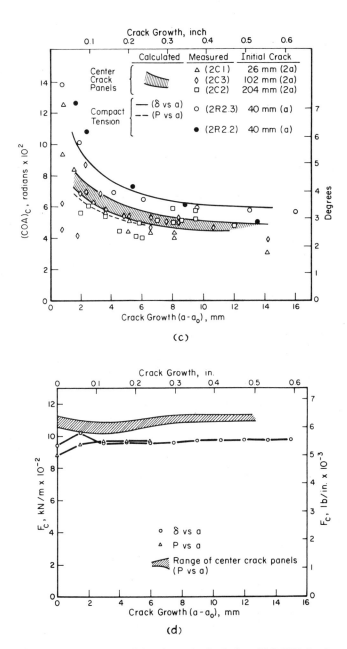

FIG. 9—*Calculated values of candidate fracture criteria for a 2219-T87 aluminum 2T compact tension specimen ($w = 102$ mm, $a_o = 40$ mm, $b = 6.35$ mm).*

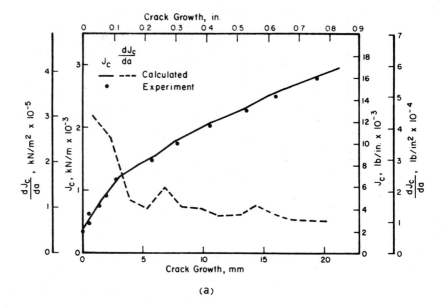

(a)

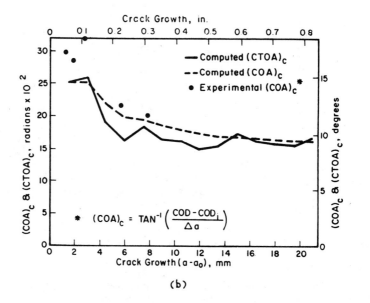

(b)

KANNINEN ET AL ON INSTABILITY PROBLEMS   143

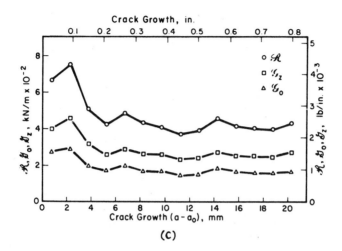

(c)

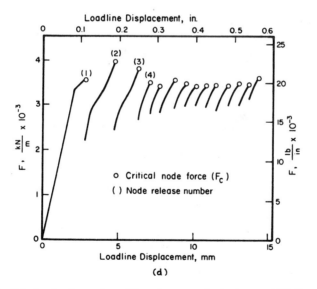

(d)

FIG. 10—*Calculated values of candidate fracture criteria for an A533-B steel, 4T, 25 percent side-grooved compact tension specimen* ($w = 203$ mm, $a_o = 134$ mm, $b = 102$ mm).

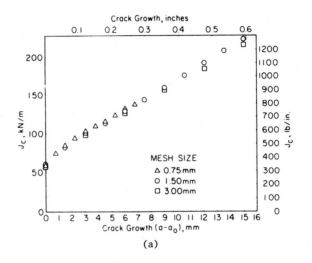

(a)

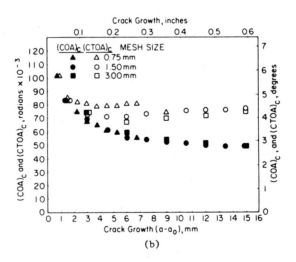

(b)

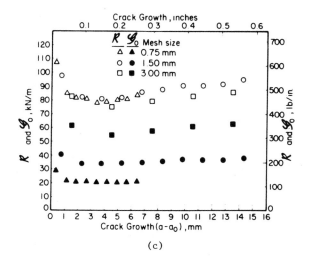

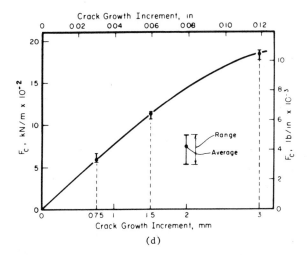

FIG. 11—*Mesh size dependence of candidate fracture criteria for a 2219-T87 aluminum center-crack panel ($2w = 305$ mm, $2a_o = 102$ mm, $b = 6.35$ mm).*

146 ELASTIC-PLASTIC FRACTURE

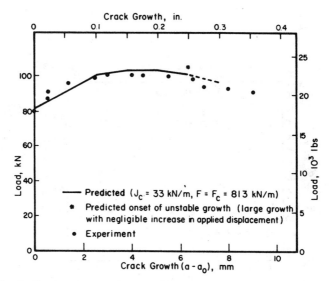

FIG. 12—*Application-phase analysis of a 2219-T87 aluminum center-crack panel using $\mathcal{R}$ as the fracture criterion* ($2w = 305$ mm, $2a_o = 102$ mm, $b = 6.35$ mm).

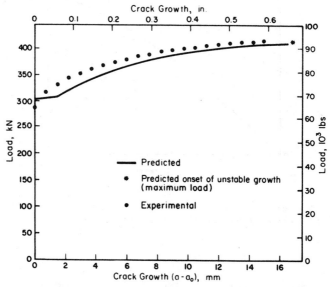

FIG. 13—*Application-phase analysis of a 2219-T87 aluminum center-crack panel using F_c as the fracture criterion* ($2w = 152$ mm, $2a_o = 102$ mm, $b = 6.35$ mm).

TABLE 2—*Appraisal of candidate fracture criteria for the basis of a plastic fracture methodology.*

Desirable Features	F_c	$\mathcal{G}_{oc}$	$\mathcal{R}$	$(COA)_c$	$(CTOA)_c$	J_c	$(dJ/da)_c$	R
Constant during stable crack growth with fixed fracture mode	yes[a]	yes	yes	no	yes	no	no	no
Independent of geometry	no	no	yes	yes	yes	no	no	no
Computer model independent	no	no	yes[a]	yes[a]	no	yes[a]	yes	yes
Possibility of direct measurement	no	no	no	yes[a]	yes	no	no	no
Computational efficiency	yes[a]	yes	no	yes[a]	yes[a]	yes[a]	yes	yes

[a] yes—Criterion is significantly better than others in this respect.

during stable crack growth means that the portion of the energy flow reaching the crack tip region diminishes with crack extension. The reduced energy flow can be thought to result from the "screening" action of the plastic zone accompanying the growing crack, as described by Broberg [9]. It is possible that the toughness increases in the case of a mixed-mode fracture with an increasing shear component. But, where this condition is not attained, the interpretation of the so-called "J-resistance" curve as a manifestation of increasing toughness of material of the process zone is fundamentally incorrect. This point is treated more fully by Hutchinson and Paris [25] and by Shih et al [27] elsewhere in this publication. They also conclude that there are departures from the Hutchinson-Rice-Rosengren singularity and, hence, the meaning of J, in a zone at the tip of a growing crack. This zone, which expands with crack extension, is "J-controlled" in the same sense as the small-scale yielding plastic zone of a stationary crack is "K-controlled." That is, there is a direct relation between the J-value measured remotely and the deformation state of the crack tip. However, this relation changes as crack growth proceeds. The rising J_c curve is a consequence of this changing relation, not of a change in the local toughness. Also, as Hutchinson and Paris have shown, the relation becomes invalid after some small amount of stable crack growth.

The present work also shows that the LEFM energy release rate concept can be generalized to elastic-plastic materials by enlarging the energy sink to include a finite computational process zone. The generalized energy release concept has several attractive features. The critical energy release rate has a well-defined physical meaning. It does not require elaborate modeling and is insensitive to mesh spacing and specimen geometry. It can reduce three-dimensional plane stress and mixed-mode crack extension to a two-dimensional problems. Also, it appears to be essentially constant from the beginning of stable crack growth. This latter feature provides the basis for the application-phase calculation, described in the previous section, which illustrates that a single constant value of $\mathcal{R}$ can predict stable crack growth and instability with precision.

The deduction that the process zone is essentially invariant during fixed-mode stable growth is important because it serves to validate other criteria, such as $\mathcal{G}_{\infty}$, F_c, and $(CTOA)_c$, which may be computationally more convenient. It follows that any parameter reflecting the state of the process zone may be independent of crack growth. More important, such parameters would not have to be measured separately. Instead, process zone parameters that are invariant with crack growth can be evaluated from a finite-element model of the state of the crack tip region at the onset of crack extension. This is illustrated by a second application-phase calculation in the previous section. Here the crack growth parameter F_c, which is determined by the value of J_{ci}, serves only an operational function in the calculation. This calculation demonstrates the feasibility of characterizing the onset of crack extension, stable crack growth, and instability for a large-scale yielding problem using a single toughness value as input.

Recently, nine different requirements for an acceptable plastic fracture criterion were identified [26]. These include that it be (1) well suited for models of three-dimensional crack fronts, (2) valid for both small and large stable crack extensions, (3) geometry independent, (4) computer model independent, (5) economical to use, (6) measurable with small test pieces, (7) able to predict crack initiation, (8) able to predict crack instability, and (9) valid for fully plastic behavior. An appraisal based on these requirements is summarized in Table 2. Shih et al [27] have examined the same group of candidate parameters (except for $\Re$). After subjecting them to essentially the same requirements as those just given, they have concluded that the J_c- and $(COD)_c$-curves are the most promising. In our view, however, such a conclusion is premature.

It should be clear from the foregoing discussion that the requirements for geometry independence, model independence, and the ability to measure the parameter are all redundant when the parameter is constant during stable crack growth. Of the parameters examined so far, the $(CTOA)_c$, F_c, and $\mathcal{G}_o$ appear to be both constant and computationally convenient operational criteria. These can be related to the value of a crack initiation parameter such as J_{1c} (or J_{ci}) or possibly $(COD)_c$, thereby satisfying requirements 2, 4, and 8 regarding initiation, stable growth, and instability. Fully plastic behavior has not posed special problems, but more work is needed to establish the utility of this approach for three-dimensional crack fronts.

Acknowledgment

This work was supported by the Electric Power Research Institute (EPRI), Palo Alto, California. The authors would like to express their appreciation to T. U. Marston and R. E. Smith of EPRI for their help and encouragement of the work. They are also indebted to John Fox of Battelle's Columbus Laboratories for his nondestructive evaluation work in this program and to F. Shih and W. Andrews of General Electric for making some of their results available. Many useful and stimulating discussions with the EPRI plastic fracture analysis group should also be acknowledged.

References

[1] Rice, J. R. in *The Mechanics of Fracture,* F. Erdogan, Ed., American Society of Mechanical Engineers Publication AMD, Vol. 19, 1976, pp. 23-53.
[2] Shih, C. F. and Hutchinson, J. W., *Journal of Engineering Materials and Technology,* Vol. 98, 1976, pp. 289-295.
[3] Amazigo, J. C. and Hutchinson, J. W., *Journal of the Mechanics and Physics of Solids,* Vol. 25, 1977, pp. 81-97.
[4] Wilkinson, J. P. D., Hahn, G. T., and Smith, R. E., "Methodology for Plastic Frac-

ture—A Progress Report" in *Proceedings*, 4th International Conference on Structural Mechanics and Reactor Technology, Aug. 15-19, 1977.
[5] Shih, C. F., de Lorenzi, H. G., Yukawa, S., Andrews, W. R., Van Stone, R. H., and Wilkinson, J. P. D., "Methodology for Plastic Fracture," Third Quarterly Report to the Electric Power Research Institute, Palo Alto, Calif., 1 Nov. 1976 to 31 Jan. 1977.
[6] Shih, C. F., de Lorenzi, H. G., Andrews, W. R., Van Stone, R. H., and Wilkinson, J. P. D., "Methodology for Plastic Fracture," Fourth Quarterly Report to the Electric Power Research Institute, Palo Alto, Calif., 1 Feb. 1977 to 30 April 1977.
[7] Paris, P. C. et al, "A Treatment of the Subject of Tearing Instability," Washington University Report to the Nuclear Regulatory Commission, NUREG-0311, 1977.
[8] de Koning, A. U., "A Contribution to the Analysis of Slow Stable Crack Growth," The Netherlands National Aerospace Laboratory Report NLR MP 75035U, 1975.
[9] Green, G. and Knott, J. F., *Journal of the Mechanics and Physics of Solids*, Vol. 23, 1975, pp. 167-183.
[10] Garwood, S. and Turner, C. E., unpublished work, Imperial College, Department of Mechanical Engineering, London, U. K., 1977.
[11] Broberg, K. B., *Journal of the Mechanics and Physics of Solids*, Vol. 23, 1975, pp. 215-237.
[12] de Koning, A. U. in *Proceedings*, 4th International Conference on Fracture, University of Waterloo Press, 1977, Vol. 3, pp. 25-31.
[13] Kfouri, A. P. and Miller, K. J. in *Proceedings*, Institution of Mechanical Engineers, Vol. 190, No. 48/87, 1976, pp. 571-584.
[14] Hellan, K., *Engineering Fracture Mechanics*, Vol. 8, 1976, pp. 501-506.
[15] Carlsson, A. J., "Progress in Nonlinear Fracture Mechanics," Presentation at the 14th International Congress of Theoretical and Applied Mechanics, Delft, The Netherlands, 30 Aug. to 4 Sept. 1976.
[16] Cotterell, B. and Reddel, J. K., *International Journal of Fracture*, Vol. 13, 1977, pp. 267-278.
[17] Andrews, E. H. and Billington, E. W., *Journal of Materials Science*, Vol. 11, 1976, pp. 1354-1361.
[18] Rice, J. R. in *Proceedings*, 1st International Conference on Fracture, Japanese Society for Strength and Fracture, Tokyo, Vol. 1, 1966, pp. 309-340.
[19] Kfouri, A. P. and Rice, J. R. in *Proceedings*, 4th International Conference on Fracture, University of Waterloo Press, Vol. 1, 1977, pp. 43-60.
[20] Kobayashi, A. S., Chiu, S. T., and Beeuwkes, R., *Engineering Fracture Mechanics*, Vol. 5, 1973, pp. 293-305.
[21] Light, M. F., Luxmoore, A. R., and Evans, E. T., *International Journal of Fracture*, Vol. 11, 1975, pp. 1045-1046.
[22] Andersson, H., *Journal of the Mechanics and Physics of Solids*, Vol. 2, 1974, pp. 285-308.
[23] Andersson, H. in *Computational Fracture Mechanics*, E. F. Rybicki and S. E. Benzley, Eds., American Society of Mechanical Engineers Special Publication, 1975, pp. 185-198.
[24] Newman, T. C. and Armen, H., Jr., "Elastic-Plastic Analysis of a Propagating Crack Under Cyclic Loading," American Institute of Aeronautics and Astronautics Paper No. 74-366, AIAA/ASME/SAE Conference, Las Vegas, Nev., 1974.
[25] Hutchinson, J. W. and Paris, P. C., this publication, pp. 37-64.
[26] Private communication, Plastic Fracture Analysis Group, T. U. Marston and R. E. Smith, chairmen, Electric Power Research Institute, Palo Alto, Calif., 12 Aug. 1977.
[27] Shih, C. F., de Lorenzi, H. G., and Andrews, W. R., this publication, pp. 65-120.

E. P. Sorensen[1]

A Numerical Investigation of Plane Strain Stable Crack Growth Under Small-Scale Yielding Conditions

REFERENCE: Sorensen, E. P., "**A Numerical Investigation of Plane Strain Stable Crack Growth Under Small-Scale Yielding Conditions,**" *Elastic-Plastic Fracture, ASTM STP 668*, J. D. Landes, J. A. Begley, and G. A. Clarke, Eds., American Society for Testing and Materials, 1979, pp. 151-174.

ABSTRACT: Plane strain crack advance under small-scale yielding conditions in elastic-perfectly plastic and power-law hardening materials is investigated numerically via the finite element method. Results indicate that the stress distribution ahead of a growing crack is essentially the same as that ahead of a stationary crack, and that the numerically evaluated steady-state crack tip profiles reflect a vertical tangent at the extending crack tip which corresponds to the theoretically predicted outline. It is found that the increment $d\delta_t$ in crack tip opening, when loads are increased at fixed crack length, seems to be uniquely related to dJ/σ_0 irrespective of the amount of previous crack growth, and for increments dl of crack advance at constant external load, the incremental crack tip opening appears related to $\ln(J/\sigma_0 r)\, dl$ when evaluated at distance r from the tip. A discussion of proposed fracture parameters for continued crack growth (as opposed to growth initiation) is included.

KEY WORDS: stable crack growth, small-scale yielding, nonhardening and power-law hardening materials, fracture criteria for continued crack advance, crack propagation

In contrast to the extensive literature pertinent to stationary cracks, for example, [1-4],[2] a paucity of solutions for the growing crack case is evident. One obvious reason for this is the added mathematical complexity inherent in a continuum formulation of the growing crack.

Analytic investigations of quasi-statically extending cracks under Mode III conditions in elastic-perfectly plastic materials are presented in [5-7], and Rice [7] extends the discussion to the form of the solution for a growing crack subject to plane strain, Mode I conditions. One conclusion of these studies is that the strain field ahead of an extending crack is dominated by

---

[1] Research assistant, Division of Engineering, Brown University, Providence, R.I. 02912. Current affiliation: Mathematics Department, General Motors Research Laboratories, Warren, Mich. 48090.

[2] The italic numbers in brackets refer to the list of references appended to this paper.

a logarithmic singularity which is weaker than the $1/r$ singularity experienced at the tip of a stationary crack (where r is the distance measured from the crack tip). The weaker strain singularity is due to the crack extending into material that has deformed plastically so that complete refocusing of the strain field at the tip of the extended crack is prevented. This reduced crack-tip strain concentration is a primary reason for stable crack growth in elastic-plastic materials.

Various aspects of the incremental solution to crack growth problems are considered by Rice [8]. His work reveals an incompatible elastic strain increment (one not derivable from a displacement field) caused by a Prandtl stress distribution traveling with the crack tip. The incompatible elastic strain increment induces plastic strain during an increment of crack advance and the additional straining promotes crack growth. Another consequence of the elastic incompatibility is that, contrary to the rigid-plastic case in which the crack advances with a finite crack tip opening angle (COA), the elastic-plastic incremental formulation results in a crack face profile exhibiting a vertical tangent at the crack tip and a corresponding ill-defined crack tip opening angle. Both the rigid-plastic case and the elastic-perfectly plastic case predict a zero crack opening displacement (COD) at the tip of an extending crack. In this respect, the Mode III asymptotic analysis presented by Chitaley and McClintock [9] is incorrect in its prediction of a nonzero COD at the tip of a steadily extending crack, and the difficulty seems to arise from their approximate numerical evaluation of an integral which should have given zero for a result [8].

Experimental observations of stable crack growth have been reported by several investigators, for example, Refs 10-12. Green and Knott [12] observe a constant increase in the nominal COD per increment of crack growth and conclude that the crack face profile associated with an extending crack tip in a ductile metal is constant. J-integral experimentation due to Clarke et al [13] and Griffis and Yoder [14] indicates a constant change of J with change of crack length following the blunting of an initially sharp crack. The implication is that the advancing crack tip experiences constant surrounding fields. These results are analogous should J and δ, the crack-tip opening displacement, remain linearly related as they are in the stationary crack case.

Finite element solutions to extending crack problems include the work of deKoning [15], Andersson [16], and Sorensen [17]. A key objective of the numerical solutions, aside from the illumination of the stress and strain distributions accompanying growing cracks, is the investigation of possible macroscopic parameters which may be correlated with the "state" at the growing crack tip. Although the meaning of the J-integral is unclear in the growing-crack case, J is known to rise monotonically, for example, Ref 13, and its use as a fracture predictor in the extending crack case is a possibility. Other proposed parameters are the crack-tip opening angle [18] and the Griffith-like separation energy rate associated with a finite crack advance step [19]. Recent work by Shih et al [20] examines and evaluates various proposed criteria for continuing fracture.

The present paper investigates the distributions of stress and deformation associated with an extending crack tip in hardening and nonhardening materials under plane strain, small-scale yielding conditions. This is accomplished via large-scale finite element calculations. The results indicate the existence of a Prandtl stress distribution traveling with the crack tip, and the crack face profiles are consistent with the logarithmic dependence noted by Rice [8]. Discussion of various proposed fracture criteria is provided and a relation characterizing continuing fracture is sketched from considerations presented in Ref 8 and the numerical results.

Numerical Considerations

Finite Element Equations

A form of the virtual work equation, valid for incremental small-strain theory in which all integrals are carried out over the reference volume and surface, is used in the derivation of the governing finite-element equations

$$\int_V \dot{\sigma}_{ij} \delta \dot{u}_{j,i} \, dV = \int_S \dot{T}_i \, \delta \dot{u}_i \, dS$$

where

u_i = displacement vector,
T_i = traction vector, and
σ_{ij} = Cauchy stress tensor.

Superimposed dots denote rates.

Let $[N]$ denote the shape functions used to represent variations of displacement within an element as interpolated from nodal displacement values, $\{u\}$, so that $[N]\{u\}$ represents the displacement field. The incremental strain-displacement relation is $\{\dot{\epsilon}\} = [B]\{\dot{u}\}$, where $[B]$ is composed of the appropriate derivatives of $[N]$. The constitutive matrix is denoted by $[C]$ such that the incremental stress is related to the incremental strain by $\{\dot{\sigma}\} = [C]\{\dot{\epsilon}\}$. Substituting the foregoing matrix relations into the governing variational equation and recognizing that arbitrary variations may not influence the resulting equilibrium equations, one obtains the well-known tangent stiffness equations

$$\int_V [B]^T [C] [B] \, dV \, \{\dot{u}\} = \int_S [N]^T \{\dot{T}\} \, dS$$

where integrals are carried out over all elements and over all externally loaded surfaces. In conventional finite-element notation this equation is written $[K]\{\dot{u}\} = \{\dot{P}\}$ with $[K]$ termed the master stiffness matrix and $\{P\}$ the forcing function or right-hand side.

Constitutive Relations

The material constitutive behavior is modeled as isotropic, elastic-perfectly plastic, and elastic power-law hardening together with the Mises yield condition and the associated Prandtl-Reuss flow law [21]. The power-law hardening relation is that used by Tracey [22], namely

$$\bar{\sigma}/\sigma_o = (\bar{\sigma}/\sigma_o + 3G\bar{e}^P/\sigma_o)^N$$

where

N = hardening exponent,
G = shear modulus,
σ_o = yield stress in tension
$\bar{\sigma} = \sqrt{\frac{3}{2} s_{ij} s_{ij}}$,
s_{ij} = deviatoric Cauchy stress tensor,
$\bar{e}^P = \sqrt{\frac{2}{3} e_{ij}^P e_{ij}^P}$, and
e_{ij}^P = plastic portion of the deviatoric strain tensor.

This power-law hardening expression is obtained from the relation $\bar{\tau}/\tau_o = (\bar{\gamma}/\gamma_o)^N$ for pure shear used by Rice and Rosengren [23] through the conversions $\bar{\tau} = \bar{\sigma}/\sqrt{3}$, $\bar{\gamma}^P = \sqrt{3}\bar{e}^P$, and $\bar{\gamma} = \bar{\gamma}^e + \bar{\gamma}^P$. No account is taken of the Bauschinger effect or possible vertex development of the yield surface during the nonradial loading experienced by material points during crack growth. These omissions must be kept in mind during the interpretation of the present results. The nonlinear problem is linearized by specifying small load increments and iterating within each increment for convergence to the best representative plastic constitutive matrices. The constitutive matrix $[C]$ at any point in the loading history may be written as

$$[C] = m[C^{el}] + (1-m)[C^{el-pl}]$$

with $0 \leq m \leq 1$. This partial-stiffness approach is due to Marcal and King [24]. $[C^{el-pl}]$ is determined from the normal to the yield surface and m depends on the amount of elastic response an element undergoes during a load increment (that is, $m = 1$ for totally elastic response and $m = 0$ for totally plastic response). The normal used in these expressions is chosen in the manner of Rice and Tracey [25] such that resulting stress states precisely satisfy the yield criterion in the elastic-perfectly plastic case and approximately satisfy the yield criterion in the power-law hardening case [22]. Typically two to three iterations are required per loading increment for convergence to an appropriate constitutive representation. Reassembly and redecomposition of the master stiffness matrix are accomplished in a cost-minimizing manner, using the efficient procedure discussed by Yang [26] and Sorensen [27] and various in- and out-of-core procedures as required [28].

Crack Growth Simulation

A nodal release technique is implemented to simulate crack advancement through the finite element mesh. This technique is used by Andersson [18] and by Kfouri and Miller [19]. As applied here, the technique proceeds as follows. Upon satisfaction of a chosen fracture criterion or specified load level, the crack is deemed ready to propagate and the boundary condition at the crack tip passes from displacement controlled to traction controlled. The reaction force corresponding to the zero displacement condition at the crack tip node is calculated and relaxed to zero in five equal increments as more steps provide minimal differences in results at significant computational expense. Following this procedure, the crack tip has advanced by one element length. The present analyses hold external loads constant during the nodal relaxation procedure. It is anticipated that due to the history dependence of the strain distribution the process of nodal force relaxation under increasing external load might result in a somewhat different strain rate ahead of the crack tip from that obtained under constant external load. The present results are interpreted as representative of crack advance under constant load or perhaps under slight increases of external load.

Element Modeling

The element used in the present analyses is the constant-strain triangle. Quadrilaterals are formed from four of these elements in the manner of Nagtegaal et al [29] to accommodate the possibility of nearly incompressible straining, and the degrees of freedom associated with the internal node are eliminated from the stiffness equations [30]. This configuration in no way accounts for the mathematical singularities encountered at the crack tip, but useful results are obtained by sufficient mesh refinement (see discussion of results). Due to the nodal release technique employed in the analyses, no use of special crack-tip singular elements, for example, Tracey [31] and Barsoum [32], is made since the crack advance would require a procedure for refocusing the mesh at the tip of the extended crack. In this context a Eulerian finite element formulation holds much promise, for then a mesh remains focused at the crack tip, and singularity elements may be employed; however, such a formulation presents other difficulties such as the convective terms which require spatial derivatives of field quantities, and an economically feasible formulation has not yet been found.

Numerical Procedure

Following an elastic increment in which the highest stressed element is scaled to cause incipient yielding, various increments of load equal to 10 or 20 percent of that in the initial solution are carried out. The nodal release procedure is implemented upon achievement of the static similarity solution of Tracey [22] and further loading is applied at the new crack length. Various steps of crack advance and external loading at constant crack length are performed. Displacement boundary conditions corresponding to the elastic singular strain dominant at the crack tip are specified on a radius which is 224 times the smallest element size and 20 times the maximum extent which the plastic zone acquires in the course of the computation. These ratios insure an appropriate boundary-layer formulation of the small-scale yielding situation [7]. The next term beyond the inverse square-root singularity in the surrounding elastic field, namely, a tension T parallel to the crack, is taken as zero. Figure 1 presents the load histories relevant to the analyses presented here; in this figure, K_o is the stress intensity factor at the first load increment, σ_o is the yield stress in tension, and $l - l_o$ is the difference between the current and initial crack lengths. These "staircase" load histories represent hypothetical cases which might

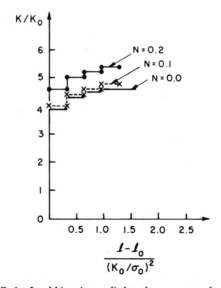

FIG. 1—*Load histories applied to the present analyses.*

be found in service and reflect the different K levels required to obtain appropriate similarity solutions prior to crack advance.

Mesh Configuration and Material Properties

The finite-element grid used in these analyses is indicated in Fig. 2 with details of the refined mesh surrounding the crack tip presented in Fig. 3. A total of 1660 elements is used together with 865 nodes and 1730 degrees of freedom. Use of static condensation reduces the number of active degrees of freedom to 946. The radius of the outer ring in Fig. 3 divided by the smallest element length is 28. The radii of the rings in Fig. 2 divided by the inner element length are 28, 34, 42, 52, 64, 80, 104, 136, 176, and 224. The radial lines are spaced at 10-deg intervals. Material properties are $\nu = 0.3$ and $E/\sigma_o = 1000$ where E is Young's modulus and σ_o the tensile yield strength. The analysis corresponding to the ideal plasticity case was carried out on the IBM 360/67 available at Brown University Computing Laboratory. The two hardening analyses were performed using the IBM 370/168 available at the Massachusetts Institute of Technology.

Results and Discussion

Crack Face Profiles

Figure 4 presents crack face profiles following the load incrementation procedure for the stationary crack and following the final crack advance

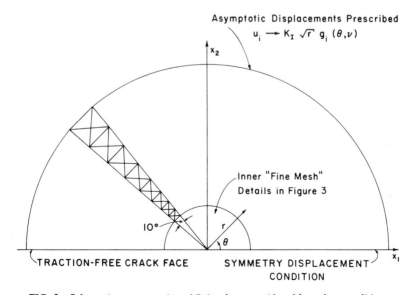

FIG. 2—*Schematic representation of finite element grid and boundary conditions.*

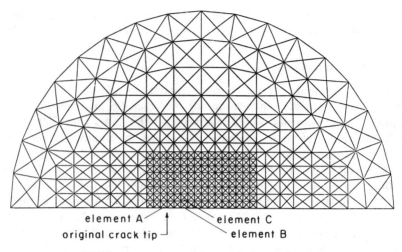

FIG. 3—*Arrangement of elements in fine-mesh region.*

step for the cases $N = 0.0$, 0.1, and 0.2. The profiles following the final crack advance step are considered representative of steady-state conditions in the vicinity of the crack tip, but not overall, as away from the crack tip the crack faces experience continuing deformation. Direct comparisons of the stationary crack profiles with those of Tracey [22] indicate maximum deviations of 6, 5, and 3 percent for $N = 0.0$, 0.1, and 0.2, respectively. Since the present analyses do not employ special singularity elements like those of Tracey [31] and do not include finite geometry changes, the crack tip opening displacements, δ, are estimated by extrapolation. For the nonhardening case, a value of 0.66 is obtained for δ nondimensionalized by J/σ_o, where J is taken equal to $(1 - \nu^2)K_I^2/E$, corresponding to the small-scale yielding situation. Tracey predicts a value of 0.54 for this ratio, but Parks [33] suggests that this number should be 0.65 due to the artificial path dependence of J that seems (through comparison with a corresponding "deformation theory" solution based on Tracey's mesh, leading to similar path-dependence) to be directly traceable to Tracey's nonhardening singularity element. For nonhardening blunting solutions, McMeeking [3] reports values for the nondimensionalized COD between 0.55 and 0.67, depending on the point of measurement. The larger of these two numbers is representative of larger values of σ_o/E, on the order of 1/100. For σ_o/E equal to 1/300 and $N = 0.1$, McMeeking reports values of the nondimensionalized COD between 0.41 and 0.44, and for $N = 0.2$ values between 0.27 and 0.30 are reported, although higher values result when measured at the elastic-plastic boundary. The present hardening analyses predict a value of 0.54 for the nondimensionalized COD when $N = 0.1$ and 0.44 when the hardening exponent equals 0.2. The good agreement of the extrap-

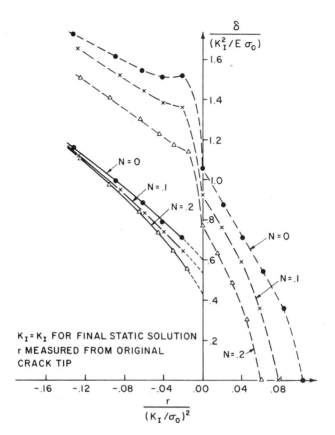

FIG. 4—*Crack surface displacements, stationary and steady-state advancing crack solutions for N = 0.0, 0.1, and 0.2.*

olated values of δ with the work of McMeeking and others lends confidence to the results of the present analyses as regards the prediction of the crack face profiles.

Figure 5 presents crack face profiles at key points in the evolution of the prescribed load history for the nonhardening case. This analysis, which models crack growth at constant load between equidistant nodal points in rate-independent materials, results in a crack face profile which as the crack advances becomes less angular with distance from the crack tip (where the "angle" is measured clockwise from a horizontal line behind the crack tip). This is also true for the hardening cases as indicated by the final crack profiles shown in Fig. 4. Rice [7,8], for quasi-static crack advance in a nonhardening material, derives a displacement distribution proportional to $r \ln r$ (where r is the radial distance measured from the crack tip) which implies a vertical tangent at $r = 0$. Nodal displacements from

160 ELASTIC-PLASTIC FRACTURE

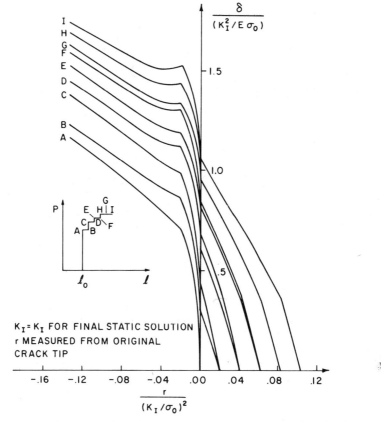

FIG. 5—*Evolution of crack surface displacements through the loading history of the non-hardening case.*

the present analyses permit curve fairing, which exhibit the vertical tangent required by the analytic solution. This infinite slope is a local phenomenon and may not overly influence the effective definition of a crack tip opening angle, defined here as the total angle between the separating crack faces behind the extending crack tip. However, this remains an open issue in need of further study. Due to the linear interpolation functions used in the present analysis, this angle is evaluated at the node immediately behind the crack tip, and resulting values are presented in Table 1. The trend toward a steady-state value is anticipated from the prescribed loading, and the numbers in Table 1 indicate the material dependent nature of the COA. As the final displacement distributions in Fig. 4 suggest, the angles would be less if based on elements farther back from the crack tip, and the parameter seems to be meaningless according to theoretical considerations in the limit of r approaching zero. To further clarify the role of the COA and its

TABLE 1—*Crack-tip opening angles, rad.*

	Release 1	Release 2	Release 3	Release 4	Release 5
$N = 0.0$	0.015	0.016	0.017	0.017	0.017
$N = 0.1$	0.016	0.017	0.018	0.018	...
$N = 0.2$	0.018	0.019	0.020	0.020	...

relation to continuing fracture, a finite strain analysis in the spirit of McMeeking [3] is desirable for the region in the vicinity of the crack tip.

Prior to any crack advance, the COD is related to the external loading according to $\delta = \alpha J/\sigma_o$, where α is dependent on material properties and hardening exponent. As previously remarked, values of α for the stationary portion of the present analyses are 0.66, 0.54, and 0.44, which agree well with published values. Following crack advance, the load histories prescribe increments of external load at constant crack length, and a relation between incremental COD and increments in external loads is

$$d\delta = \alpha \frac{dJ}{\sigma_0} \qquad (1)$$

Values of α in this relation are presented in Table 2; increments in COD are measured at the node immediately behind the current crack tip and the original crack tip position. The numbers in Table 2 indicate that the nominal crack tip opening displacement continues to be effectively characterized by increments in J for external loading at fixed crack length following crack advance steps. For each of these analyses, α seems to be constant, namely, 0.66, 0.54, and 0.44 for $N = 0.0, 0.1$, and 0.2, respectively.

The present analyses prescribe crack advance at constant external load, and here the resulting incremental crack surface displacements are related to the crack advance step. Rice [8] obtains a relation between incremental displacements resulting from an increment of crack advance, dl, and the increment of crack advance. This relation may be written as

$$du_i = \frac{\sigma_o}{E}\left[A_i(\theta) + B_i(\theta) \ln \frac{R}{r}\right] dl$$

where $A_i(\theta)$ and $B_i(\theta)$ are dimensionless functions and R is a characteristic length of the plastic zone. Noting that for small-scale yielding, R scales with EJ/σ_o^2, and specializing the above expression to the crack surface ($\theta = \pi$), the following is obtained

$$d\delta = \frac{\sigma_o}{E} \beta \ln \left(\lambda \frac{EJ}{\sigma_o^2 r}\right) dl \qquad (2)$$

TABLE 2—Values of α in the correlation of dδ and dJ/σ₀.

	Initial Value	Measured Immediately Behind Current Tip			Measured at Original Tip Position		
		After Release 1	After Release 2	After Release 3	After Release 1	After Release 2	After Release 3
N = 0.0	0.66	0.65	0.66	0.66	0.65	0.60	0.60
N = 0.1	0.54	0.54	0.53	0.52	0.54	0.51	0.54
N = 0.2	0.44	0.43	0.43	0.41	0.43	0.41	0.46

Here, the proportionality constants have been lumped into β and λ, and $d\delta$ is the COD increment. Equation 2 may be integrated to obtain an expression for the additional COD resulting at a fixed point X as the crack advances under constant J, from l_1 to l_2, two crack lengths such that $l_2 > l_1 \geq X$. The integration results in

$$\delta(l_2, X) - \delta(l_1, X) = \beta \frac{\sigma_o}{E}\left[(l_2 - X) \ln \frac{\lambda e EJ}{\sigma_o^2(l_2 - X)} \right.$$
$$\left. - (l_1 - X) \ln \frac{\lambda e EJ}{\sigma_o^2(l_1 - X)} \right] \quad (3)$$

For the node immediately behind the advancing crack tip, $l_1 = X$ and the second term on both the left side and right side of Eq 3 is zero. Figure 6 presents a plot of crack surface displacement values for the node immediately behind the crack tip taken from the present hardening and nonhardening analyses. The plot indicates a unique pair (β, λ) which satisfies Eq 3, namely, $\beta = 9.5$ and $\lambda = 0.04$. These values of β and λ do not correctly predict the incremental COD due to crack advance at the other nodes behind the advancing tip, indicating that Eq 2 may apply only within a certain distance from the crack tip. This latter point is currently under further investigation. However, the good correlation provided by Eq 2 for incremental crack opening near the advancing crack tip due to an increment in crack advance at constant external load indicates the possibility of using Eq 2 in conjunction with Eq 1 to provide a relationship for the characterization of incremental COD's with increments of external loads as well. That is

$$d\delta = \alpha \frac{dJ}{\sigma_o} + \beta \frac{\sigma_o}{E} \ln \frac{\lambda EJ}{\sigma_o^2 r} dl \quad (4)$$

This equation appears useful in the study of continuing fracture, and its use in developing a fracture criterion based on near-tip crack opening is more fully explored in Ref *40*.

Stress Distributions Ahead of the Crack Tip

The stress fields associated with hardening and nonhardening plane strain stationary cracks under small-scale yielding conditions are provided by Tracey [22]. Figure 7 presents curves for the opening stress ahead of a stationary crack together with corresponding centroidal stress values of elements directly ahead of the crack tip taken from the present analyses. The distance from the crack tip to the centroid is used for the position on the abscissa of the plot. Maximum deviations of the present results from

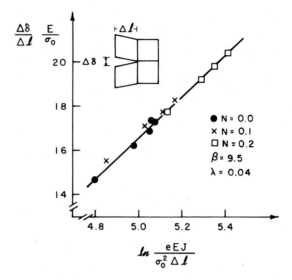

FIG. 6—*Correlation between increments of crack opening displacement and J.*

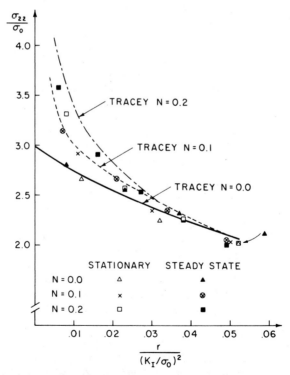

FIG. 7—*σ_{22} stress distributions ahead of the crack tip.*

Tracey's curves are 5, 5, and 7 percent for $N = 0.0$, 0.1, and 0.2, respectively. The good agreement of these values is a consequence of the fine mesh employed in the solution.

As the stress gradients near the crack tip become steeper with increasing hardening exponent, the deviations of the present results from those of Tracey are expected to increase since these analyses make no use of singular elements but rely on fine-mesh gradation to capture the appropriate stress distributions. This formulation provides little information on the angular stress distribution as r approaches zero and does not precisely obtain the factor of 2.97 in σ_{22} stress elevation over σ_o as the Prandtl solution demands in the nonhardening case.

Figure 7 also indicates points corresponding to the apparently steady-state stress distribution predicted in the present analyses. Following the final crack advance step, there are minor elevations of σ_{22} ahead of the tip with maximum deviations from Tracey's results for a stationary crack of 4, 3, and 7 percent for $N = 0.0$, 0.1, and 0.2, respectively. Similar plots of the σ_{22} stress distribution ahead of the crack tip, following intermediate crack advance steps, reveal points between the static and steady-state points presented in Fig. 7. The conclusion is that under small-scale yielding conditions for both hardening and nonhardening materials, the σ_{22} stress distribution ahead of a growing crack is effectively the same as the corresponding stress distribution for a stationary crack.

Material Stress Histories

Three material point stress histories are now described. The points under scrutiny are the element immediately behind the initial crack tip, an element ahead of the initial crack tip and on the prospective fracture plane, and an element ahead of the initial crack tip but removed from the fracture plane. These material points are designated Elements A, B, and C, respectively, and are indicated in Fig. 3. The stresses described are centroidal stresses associated with the constant-strain triangles used in the analyses.

Element A—Following the final static load incrementation step, values of σ_{11}/σ_o are 1.14, 1.29, and 1.51; values of σ_{22}/σ_o are 1.27, 1.44, and 1.69; and values of σ_{12}/σ_o are -0.57, -0.69, and -0.86 for $N = 0.0$, 0.1, and 0.2, respectively. Upon subsequent crack advance and load incrementation steps, both σ_{12} and σ_{22} become small in contrast with σ_{11}, which dominates the later stress history. After the last crack advance step, nondimensionalized values of σ_{11} are 1.07, 1.13, and 1.06, respectively, for the three analyses. The final value of σ_{11} seems to represent a residual tensile stress in a plastic wake region. For the nonhardening case there is some continuing plastic flow in this wake, but not in the hardening cases. This is due to the isotropic model of strain hardening used in these analyses and underlines the sensitivity of the results to the constitutive model used. The Prandtl stress distribution suggests that $\sigma_{11} = 1.15\,\sigma_o$ on the crack surfaces immediately behind the crack tip, and the final value of σ_{11} in the nonhardening case is 7 percent below this number.

Element B—This material point lies ahead of the initial crack tip but behind the final crack tip position in these analyses. This material experiences continued increases in σ_{11} and σ_{22} during the first two nodal release steps. Values of σ_{22}/σ_o following the second nodal release step are 2.74, 3.03, and 3.45 for $N = 0.0, 0.1,$ and 0.2, respectively. Then, during the remaining nodal release steps, σ_{22} drops to a small value and σ_{11} becomes the dominant stress at the point. It is not surprising that σ_{22} should drop drastically, because the traction-free boundary condition imposed on the open crack face requires that σ_{22} be zero there (assuming negligible deviation of the surface normal from the X_2 direction). Also, following the third nodal release step during which this element becomes part of the material behind the crack tip, values of σ_{12}/σ_o are $-0.56, -0.67,$ and -0.80, respectively, for the three hardening exponents. For the nonhardening case, σ_{11} and σ_{22} are nearly equal and this is consistent with the notion that at this stage of its history the element is part of a centered fan above the crack tip, requiring a hydrostatic stress state coupled with plastic shearing. Then, as the crack advances farther, this element passes from the centered fan region to a residually stressed wake region as discussed above.

Element C—This element is farther removed from the plane of fracture than Element B. Figure 8a presents a plot of its stress history versus crack advance step for the nonhardening case. This plot is also representative of the hardening results but with an appropriate shift of the vertical axis. Figure 8a also presents stresses as predicted from a Prandtl stress distribution traveling with the crack; the Prandtl slipline construction is indicated in Fig. 8b. The material point under discussion is imagined enveloped by Region 1 in Fig. 8b following Release 1, in the centered fan of Region 2 following Releases 2 and 3, and in constant-state Region 3 following the final two steps of crack advance. For the stress values plotted in Fig. 8a, θ is taken as the angle between a horizontal line and a line joining the centroid of Element C to the actual crack tip. The numerically calculated stress distribution reflects a pattern similar to that predicted by the Prandtl field. However, the fuzzy crack-tip phenomenon is also evident. This terminology describes the consequences of a finite element discretization which cannot exactly reproduce appropriate strain singularities at the crack tip so that elements surrounding the crack tip respond to an ill-defined crack tip with a corresponding vagueness in the definition of θ and r. This effect is minimized as the mesh is refined.

These brief descriptions of material stress-point histories reinforce the previously made point that the steady-state σ_{22} stress distribution is effectively the same as that prior to any crack extension. Additional conclusions are as follows. (1) Following crack advance steps, points positionally similar

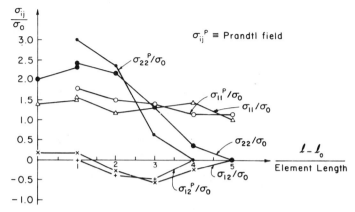

FIG. 8a—*Stress history of Element C plotted versus crack advance for the nonhardening case.*

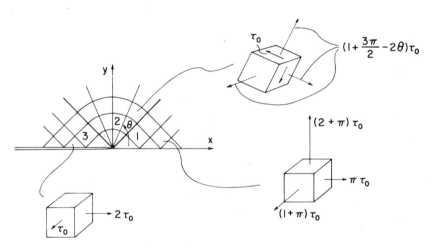

FIG. 8b—*Prandtl slipline field stress distribution.*

with respect to the crack tip experience similar stress histories of loading and unloading, which suggests that a steady state prevails near the advancing crack tip. Positionally similar stress histories are a tacit assumption of the ideally plastic theoretical analysis of plastic strain singularities for growing cracks [7,8] discussed earlier. (2) Prior to any given nodal release step, the elements surrounding the crack tip experience essentially the same stress field as those surrounding a stationary crack. This stress field, within the limitations of the present analyses, resembles the Prandtl slipline solution for the nonhardening situation. (3) The wake material is dominated by a residual, tensile σ_{11} stress which results in continued yielding in the nonhardening case, but not in the hardening cases due to the isotropic hardening model used.

Plastic Zone Shapes

The shapes of the active plastic zones, corresponding to the final static load increment and the final step of crack advance, nondimensionalized by the similarity quantity $(K_I/\sigma_o)^2$, are presented in Fig. 9a-c for the hardening exponents 0.0, 0.1, and 0.2, respectively. The plastic zone shapes corresponding to the static crack solutions are in good agreement with appropriate cases documented in Refs 22, 25, and 34. The effect of the crack advance on the shape of the plastic zone is to constrict its width and to tilt the inclination of the zone toward the symmetry axis ($X_2 = 0$). Between the nodal release steps and depending on the amount of load incrementation, the plastic zone attempts to restore the butterfly shape familiar from static crack analysis. The achievement of a steady-state solution is particularly evident from results of the nonhardening case, which includes two consecutive nodal release steps; there is negligible difference between the plastic zone shapes following these crack advance steps. The small, actively plastic wake region in the nonhardening case contrasts sharply with the lack of this region in the hardening cases. As has been remarked, this is attributed to the isotropic constitutive theory used in these analyses. Also, the angular tilt and plastic zone constriction are reminiscent of similar growth effects encountered in Mode III analyses, for example, Refs 9 and 17.

Material Strain Histories

The equivalent plastic strain history of a material point in the slipline fan above the crack tip before crack advance indicates a high amount of straining prior to crack advance, further straining during the first nodal release step, and negligible further straining. The strain incurred during the first crack advance step is due to the fan region sweeping by the material point. A material point positionally similar with respect to the crack tip before the third crack advance step is not plastically strained until the crack has advanced sufficiently to engulf this point with its accompanying plastic zone. This element is then strained irreversibly, but less than the corresponding element in the stationary crack case. Ratios of the plastic strain at the second material point prior to Nodal Release 3 divided by the plastic strain at the positionally similar material point prior to the first step of crack advance are 0.95, 0.92, and 0.87 for $N = 0.0, 0.1$, and 0.2, respectively. These values corroborate the observation of Rice [8] that the strain field associated with an extending crack sustains a weaker singularity than that associated with a stationary crack, r^{-1} versus ln r for

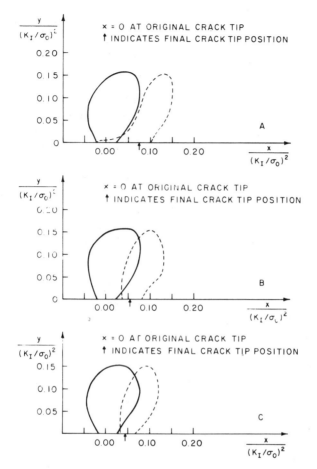

FIG. 9—*Stationary and steady-state plastic zone shapes for* (a) $N = 0.0$, (b) $N = 0.1$, *and* (c) $N = 0.2$.

the nonhardening case. The foregoing ratios also imply that, at least for strain-controlled ductile rupture mechanisms, the extent of stable crack growth is greater for a hardening material than for a nonhardening material.

Separation Energy Rates

Each of the nodal release steps provides a record of vertical displacement versus reaction force. Calculating the area under this curve and dividing by the finite crack advance step, one obtains a separation energy rate, G^Δ in the notation of Kfouri and Rice [*35*]. It is the finite value of the crack advance step that renders this calculation nontrivial, for in the limit of growth step approaching zero and for materials which exhibit a finite stress level at the crack tip, such a calculation yields zero for G^Δ [36]. G^Δ values taken from the work of Kfouri and Miller [19] and McMeeking [3] together with values from the present analyses are plotted in Fig. 10. The points of Kfouri and Miller result from the plane strain analysis of the tensile and equibiaxial loading of a finite plate containing a crack. The ratio of crack length to plate width is 0.125 and the ratio of Young's modulus to initial yield stress is 667.7. Their analyses model the material as linear hardening with a tangent modulus equal to 0.023 times the elastic modulus and Poisson's ratio equal to 0.3. The points of McMeeking are taken from separation energy rate calculations for the small-scale yielding analysis of a blunted notch; material properties are $E/\sigma_o = 300$, Poisson's ratio $= 0.3$, and a power-law hardening exponent N equal to 0.1. McMeeking's work includes finite strain effects at the blunted notch and employs crack-growth steps on the order of the crack opening displacement, whereas the growth steps employed by Kfouri and Miller and the author are much larger. The "steady state" point of McMeeking corresponds to the final growth step calculated and it is presented for completeness although his analysis does not indicate that steady-state conditions are achieved.

The explanation for the separate pattern of points due to Kfouri and Miller is thought to be the "T effect", which is explored by Larsson and Carlsson [34] and Rice [37]. The origin of the effect is the presence of nonvanishing, nonsingular terms in the eigenvalue expansion of the elastic

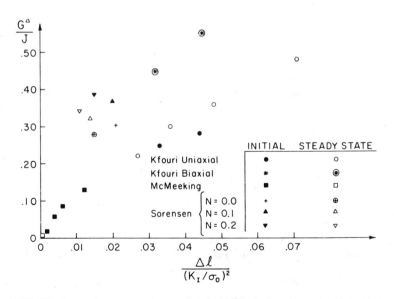

FIG. 10—*Separation energy rates correlated with J and plotted versus growth step.*

stress tensor at the crack tip in plane strain. The points corresponding to the equibiaxial tensile loading case of Kfouri and Miller provide the best fit with points from the present analyses. This is because the present formulation has $T = 0$ and, for an infinite plate under equibiaxial loading, $T = 0$ (for a finite plate $T \approx 0$). Kfouri and Rice [35] present a relation between J and G^Δ for the tensile load case in an effort to correlate the two quantities, but a subsequent communication with Kfouri has indicated a different relation for the equibiaxial loading case. The conclusion is that the use of J as a correlator of the separation energy during a finite growth step of the size they explored is sensitive to the nonzero, nonsingular stress terms present at the crack tip in plane strain. Figure 10 also indicates that the value of G^Δ is sensitive to the degree of strain hardening. But, these observations are made for points corresponding to growth step sizes far in excess of values comparable to the COD. Since it is in a region of linear extent on the order of COD that ductile fracture mechanisms such as void coalescence and localization of shear dominate, it would seem to be step sizes of this order that are of greatest interest. By using such small crack advance steps it may be investigated whether J correlates with G^Δ independently of T, the value of N, and the extent of yielding. That is, do all the curves which are different for larger crack advance steps merge into a single curve for step sizes on the order of the COD? At present, only the values reported by McMeeking have growth steps in this range. Of course, in such analyses, finite-strain considerations must be properly treated by McMeeking and Rice [38]. Due to the elastic unloading that occurs during crack advance, J should be correlated with G^Δ values for the initial nodal release; similar correlations with G^Δ values for subsequent nodal releases must be interpreted carefully due to the clouded meaning of J following crack extension.

Finally, a discussion of the validity of the G^Δ quantity is warranted. The nodal reaction force to be relaxed to zero is related to the stress field surrounding the crack tip. As this force is relaxed to zero, the appropriate nodal displacement is monitored so the work expended in the relaxation process may be calculated. The elastic unloading of the body and the accumulated strain at the crack tip influence this displacement. However, no size scale is inherent in this calculation except that imposed by the finite-element mesh, and, as such, the fundamental significance of G^Δ is obfuscated unless a direct correlation of the step of crack advance may be made to a microstructurally significant dimension such as the crack opening displacement. Although their model involves failure by cleavage, Ritchie et al [39] emphasize the necessity of an appropriate size scale in a fracture criterion and correlate the fracture toughness of mild steel with the achievement of a critical tensile stress over a distance on the order of the spacing of crack nucleating carbides.

Conclusions

The following conclusions are drawn from the present analyses:

1. The advancing crack profiles are consistent with the theoretical result of a vertical tangent at the crack tip. Since this is a local effect, it may yet be possible to sensibly define a crack opening angle.

2. Extending cracks in hardening and nonhardening materials are subject to effectively the same stress distributions as geometrically similar stationary cracks. The strains accumulated ahead of a moving crack tip are less than those of a corresponding stationary crack in corroboration of the analytical work of Rice [7,8].

3. The active plastic zone ahead of a growing crack constricts and tilts, paralleling the behavior predicted from analytic and numerical investigations of Mode III cracks.

4. For separation energy rates calculated for crack growth steps much greater than the nominal crack opening displacement, the use of J as a correlator is highly sensitive to strain-hardening properties and the details of external loading.

5. The incremental opening at the crack tip, due to load increase at fixed crack length, seems to be given by $d\delta = \alpha \, dJ/\sigma_0$; the value of α depends on material properties but is the same regardless of the extent of crack growth. Increments in crack surface displacement may be correlated with increments of crack growth at constant external load through the expression

$$d\delta = \frac{\sigma_0}{E} \beta \ln\left(\lambda \frac{EJ}{\sigma_0^2 r}\right) dl$$

The parameters β and λ are constant in these analyses.

Considerations which remain to be addressed are the effects of crack advance under increasing external load, the effects of different plasticity models on the present small-scale yielding solutions, and the effects of large-scale, fully plastic specimen behavior. Also, the role of the finite strain at the root of a blunted crack should be further investigated in the extending-crack situation.

Acknowledgments

This study was supported by the Energy Research and Development Agency under Contract EY-76-S-02-3084 and by the National Science Foundation Materials Research Laboratory at Brown University. The author expresses his gratitude to Professor James R. Rice for his guidance in this study and his patience in reviewing this manuscript.

References

[1] Rice, J. R. in *The Mechanics of Fracture*, F. Erdogan, Ed., American Society of Mechanical Engineers, AMD-Vol. 19, 1976, pp. 23-53.
[2] Rice, J. R. and Johnson, M. A. in *Inelastic Behavior of Solids*, M. F. Kanninen et al, Eds., McGraw-Hill, New York, 1970, pp. 641-672.
[3] McMeeking, R. M., *Journal of the Mechanics and Physics of Solids*, Vol. 25, 1977, pp. 357-381.
[4] Shih, C. F. in *Fracture Analysis, ASTM STP 560*, American Society for Testing and Materials, 1974, pp. 187-210.
[5] McClintock, F. A., *Journal of Applied Mechanics*, Vol. 25, 1958, pp. 582-588.
[6] McClintock, F. A. and Irwin, G. R. in *Fracture Toughness Testing and Its Applications, ASTM STP 381*, American Society for Testing and Materials, 1965, pp. 84-113.
[7] Rice, J. R. in *Fracture: An Advanced Treatise*, H. Liebowitz, Ed., Vol. 2, Academic Press, New York, 1968, pp. 191-311.
[8] Rice, J. R. in *Mechanics and Mechanisms of Crack Growth* (*Proceedings*, Conference at Cambridge, England, April 1973), M. J. May, Ed., British Steel Corporation Physical Metallurgy Centre Publication, 1975, pp. 14-39.
[9] Chitaley, A. D. and McClintock, F. A., *Journal of the Mechanics and Physics of Solids*, Vol. 19, 1971, pp. 147-163.
[10] Broek, D., *International Journal of Fracture Mechanics*, Vol. 4, 1968, pp. 19-29.
[11] Green, G., Smith, R. F., and Knott, J. F. in *Mechanics and Mechanisms of Crack Growth* (*Proceedings*, Conference at Cambridge, England, April 1973), M. J. May, Ed., British Steel Corporation Physical Metallurgy Centre Publication, 1975, pp. 40-54.
[12] Green, G. and Knott, J. F., *Journal of the Mechanics and Physics of Solids*, Vol. 23, 1975, pp. 167-183.
[13] Clarke, G. A., Andrews, W. R., Paris, P. C., and Schmidt, D. W. in *Mechanics of Crack Growth, ASTM STP 590*, American Society for Testing and Materials, 1976, pp. 27-42.
[14] Griffis, C. A. and Yoder, G. R., *Transactions*, American Society of Mechanical Engineers, *Journal of Engineering Materials and Technology*, Vol. 98, 1976, pp. 152-158.
[15] de Koning, A. U., "A Contribution to the Analysis of Slow Stable Crack Growth," presented at the 14th International Congress of Theoretical and Applied Mechanics, Delft (also Report NLR MP 75035 U, National Aerospace Laboratory NLR, Amsterdam), The Netherlands, 1976.
[16] Andersson, H., *Journal of the Mechanics and Physics of Solids*, Vol. 22, 1974, pp. 285-308.
[17] Sorensen, E. P., *International Journal of Fracture*, Vol. 14, 1978, pp. 485-500.
[18] Andersson, H., *Journal of the Mechanics and Physics of Solids*, Vol. 21, 1973, pp. 337-356.
[19] Kfouri, A. P. and Miller, K. J. in *Proceedings*, Institution of Mechanical Engineers, Vol. 190, 1976, pp. 571-584.
[20] Shih, C. F., de Lorenzi, H. G., and Andrews, W. R., this publication, pp. 65-120.
[21] Hill, R., *The Mathematical Theory of Plasticity*, Oxford University Press, Oxford, England, 1950.
[22] Tracey, D. M., *Transactions*, American Society of Mechanical Engineers, *Journal of Engineering Materials and Technology*, Vol. 98, 1976, pp. 146-151.
[23] Rice, J. R. and Rosengren, G. F., *Journal of the Mechanics and Physics of Solids*, Vol. 16, 1968, pp. 1-12.
[24] Marcal, P. V. and King, I. P., *International Journal of Mechanical Sciences*, Vol. 9, 1967, pp. 143-155.
[25] Rice, J. R. and Tracey, D. M. in *Numerical and Computer Methods in Structural Mechanics*, S. J. Fenves et al, Eds., Academic Press, New York, 1973, pp. 585-623.
[26] Yang, W. H., *Computer Methods in Applied Mechanics and Engineering*, Vol. 12, 1977, pp. 281-288.
[27] Sorensen, E. P., *Computer Methods in Applied Mechanics and Engineering*, Vol. 13, 1978, pp. 89-93.
[28] Sorensen, E. P., "Some Numerical Studies of Stable Crack Growth," Ph.D. dissertation, Brown University, Providence, R.I., 1977.

[29] Nagtegaal, J. C., Parks, D. M., and Rice, J. R., *Computer Methods in Applied Mechanics and Engineering,* Vol. 4, 1974, pp. 153-177.
[30] Guyan, R. J., *Journal of the American Institute of Aeronautics and Astronautics,* Vol. 3, 1965, p. 380.
[31] Tracey, D. M., *Engineering Fracture Mechanics,* Vol. 3, 1971, pp. 255-265.
[32] Barsoum, R. S., *International Journal for Numerical Methods in Engineering,* Vol. 11, 1977, pp. 85-98.
[33] Parks, D. M., "Some Problems in Elastic-Plastic Finite Element Analysis of Cracks," Ph.D. dissertation, Brown University, Providence, R.I., Chapter 3, 1975.
[34] Larsson, S. G. and Carlsson, A. J., *Journal of the Mechanics and Physics of Solids,* Vol. 21, 1973, pp. 263-277.
[35] Kfouri, A. P. and Rice, J. R. in *Fracture 1977,* D. M. R. Taplin et al, Eds., Solid Mechanics Division Publication, University of Waterloo Press, Waterloo, Ont., Canada, Vol. 1, 1977, pp. 43-59.
[36] Rice, J. R. in *Proceedings,* 1st International Congress on Fracture, Sendai, Japan, T. Yokobori et al, Eds., Japanese Society for Strength and Fracture, Vol. 1, 1965, pp. 309-340.
[37] Rice, J. R., *Journal of the Mechanics and Physics of Solids,* Vol. 22, 1974, pp. 17-26.
[38] McMeeking, R. M. and Rice, J. R., *International Journal of Solids and Structures,* Vol. 11, 1965, pp. 601-616.
[39] Ritchie, R. O., Knott, J. F., and Rice, J. R., *Journal of the Mechanics and Physics of Solids,* Vol. 21, 1973, pp. 395-410.
[40] Rice, J. R. and Sorensen, E. P., *Journal of the Mechanics and Physics of Solids,* Vol. 26, 1978, pp. 163-186.

R. M. McMeeking[1] and D. M. Parks[2]

On Criteria for J-Dominance of Crack-Tip Fields In Large-Scale Yielding

REFERENCE: McMeeking, R. M. and Parks, D. M., "**On Criteria for J-Dominance of Crack-Tip Fields in Large-Scale Yielding,**" *Elastic-Plastic Fracture, ASTM STP 668,* J. D. Landes, J. A. Begley, and G. A. Clarke, Eds., American Society for Testing and Materials, 1979, pp. 175-194.

ABSTRACT: Very detailed finite-strain/finite-element analyses of deeply cracked plane-strain center-notch panel and single-edge crack bend specimens were generated using nonhardening and power-law-hardening constitutive laws. The deformation was followed from small-scale yielding into the fully plastic range. The objective was to provide insight as to the minimum specimen size limitations, relative to the characteristic crack-tip opening dimension J/σ_0, necessary to assure a J-based dominance of the crack-tip region. The criterion used to judge the degree of dominance was the extent of agreement of the present stress and deformation fields at the blunted crack tips with those calculated by McMeeking for small-scale yielding. For deeply cracked bend specimens, we find very close agreement of the near-tip fields with those of small-scale yielding up to J values of $\sigma_0 L/25$, where L represents the remaining uncracked ligament (and in the deeply cracked case, the only pertinent specimen dimension). This value is consistent with previously proposed J testing size limitations. However, we find that quite detectable deviation from the small-scale yielding fields occurs in both hardening and nonhardening center-crack specimens at considerably smaller J values relative to ligament dimension. This suggests that minimum specimen size requirements necessary to ensure a J-based characterization of the crack tip region may well be more stringent for center-crack or other low plastic constraint configurations than in bend-type specimens. A perhaps overly conservative value of 200 is proposed as the minimum ligament-to-J/σ_0 ratio which ensures a sensible J-based characterization of the crack-tip region in center-crack specimens of materials exhibiting moderate to low strain hardening.

KEY WORDS: crack propagation, J-integral, plasticity, large-scale yielding, fracture (materials), fracture toughness testing, tip field dominance

---

[1]Formerly, acting assistant professor, Division of Applied Mechanics, Stanford University, Stanford, Calif. 94305; currently, assistant professor, Department of Theoretical and Applied Mechanics, University of Illinois, Urbana, Ill. 61801.
[2]Formerly, assistant professor, Department of Engineering and Applied Science, Yale University, New Haven, Conn. 06520; currently, assistant professor, Department of Mechanical Engineering, Massachussetts Institute of Technology, Cambridge, Mass. 02139.

In 1971 Begley and Landes [1,2][3] and, independently, Broberg [3], proposed that the J-integral [4] could be used as a ductile fracture criterion, subject to certain not-well-defined limitations. In addition, Begley and Landes provided experimental evidence suggesting that fracture initiation could be correlated with attainment of a critical J value in a wide, but not unlimited, range of specimen sizes and configurations.

In the original papers, Begley and Landes discussed the possible influence of specimen geometry on the suitability of a one-parameter ductile fracture criterion. They noted McClintock's [5] observations concerning the widely varying crack-tip stress and deformation states, as calculated from nonhardening plane-strain slipline solutions, in different specimen (or structural) geometries. For example, deep double-edge notched (DEN) tension and edge-cracked bend (ECB) type specimens exhibit high triaxial tension on the plane ahead of the crack, while the center-cracked panel (CCP) develops no such elevation of triaxiality, as straight 45-deg sliplines proceed from the crack tips to the free surfaces. McClintock noted that the radically different near-tip stress states in these specimens could presumably affect the microstructural mechanisms of ductile cracking, notably void growth and coalescence. The nonuniqueness in crack-tip fields determined from small strain formulations is a consequence of the nonhardening idealization. When any strain hardening is included, asymptotic analysis [6,7] leads to crack-tip singular fields which are unique to within a scalar amplitude factor, and the J-integral serves as a measure of this amplitude. Begley and Landes argued that since virtually all materials exhibit some strain hardening, J should characterize the near-tip fields at least up to the inception of crack extension.

However, Rice's [8] complete solutions in antiplane strain for large-scale, but contained, yielding in power-law hardening materials ($\sigma \propto \epsilon^N$ in the plastic range) show that the size over which the J-controlled term actually *dominates* higher-order terms in the crack-tip region is a decreasing function of strain hardening exponent N, and vanishes in the nonhardening limit of N going to zero. If the unique Hutchinson-Rice-Rosengren (HRR) [6,7] fields are to dominate the crack tip fields over a physical size scale relevant to microstructural fracture processes, it is evident that the *degree* of hardening is important as well.

Begley and Landes also argued that the finite geometry changes associated with crack-tip plastic blunting should contribute to a J-characterized uniqueness. However, as the plastic deformation leading to the blunted configuration is itself a response to the changing crack-tip fields, it is not clear that blunting *per se* should be assigned a casual role as to maintaining a J-characterized uniqueness. Furthermore, the degree of crack-tip uniqueness associated with blunting must itself depend upon strain harden-

---

[3]The italic numbers in brackets refer to the list of references appended to this paper.

ing. For example, in the nonhardening case, the blunted crack configuration may be smooth [9] or contain any number of sharp vertex features [10].

The finite geometry changes associated with blunting, however, do provide a framework for assessing the degree to which a single parameter characterization of the crack tip is appropriate. The reason is that the blunted crack opening, given roughly by J/σ_{flow}, where σ_{flow} is a representative tensile stress level for plastic deformation, sets the local size scale over which large strain and high triaxiality develop [9,11] and, consequently, the size scale on which microscopic ductile fracture processes may be presumed to act. Indeed, for ductile fracture initiation in small-scale yielding, calculated crack opening displacements correlate well with the spacing of void-nucleating second-phase particles [9,11].

For this blunted region to be uniquely characterized by a single parameter such as J, but otherwise independent of geometry and loading, it is evident that its size must be small compared to any other specimen dimension. Or, to put it the other way, for operative microstructural fracture processes to be embedded within a blunted region characterized by J, a specimen must meet certain minimum size requirements. Paris [12] suggested that, in addition to generating plane-strain constraint along the crack front, all specimen dimensions in a valid J test be chosen to exceed some multiple M of $J_{\text{Ic}}/\sigma_{\text{flow}}$, where J_{Ic} is the value of J identified with the initiation of stable crack growth. Values of 25 or 50 for the coefficient M have been found to give rise to essentially size-independent J_{Ic} and J-Δa curves in bend or compact tension (CT) specimens [13]. However, Begley and Landes [14] report that the resistance curve determined for a CCP specimen was quite different from that of a CT specimen, even though the remaining ligament of the CCP did exceed 25 times the inferred $J_{\text{Ic}}/\sigma_{\text{flow}}$. In fact, the direction of macroscopic crack growth in this CCP specimen followed the 45-deg sliplines of the nonhardening idealization. This suggests the likelihood of a breakdown, due to the remote plastic flow field, of a J-dominated crack-tip region in this specimen at a ratio of ligament L over J/σ_{flow} somewhat greater than 25. In discussing Ref *14*, Rice [15] suggested that the size, relative to remaining ligament, over which J dominates crack-tip fields may well be considerably smaller in fully plastic CCP specimens than in bend configurations. Consequently, the numerical factor M in the minimum-size requirement may be considerably larger than 50 in the case of CCP specimens.

In view of the uncertainties just noted regarding the limits of validity of a J-characterized fracture process zone, the present work was undertaken. The objective was to provide some insight into the specimen geometry and strain-hardening dependence of the scalar M which, for fixed specimen dimensions, defines the limit of deformation (as measured by J) at which uniqueness of the crack-tip fields breaks down.

In the following sections, the basic computational procedures are outlined and certain of the results are presented. Finally, the results are discussed with special consideration of the possible implications for future experimental work in J_{1c} and J resistance curve testing.

Details of the Computer Analysis

The updated Lagrangian finite-deformation finite-element method suggested by McMeeking and Rice [16], modified according to Appendix 2 of Nagtegaal, Parks, and Rice [17] to free the mesh of artificial constraint, was used to analyze plane-strain precracked specimens. The method is based on Hill's [18] variational principle phrased in terms of a current reference state. In yielded elements the partial stiffness approach of Marcal and King [19], as modified by Rice and Tracey [20] and Tracey [21,22], is used to calculate the tangent stiffness, based on the Prandtl-Reuss equations for isotropically hardening materials as in Hill [23, pp. 15-39], but generalized to account for material spin.

In some of the calculations a power law for hardening of the uniaxial Kirchhoff stress $\bar{\tau}$ [16] versus logarithmic plastic strain $\bar{\epsilon}^p$ curve was used which had the form

$$(\bar{\tau}/\sigma_o)^{1/N} = \bar{\tau}/\sigma_o + 3G\bar{\epsilon}^p/\sigma_o \tag{1}$$

where σ_o is the tensile stress and G is the elastic shear modulus. In other calculations a nonhardening law for $\bar{\tau}$ versus plastic strain was used which will be designated $N = 0$.

An undeformed finite-element mesh representing one quarter of a center-cracked panel in tension or one half of an edge-cracked bend specimen is shown in Fig. 1, with the detail of the undeformed near-tip mesh in Fig. 2. The mesh shown has an a/w ratio of 0.9 (see Fig. 3) where a is measured to the center of the semicircular notch tip, $h/w = 3$, and the ratio of undeformed notch width to ligament, b_o/L, is 2×10^{-3}. There were also two other meshes, one with $a/w = 0.9$ and $b_o/L = 2 \times 10^{-4}$ and one with $a/w = 0.5$ and $b_o/L = 2 \times 10^{-4}$. All elements were 4-node quadrilaterals. To model the precracked specimens, traction-free boundary conditions were applied on the notch surface, while the nodes ahead of the tip on the crack line were restrained to remain on the crack line. For the center-cracked panel in tension, the appropriate nodes on the vertical axis of symmetry were restrained to remain on the axis of symmetry and uniform vertical displacements were applied across the top of the specimen. In the case of the edge-cracked bend specimen, traction-free conditions were applied on the sides while the nodes across the top were constrained to lie on a straight line rotating around the center of the top of the specimen. This was done in such a way as to assure that there was no constraint of nodes parallel to the straight line and that the sum of nodal force increments

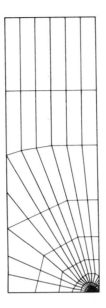

FIG. 1—*Finite-element mesh in undeformed configuration.*

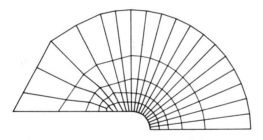

FIG. 2—*Crack-tip detail of finite element mesh.*

normal to the line was zero. Table 1 summarizes the specimen configurations and material properties.

The crack tips were blunted out to maximum openings between 25 and 55 times their undeformed openings. These openings were accommodated partly by arranging the near-tip mesh in a way that would lead to lengthening of the short sides of elements as the tip opened up. The elements on the crack-tip surface ultimately became extremely long in the circumferential direction and, through plastic incompressibility, extremely thin in the radial direction. The results were considered to be sufficiently accurate because the blunted crack tip was always defined by 13 fairly regularly spaced nodes and the region within a few crack-tip openings of a crack tip was always composed of elements quite small compared to the current crack

180 ELASTIC-PLASTIC FRACTURE

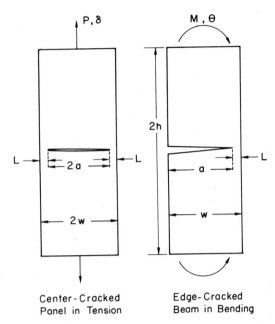

FIG. 3—*Schematic representation of specimen geometries and loadings considered.*

tip opening. Furthermore, the long thin elements are oriented in a way that is quite favorable for modeling the deformation near the tip surface, which is predominantly one that stretches fibers parallel to the tip surface.

Results

The current crack-tip opening to undeformed crack-tip opening ratio δ_t/b_o has been plotted versus $J/(\sigma_o b_o)$ in Fig. 4 for all seven specimens analyzed. The current crack opening δ_t was arbitrarily measured in all

TABLE 1—*Summary of finite-element solutions generated.*

Specimen	a/w	b_o/L	N
ECB[a]	0.9	2×10^{-4}	0
ECB	0.9	2×10^{-3}	0
ECB	0.9	2×10^{-4}	0.1
CCP[b]	0.9	2×10^{-4}	0
CCP	0.9	2×10^{-4}	0.1
CCP	0.5	2×10^{-4}	0
CCP	0.5	2×10^{-4}	0.1

[a] ECB = edge-cracked bend specimen.
[b] CCP = center-cracked panel in tension.

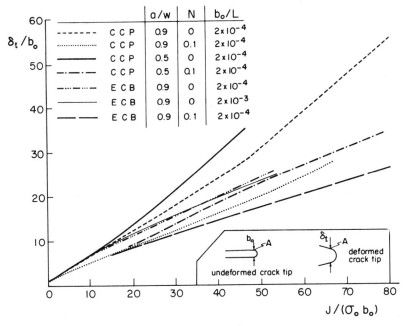

FIG. 4—*Crack-tip opening δ_t versus applied J value for all solutions generated. Crack-tip opening is normalized with respect to undeformed notch root diameter b_o and J is normalized with respect to $\sigma_o b_o$.*

cases as the distance between the nodes that, in the undeformed configuration, lay at the intersection of the straight flank and the semicircular tip as shown in the inset of Fig. 4. The value of J was determined from a load deflection curve of the specimen through the formulas of Rice, Paris, and Merkle [24]. In addition, the virtual crack extension (VCE) method of Parks [25] was used to obtain a numerical analog of the contour value of J. The contour integral definition of J appropriate to finite deformation was given by Eshelby [26] and has been discussed by McMeeking [11,27]. For constant-strain triangular elements, the VCE method exactly computes the contour integral J value on the path connecting midpoints of the sides of the distorted elements and, for isoparametric elements, agrees closely with line integral values [25]. For each analog contour remote from the crack tip, the VCE method gave results for J in agreement with the load-deflection curve method of Rice et al [24]. However, the VCE method showed that J decreases as the contour on which it is computed approaches the crack tip to within a few current crack-tip openings both in small-scale and large-scale yielding. This is in agreement with line integral calculations near blunted crack tips for both small-scale yielding in hardening and non-hardening elastic-plastic materials [11] and the fully plastic deformation of a nonhardening DEN specimen [27].

Returning to Fig. 4, we note that each specimen is subject to three different stages of deformation. At first, the plastic zone is small, or comparable in size to b_o and the ratio δ_t/b_o is not much greater than unity. Later, small-scale yielding occurs as the plastic zone develops into a size that is very much larger than δ_t, and δ_t is only a few times larger than b_o. During small-scale yielding, the slope of δ_t/b_o versus $J/(\sigma_o b_o)$ is close to constant in each specimen with the value of constant depending only on material properties rather than on specimen configuration. The values of the constants are in agreement with those found earlier for small-scale yielding of a sharp crack by McMeeking [11], whose work indicates that the near-tip fields calculated here for a crack of width δ_t are very close to that around an initially sharp crack in the same material blunted to the same width. Since the influence of original notch geometry is lost as far as both $d\delta_t/dJ$ and the near-tip stress and deformation are concerned, the curves of δ_t versus J/σ_o for an initially sharp crack may be obtained from Fig. 4. First, regard b_o as an arbitrary length measure, then extrapolate from the small-scale yielding regime a straight line down and to the left, with the appropriate small-scale yielding slope $d\delta_t/dJ$ for the material as in Ref 11. For an initially sharp crack, the value of J/σ_o in arbitrary length units b_o for a crack-tip opening δ_t in arbitrary length units b_o is then measured from the intercept of the extrapolated line with $\delta_t = 0$. This value will be quoted in future references to J.

In large-scale yielding, the value of $d\delta_t/dJ$ depends not only on material properties but also on specimen configuration, as can be seen in Fig. 4. This has been pointed out for nonhardening materials by Rice [28], who noted that the center-cracked panel in fully plastic tension has a larger value of $d\delta_t/dJ$ than the fully plastic bend specimen. The nonhardening specimens were all deformed until the loads reached those corresponding to slipline solutions. In particular, at very large deformations, the moment on the nonhardening bend specimen attained a value 25 percent in excess of the limit load for the initial geometry. In the final increments of computation, however, the relationship of end rotation rate to rate of change of δ_t was appropriate to the deformation of the fully plastic specimen [28]. It may be that a part of the excess load can be attributed to having too few elements across the ligament, although the precise meaning of a limit load is unclear when finite geometry change is considered.

The near-tip tensile stress normal to the crack line for the nonhardening edge-cracked bend specimen has been plotted in Fig. 5 against the position R of the material in the undeformed configuration for material points lying along the crack line ahead of the tip. The stress is normalized by yield stress and the position is normalized by J/σ_o. The results are taken from the later stages of deformation as can be seen from the key, where r_p is the maximum extent of the plastic zone from the crack tip (—indicates that the zone has reached the specimen edge). The mesh with $b_o/L = 2 \times 10^{-4}$

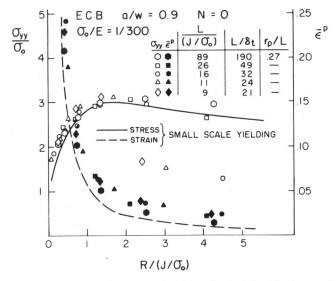

FIG. 5—*Normal stress distribution on the plane ahead of the blunting crack tip versus distance R of material points ahead of the notch root in the undeformed configuration for the nonhardening edge-cracked bend specimen. Distance axis is normalized by J/σ_0. Solid curve is McMeeking's [11] solution for small-scale yielding. Also shown is equivalent plastic strain $\bar{\epsilon}^p$ versus distance from crack tip for material points at 45 deg from the plane ahead of the crack tip in the undeformed configuration. Dashed line is this strain distribution is small-scale yielding [11].*

was used for the hexagons and these are in fact the results at the largest J/σ_0 for this specimen. The remaining points in Fig. 5 were taken from the result for the mesh with $b_0/L = 2 \times 10^{-3}$, and it can be seen that much larger values of δ_t and J/σ_0, measured in terms of L, were achieved in this solution. The full line is the small-scale yielding result of McMeeking [11], who determined the near-tip deformations and stresses around a blunting crack in small-scale yielding by enforcing at a distance remote from the tip an asymptotic dependence of the deformation on the singular term of the crack-tip elastic displacements. Note that this last-mentioned result is self-similar when lengths are measured in terms of δ_t or J/σ_0, since these two quantities are proportional in small-scale yielding. Remarkably, the normal stress distribution on the plane ahead of the crack in the ECB specimen is nearly identical to that of small-scale yielding even for large J values. For $L/(J/\sigma_0) = 26$, the stress state agrees closely over a distance of eight blunted openings. At the three largest J values shown, the stress points farthest from the blunted crack tip in Fig. 5 lie considerably below the small-scale yielding curves at distances from six to nine blunted openings away. These can be explained by noting that for these points the blunted opening δ_t corresponds to between 3 and 5 percent of the uncracked liga-

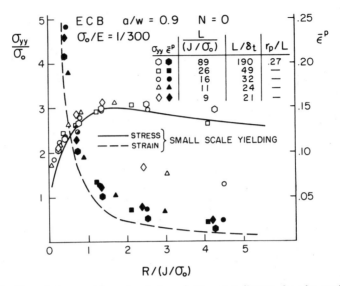

FIG. 6—*Normal stress on plane ahead of crack tip versus distance for edge-cracked bend with strain hardening exponent $N = 0.1$. Also shown is equivalent plastic strain versus distance from crack tip at 45 deg from plane ahead of crack tip.*

ment. Since there is no net force on the ligament, the normal stress must go into compression at a point somewhat nearer to the crack tip than to the back face, and the three low data points are in accord with this. Figure 5 also shows the equivalent plastic strain at 45 deg from the plane ahead of the crack in this specimen and in small-scale yielding [*11*] (dashed line). The fully plastic bend specimen shows somewhat higher strains than small-scale yielding, but the data appear to be rather closely clustered over an order of magnitude range of J.

Figure 6 shows the normal stress distribution in the edge-cracked beam with $N = 0.1$ and $a/w = 0.9$ at three deformation levels in the fully plastic range. Also shown is the small-scale yielding result which, for $N = 0.1$, reaches a maximum value of σ_{yy} approximately equal to $3.8\sigma_o$ as opposed to $3\sigma_o$ for the nonhardening case shown in Fig. 5. Again, at values of $L/(J/\sigma_o)$ as small as 50, the characteristic near-tip field is as in small-scale yielding. The plastic strains at 45 deg are also shown in Fig. 6, and are again close to but somewhat larger than those in the small-scale yielding.

The equivalent plastic strain distribution on the plane ahead of the crack is fairly insensitive to specimen geometry or hardening behavior. This is in accord with the results of McMeeking [*11*], who found that the plastic strain distribution ahead of the blunted crack was virtually identical for strain hardening exponents $N = 0$, 0.1, and 0.2.

Figure 7 shows the normal stress acting on the plane ahead of the crack tip for the nonhardening CCP specimen with $a/w = 0.9$, and the solid line

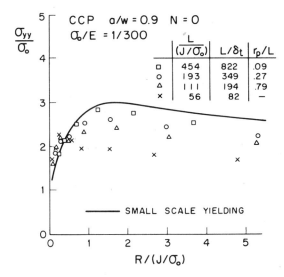

FIG. 7—*Normal stress on plane ahead of crack tip versus distance for the nonhardening center-crack panel, a/w = 0.9. Note the falloff of maximum attained triaxiality at the larger J values.*

is again for small-scale yielding. Data from a wider range of J values, from contained plasticity to fully plastic conditions, are shown. As can be seen, the stress distribution deviates sharply from the small-scale yielding result as fully plastic conditions are approached. At $L/(J/\sigma_o) = 56$, the maximum stress ahead of the crack is only $2.3\sigma_o$ as opposed to $3\sigma_o$ for contained yielding.

The stress state at the smallest J shown lies somewhat below the curve, and it may be significant to consider this fact. We may roughly identify an ASTM-like limit for a valid fracture toughness test in this very deeply cracked specimen as $K_I = 0.6\sigma_o \sqrt{L}$, or by using the small-scale yield relationship $J = (1 - \nu^2)K_I^2/E$, $L/(J/\sigma_o) \approx 900$ for $\nu = 0.3$, and $E/\sigma_o = 300$. Thus the smallest J value shown is well beyond small-scale yielding. Furthermore, Larsson and Carlsson [29] showed that center-cracked specimens exhibited larger overall plastic zones at the ASTM limit than do other geometries. They attribute the specimen dependence of overall plastic zone size, and of stress-state variations within the plastic zone, to the effect of the normal stress parallel to the crack plane. For center-cracked geometries, because this term is negative, Rice [30] noted that the expected effect should be to reduce triaxial tension, and hence maximum normal stress, ahead of the crack tip. Indeed, Larsson and Carlsson noted considerably lower stresses on the plane ahead of crack tip at the ASTM limit in the center-cracked geometry than in bend, double-edge notched or compact-tension specimens. Their mesh was not nearly so detailed as in the present investigation, however, and our observed effects at very small J levels of order $\sigma_o L/1000$ indicated a maximum normal stress ahead of the crack of value $2.89\sigma_o$, which is within 3 percent of the expected Prandtl value. This may well account for the somewhat lower contained yielding stress distributions shown in Fig. 7 and later in Figs. 9 and 11. This subject could be further investigated by applying the modified boundary-layer analysis reported by Larsson and Carlsson [29] to the small-scale yielding blunting solution method used by McMeeking [11].

Figure 8 shows the equivalent plastic strain at 45 deg from the plane ahead of the crack tip for the nonhardening CCP specimen with $a/w = 0.9$. For well-contained plasticity the results are virtually identical to small-scale yielding values. It is quite apparent, however, that for larger deformations crack-tip plastic strain on this ray is considerably larger than for contained yielding. Furthermore, although not shown in Fig. 8, the results are consistently drifting farther from the solid curve at each of the last few deformation increments computed. This means that the intense global deformation on the 45-deg sliplines is intruding on the near tip field, substantially amplifying the deformation on this ray far beyond the J-controlled value. The crack tip characterizing property of J is breaking down.

One may reasonably argue that the dramatic decrease of triaxiality ahead of the crack, and amplification of plastic deformation at 45 deg at general yield shown in Figs. 7 and 8, could be expected from the nonhardening

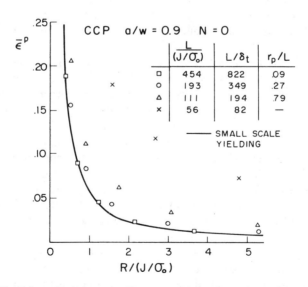

FIG. 8—*Equivalent plastic strain at 45 deg from plane ahead of crack tip versus distance for the nonhardening center-cracked panel with* $a/w = 0.9$. *Note the amplification of plastic deformation for these points at the higher* J *values shown.*

idealization. However, the same trend is shown in Figs. 9 and 10 for the CCP specimen with $a/w = 0.9$ and strain hardening exponent $N = 0.1$. Again, the maximum small-scale yield triaxiality of 3.8 is not quite reached in well-contained yielding, and the stress distribution for moderately contained plasticity again lies somewhat below the reference curve. However, it is apparent that maximum achieved triaxiality is steadily dropping as large plastic deformation ensues. Furthermore, Fig. 10 shows that the plastic strain at 45 deg is also drifting away from the J-characterized field. Although calculations were terminated $L/(J/\sigma_o) = 72$, it would seem that the trend indicated in Figs. 9 and 10 would result in crack-tip fields considerably far from J-dominance if extrapolated to values of $L/(J/\sigma_o)$ equal to 50 or 25.

It is realized that the extreme a/w value of 0.9 used in the previous CCP calculations is likely to be of little value in actual J testing because of difficulties in machining, fatigue sharpening, instrumentation, etc. Rather, they were performed so as to isolate a single characteristic specimen dimension, the ligament, for comparison with J/σ_o. As was noted earlier, the CCP specimen was also solved with $a/w = 0.5$ so that $a = L = w/2$.

Figures 11 and 12 show the stress and plastic strain distributions, respectively, plotted for the CCP specimen with $a/w = 0.5$ and $N = 0.1$. Again, a substantial deviation from J-dominance occurs as the macroscopic plastic deformation field impinges on the blunting crack-tip region. A similar

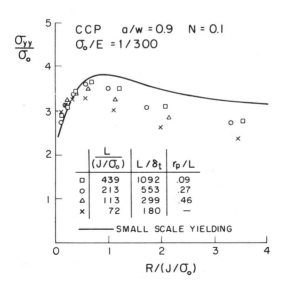

FIG. 9—*Normal stress on plane ahead of crack tip versus distance for center-cracked panel with* a/w = 0.9 *and* N = 0.1. *Note that the maximum attained normal stress is decreasing at the larger* J *values.*

188   ELASTIC-PLASTIC FRACTURE

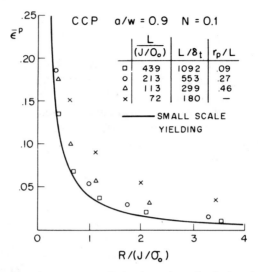

FIG. 10—*Equivalent plastic strain at 45 deg from plane ahead of crack tip versus distance for the center-cracked panel with* a/w = 0.9 *and* N = 0.1. *Plastic deformation on this plane is intensifying at the larger J values.*

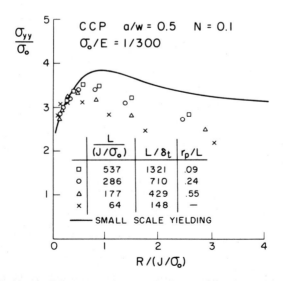

FIG. 11—*Normal stress distribution on plane ahead of crack tip versus distance for center-cracked panel with* a/w = 0.5 *and* N = 0.1.

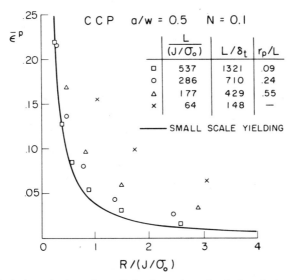

FIG. 12—*Equivalent plastic strain at 45 deg from plane ahead of crack tip versus distance for center-cracked panel with* a/w = 0.5 *and* N = 0.1.

marked deviation from *J*-dominance also occurred in the nonhardening solution of this specimen.

Discussion

We believe that the calculations presented in the previous section have considerable relevance to the development of elastic-plastic fracture testing and evaluation. In this section tentative conclusions drawn from the numerical results are discussed, as well as factors which may tend to modify these conclusions. The utility of this type of investigation in assessing the validity of experimental results is noted, as well as a brief consideration of the implications for micro-mechanical modeling.

A major result of this work is the demonstration in fully plastic bend specimens of unique fields in the blunted crack-tip region, when scaled by J/σ_o. Indeed, Fig. 5 shows stress distributions ahead of the crack at *J* values as large as $\sigma_o L/25$ which are virtually identical to small-scale yielding within a region of eight blunted openings. Similarly, the plastic strain distributions are also close to those in small-scale yielding, as shown in Figs 5 and 6. This strongly suggests that current suggested minimum specimen size requirements of 25 to 50 J/σ_o reasonably assure *J*-dominance of the crack-tip region in bend specimens.

On the other hand, calculations for the CCP geometry, even with a moderate amount of strain hardening, suggest that more stringent size requirements may be necessary to assure a sensible J characterization of the crack tip region. Indeed, the maximum tensile stress ahead of the crack tip for $N = 0.1$ has decreased by 15 percent from its contained yield maximum at a J level of roughly $\sigma_o L/70$ in both the CCP problems of $a/w = 0.5$ and 0.9. Further, the marked intensification of plastic straining at 45 deg to the crack tip, as compared to that in small-scale yielding, is quite apparent at this J level, as shown in Figs. 10 and 12. While these results show that near-tip fields are drifting from those of well-contained yielding, they do not provide a clear-cut indication of when to declare the end of J-dominance. It seems likely that any such explicit criterion must involve a certain degree of arbitrariness, although perhaps consideration of criteria for driving the microscopic fracture mechanisms may provide a rationale for imposing restrictions. In any event, it would appear from the present work that a value of $M = 200$ would be a (perhaps unduly) conservative estimate for minimum specimen size as compared with J/σ_o.

As an example of the potential utility of such calculations, we can consider the experiments of Märkstrom [*31*] and Begley and Landes [*14*]. Märkstrom considered two alloys in CT, CCP, and DEN configurations. He reported essentially constant values of J at crack initiation in each specimen geometry for both materials. Because the CT specimen has a fully plastic flow field similar to that of pure bending, it is presumed that J_c values determined by this test were valid, since specimen dimensions exceeded 25 J_c/σ_o, where σ_o is the 0.2 percent offset yield strength. Let us now ask if the more stringent size requirements we propose for a valid J_c test in the CCP specimens were met. The material Domex 400 had tensile yield and ultimate tensile strength values of 460 and 620 MN/m², respectively. Although no explicit value of a strain-hardening exponent is given, it is inferred from the ratio of the previous numbers to be near $N = 0.1$. The reported J_c value for this material is roughly 0.1 MN/m, so that $J_c/\sigma_o = 0.22$ mm. Although test-by-test data are not reported, the minimum ligament size over all tests was 200 mm. Consequently, the minimum ratio of $L/(J_c/\sigma_o)$ is about 910. This corresponds to well-contained plasticity in the numerical solution with $a/w = 0.5$ and $N = 0.1$, so we would expect J-dominance at the tip, and hence a specimen-independent J_c. For the other material, Ox 802, the tensile yield and ultimate tensile strengths were 800 and 836 MN/m², respectively, and J_c is roughly 0.2 MN/m. This gives a minimum ratio of $L/(J_c/\sigma_o) = 800$. For $a/w = 0.5$, $N = 0$, this J level also corresponds to contained yielding, and again a J-characterized crack tip would be expected. As a contrasting interpretation, we may cite the tests of Begley and Landes [*14*]. They infer a specimen-independent J_{Ic} value at crack extension of 0.105 MN/m in a nickel-chromium-molybdenum-vanadium steel at 394 K. The yield and ultimate strengths are 855 and 957 MN/m², respectively. The CCP specimen had an a/w ratio of 0.6 and a remaining ligament dimension L of 5.08 mm, corresponding to $L/(J_{Ic}/\sigma_o)$

= 41. Our calculations for $N = 0.1$, $a/w = 0.5$ strongly suggest that J characterization of the crack tip has broken down by this point, so a geometry-independent J_{Ic} would not necessarily be expected. Although Begley and Landes report specimen-independent J_{Ic} values, obtained by extrapolation to zero crack extension on the $J - \Delta a$ curve, the initial stable growth in the CCP specimen was evidently influenced by the remote flow field as noted earlier. We take this as indirect evidence supporting our prediction that J-dominance of the crack-tip fields has broken down.

We return to the point that even in this CCP specimen, the inferred J at the initiation of crack growth was apparently the same as in the CT configuration, even though our calculations would suggest rather different crack-tip states. Of course, the agreement could be merely coincidental. On the other hand, some rationalization of the consistency of the results may be attempted. In the earlier stages of loading, the CCP specimen exhibited high triaxiality at the crack tip, and, if this constraint was generated over a significant microstructural dimension, it may be that voids were at least initiated by decohesion or cracking of second-phase particles. As deformation increased, however, the high triaxiality relaxed. This relaxation, by itself, would be expected to decrease the void growth rate. However, there is another factor to consider. As seen in Fig. 4, the CCP specimens tend, in the fully plastic range, to exhibit larger crack opening displacements than do bend specimens at equivalent J values. Consequently, the larger overall deformations in CCP, as opposed to bend specimens, could tend to offset the effect of reduced triaxiality on void growth and coalescence. While this discussion has been of a rather qualitative (and speculative) nature, it does seem to emphasize the importance of obtaining a clearer understanding of the microscopic processes of ductile fracture in developing rational macroscopic criteria which characterize crack extension.

Another area which requires further study is the area of constitutive equations. We used a straightforward generalization of the isotropic hardening Prandtl-Reuss equations based on a smooth yield surface. Clearly this idealization does not precisely model certain aspects of the constitutive behavior of polycrystalline materials at large deformation, so it is reasonable to ask if our constitutive law contributes substantially to the observed breakdown of a J-dominated crack tip in fully plastic CCP specimens. In the nonhardening limit $N = 0$, this must be true, because in some sense *no* hardening brings us back to slipline theory, regardless of the proper treatment of changing geometry. Consequently, we feel that the dramatic loss of triaxiality and amplification of plastic deformation at 45 deg in the fully plastic nonhardening CCP specimens is indeed a direct consequence of the constitutive law. On the other hand, the less dramatic, but quite observable, breakdown with $N = 0.1$ is not so readily attributable to a constitutive inadequacy.

There are at least two important features of stress/strain behavior which have not been modeled here, but they would probably have opposing effects regarding the continuation of uniqueness at the crack tip. The first neglected feature is the possibility that subsequent yield surfaces may develop vertices, or at least very high local curvatures, at the current stress state on the yield surface in stress space. The effect of such features is to make the material behave a bit more like a nonlinear elastic material, as in deformation theory plasticity, thus tending to promote a continued crack-tip uniqueness, as discussed by Rice [32]. On the other hand, for nonzero strain hardening exponent N, the present stress/strain law (Eq 1) never saturates in terms of Kirchhoff stress for arbitrarily large deformation. If, say, the Kirchhoff stress for continuing plastic flow in a particular material saturates after some finite amount of deformation, perhaps identifiable with an equilibrium of dislocation generation and annihilation, then our proposed limitations for J-dominance should be conservative since, as discussed previously, this type of behavior does not tend to promote continuing near-tip uniqueness. These observations suggest that further consideration should be given to the influence of constitutive equations on computed crack-tip fields in both contained and large-scale yielding.

Although the present calculations were not directed toward the problem of stable crack growth, the procedures can possibly be modified to assess the likelihood of J-dominance of crack growth over small distances, perhaps of the order of a few times the blunted crack opening at initiation [33].

Conclusions

On the basis of the finite-element solutions generated here, the following conclusions are drawn:

1. The proposed specimen size limitations [13] for J testing requiring all specimen dimensions to exceed MJ/σ_o, where M is typically 25 to 50, sensibly assure a J-based characterization of the crack-tip region at the initiation of crack extension in pure bending specimens. Because of their similar fully plastic flow fields, this conclusion is presumed to apply to compact tension and three-point bend specimens as well.

2. More stringent specimen size limitations seem necessary to assure a similar J-based characterization of the crack-tip fields in center-cracked panel test configurations, at least in lightly to moderately strain-hardening materials. Based on the present calculations, we would propose a conservative estimate of $M = 200$ as a size limitation which seems to assure the validity of J-characterized crack tip fields.

3. Because the loss of J-dominance of the crack-tip fields in the center-crack geometries is gradual rather than abrupt, there is an arbitrariness in the imposition of size or deformation limitations beyond which J, or other single-parameter characterizations of the crack-tip region, should be deemed invalid. This arbitrariness can be removed, and rational guidelines adopted,

only from the results of careful and systematic experimental investigations of these geometries.

4. Even when a substantial body of such experimental results does become available (there is currently nothing comparable to the extensive results in bend-type specimens) it may be that necessary size limitations to obtain J_{Ic} results consistent with those obtained in bend-type geometries will vary from material to material depending upon aspects of the microstructural ductile fracture mechanisms involved.

Acknowledgment

This work was supported by the National Science Foundation's Center for Materials Research at Stanford University. We are pleased to acknowledge the encouragement received from Professor W. D. Nix in pursuing this topic.

References

[1] Begley, J. A. and Landes, J. D. in *Fracture Toughness, ASTM STP 514*, American Society for Testing and Materials, 1972, pp. 1-23.
[2] Landes, J. D. and Begley, J. A. in *Fracture Toughness, ASTM STP 514*, American Society for Testing and Materials, 1972, pp. 24-39.
[3] Broberg, K. B., *Journal of the Mechanics and Physics of Solids*, Vol. 19, 1971, pp. 407-418.
[4] Rice, J. R., *Journal of Applied Mechanics*, Vol. 35, 1968, pp. 379-386.
[5] McClintock, F. A. in *Fracture: An Advanced Treatise*, H. Leibowitz, Ed., Vol. 3, Academic Press, New York, 1971, pp. 47-225.
[6] Hutchinson, J. W., *Journal of the Mechanics and Physics of Solids*, Vol. 16, 1968, pp. 13-31.
[7] Rice, J. R. and Rosengren, G. F., *Journal of the Mechanics and Physics of Solids*, Vol. 16, 1968, pp. 1-12.
[8] Rice, J. R., *Journal of Applied Mechanics*, Vol. 34, 1967, pp. 287-298.
[9] Rice, J. R. and Johnson, M. A. in *Inelastic Behavior of Solids*, M. F. Kanninen et al, Eds., McGraw-Hill, New York, 1970, pp. 641-671.
[10] McMeeking, R. M., *Transactions, American Society of Mechanical Engineers, Journal of Engineering Materials Technology*, Vol. 99, 1977, pp. 290-297.
[11] McMeeking, R. M., *Journal of the Mechanics and Physics of Solids*, Vol. 25, 1977, pp. 357-381.
[12] Paris, P. C., discussion in *Fracture Toughness, ASTM STP 514*, American Society for Testing and Materials, 1972, pp. 21-22.
[13] Landes, J. D. and Begley, J. A. in *Fracture Analysis, ASTM STP 560*, American Society for Testing and Materials, 1974, pp. 170-186.
[14] Begley, J. A. and Landes, J. D., *International Journal of Fracture Mechanics*, Vol. 12, 1976, pp. 764-766.
[15] Rice, J. R. in *The Mechanics of Fracture*, F. Erdogan, Ed., Applied Mechanics Division, American Society of Mechanical Engineers, Vol. 19, 1976, pp. 23-53.
[16] McMeeking, R. M. and Rice, J. R., *International Journal of Solids and Structures*, Vol. 11, 1975, pp. 601-616.
[17] Nagtegaal, J. C., Parks, D. M., and Rice, J. R., *Computer Methods in Applied Mechanics and Engineering*, Vol. 4, 1974, pp. 153-177.
[18] Hill, R., *Journal of the Mechanics and Physics of Solids*, Vol. 7, 1959, pp. 209-225.
[19] Marcal, P. V. and King, I. P., *International Journal of Mechanical Sciences*, Vol. 9, 1967, pp. 143-155.

[20] Rice, J. R. and Tracey, D. M. in *Numerical and Computer Methods in Structural Mechanics*, S. J. Fenves et al, Eds., Academic Press, New York, 1973, pp. 585-623.
[21] Tracey, D. M., "On the Fracture Mechanics Analysis of Elastic-Plastic Materials Using the Finite Element Method," Ph.D. dissertation, Brown University, Providence, R.I., 1973.
[22] Tracey, D. M., *Transactions* ASME, *Journal of Engineering Materials Technology*, Vol. 98, 1976, pp. 146-151.
[23] Hill, R., *The Mathematical Theory of Plasticity*, Oxford University Press, London, U.K., 1950.
[24] Rice, J. R., Paris, P. C., and Merkle, J. G. in *Progress in Flaw Growth and Fracture Toughness Testing, ASTM STP 536*, American Society for Testing and Materials, 1973, pp. 231-245.
[25] Parks, D. M., *Computer Methods in Applied Mechanics and Engineering*, Vol. 12, 1977, pp. 353-364.
[26] Eshelby, J. D. in *Inelastic Behavior of Solids*, M. F. Kanninen et al, Eds., McGraw-Hill, New York, 1970, pp. 77-115.
[27] McMeeking, R. M. in *Flaw Growth and Fracture, ASTM STP 631*, American Society for Testing and Materials, 1977, pp. 28-41.
[28] Rice, J. R. in *Mechanics and Mechanisms of Crack Growth* (*Proceedings*, Conference at Cambridge, England, April 1973), M. J. May, Ed., British Steel Corporation Physical Metallurgy Centre Publication, 1975, pp. 14-39.
[29] Larsson, S. G. and Carlsson, A. J., *Journal of the Mechanics and Physics of Solids*, Vol. 21, 1973, pp. 263-277.
[30] Rice, J. R., *Journal of the Mechanics and Physics of Solids*, Vol. 22, 1974, pp. 17-26.
[31] Mårkstrom, K., *Engineering Fracture Mechanics*, Vol. 9, 1977, pp. 637-646.
[32] Rice, J. R. in *Numerical Methods in Fracture Mechanics*, A. R. Luxmoore and D. R. J. Owen, Eds., *Proceedings*, International Symposium on Numerical Methods in Fracture Mechanics, Swansea, Wales, Jan. 1978.
[33] Hutchinson, J. W. and Paris, P. C., this publication, pp. 37-64.

M. Nakagaki,[1] W. H. Chen,[1] and S. N. Atluri[1]

A Finite-Element Analysis of Stable Crack Growth—I

REFERENCE: Nakagaki, M., Chen, W. H., and Atluri, S. N., "**A Finite-Element Analysis of Stable Crack Growth—I,**" *Elastic-Plastic Fracture, ASTM STP 668,* J. D. Landes, J. A. Begley, and G. A. Clarke, Eds., American Society for Testing and Materials, 1979, pp. 195-213.

ABSTRACT: A finite-element methodology is developed to study the phenomenon of stable crack growth in two-dimensional problems involving ductile materials. Crack growth is simulated by (1) the translation in steps, of a core of sector elements, with embedded singularities of Hutchinson-Rice-Rosengran type by an arbitrary amount, Δa in each step, in the desired direction; (2) reinterpolation of the requisite data in the new finite-element mesh; and (3) incremental relaxation of tractions in order to create a new crack face of length Δa. Steps 1 and 2 were followed by corrective equilibrium-check iterations. A finite deformation analysis based on the incremental updated Lagrangian formulation of the hybrid-displacement finite-element method is used.
  The present procedure is used to simulate available experimental data on stable crack growth, and thus to study the variation during crack growth of certain physical parameters that may govern the stability of such growth and the subsequent onset of rapid fracture. Attention is focused in this study on the following parameters: $G^{*\Delta}$, the energy release to the crack tip per unit crack growth, for growth in finite steps Δa, calculated from global energy balance considerations; G_{pz}^{Δ} the energy release to a finite "process zone" near the crack tip per unit crack growth, for growth in finite steps Δa, calculated again from global energy balance considerations; and the crack opening angles. However, the work reported here is limited to the first phase of our study, that is, to simulation of available experimental data.

KEY WORDS: ductile fracture, stable crack growth, translation of singularities, finite-element method, J-integral, crack-tip energy release rate, process zone energy release rate, crack opening angle, aluminums, steels, crack propagation.

Considerable research has been reported in recent literature on elastic-plastic fracture mechanics that deals with the use of the characteristic parameters such as the J-integral and crack opening displacement (COD) to define the conditions of incipient growth of preexisting cracks in ductile materials. Only currently are efforts underway to identify characteristic

---

[1] Post-doctoral fellow, graduate student (presently, associate professor, National Tsing Hua University, Taiwan), and professor, respectively, School of Engineering Science and Mechanics, Georgia Institute of Technology, Atlanta, Ga. 30332.

parameters, if any, to deal with the often-observed phenomenon of stable crack growth under rising load, prior to final fracture, in ductile materials. In the present paper we describe attempts to analyze such crack growth. The study is conducted in two phases: the first is a finite-element simulation of available experimental data to understand the variation of several parameters of interest during stable crack growth; and the second phase involves application of criteria chosen from the first phase to predict, using finite-element methodology, stable crack growth and final failure in different cases and to check the accuracy of the prediction against available experimental data. The work reported herein, however, is limited to the first phase of our study.

Scope of Analysis

We consider a two-dimensional stable crack growth situation in an elastic-plastic material and consider an instant of time during which a crack extension by an increment Δa is taking place. During this incremental growth, let ΔW_f be the work performed by the forces applied to the structure; ΔW_e be the change in elastic internal energy of the structure, ΔW_p be the change in plastically dissipated energy in the structure; ΔT the change in kinetic energy, if any, in the structure; and ΔW_c be the work done in quasi-statically and proportionally erasing the tractions (holding the length Δa of the crack together) in order to create a new crack surface of length Δa. Neglecting any thermal input to the structure, one can write an energy balance equation for the entire structure as

$$\frac{\Delta W_f}{\Delta a} = \frac{\Delta W_e + \Delta W_p + \Delta W_c + \Delta T}{\Delta a} \tag{1a}$$

or

$$\frac{\Delta W_f - \Delta W_e - \Delta W_p}{\Delta a} = \frac{\Delta W_c + \Delta T}{\Delta a} \tag{1b}$$

or more conveniently

$$\frac{\Delta(W_f - W_e - W_p)}{\Delta a} = \frac{\Delta W_c}{\Delta a} + \frac{\Delta T}{\Delta a} \tag{2}$$

It is noted to start with, that the increment of crack growth, Δa is postulated to be finite; and in the following we discuss the ramifications when the limit $\Delta a \to 0$ is considered and accordingly the incremental symbol Δ is replaced by the symbol ∂, denoting partial differentiation. The left-hand-side term of Eq 2 is the rate of energy release per unit crack growth (for growth in a finite increment of Δa) from the structure to the crack tip and will be denoted as $G^{*\Delta}$. Thus

$$G^{*\Delta} = \frac{\Delta(W_f - W_e - W_p)}{\Delta a} \quad (3)$$

Thus $G^{*\Delta}$ can be interpreted as the rate of energy "available" to create a new crack surface. By simulating a stable crack extension of amount Δa in the finite-element computations, $G^{*\Delta}$ is computed in the present study by directly computing the terms ΔW_f, and ΔW_e and ΔW_p, in the entire body during such crack extension. It is noted that it has been indicated by Rice [1][2], for nonhardening materials, that $G^{*\Delta} \to 0$ as $\Delta a \to 0$, and thus, according to Ref 1

$$\lim_{\Delta a \to 0} \frac{\Delta(W_f - W_e - W_p)}{\Delta a} = 0 \quad (4)$$

Later in the present paper, the validity of Eq 4 is numerically examined. If Eq 4 is valid, then the dependence of $G^{*\Delta}$ on Δa must be studied, and possible guidelines for selecting "finite" growth step size Δa in finite-element simulations of stable growth must be arrived at. These issues are addressed later in the present paper.

The first term on the right-hand side of Eq 2, on the other hand, is the rate of work "needed" to quasi-statically and proportionally release the cohesive tractions holding the crack surfaces of length Δa, and this term is denoted by G^{Δ}. Thus

$$G^{\Delta} = \frac{\Delta W_c}{\Delta a} \quad (5)$$

Calculations of G^{Δ} for a center-cracked plate in plane strain were presented recently by Kfouri and Miller [2] using an elastic-plastic finite-element analysis. In the analysis of Ref 2, the crack-growth increment, of a finite size Δa, is in fact the distance between two neighboring finite-element nodes on the crack axis. Thus, let Node A be the "current" crack tip, and let the next immediate node, located at distance Δa from A, be Node B. Then crack growth is simulated in the procedure of Ref 2 by proportionally reducing to zero the restraining "nodal force" at Node A. The work done in this nodal force release process is considered to be ΔW_c, and G^{Δ} was computed from Eq 5. In the study of Ref 2, the center-cracked panel was loaded to different levels of far-field tensile stress and, at each load level, a crack-growth increment of Δa was simulated and the attendant G^{Δ} computed. However, it should be noted that the growth increment Δa was considered from the same virgin crack length a_0 at all load levels; and, since the finite-element mesh was kept constant in each load-level case,

---

[2] The italic numbers in brackets refer to the list of references appended to this paper.

the growth increment Δa itself was of constant magnitude at each load level. Kfouri and Rice [3] later concluded that the results of Ref 2 show that, at the same value of the applied load, G^Δ decreases and eventually vanishes as a certain parameter $S = (\sigma_y/K_I)^2 \Delta a$ (where σ_y is the uniaxial yield stress and K_I is the Mode I linear elastic stress-intensity factor for crack of length a_0 at the applied load) tends to zero. This conclusion in Ref 3 is examined in some detail in the present study.

In the present study, as shown later, crack growth is simulated by translating the finite elements near the crack tip in the direction of growth by an arbitrary amount Δa which is not related to the 'distance' between two adjacent nodes; thus the "new" crack tip need not be a 'node' in the original finite-element mesh prior to translation. Thus, the dependence of $G^{*\Delta}$ and G^Δ on Δa at the same load level is more easily studied. Knowing this dependence and postulating a finite-growth increment Δa to be about three to five times the COD at incipient growth condition, available experimental data for fracture test specimens in the form of load versus crack growth or J-integral versus crack growth are simulated in the finite-element model. Thus the variation of $G^{*\Delta}$ and G^Δ during stable crack growth and at the point of unstable fracture is obtained.

Also in the present study, attention is focused on the rate of energy flow, per unit crack growth (for growth in finite increments), into a "process zone" from the rest of the structure. Even though the dimensions of this process zone are arbitrary, several choices can be made. In the present study, the process zone is considered to be a circle Γ of radius equal to $a_0/10.0$, where a_0 is the initial crack length, and, for the problem treated, this size is roughly 40 percent of the plastic zone size at the onset of stable growth. This energy flow rate to the process zone, denoted as $G_\Gamma^{*\Delta}$, is given by the relation

$$G_\Gamma^{*\Delta} = \frac{\Delta W_f}{\Delta a} - \frac{(\Delta W_e + \Delta W_p)}{\Delta a}\bigg|_{\Omega-\Gamma} \qquad (6)$$

In Eq 6 the changes of elastic and plastic energies, $(\Delta W_e)|_{\Omega-\Gamma}$ and $(\Delta W_p)|_{\Omega-\Gamma}$, are evaluated in the rest of the structure, excluding the process zone (Ω is the domain of the structure and Γ is the domain of the process zone). Equivalently, $G_\Gamma^{*\Delta}$ can be evaluated as

$$G_\Gamma^{*\Delta} = \int_{\partial \Gamma} T_i \frac{\Delta u_i}{\Delta a} d\partial\Gamma \qquad (7)$$

where T_i are the tractions at the boundary $\partial\Gamma$ of the process zone Γ, and Δu_i are increments of displacements of $\partial\Gamma$. The variations of $G_\Gamma^{*\Delta}$ during finite-element simulation of experimental data of stable crack growth in a test specimen are also studied.

Further, the crack-tip opening angle (CTOA), which reflects the angle between crack faces at the tip of the advancing crack, and an average value, designated COA, based on the crack opening displacement at the original crack tip position, are studied during the simulated stable crack growth.

Finally, it is perhaps worth noting that, in the present finite-element formulation, the well-known Hutchinson-Rice-Rosengren (HRR) singularities for stresses/strains are built into the elements near the crack tip while analyzing a stationary crack. However, during the translation of the near-tip singular elements to simulate crack growth, the same singularities are assumed to be present at the new crack-tip location. Notice is taken of existing attempts in the literature to obtain analytical solutions to the problem of "steadily" moving cracks in perfect-plastic materials [4,5] which show a logarithmic strain-singularity at the crack tip. However, one of these solutions [4] has been recently argued by Broberg [6] to be in error. For this reason, and for lack of criteria to define "steady"-state conditions, a priori, the HRR singularities are allowed, in the present simulation, to be translated with the advancing crack tip. This may, however, be viewed as an approximation in the general context of the finite-element method.

Brief Description of the Finite-Element Method

To account for the previously mentioned plastic singularities of the HRR type near the crack tip, several special elements called "sector-core" elements or "singularity" elements are designed to model the crack tip region. Each of the core elements has six or eight nodes with the apex node being the crack tip. A hybrid displacement finite-element model is considered only in these core elements while a conventional displacement model is assumed in the rest of the domain, where eight-noded quadrilateral elements are used. Displacement compatibility and traction reciprocity between these two regions are still maintained through the variational principle governing the hybrid displacement model.

In dealing with the geometric as well as material nonlinearities, the present finite-element procedure is based on a large deformation theory, using an updated Lagrangian coordinate formulation, and a tangent modulus incremental plasticity approach with iterative equilibrium corrections. In each step or iterative loading a constitutive relation between the incremental stress $\Delta^* S_{ij}$ and the incremental strain $\Delta^* e_{ij}$ is assumed as

$$\Delta^* S_{ij} = E_{ijkl}{}^t (\tau_{mn}{}^N) \Delta^* e_{kl} \tag{8}$$

where $E_{ijkl}{}^t$ is the current elastic-plastic constitutive property and is a function of current true stress $\tau_{mn}{}^N$ in Nth state of increment.

For defining the yield function f, the Huber-Mises-Hencky type initial yield criterion is considered together with the kinematic hardening rule, modified by Ziegler, which best describes Bauschinger effects. Thus the yield surface in the general condition may be written as

$$\frac{3}{2}(\tau'_{ij} - \alpha_{ij})(\tau'_{ij} - \alpha_{ij}) = \sigma_Y^2 \tag{9}$$

where

α_{ij} = translation of the yield surface in the stress space,
σ_y = yield strength, and
τ'_{ij} = deviatoric part of the stress.

The stress increment $\Delta^* S_{ij}$ in Eq 8 is added to the true stress of the previous state to update the Kirchhoff-Trifftz stress

$$S_{ij(N)}^{N+1} = \tau_{ij}^N + \Delta^* S_{ij} \tag{10}$$

Then $S_{ij(N)}^{N+1}$ are converted to the Euler true stress of the current state by the relation

$$\tau_{ij}^{N+1} = \frac{1}{D} \frac{\partial x_i^{N+1}}{\partial x_k^N} \frac{\partial x_j^{N+1}}{\partial x_l^N} S_{kl(N)}^{N+1} \tag{11}$$

where D is the determinant of the Jacobian

$$\left[\frac{\partial x_i^{N+1}}{\partial x_j^N} \right]$$

and X_i^N is the updated material coordinate at the Nth state.

The variational functional π_{HD} Euler-Lagrange equations derived from the principle $\delta \Delta \pi_{HD} = 0$, and the details of the finite-element formulations leading to the incremental (tangent-modulus type) stiffness relations for the cracked structure, are elaborated in Ref 7. For purposes of present interest, we briefly discuss the relevant field assumptions in the sector-shaped "core" (singularity) elements. The three field variables in each sector element are assumed as

$$\Delta \mathbf{u} = U\beta; \quad \Delta \mathbf{v} = L\Delta \mathbf{q}; \quad \mathbf{T}_L = R\alpha \tag{12}$$

Where β and α are undetermined independent parameters, and $\Delta \mathbf{q}$ are incremental nodal displacements. The interpolation function for the displacements, U in Eq 12 includes regular polynomial modes and a singularity mode such as

$$r^{1/(n+1)}(\beta_1 + \beta_2\theta + \beta_3\theta^2) \tag{13}$$

Where (r, θ) are polar coordinates at the crack tip and n is an exponent coefficient in the strain hardening law. The functions L, which interpolate the displacements at the interelement radial boundary between two sector elements in terms of nodal displacements, also contain the singularity behavior of the type

$$a_1 r^{1/(n+1)} + a_2 r + a_3 \tag{14}$$

whereas at the interface between the singular sector elements and the far-field regular elements, they contain a variation of the type

$$a_1\theta^2 + a_2\theta + a_3 \tag{15}$$

The traction interpolation function R on the element boundary is derived from a self-equilibrated stress field derived from a set of Airy stress functions. R consists of regular polynomials as well as

$$r^{-1/(n+1)}(\alpha_1 + \alpha_2\theta + \alpha_3\theta^2) \tag{16}$$

type singularity behavior.

At the end of each incremental load step, the equilibrium of the total structure is examined in such a way that the Euler stresses are converted to equivalent forces concentrated at the element nodes, to check whether a total norm of the resultant residual force vector at the nodes, except those on constrained or loaded boundary, is small enough or not. Otherwise, the unequilibrated nodal forces are reapplied using an updated stiffness matrix. The Newton-Raphson type iteration is continued until the residual norm reaches below a certain specified tolerance level.

In each element, 5 by 5 (7 by 7 for a singularity element) product Gaussian integration points are used, with data of loading history necessary to determine the current state of deformation and plasticity being stored at each of these Gaussian points. Therefore, a more precise distribution of plasticity conditions is defined than by using constant-stress triangular elements. This makes it possible to use comparatively large quadrilateral elements even adjacent to the core elements as can be seen in Fig. 1. Once the effective stress on the Gaussian point reaches the yield strength, the point is claimed as yielded. The subsequent yield condition is determined by checking $(\partial f/\partial \tau_{ij}) \cdot d\tau_{ij}$ as loading, neutral loading, or unloading depending on whether it is positive, zero, or negative, respectively. However,

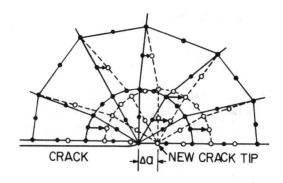

FIG. 1—*Scheme of translation of core elements and reinterpolation of data.*

when an elastic or unloaded part of the domain which is about to yield becomes plastic, it is difficult to follow a uniaxial stress-strain curve with sharp transition point. This difficulty is overcome by using a "knee-correction" procedure, the mathematical details of which are omitted here.

Finite-Element Modeling of Crack Growth

The finite-element simulation of crack growth may be described as (1) geometrical change in the crack surface boundary, (2) translation of the crack tip singularities to the advanced crack tip, and (3) release of surface tractions on the newly created crack surface.

The change in the crack surface boundary is made by translating the whole set of crack tip core elements, as shown in Fig. 1, by an arbitrary distance Δa in the direction of intended crack extension; thus the new crack tip node, which is designated by the center of the sector-shaped core elements, need not be coincident with any previously existing finite-element node before extension. Thus for the Mode I case, even though the fixed boundary in the uncracked ligament of the structure is changed, the constraining condition of the nodes need not be altered. Elements immediately adjacent to the core must be readjusted to fit the translated core. This process of translating the core mesh also moves the embedded singularity in the elements to the new crack tip area, leaving no singularities but large deformations and strains in the wake of the advanced crack tip. All the 7 by 7 Gaussian data points in each of the translated core elements (5 by 5 points for the conventional elements) may generally not coincide with those before translation, for which plastic history data such as current stresses, plastic strains, plastically dissipated work, and yield surface translation are available. Therefore the data at points in the new mesh are estimated by linearly interpolating data on four Gaussian points in the old mesh that are nearest to the point under question in the new mesh. For the sake of brevity, the mathematical details of this interpolation and smoothing process are omitted here. With the fitted plastic data and the new element geometry, element stiffness matrices are recalculated for the core elements as well as for the surrounding rearranged elements, and the global stiffness is appropriately modified. Subsequent equilibrium check iterations using the new stiffness of the structure correct fitting errors, if any, of the plastic data in the new mesh. At the same time, the tractions over the distance AB (Δa as shown in Fig. 1) are incrementally removed, with equilibrium check iterations at each step, to create a new traction-free crack surface of length Δa. The finite-element simulation of crack extension is now completed. The mathematical details of the steps just described are omitted here. During the foregoing extension process, increments for externally supplied energy, elastic strain energy, and plastically dissipated energy in the structure and energy flow into the process zone are calculated to estimate the previously defined quantities $G^{*\Delta}$ and $G_\Gamma^{*\Delta}$. The work done in releasing the preexisting tractions to create a new crack surface of length Δa is computed, and labeled G^Δ. The angles COA and CTOA are computed according to the previously cited definitions.

Problems and Results

First, to understand the dependence of $G^{*\Delta}$ and G^Δ on Δa, the problem of a center-cracked plate under plane strain, identical to the case solved by Kfouri and Miller [2], is analyzed. As in Ref 2, the plate is dimensioned as 40.6 by 40.6 mm and the original half-crack length a_0 is 2.60 mm. Material properties are described as Young's modulus, $E = 207$ KN/mm²; Poisson's ratio, $\nu = 0.3$; yield stress, $\sigma_y = 310$ N/mm²; and a linear strain hardening with a tangent modulus of 4830 N/mm². In the present finite-element analysis, a total of 51 elements and a total number of 148 deg of freedom are used. The hybrid displacement model is used only in the six 6-noded sector-shaped core elements surrounding the crack tip, while the rest of the domain is divided into 8-noded quadrilateral elements wherein a conventional displacement model is used. Three J-integral paths encircling the crack tip, as indicated by broken lines in Fig. 2, are used. The average of J values integrated over the three paths is used in the present analysis, where differences between the paths are within ±0.4 percent. Keeping

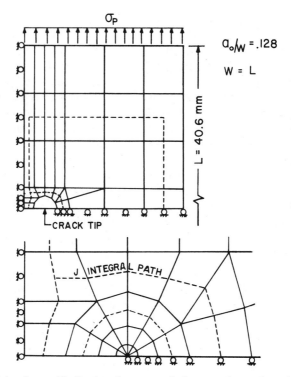

FIG. 2—*Finite-element idealization of center-cracked square plate under uniaxial tension.*

the crack extension Δa as a constant (5 percent of the original crack length a_0), the global energy release rate $G^{*\Delta}$ is estimated for single-step extensions at various load levels and is plotted in Fig. 3 against the parameter S, which is the ratio of Δa divided by the plastic zone size. At about $S = 0.54$, discrepancies among $G^{*\Delta}$, G, and J are negligible, representing a small-scale yielding stage of the loaded specimen, where G is a strain-energy release rate defined by $G = [(1 - \nu^2)K_I^2/E]$. (Note that the normalizing factor G used in Fig. 3 is not a constant, but varies with the load level as per the given definition.) However, remarkable discrepancies are noticed between J and $G^{*\Delta}$ under large-scale yield conditions, as shown in the Fig. 3, wherein $J/G^{*\Delta}$ attains large values as S tends to zero. This result indicates that unless the yielded region is so small that a linear elastic analysis is approximately applicable, J cannot be used to characterize the growth process even at the beginning of the stable crack growth. This would contradict Brogerg's hypothesis, which states that the energy flow Φ to the end region per unit of crack growth is represented by

$$\Phi = J[\alpha + (1 - \alpha)\exp\{-\beta\sigma_y^2(a - a_0)/EJ\}]$$

where α and β are some constants; therefore, $\Phi = J$ at the start of the crack growth.

Also plotted in Fig. 3 is the variation of $G^{*\Delta}$ normalized by the variable G, which is compared with Kfouri and Miller's result for the crack surface energy release rate G^Δ. It is noted that Kfouri and Miller obtained G^Δ at various load levels for a constant $\Delta a = 0.05\ a_0$, as in the present case. The present results are 5 to 10 percent higher than the G^Δ value reported in Ref 2 up to the value of $S \simeq 0.02$. Even though no numerical results for G^Δ/G were obtained by Kfouri and Miller [2] for values of $0 < S < 0.02$, it was surmised in Refs 2,3 that G^Δ/G may tend to zero as $S \rightarrow 0$. As shown in Fig. 3, however, the present results indicate that in fact $G^{*\Delta}/G$ does not monotonically tend to zero as $S \rightarrow 0$, but reverses its trend and raises again. The results of Kfouri and Miller [2] for G^Δ/G up to $S \simeq 0.02$ were in fact considered to be numerical evidence for the paradox that the global energy release rate $G^{*\Delta}$ tends to zero as $\Delta a \rightarrow 0$, as originally predicted by Rice [1]. To understand this further, the absolute value of $G^{*\Delta}$, for constant value of growth increment ($\Delta a = 0.05 a_0$), is shown in Fig 4 for a single-step growth at various values of applied stress, σ_p. It is seen

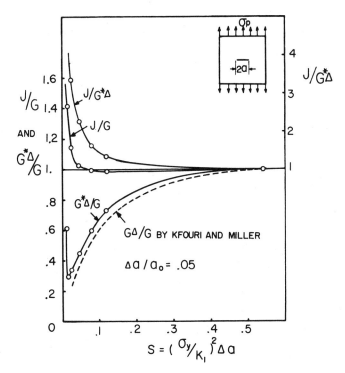

FIG. 3—*Crack separation energy release rates: constant Δa at various load levels.*

from Fig. 4 that $G^{*\Delta}$ in fact increases monotonically with increasing σ_p, at least for the case of $\Delta a/a_0 = 0.05$, even though the normalized values $G^{*\Delta}/G$ (with G as shown in Fig. 4) may decrease with σ_p.

Thus, to numerically study the original hypothesis of Rice [1] (that $G^{*\Delta} \to 0$ as $\Delta a \to 0$), the calculations were repeated for various values of Δa, ranging from $\Delta a/a_0 = 0.07$ to 0.001, while keeping the load constant at $\sigma_p = 245$ N/mm², at which load large-scale yielding conditions are numerically found to exist. Precisely speaking, keeping the load constant, single-step growth increments were simulated for various values of growth increments, Δa. The results are shown in Fig. 5, where $G^{*\Delta}$ is normalized with respect to constant $G = (1 - \nu^2)K_I^2/E$, K_I being the elastic intensity factor at $\sigma_p = 245$ N/mm². It is seen from Fig. 5 that even for the present slightly hardening material, $G^{*\Delta}$ tends to zero as $\Delta a \to 0$ while the load is kept constant. The result in Fig. 5 may then be considered as a direct numerical proof of Rice's original hypothesis [1].

Thus, even though $G^{*\Delta} \to 0$ as $\Delta a \to 0$ at constant load, as seen from Fig. 5, it is finite for all finite growth step values of Δa. Thus, postulating a finite growth step to be of the order of $\Delta a/a = 0.01 \sim 0.02$ (to avoid possible numerical difficulties in the region $\Delta a/a_0 < 0.005$ as in Fig. 5),

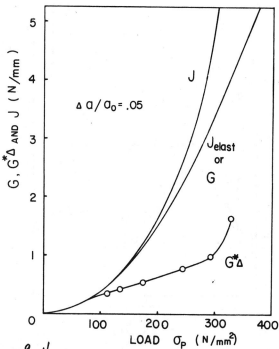

FIG. 4—*Crack separation energy release rate: constant Δa at various load levels.*

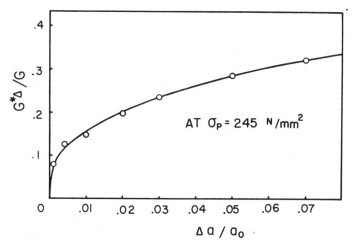

FIG. 5—*Crack separation energy release rate: for various values of Δa at constant load level corresponding to large-scale yielding.*

one may meaningfully simulate experimental data on stable crack growth to study the behavior of $G^{*\Delta}$, G^{Δ}, $G_{\Gamma}^{*\Delta}$, during such growth. This is done in the next problem.

Using the foregoing concepts, experimental data on stable crack growth in two 3-point bend specimens are simulated using the present finite-element procedure. Such experimental data are reported by Griffis and Yoder [8]. The geometry of the first specimen is: width $W = 1.91$ cm, thickness $B = 0.64$ cm, original crack length $a_0 = 0.89$ cm, and $S/W = 4$, where S is the roll-span of the specimen. The initial crack length in the second specimen was $a_0 = 1.14$ cm and all the other dimensions were identical to those of the first specimen. The material of the specimen is an intermediate-strength aluminum alloy, 2024-T351, and the properties are: Young's modulus $E = 80\,300$ MPa, yield strength $\sigma_y = 338$ MPa, ultimate tensile strength $\sigma_u = 492$ MPa, and elongation 21.5 percent. This uniaxial data are fitted by a Ramberg-Osgood type strain hardening law, $\epsilon = (\sigma/E) + (\sigma/B_0)^n$, where the hardening coefficients are $B_0 = 768.6$ MPa and $n = 7.57$. A finite-element mesh breakdown analogous to that of the centrally cracked square plate is considered for both of the present specimens, with a total of 43 elements and a total number of 143 degrees of freedom.

This present two-dimensional analysis does involve the thickness of the specimen, with only an "either-or" choice of plane stress versus plane strain. Based on earlier experience [9], a plane-stress condition was invoked.

A constant crack growth step of $\Delta a = 0.128$ mm is assumed, which is roughly four times the crack opening displacement and about $1/15$ of the plastic zone radius at the loading level corresponding to crack-growth initiation in both specimens. The crack is extended in a total of nine equal steps so that the overall crack extension is about 10 to 13 percent of the original crack length in the specimens. The crack tip is kept unopened until the externally applied load reaches to the level that crack growth initiation is observed in Griffis and Yoder's experiment, that is, at $P/2B = 0.2125$ MN/m for the first specimen and $P/2B = 0.1365$ MN/m for the second. Thereupon, a load increment ΔP corresponding to Δa as in the experimental P versus a (load versus crack growth) record is applied to the structure with an extended crack length $a + \Delta a$. It is noted therefore that the incremental loading and incremental crack opening are carried out simultaneously in the finite-element analysis; thus the present finite analysis results in a piecewise linear approximation to the actual experimental data rather than in a "staircase"-type approximation. This incremental crack extension process is continued to about nine steps (with crack extension of about 13 percent in the first specimen and about 10 percent in the second), when the experimental load records terminate. During this process, the load-point displacement (δ) is computed, and the computed load versus δ-curve is plotted in Fig. 6 for the first specimen ($a_0 = 0.89$ cm), where Griffis and Yoder's experimental data are also included. The excellent correlation of the present results with the cited experimental data suggests the validity of the present finite-element procedure and the correctness of assuming plane-stress conditions in analyzing the problem. Similar correlation for the second specimen has also been noticed but, for want of space, the corresponding curve is not included here. The averages of the three J-contour integral values obtained on three paths for both specimens are plotted against the crack extension ($a - a_0$) in Fig. 7 along

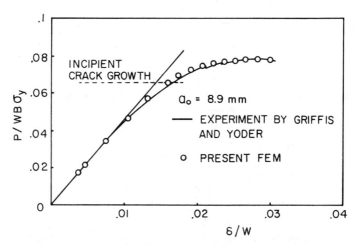

FIG. 6—*Load versus load-point displacement crack growth curve.*

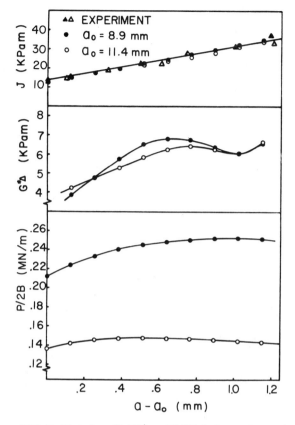

FIG. 7—Variation of J, $G^{*\Delta}$, and P/2B during crack growth.

with the experimental J obtained by Griffis and Yoder [8] using a method established by Begley and Landes [10]. At incipient crack growth in the first specimen, the presently computer J-integral average value is 12.4 kPa·m and 12.6 kPa·m in the second, whereas that reported by Griffis and Yoder is $J_{Ic} = 13.7$ kPa·m. Again, a good correlation is observed between the present numerical J-integral values and the experimental J results at growth initiation. Crack surface profiles in the crack tip region for each step of extension are demonstrated in Fig. 8, showing the characteristic shape of the extended crack surface.

From Fig. 8 it can be seen that at growth initiation the crack tip is blunted, thus reflecting the behavior of asymptotic crack surface deformations of the HRR type that are embedded in the core elements in the present analysis. It is also seen from Fig. 8 that the crack profile becomes much sharper after crack extension; this suggests the possibility of a change in the order and nature of strain singularities at the tip of an extending

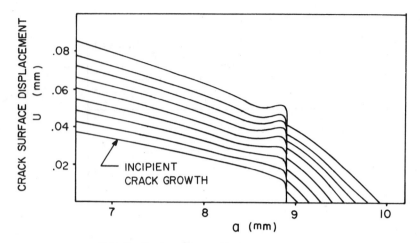

FIG. 8—*Crack surface profile during crack growth.*

crack. It is interesting to note that in the present analysis the HRR singularities are translated with the extending crack tip without any further modification. The fact that in spite of this the crack profile tends to become sharper after extension is surprising. However, if the nature of singularities near the advancing crack tip both in the "transitory" as well as in the "steady"-state conditions is clarified analytically, it is, in principle, possible to effect the appropriate changes in the finite-element modeling. For the present, in view of the foregoing observations concerning Fig. 8, it appears that the assumption of HRR singularities at the tip of an advancing crack may be viewed as an "approximation" in the general context of the finite-element method, in the sense that the hypothetical "exact" solution is approximated by a set of assumed basis functions.

Also shown in Fig. 7 are the variations of $G^{*\Delta}$ during crack growth, for both specimens. It is noted that, for both specimens, $G^{*\Delta}$ exhibits marked variations during the crack extension process, from about 4 kPa·m at initiation to about 7 kPa·m at final fracture. Further, $G^{*\Delta}$ is seen to increase almost monotonically during the crack extension process, except for a slight dip near the point of unstable crack propagation, for both test cases. Also shown in Fig. 7 are the load versus crack-growth curves for both specimens which, of course, are also the experimental curves used in the present simulation.

In Fig. 9, the rate of energy flow to the process zone Γ for finite growth steps, designated as $G_\Gamma^{*\Delta}$, is plotted for both of the simulated cases. Once again it is seen that, for both cases, $G_\Gamma^{*\Delta}$ increases monotonically almost up to the point of final fracture. The rate of energy dissipated in the process zone, which is the difference between $G_\Gamma^{*\Delta}$ and $G^{*\Delta}$, is also shown in Fig. 9 for both test cases. Once again this energy dissipation rate in the process zone, which in the present analysis is completely embedded in the plastic zone near the crack tip, is seen to increase monotonically during crack extension, but is seen to level off or start decreasing near the point of unstable fracture as observed in the experiment.

Finally, the variations of the crack tip opening angles during crack extension, for both of the simulated cases, are shown in Fig. 10. It is observed that this variation of CTOA is analogous to that of $G^{*\Delta}$ as shown in Fig. 7.

Conclusions

It is recognized that formulating any criterion or criteria governing the loss of stability of crack growth, based on numerical simulation of a few experimental data, is, at best, a risky proposition. Thus, we defer any conclusions regarding such criteria until the completion of the second phase of our research. The results reported herein, however, lead to the following conclusions that may be germane to the problem of stable crack growth in ductile materials.

1. A direct numerical proof is provided for the original hypothesis of Rice [1] that $G^{*\Delta} \to 0$ as $\Delta a \to 0$ for those materials for which the flow stress saturates at a finite value of large strain. Thus, for any meaningful numerical study of stable crack growth, a finite growth step must be postulated. Results similar to those in Fig. 5 may be useful in providing guidelines for choosing Δa such that the numerically computed $G^{*\Delta}$ is not

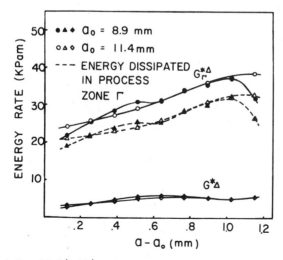

FIG. 9—*Variation of* $G_\Gamma^{*\Delta}$, $G^{*\Delta}$, *and energy dissipation in process zone during crack growth.*

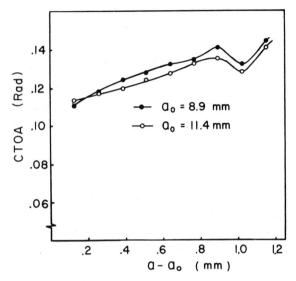

FIG. 10—*Variation of crack-tip opening angle during crack growth.*

sensitive to the errors inherent in numerical processes such as in the finite-element method.

2. Since the magnitudes of $G^{*\Delta}$ and $G_\Gamma^{*\Delta}$ are clearly shown to depend on the postulated magnitudes of "finite" growth steps Δa, in the finite-element modeling it is clear that any criteria governing the loss of stability of crack growth cannot be based on the absolute magnitudes of these quantities. Any such criteria can be based only on the relative qualitative behavior of $G^{*\Delta}$ and $G_\Gamma^{*\Delta}$ for the postulated growth step Δa.

3. In view of the previous observation and the fact that $G^{*\Delta}$ and $G_\Gamma^{*\Delta}$ vary substantially during the crack-extension process, as in Figs. 7 and 9, it is clear that the generalizations of Griffis's approach, in the sense that a ductile material has some characteristic work of separation per unit of new crack area, and that this is to be equated at the critical condition to the rate of surplus work done on the material, cannot be made in situations of stable growth under large-scale yield conditions.

4. The only discernible trend near the points of fracture, as observed experimentally for both of the cases studied, is the marked change in the behavior of $G^{*\Delta}$, CTOA, and to an extent in $G_\Gamma^{*\Delta}$, near these points, when these quantities reverse their monotonically increasing trend during the prior extension process. While a theoretical argument explaining this is lacking, it remains to be seen whether these observations can be used to numerically predict loss of stability of growth in different specimens of the same material. This is the object of our work in progress.

Acknowledgment

The results presented here were obtained during the course of an investigation sponsored by Air Force Office of Scientific Research under Grant AFOSR-74-2667 and by the National Science Foundation under Grant NSF-ENG-74-21346. These and the supplemental support from the Georgia Institute of Technology are gratefully acknowledged. The authors also appreciate the thoughtful comments offered by the reviewers of this manuscript.

References

[1] Rice, J. R. in *Proceedings,* 1st International Congress on Fracture, T. Yokobori et al, Eds., Sendai, Japan, 1965, Japanese Society for Strength and Fracture, Tokyo, Vol. 1, 1966, pp. 309-340.
[2] Kfouri, A. P. and Miller, K. J. in *Proceeding,* Institution of Mechanical Engineers, London, U. K., Vol. 190, 1976, pp. 571-586.
[3] Kfouri, A. P. and Rice, J. R. in *Proceedings,* 4th International Conference on Fracture, D. M. R. Taplin, Ed., Waterloo, Ont., Canada, June 1977.
[4] Chitaley, A. D. and McClintock, F. A., *Journal of the Mechanics and Physics of Solids,* Vol. 19, 1971, pp. 147-163.
[5] Rice, J. R. in *Mechanics and Mechanisms of Crack Growth* (Proceedings, Conference at Cambridge, England, April 1973), M. J. May, Ed., British Steel Corporation Physical Metallurgy Center Publication, 1975, pp. 14-39.
[6] Broberg, K. B., *Journal of the Mechanics and Physics of Solids,* Vol. 23, 1975, pp. 215-237.
[7] Atluri, S. N., Nakagaki, M., and Chen W. H. in *Flaw Growth and Fracture, ASTM STP 631,* American Society for Testing and Materials, 1977, pp. 42-61.
[8] Griffis, C. A. and Yoder, G. R., *Transactions,* American Society of Mechanical Engineers, *Journal of Engineering Materials and Technology,* Vol. 98, 1976, pp. 152-158.
[9] Atluri, S. N. and Nakagaki, M., *American Institute of Aeronautics and Astronautics Journal,* Vol. 15, No. 7, 1977, pp. 923-931.
[10] Begley, J. A. and Landes, J. D. in *Fracture Toughness, ASTM STP 514,* American Society for Testing and Materials, 1972, pp. 1-20.

K. J. Miller[1] and A. P. Kfouri[1]

A Comparison of Elastic-Plastic Fracture Parameters in Biaxial Stress States

REFERENCE: Miller, K. J. and Kfouri, A. P., "**A Comparison of Elastic-Plastic Fracture Parameters in Biaxial Stress States,**" *Elastic-Plastic Fracture, ASTM STP 668*, J. D. Landes, J. A. Begley, and G. A. Clarke, Eds., American Society for Testing and Materials, 1979, pp. 214–228.

ABSTRACT: Parameters used in fracture predictions in elastic-plastic materials will differ from those used when the material is elastic. However, the first set of parameters may reduce to the latter in the limiting case of brittle behavior involving minimal plasticity. Analyses of the stress and plastic strain fields in the region of the crack tip and evaluations of energy release rates are therefore relevant to studies on fracture processes in engineering materials. It is becoming generally recognized that more than one parameter is needed in the formulation of a realistic fracture criterion applicable to elastic-plastic materials. In particular, such a criterion must take into account the possible biaxial nature of the applied stress.

This paper presents some of the results of extensive elastic-plastic finite-element analyses on a center-cracked plate. Information is provided, and comparisons made, on such features as crack-tip plastic zone sizes, intensities of plastic strain near the tip, the major principal stress in the crack-tip region, crack opening displacements, values of the J contour integral, and crack separation energy rates G^Δ—all corresponding to different biaxial stress states.

KEY WORDS: biaxiality, center-cracked plate, compact tension specimen, crack growth step, crack-tip opening displacement, crack-tip plasticity, crack tip plastic zone sizes, crack-tip stresses and strains, crack separation energy rate, elastic-plastic fracture mechanics, finite-element method, Griffith's energy release rate, incremental-load initial-stress elastic-plastic finite-element analysis, J contour integral, small-scale yielding, crack propagation

Nomenclature

A Crack-tip plastic zone size factor
a Half crack length of center-cracked plate
E Modulus of elasticity
G Griffith's energy release rate for a linear elastic material

[1] Professor and research fellow, respectively, Faculty of Engineering, University of Sheffield, Sheffield, U. K.

G_0 Value of G at incipient yielding in a plane-strain uniaxial finite-element analysis of the center-cracked plate
G^Δ Crack separation energy rate
H Tangent modulus in linear hardening stress-strain law
J Rice's contour integral for a path through unyielded material
J_0 G_0
K_I Irwin's Mode I stress intensity factor
$K_I{}^*$ Non-dimensional Mode I stress intensity factor
r_p Maximum dimension of the crack-tip plastic zone
V_A Displacement normal to the crack plane of Node A on the crack surface, nearest to the crack-tip node
Δa Crack growth step = side of leading element at the tip of the crack
ΔW Work absorbed during the stress release for the growth step Δa
$\bar\epsilon_0$ Calculated equivalent strain at the crack-tip node before crack extension
$\bar\epsilon_1$ Calculated equivalent strain at the crack-tip node after the consecutive release of three crack-tip nodes
η_0 20 000 $V_A/\Delta a$ before crack extension
η_1 20 000 $V_A/\Delta a$ after the consecutive release of three crack-tip nodes
ψ Load parameter = G/G_0
λ Biaxiality parameter = σ_P/σ_Q
ν Poisson's ratio for the material in the elastic state
ν^* Effective value of Poisson's ratio for the elastic-plastic material, $\nu < \nu^* < 0.5$
σ_P Applied stress normal to the crack plane, on the boundaries of the center-cracked plate parallel to the crack plane
σ_{P0} Value of σ_P at the start of the incremental load elastic-plastic analysis
σ_Q Applied stress parallel to the crack plane, on the boundaries of the center-cracked plate normal to the crack plane
σ_{10} Major principal stress at the center of the leading element ahead of the crack tip before crack extension
σ_{11} Major principal stress at the center of the leading element ahead of the crack tip after the consecutive release of three crack tip nodes
σ_y Yield stress

Defects in engineering structures are usually situated in a complex stress field and so it is of interest to ascertain the extent to which currently used fracture parameters are affected by the biaxiality of the mode of loading.

Several fracture parameters are known to be dependent on the size of the crack-tip plastic zone. The size and, to a lesser extent, the shape and orientation of the crack-tip plastic zone depend on the biaxiality of the applied load [1][2] or on the type of specimen used [2]. The suggestion has also been made

[2]The italic numbers in brackets refer to the list of references appended to this paper.

[3,4] that even for small-scale yielding it is not sufficient to consider only the first singular term in a series expansion for determining the crack tip stresses in a boundary-layer approach. A second nonsingular term reflecting an additional load parallel to the crack must be taken into account to obtain a realistic assessment of the shape and size of the plastic enclave. This is not surprising if one considers that at the elastic-plastic boundary the equivalent stress in a von Mises material must be equal to the yield stress and therefore stresses of the order of the yield stress must influence the exact location of the boundary.

This paper collates unpublished data obtained from elastic-plastic finite-element analyses on a center-cracked plate [5] in order to assess and compare the effect of load biaxiality on crack-tip plastic zone size, crack-tip opening displacement, crack-tip stresses and strains, Rice's J-integral, and the crack separation energy rate, G^Δ.

The Analyses

The finite-element analyses were carried out on the center-cracked plate of a von Mises material in plane strain of which the finite-element idealization of one quadrant is shown in Fig. 1. The idealization of the region near the tip of the crack is shown in Fig. 2. Material properties were $E = 207$ GN/m^2, Poisson's ratio ν (elastic) $= 0.3$, yield stress $\sigma_y = 310$ MN/m^2, and linear strain hardening with tangent modulus H of 4830 MN/m^2. The elastic-plastic finite-element program is based on the initial stress approach and is described elsewhere [5–7]. Simple isoparametric quadrilateral elements with four corner nodes are used. The effective in-plane Poisson ratio ν^* varies from 0.3 on the elastic-plastic boundary to a maximum of 0.5 in regions which have incurred substantial plastic flow, the actual value being dictated by the extent of the plastic flow and the plane-strain constraint. A consequence of this is that the plastic zone size is somewhat larger than would be the case if a constant value of ν^* equal to 0.5 were used throughout the plastic region.

A measure of the load applied to the crack-tip region is given as G equal to $K_I^2(1 - \nu^2)/E$, that is, equal to the Griffith energy release rate for the equivalent unyielding elastic material. Here K_I is Irwin's Mode I elastic stress intensity factor. At incipient yielding, the largest equivalent stress occurring at any node of any element of the structure is equal to σ_y. The load G will often be normalized with respect to G_0, that is, $\psi = G/G_0$, where G_0 is the value of G required to cause incipient yielding in the finite-element analysis for uniaxial loading. Note that G_0 is proportional to $\sigma_y^2 \Delta a$. This follows from the expression $\sigma_1 = K_I(2\pi r)^{-0.5}$ for the normal stress at a point situated at a small distance r ahead of the crack tip in the plane of the crack in a linear elastic material. Alternatively, as a load characterization parameter, Rice's J-integral, calculated along a contour running entirely through elastic

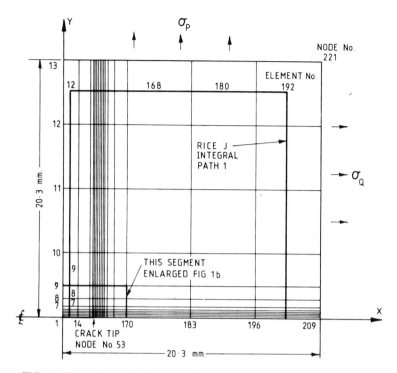

FIG. 1—*Finite-element idealization of top right quadrant of center-cracked plate.*

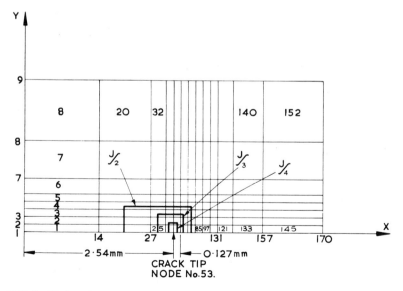

FIG. 2—*Nodes, element numbers, and J-contours in the neighborhood of the crack tip.*

material, is used and normalized with respect to J_0 equal to G_0. The crack growth step Δa is equal to the side of the leading element ahead of the crack tip, that is, 0.127 mm in the mesh shown in Fig. 2. For convenience the load biaxiality parameter $\lambda = \sigma_Q/\sigma_P$ is used where σ_P and σ_Q are the applied boundary stresses normal and parallel to the crack, respectively. Values of λ corresponding to the uniaxial, equibiaxial, and shear modes are 0, 1, and -1, respectively.

Incremental Load Analyses and Crack-Tip Node Releases

Details of the incremental load elastic-plastic finite-element analyses and of the crack-tip node release technique used in the determination of the crack separation energy rate can be found in Refs 5, 8, and 9. For each loading mode the analyses were carried out as follows:

1. An initial elastic analysis adjusted the load σ_P to a value σ_{P0} equal to 95 percent of the value required for incipient yielding.

2. For the incremental load elastic-plastic analysis the load was increased from σ_{P0} in steps of 0.08 σ_{P0}, causing plastic zones to develop at the crack tip, without crack extension. Now at each of the loads σ_P equal to 1.56 σ_{P0}, 2.12 σ_{P0}, 2.68 σ_{P0}, 3.24 σ_{P0}, 3.80 σ_{P0} ... corresponding to different sizes of crack-tip plastic zones, four crack-tip nodes were released consecutively by the following procedure.

3. The equivalent nodal reaction at the crack-tip node is calculated. This force is in a direction normal to the plane of the crack.

4. The constraint on the corresponding displacement of the crack-tip node is relaxed and an external equivalent nodal force equal to the reaction is substituted. Thus the crack is extended, but not opened to the next node along the plane of the crack.

5. Maintaining σ_P constant, the equivalent nodal force at the released node is gradually reduced to zero in six incremental release steps and the displacement of the released node is calculated for each release. This enables the evaluation of the work absorbed during the release, ΔW, to be carried out, and G^Δ is determined by dividing ΔW by Δa.

6. Steps 3, 4, and 5 of the foregoing are repeated on the new crack tip node until four nodes have been released.

Results

The results of this work are given in Table 1 for ease of reference.

Crack-Tip Plastic Zone Sizes

In small-scale yielding, the maximum dimension of the crack-tip plastic zone is usually related to the applied load and yield stress by

$$r_p = A \left(\frac{K_1}{\sigma_y}\right)^2 \qquad (1)$$

where A is assumed to take a constant value approximately equal to 0.175 from an analysis based on the boundary-layer approach and small-scale plasticity [10]. In fact the finite-element results reveal that only in the equibiaxial mode can A be taken as a constant. Note that the boundary-layer approach approximates the equibiaxial mode. Values of A equal to $r_p/(K_I/\sigma_y)^2$ against ψ for the different biaxial modes of loading are shown in Fig. 3. For the equibiaxial mode, A takes the constant value of 0.19. In the uniaxial mode, A appears to increase linearly with ψ from an initial value of approximately 0.24. The variation of A is quite large in the case of the shear mode. Also shown in Fig. 3 are some spot checks carried out in the uniaxial mode on a finer mesh having tip elements of side equal to 0.0635 mm, that is, one half of that for the mesh of Fig. 1. The points referring to the finer mesh are marked by triangles in Fig. 3 and by asterisks in Table 1. Recent work, reported later in this paper, gives a value of A equal to 0.149 for the compact tension specimen.

In Fig. 4, ψ and J/J_0 are plotted against the plastic zone size normalized with respect to Δa. For a given value of ψ, the value of $r_p/\Delta a$ is smallest for the equibiaxial mode and largest for the shear mode while the value corresponding to the uniaxial mode occupies an intermediate position. Note that for small values of $r_p/\Delta a$ the values of J are approximately equal to those of G.

Since G and K_1^* depend only on σ_P and not on σ_Q, that is

$$K_1 = \sigma_P K_1^* \sqrt{a} \qquad (2)$$

(where K_1^*, approximately equal to $\sqrt{\pi}$, is the non-dimensional stress intensity factor and a is the half crack length), G and J do not depend on the mode of biaxiality of the load in small-scale yielding situations. Note that ψ is approximately equal to $0.51 \Delta a^{-1} (K_1/\sigma_y)^2$; that is, ψ is equal to $r_p/\Delta a$ if the constant A in Eq 1 assumes the same value of 0.51 for all modes of biaxiality. This means that Fig. 3 is a plot of actual plastic zone size against that predicted by Eq 1.

Crack Tip Opening Displacement (CTOD)

A nondimensional measure of the CTOD is obtained from the value of η equal to $2V_A/\Delta a \times 10^4$ where V_A is the distance from the crack plane to the node A on the crack surface immediately before the crack tip. As the load is applied without crack extension, the crack-tip region incurs permanent plastic deformation and attains a value η_0. A distinction is made between

220 ELASTIC-PLASTIC FRACTURE

TABLE 1—*Results of elastic-plastic finite-element analyses on a center-cracked plate.*

λ	$r_p/\Delta a$	G/G_0	A	η_0	η_1	$\bar{\epsilon}_0 \times 10^4$	$\bar{\epsilon}_1 \times 10^4$	J/J_0	σ_{10}/σ_y	σ_{11}/σ_y	G^Δ/G
−1	1.7	1.9	0.45	86.0	…	27	…	1.91	1.06	…	0.79
	…	2.2	…	…	…	…	22	…	…	1.12	…
	4.0	3.5	0.59	136	79	68	…	3.56	1.18	…	0.55
	…	4.0	…	…	…	…	39	…	…	1.32	…
	10.0	5.6	0.92	215	97	120	…	5.97	1.18	…	0.37
	…	6.4	…	…	…	…	48	…	…	1.32	…
	…	8.2	…	…	106	203	…	…	1.18	1.32	0.28
	27.9	9.4	1.75	358	120	…	58	10.4	…	1.32	…
0	…	*2.0	0.28	…	…	…	…	…	…	…	…
	1.3	2.2	0.26	86.7	…	24	…	2.2	1.17	…	0.84
	…	2.5	…	…	86.1	…	23	…	…	1.23	…
	…	*3.7	0.30	…	…	…	…	…	…	…	…
	2.2	4.1	0.29	137	…	60	…	4.1	1.45	…	0.66
	…	4.7	…	…	114	…	41	…	…	1.57	…
	…	*5.9	0.32	…	…	…	…	…	…	…	…
	4.1	6.5	0.33	196	…	90	…	6.64	1.59	…	0.51
	…	7.5	…	…	135	…	52	…	…	1.77	…
	6.5	9.5	0.35	269	…	126	…	9.9	1.66	…	0.39
	…	10.9	…	…	150	…	61	…	…	1.81	…
	11.0	13.0	0.43	361	…	168	…	14.2	1.71	…	0.32
	…	15.0	…	…	164	…	69	…	…	1.93	…
	16.8	17.2	0.50	474	…	215	…	19.7	1.76	…	0.26
	…	19.8	…	…	176	…	76	…	…	1.97	…
+1	0.9	2.5	0.19	94	…	23	…	2.5	1.26	…	0.87
	…	2.9	…	…	92	…	21	…	…	1.32	…
	1.7	4.5	0.19	143	…	49	…	4.6	1.62	…	0.70
	…	5.2	…	…	126	…	39	…	…	1.70	…
	2.9	7.3	0.20	208	…	75	…	7.41	1.90	…	0.59
	…	8.4	…	…	159	…	51	…	…	2.05	…
	4.1	10.6	0.20	261	…	101	…	10.92	2.13	…	0.50
	…	12.2	…	…	190	…	65	…	…	2.30	…
	5.3	14.6	0.19	330	…	126	…	15.2	2.31	…	0.43
	…	16.8	…	…	219	…	78	…	…	2.48	…

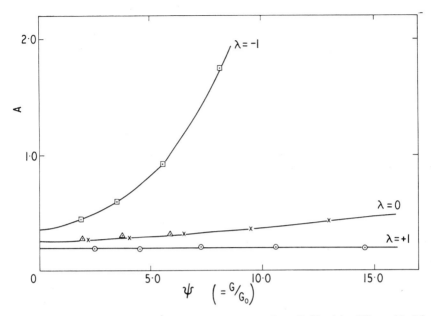

FIG. 3—*Variation of crack-tip plastic zone size factor with applied load for different biaxial loading modes.*

the case mentioned in the foregoing and that of crack extension under constant load by the gradual and consecutive release of successive crack-tip nodes. The value of η after three tip nodes have been released will be referred to here as η_1. In the idealization shown, η_0 will therefore refer to the vertical displacement of Node 40 and η_1 to the vertical displacement of Node 79; see Fig. 5. A change in crack profile is known to occur as the crack begins to extend during the initial stages of subcritical crack growth [11]. In Fig. 6 the top three curves give the values of η_0 and the bottom three those of η_1, for the three biaxial modes of loading. The values of η_0 diverge considerably with different values of λ, the values for the shear mode being greater than those for the other two modes. The values of η_1 are all smaller than those of η_0 but the order of the relative magnitudes is reversed in the case of the shear and equibiaxial modes. This is probably due to the residual stress pattern developed in the wake of the crack tip which is an effect of the size, shape, and position of the plastic zone prior to crack extension, all of which are functions of stress biaxiality [1]. In all cases the variation of η with ψ seems to stabilize into a near linear relationship with increasing ψ but the slope is of course different for each mode.

Crack-Tip Strains and Stresses Ahead of Tip

Figure 7 shows the calculated equivalent strains at the crack tip plotted

222 ELASTIC-PLASTIC FRACTURE

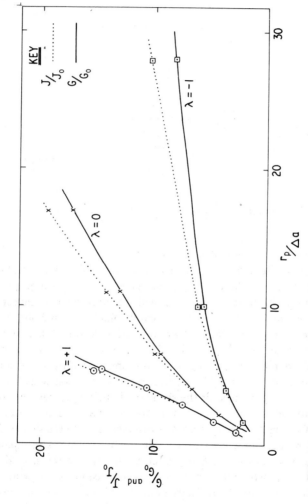

FIG. 4—*Applied load parameters G and J against crack-tip plastic zone size for different biaxial loading modes.*

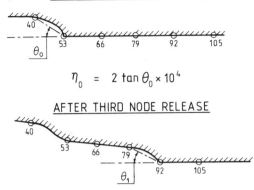

FIG. 5—*Node numbers on the crack profile and beyond the crack tip before and after the release of the crack-tip nodes.*

against ψ for the same situations as described for CTOD, that is, $\bar{\epsilon}_0$ applying to a stationary crack being loaded and $\bar{\epsilon}_1$ applying to the equivalent strain after the release of three tip nodes. The crack-tip node is 53 in the case of $\bar{\epsilon}_0$ and 92 in the case of $\bar{\epsilon}_1$. The results are of only qualitative interest since the exact strain at the crack tip cannot be known with any precision, but they do give some indication of the relative strains incurred in the tip region under different biaxial modes of loading. The general pattern in Fig. 7 is not very different from that of Fig. 6. However, the grouping of the lower three curves giving $\bar{\epsilon}_1$ is much closer than the corresponding grouping in Fig. 6 for η_1. This suggests that plasticity intensity at the tip of a growing crack is not greatly influenced by load biaxiality.

Figure 8 gives the magnitude of the main principal stress at the center of the leading element ahead of the crack tip, normalized with respect to the yield stress for the three modes of load biaxiality. The broken lines refer to the case of the stationary crack and the solid curves to the extended crack after three tip nodes have been released. The element number is 49 in the first case and 85 in the second. The stresses are highest for the equibiaxial mode and lowest for the shear mode. The stresses for the stationary crack are lower than those occurring after the release of the three tip nodes. The difference between the three modes is attributed to the hydrostatic component, which is highest in the case of the equibiaxial mode and lowest in the case of the shear mode, and also to the different plastic zone sizes. On the whole the curves in Fig. 8 seem to follow a somewhat similar pattern to the η_1 curves in Fig. 6. Distributions of normal principal stresses ahead of the crack tip have also been given in Ref *12*.

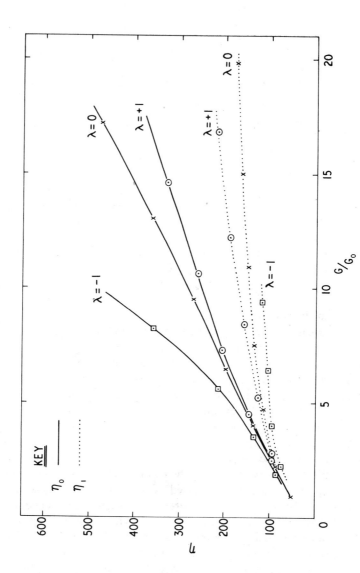

FIG. 6—*Crack tip displacements before crack extension (η_0) and after three tip node releases (η_1) for different biaxial loading modes.*

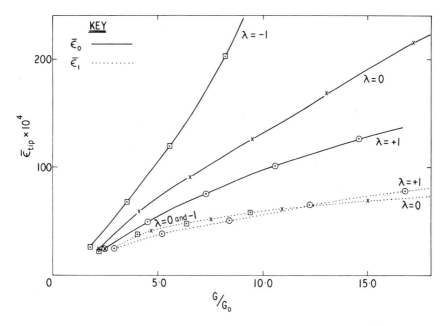

FIG. 7—*Calculated equivalent strains at the crack tip before crack extension ($\bar{\epsilon}_0$) and after three tip nodes have been released ($\bar{\epsilon}_1$) against applied load for different biaxial loading modes.*

Crack Separation Energy Rate

For a finite growth step Δa, let the stresses holding the crack surfaces together, over the distance Δa beyond the crack tip, be quasi-statically and proportionally reduced to zero, causing the surfaces to separate. Calling ΔW the work absorbed during the release of the stresses, then the crack separation energy rate, G^Δ, is defined as $\Delta W/\Delta a$.

The effect of load biaxiality on G^Δ has already been presented in Ref *12*, giving values of G^Δ/G and J/G against ψ for different modes of load biaxiality. The values of G^Δ/G converge from unity, when G equals G_0, to zero for large values of G. However, the paths are different for different modes of biaxiality, the largest value of G^Δ corresponding to the equibiaxial mode. On the same figure of Ref *12* is shown the lack of dependence of J/G on load biaxiality for small-scale yielding as previously indicated.

Figure 9 shows that when G^Δ/G is plotted against actual normalized plastic zone size ($r_p/\Delta a$), the dependence of G^Δ/G on the mode of load biaxiality is very much reduced. Hence G^Δ/G appears to be a function of plastic zone size only and independent of λ. As Fig. 4 shows, however, $r_p/\Delta a$ depends upon the mode of biaxiality λ.

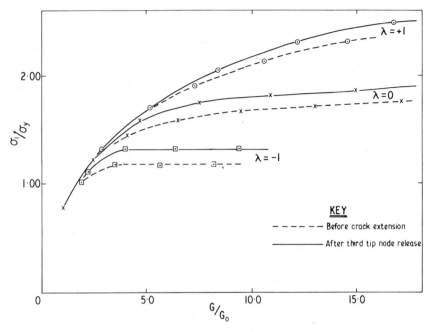

FIG. 8—*Major principal stress at the center of the element ahead of the crack tip against applied load for different biaxial loading modes.*

Discussion

Most of the quantities investigated here show a more or less marked dependence on the biaxial mode of loading. Exceptions are the crack-tip equivalent strain for the moving crack $\bar{\epsilon}_1$ and the relation between G^Δ/G and $r_p/\Delta a$. Very little is known about the growth step Δa and its dependence on biaxiality. If Δa is dependent on the intensity of plasticity in the crack-tip region, it is plausible to suppose that Δa is not affected by the mode of load biaxiality.

When considering parameters such as η and $\bar{\epsilon}_{\text{tip}}$, a distinction must be made between initiation and propagation. Generally for a stationary crack η_0 and the crack-tip plastic strain, $\bar{\epsilon}_0$ would appear to be more relevant to the initiation stage than to incipient unstable crack propagation, while G^Δ is intended as a propagation parameter. The CTOD for the moving crack η_1 would appear to be more relevant to propagation than initiation.

If Δa can be taken to be independent of the mode of biaxiality, Fig. 9 shows that the crack-tip plastic zone size is almost uniquely related to G^Δ and can therefore be used as a propagation parameter independent of λ. However, it must be noted that actual values of r_p depend on λ. It follows that crack propagation cannot be uniquely determined by J. The values of G^Δ used here are those corresponding to the third release of the crack-tip node.

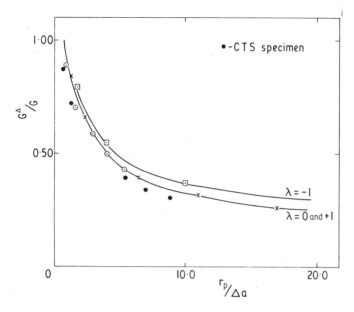

FIG. 9—*Variation of crack separation energy rate with crack-tip plastic zone size for different biaxial loading modes.*

Now cracks in engineering components may have any orientation. Our results show that in symmetric cases, where the sizes and shapes of the crack-tip plastic zones above and below the crack tip are equal (hence no ambiguity exists about the direction of crack propagation in isotropic materials), an approximate evaluation of G^Δ can be obtained from a knowledge of r_p. For the record, some recent elastic-plastic finite-element analyses on a CTS specimen are included in Fig. 9 [13].

In the case of inclined cracks, a more comprehensive analysis is required since the symmetry is lost on account of the applied shear load component additional to σ_P and σ_Q. This problem is currently being studied and is not a simple matter since elastic-plastic finite-element analyses show that unequal plastic zones are produced at the tip and the direction of crack propagation must be ascertained for each case. Furthermore, there may be some significant interaction between Mode I and Mode II propagation in some cases.

Finally, it is interesting to note that fatigue growth studies under biaxial load conditions of λ equal to -1, 0, and $+1$ indicate that crack growth rates could be correlated on a basis of similar plastic zone sizes albeit that elastic stress intensity factors are different [14]. Reference 14 concluded with a statement that a two-parameter description for fatigue fracture was required. Likewise, the present work suggests that a two-parameter relationship, that is, J and λ, is necessary to describe brittle fracture in real engineering situations.

Conclusion

Brittle crack propagation for various degrees of load biaxiality can be correlated for a given material on a basis of similar plastic zone size, which is a function of applied stress and loading mode.

References

[1] Miller, K. J. and Kfouri, A. P., *International Journal of Fracture*, Vol. 10, No. 3, 1974, pp. 393-404.
[2] Larsson, S. G. and Carlsson, A. J., *Journal of the Mechanics and Physics of Solids*, Vol. 21, 1973, pp. 263-277.
[3] Rice, J. R., *Journal of the Mechanics and Physics of Solids*, Vol. 22, 1974, pp. 17-26.
[4] Eftis, J., Subramonian, J., and Liebowitz, H., *Engineering Fracture Mechanics*, Vol. 9, 1977, pp. 189-210.
[5] Kfouri, A. P. and Miller, K. J. in *Proceedings*, Institution of Mechanical Engineers, Vol. 190, No. 48/76, 1976, pp. 571-584.
[6] Owen, D. R. J., Nayak, G. C., Kfouri, A. P., and Griffiths, J. R., *International Journal for Numerical Methods in Engineering*, Vol. 8, 1973, pp. 63-73.
[7] Hellen, T. K., Galluzzo, N. G. and Kfouri, A. P., *International Journal of Mechanical Sciences*, Vol. 19, 1977, pp. 209-221.
[8] Kfouri, A. P. and Miller, K. J., *International Journal of Pressure Vessels and Piping*, Vol. 2, 1974, pp. 179-191.
[9] Kfouri, A. P. and Miller, K. J., "Separation Energy Rates in Elastic Plastic Fracture Mechanics," Technical Report CUED/C-MAT/TR18, Engineering Department, University of Cambridge, Cambridge, England, 1974.
[10] Rice, J. R. and Johnson, M. A. in *Inelastic Behavior of Solids*, M. F. Kanninen, W. F. Adler, A. R. Rosenfield, and R. I. Yafee, Eds., McGraw-Hill, New York, 1970, pp. 641-672.
[11] Rice, J. R. in *Proceedings*, Conference on the Mechanics and Mechanisms of Crack Growth, April 1973; British Steel Corp. Physical Metallurgy Centre Report, M. J. May, Ed., 1975, pp. 14-39.
[12] Kfouri, A. P. and Miller, K. J., *Fracture*, Vol. 3, ICF 4, Waterloo, Canada, 19-24 June 1977.
[13] Kfouri, A. P., "An Elastic-Plastic Finite Element Evaluation of G^Δ in a Compact Tension Specimen," to be published.
[14] Miller, K. J., "Fatigue Under Complex Stress" in *Proceedings*, "Fatigue 1977" Conference, Cambridge, England; *Metal Science Journal*, Vol. 11, Nos. 8 and 9, 1977, pp. 432-438.

Y. d'Escatha[1] *and J. C. Devaux*[2]

Numerical Study of Initiation, Stable Crack Growth, and Maximum Load, with a Ductile Fracture Criterion Based on the Growth of Holes

REFERENCE: d'Escatha, Y. and Devaux, J. C., "**Numerical Study of Initiation, Stable Crack Growth, and Maximum Load, with a Ductile Fracture Criterion Based on the Growth of Holes,**" *Elastic-Plastic Fracture, ASTM STP 668,* J. D. Landes, J. A. Begley, and G. A. Clarke, Eds., American Society for Testing and Materials, 1979, pp. 229-248.

ABSTRACT: Considering a material at the ductile plateau, fracture tests on small specimens in generalized yielding conditions are common. We need to extract from them adequate information to characterize the fracture resistance properties of the material. We also need to predict initiation, crack growth, and maximum load for a crack found in a ductile structure. But the real problems are three-dimensional; for instance, semi-elliptical surface cracks or through-cracks in "small" thicknesses (tunneling and mixed-mode fracture). Moreover, they are not only in the symmetrical Mode I case (angled crack extension).

The common denominator of all these phenomena is the ductile fracture processes in the material at the crack border; these micromechanisms extend over some characteristic length which needs to be introduced at a crack tip because of the very intense strain gradient. We thus need a ductile fracture damage function belonging to the continuum mechanics frame and related to the history of stresses and strains averaged over such a characteristic volume. In this numerical feasibility study, using elastic-plastic finite-element computations and guided by a ductile fracture model in three stages—void nucleation, void growth, and coalescence—we tried such a differential damage history, in a most simplified form. We integrated this during the whole stress and strain history in each finite element along the crack path, and we studied the influence of the mechanical and numerical parameters playing a role in this methodology. We describe herein the evolution, which results from this criterion, of some parameters used in the literature as initiation and crack growth criteria.

KEY WORDS: ductile fracture, void growth, finite elements, elastic-plastic deforma-

---

[1] Maître de Conférences à l'Ecole Polytechnique et à l'Ecole des Mines, Laboratoire de Mécanique des Solides, Ecole Polytechnique, 91128 Palaiseau Cedex, France.
[2] Adjoint au Chef du Département Calcul de la Division des Fabrications, Framatome, B.P. 13, 71380 Saint Marcel, France.

tion, generalized yielding, crack initiation, stable crack growth, instability, crack propagation

The necessity to develop physical and numerical models of ductile fracture arises from a twofold need:

1. We must learn to read deeply into fracture tests, at the ductile plateau, of small specimens which are in more or less extensive yielding conditions, in order to extract adequate information about the fracture resistance of the studied materials, and thereby to characterize them. In effect, tests on small specimens which are in generalized yielding conditions are common, and as these tests seek the fracture resistance properties of the material, we must develop the ability to characterize the properties from the observation of initiation, crack growth, and maximum load (measuring load, displacement, crack opening, and crack growth).

2. We must develop the ability to predict, at least approximately, initiation, stable crack growth, and maximum load for a real flaw found in a ductile structure, in order to estimate safety margins.

Industry is now being urged to answer these questions. But the real problems in the desired applications are, as a matter of fact, very complicated. Most of them are essentially three-dimensional—for instance, semi-elliptical surface cracks or through-cracks in "small" thicknesses when the stress and strain state at the crack border depends on the through-thickness coordinate. Moreover, the real problems may be in complex loading conditions, that is to say, not only in the symmetrical (Mode I) case.

It does not seem that the elastic-plastic fracture mechanics criteria proposed today in the literature can be suitably extended to account for initiation, stable crack growth, and maximum load in three-dimensional cases or in complex loading cases. That is why we looked toward the only common denominator of all these various situations, namely, the damage corresponding to the ductile fracture mechanisms, undergone during the loading history by the material situated at the crack tip.

Initiation, crack growth, change in shape of a semi-elliptical surface crack, tunneling, mixed-mode fracture (flat fracture and shear lips) in "small" thicknesses, and angled crack extension in complex loading—all these phenomena can be thought of as resulting from the distribution of stresses and strains, and of the damage they create in the material situated at the crack tip, along the successive positions of the crack border, during the loading history and the crack growth.

In the case of ductile fracture, this damage can be imagined, generally speaking, as the following three stages [1,2][3]. First, included particles

---

[3] The italic numbers in brackets refer to the list of references appended to this paper.

either break or separate from the matrix, thus nucleating voids [3,4]. Then comes a void growth stage [5,6] until finally coalescence of voids occurs, meaning that localized internal necking between the voids happens. These micromechanisms of ductile fracture depend on the material and extend over a characteristic length scale, for instance, the mean distance between void nucleation sites. This characteristic length scale gives the size effect.

If we consider the ductile fracture damage as it develops in the material situated at a crack tip, where the strain gradient is very intense, we recognize the necessity to consider the damage developing globally in a volume of material situated at the crack tip and having dimensions of the order of magnitude of the characteristic length of the microscopic ductile fracture processes [7-13].

We are thus looking for a macroscopic damage function which we want to belong to the frame of continuum mechanics and whose evolution would be related to the history of stresses and strains averaged over such characteristic volumes of material representative of the length scale of the ductile fracture micromechanisms in the considered material.

To avoid being completely arbitrary in the choice of this continuum mechanics damage history, we can find some guidance by considering the foregoing three general stages of ductile fracture. Therefore, this model is restricted to fracture in the fully ductile range of temperature when purely ductile fracture is caused by micromechanisms of the general type just described. In particular, there should be no cleavage.

Ignoring completely whether this way of handling the problem could produce the general trends of the usual experimental observations, and could open onto an industrially compatible numerical tool, we began a simple and rapid feasibility study. We chose the two-dimensional plane-strain case, in symmetrical conditions (pure Mode I), taking three-point bend specimens. We used a most simplified damage history model [12], added to classical incremental elastic-plastic finite-element computations, made with the TITUS program, in the "small geometry changes" approximation, using the initial stress and tangent stiffness method, with the Von Mises yield criterion, Hill's Maximum Work Principle, and the perfectly plastic case. This ductile fracture methodology is thus "noninteraction" in the sense that the void growth does not alter the element stiffness.

Present Ductile Fracture Model

Considering a characteristic volume of material, we assume that very early in its stress and strain history, included particles either break or separate from the matrix. We thus neglect for this material the nucleation stage. Then comes a void growth stage, and we use the void growth rate formula obtained by Rice and Tracey [6] for a single spherical void in an infinite rigid-perfectly plastic material

$$d\left(\log \frac{R}{Ro}\right) = 0.28 \text{ (sign } \sigma_m^\infty) \, d\epsilon_{eq}^\infty \exp\left(1.5 \frac{|\sigma_m^\infty|}{\sigma_Y}\right) \quad (1)$$

where

Ro = initial void radius,
R = present mean radius,

$$\sigma_m^\infty = \frac{\sigma_{xx}^\infty + \sigma_{yy}^\infty + \sigma_{zz}^\infty}{3} = \text{mean stress at infinity, and}$$

$d\epsilon_{eq}^\infty = (\frac{2}{3} de_{ij}^\infty \, de_{ij}^\infty)^{1/2}$ = Von Mises equivalent incremental strain at infinity [with $de_{ij} = d\epsilon_{ij} - (d\epsilon_{kk}/3) \delta_{ij}$].

This formula does not take into account void interaction or work-hardening.

Here the values of stress and strain at infinity must be understood as the averaged values over the characteristic volume.

The finite elements are there only to calculate an approximate solution of the partial differential equations, and then this approximate solution must be averaged over the characteristic volumes. Here we simplify by taking as the characteristic volume one finite element and we use the mean values of stresses and strains in the element.

By symmetry about the crack plane, we have only to study one half of the three-point bend specimens and, in the finite-element mesh, we put along the crack path (symmetry axis), which is known a priori, a layer of identical quadrilateral elements which will be the successive characteristic material volumes at the crack tip during the crack growth.

Here we chose to represent the characteristic volume of material in front of the crack tip by a square element of side $\Delta a = 0.2$ mm. We also tried once $\Delta a = 0.4$ mm for comparison.

We thus follow, in each characteristic element along the crack path, the stress and strain history and the corresponding evolution of the R/Ro ratio by integrating Eq 1 step by step during the elastic-plastic incremental process.

We assume that initially there is a constant distance lo between void centers, the same in the x, y, and z directions, for instance, the x, y, and z axes being defined in Fig. 1, and we assume that these distances change with the mean strains in the elements along the crack path according to

$$\frac{li}{lo} = \exp(\epsilon_i) \quad i = xx, yy, zz$$

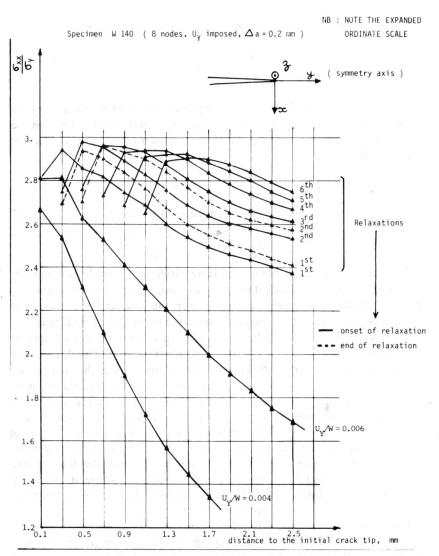

FIG. 1—σ_{xx}/σ_Y along y-axis, during crack growth, Case 1.

and we define

$$l = \min_i l_i$$

About coalescence, here we simply assume that it occurs when

$$\frac{R}{l} = \alpha_c$$

Thus the very simplified present criterion for ductile fracture of the characteristic volume at the crack tip, giving separation and thus an elemental crack growth Δa along the a priori known crack path, is (defining $\alpha_o = Ro/lo$)

$$\frac{\frac{R}{Ro}}{\frac{l}{l_o}} = \frac{\alpha_c}{\alpha_o} = \text{given material property} \qquad (2)$$

Of course, it would be very easy to incorporate in this model and numerical process any void nucleation criterion operating on continuum mechanics stresses and strains averaged over the characteristic volume of material at the crack tip, if such a criterion is known with enough reliability for the considered material (for instance, void nucleation at a manganese sulfide inclusion by decohesion of the inclusion/matrix interface). Likewise, void interaction and work-hardening could be included in the void growth rate [5,14,15], and a more realistic coalescence criterion should be introduced to characterize the localized internal necking which takes place between some voids and the crack tip and which leads to separation and incremental crack growth (for instance, void coalescence by void sheet formation, where a second population of smaller voids nucleates at carbides within a band of intense shear between two inclusion nucleated larger voids, giving a duplex distribution of dimple sizes on the fracture surfaces). In the same way, an interaction between void growth and the constitutive law of the material at the crack tip should be introduced.

For us, initiation is thus defined as the moment when the criterion, Eq 2, is met in the first element at the crack tip. We then simulate separation at the crack tip by relaxing step by step the corresponding nodal force from f_{max} to 0 for four-node quadrilateral elements. The normal displacement of this node then grows from 0 to v_{max} (Fig. 2). Note that one must take care to do this relaxation with small enough steps, especially at the end of the process.

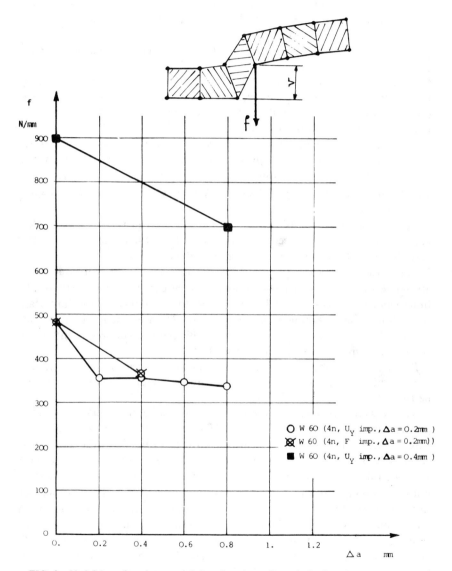

FIG. 2—*Nodal force f at the onset of the relaxations, Cases 3-5, plotted versus corresponding crack growth.*

During this relaxation process, the R/Ro ratio in the following element increases and l/lo decreases, so that we have to check whether the criterion has been reached at the end of the relaxation process in this new crack tip element: if it has been, we proceed to relax the nodal force of the new crack tip, and if not, we proceed to load further the specimen until the criterion is reached again, and so on.

For eight-node quadrilateral elements, the principle is the same except that we have two nodes to relax when the criterion is reached in the crack tip element: the corner node at the tip and the mid-side node (Fig. 3). Here we relax step by step the forces at these two nodes simultaneously and proportionally. We also made one try relaxing them one after the other, which is somewhat daring, to compare.

We studied the three-point bend specimens in the displacement-controlled case, imposing upon them a growing displacement Uy of the roller and keeping this displacement constant during the relaxations. We also made one try with the load-controlled case (imposing a growing load F and keeping it constant during the relaxations).

The foregoing oversimplified model can be used in the present Mode I case because we do not need any directionality in the criteria since in this case the crack is known a priori to grow in its own plane.

The present approach to the problems presented in the introduction is attractive because it can be incorporated very easily in any elastic-plastic finite-element program, and because it could be applied to the important three-dimensional, symmetrical (pure Mode I) cases—semi-elliptical surface cracks, or through-cracks in "small" thicknesses—to predict initiation, stable crack growth, and maximum load.

By an adequate adaptation of the mesh and of the criteria (directionality), the angled crack extension problem in complex loading could, it is hoped, be treated.

The calculations of the feasibility study are made in the elastic-perfectly plastic case (yield stress $\sigma_Y = 520$ MPa, Von Mises yield criterion), on three-point bend specimens of various widths W, but with the same a/W ratio (0.475); they are in-plane homothetic. The $(F/BW\sigma_Y, Uy/W)$ normalized load-displacement curves are the same for these various specimens until initiation, but the initiation point and crack growth behavior are different when the size of the specimen is changed (size effect) (Fig. 4). A large enough specimen will be in small-scale yielding conditions when initiation occurs, whereas a small enough specimen will be in generalized yielding conditions. This, of course, comes from the characteristic volume of material over which the ductile fracture micromechanisms extend. Since the damage results from the history of stresses and strains averaged over this characteristic volume (independent of specimen size), and since the stresses and strains are identical within the homothesis acting upon the in-plane coordinates, it is obvious that at homologous loads the damage

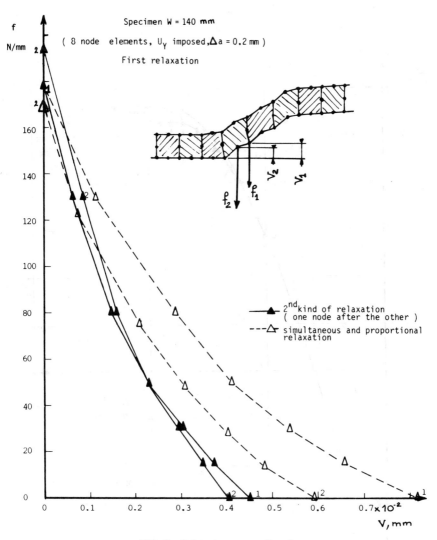

FIG. 3—*Relaxation curves, Case 1.*

will be lower in the smaller specimen. so that its initiation point will be later.

Results

The first calculations of the feasibility study showed that this model, though very simplified, reproduced the general trends of the usual experimental observations and gave no results in contradiction to them. More-

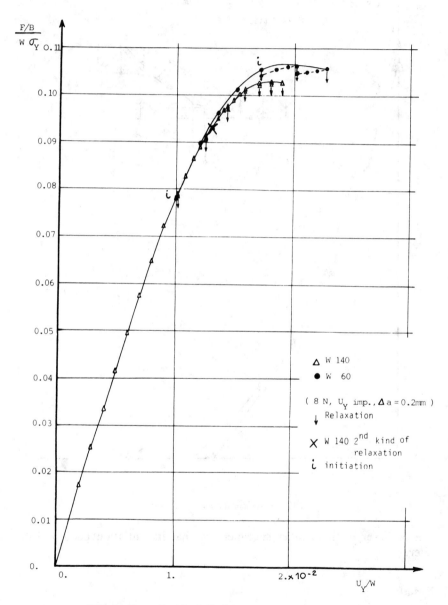

FIG. 4—*Normalized load-displacement curves, Cases 1 and 2.*

over, an industrial tool was built. As these results had only a general trend value because the whole numerical treatment was not refined enough, and as they were very encouraging, we made the present new calculations, reported here, with the necessary refinements to study the influence of the various parameters playing a role in this methodology: mesh, size of elements, type of elements, convergence precision, steps in the incremental process (especially in the node relaxations), type of relaxation, and precision in the criterion fulfillment. We get smooth solutions, with no oscillations, the principal results being given in the following, and showing in particular the behavior, which results from the present criterion, of some parameters used in the literature [7,16-20] as initiation and crack growth criteria. The results reported here deal with the following cases:

1. $W = 140$ mm, 8-node elements, $\Delta a = 0.2$ mm, displacement-controlled process, nodal forces relaxed simultaneously and proportionally. In order to test the influence of the relaxation technique for the 8-node elements, however, we made one try for this specimen where we relaxed the two nodes one after the other. Starting again from the solution obtained at the initiation point, we relaxed the two nodes one after the other, and then we loaded again the specimen until the criterion was met in the new crack tip element. This calculation is labelled "2nd kind of relaxation" in the figures, which show that, as expected, the effect is much smaller on global quantities (Figs. 4, 5, 6) than on local quantities (Figs. 7, 8, 3, and 9).

2. $W = 60$ mm, 8-node elements, $\Delta a = 0.2$ mm, displacement-controlled process, nodal forces relaxed simultaneously and proportionally.

3. $W = 60$ mm, 4-node elements, $\Delta a = 0.2$ mm, displacement-controlled process.

4. $W = 60$ mm, 4-node elements, $\Delta a = 0.2$ mm, load-controlled process.

5. $W = 60$ mm, 4-node elements, $\Delta a = 0.4$ mm, displacement-controlled process.

In the first four cases, where $\Delta a = 0.2$ mm, we took the same critical value in the criterion (1.286). In the fifth case, where $\Delta a = 0.4$ mm, we took a lower critical value (1.184) chosen to give initiation at approximately the same value of J-integral.

The normalized load-displacement curves are shown for Cases 1 and 2 on Fig. 4; as expected, they are found identical before crack growth initiation for the large and for the small specimen, and initiation (i) occurs "later" for the small one. The arrows point out the relaxations (made here with U_y kept constant). Since we were interested here in initiation, stable crack growth, and maximum load, we stopped the calculations when the load began to decrease.

The details of initiation, stable crack growth, and maximum load are shown in the same way for Cases 3-5 in Fig. 10. In the load-controlled case (No. 4), we have at initiation two successive relaxations, and it can be

240 ELASTIC-PLASTIC FRACTURE

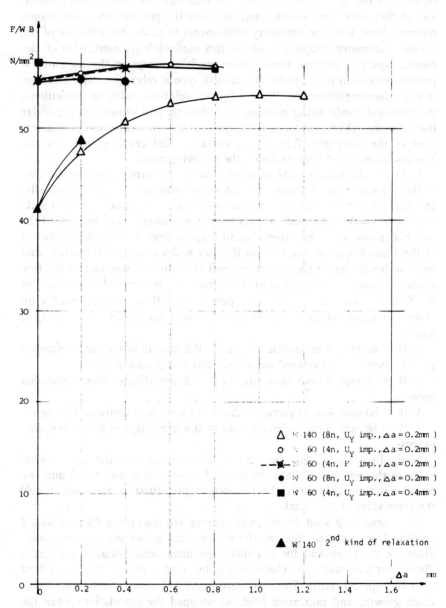

FIG. 5—*Load at the onset of the relaxations, plotted versus corresponding crack growth.*

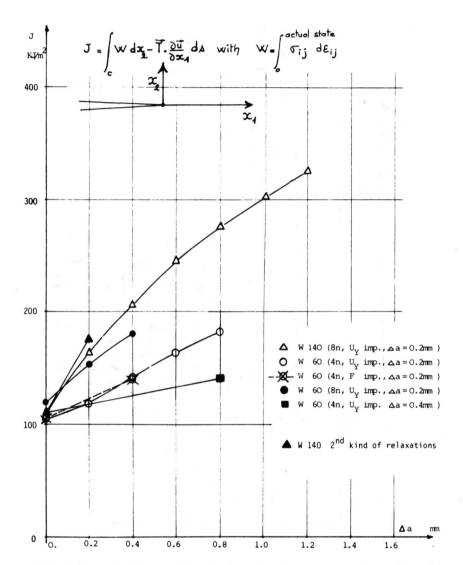

FIG. 6—*J-integral at the onset of the relaxations, plotted versus corresponding crack growth.*

242 ELASTIC-PLASTIC FRACTURE

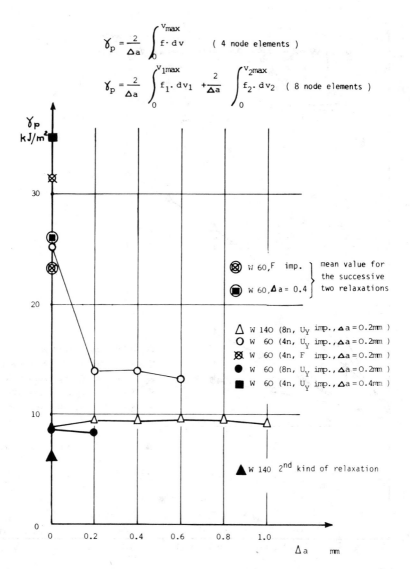

FIG. 7—γ_p for the relaxations, plotted versus corresponding crack growth (γ_p = relaxation work per unit crack extension area).

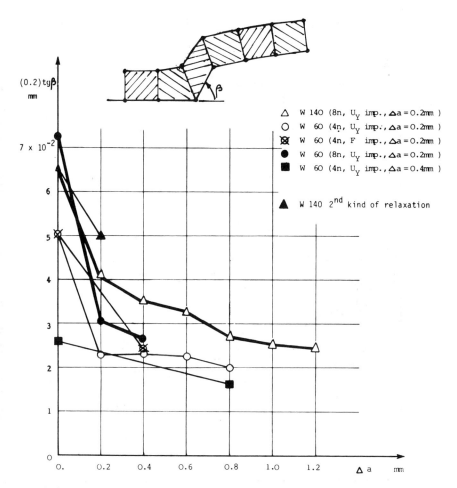

FIG. 8—*Conventional crack opening angle β, given by (0.2) tgβ, mm, at the onset of the relaxations, plotted versus corresponding crack growth.*

seen that, though they correspond to two different loading histories, the load-controlled case (No. 4) and the displacement-controlled case (No. 3) give quite close results. In Case 5, where $\Delta a = 0.4$ mm, we have also two successive relaxations at initiation, giving a crack growth of 0.8 mm, which is larger than the crack growth to maximum load obtained with $\Delta a = 0.2$ mm. This can be paralleled with the fact that the load obtained at the next relaxation of Case 5 is lower than the initiation one.

In Fig. 5, the load F/BW at the onset of each relaxation is plotted versus the corresponding crack growth Δa; the upper group of points represents the corresponding behavior from initiation to maximum load

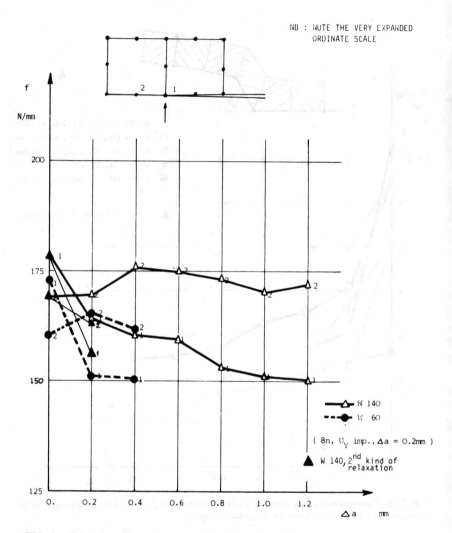

FIG. 9—*Nodal forces* f *at the onset of the relaxations. Cases 1, 2, plotted versus corresponding crack growth.*

for the small specimen. The lower curve represents this behavior for the large specimen.

Figure 7 gives the relaxation work per unit crack extension area γ_p for the relaxations, plotted versus corresponding crack growth. γ_p is defined on the figure for 4-node elements and 8-node elements. We note that, in the 8-node element cases (1 and 2), γ_p is found almost constant during crack growth, initiation included, and almost the same for the large and for the small specimen. This result is interesting since the criterion used here to recognize whether a crack tip element has reached a critical state with

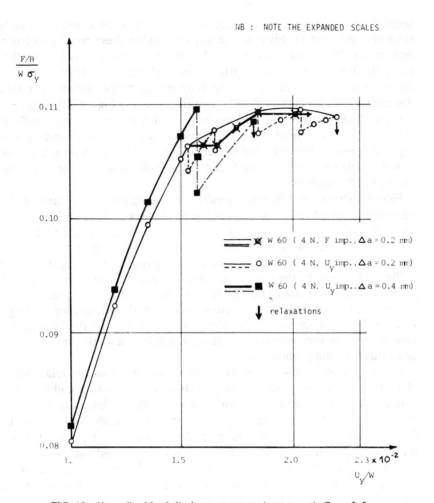

FIG. 10—*Normalized load-displacement curves* (upper part), *Cases 3-5.*

respect to ductile fracture is a criterion which integrates the whole growth process all along the elastic-plastic incremental stress and strain history, and thus is not directly linked to this energetic criterion. When we had two successive relaxations (Cases 4 and 5), we plotted the γ_p value corresponding to the first relaxation and also the mean value on the two successive relaxations.

We note, among all parameters which play a role in the approximate numerical solution of this elastic-plastic incremental problem combining specimen loading and node relaxation, differences in behavior of the 4-node and 8-node elements around the crack tip, where the mesh is the same. These differences appear on figures representing local quantities and local effects at the crack tip, for instance, Figs. 7 and 8. Recall that there is one nodal force to relax for 4-node elements and two nodal forces for 8-node elements, and that in this case the way of relaxing these two forces has a local effect. Note that there are in the present solution, which is mathematically singular, high plastic strains and high strain gradients, and that the specimens are loaded up to the fully plastic range. Effects of mesh, size and type of elements on global quantities appear in Figs. 4, 10, 5, and 6.

Figure 8 shows the crack opening angle, which is conventionally defined in the figure, at the onset of the relaxations, plotted versus the corresponding crack growth. We note that it decreases after initiation with a tendency to stabilization with crack growth and comparable trends for the large and for the small specimen.

Figure 6 shows the J-integral, recalled on the figure, at the onset of the relaxation, plotted versus corresponding crack growth. J was computed on paths distant from the crack tip, where it was found almost path-independent (within a few percent). We note an important effect of the mesh, size, and type of elements (see second-last paragraph in the foregoing).

Figure 3 compares the relaxation curves obtained at initiation in Case 1 for the two kinds of relaxation. Besides, we noticed that, for Cases 1 and 2 (8-node elements), the relaxation curves obtained at initiation for the large and for the small specimen were almost identical, though the large specimen is then in very contained yielding conditions and the small one in generalized yielding conditions.

Figures 9 and 2 show the nodal forces at the onset of the relaxations, plotted versus corresponding crack growth, in Cases 1 and 2 and 3-5. We note a tendency to stabilization with crack growth and comparable trends for the large and for the small specimen (Fig. 9). Besides, we noticed in Cases 1 and 2 that the nodal normal displacements v at the end of the relaxations were almost constant during crack growth and almost the same for the large and for the small specimen. There is thus, for the relaxation curves, a tendency to stabilization with crack growth and comparable trends for the large and for the small specimen.

Figure 1 shows the opening normal stress (over yield stress) ahead of the crack tip during crack growth, in Case 1. We note that, in this specimen, the stable crack growth takes place under a high triaxiality stress field.

Further Developments

We think that the results of this feasibility study are quite encouraging. Moreover, we are now equipped with a completely automatic tool whose parameter dependence and sensibility have been studied. Thus, we proceeded recently to the most important stage—the comparison between calculation predictions and experiments. Tests are being made on compact tension specimens, on round bars with an external circular notch in tension, and on single-edge notched specimens in tension and in bending, with different values of W, of a/W, of the notch angle, and of the notch root radius, in order to have, for the ductile fracture processes, different conditions of stress triaxiality in the characteristic volume at the notch root. The material is A 508 Cl 3 steel in fully ductile conditions. Adjusting the various parameters once and for all, we shall see to what extent this model and methodology can reproduce the experimental results for load, displacement, crack opening, and crack growth, for these very different conditions of ductile fracture, especially for stress triaxiality.

Acknowledgments

We wish to acknowledge the financial support of the French "Service Central de Sûreté des Installations Nucléaires," and J. Devaux, G. Mottet, and C. Vouillon for their precious help with the calculations.

References

[1] Bluhm, J. I. and Morrissey, R. J., *Transactions*, 1st International Conference on Fracture, Sendai, Japan, Vol. 3, 1966, pp. 1739-1780.
[2] Hodgson, D. E., "An Experimental Investigation of Deformation and Fracture Mechanisms in Spheroidized Carbon Steels," Ph.D. Thesis, Stanford University, Stanford, Calif., 1972.
[3] Argon, A. S., Im, J., and Safoglu, R., *Metallurgical Transactions*, Series A, Vol. 6A, April 1975, pp. 825-837.
[4] Tanaka, K., Mori, T., and Nakamura, T., *Transactions*, Iron and Steel Institute of Japan, Vol. 11, 1971, pp. 383-389.
[5] McClintock, F. A., *Transactions*, American Society of Mechanical Engineers, *Journal of Applied Mechanics*, June 1968, pp. 363-371.
[6] Rice, J. R. and Tracey, D. M., *Journal of the Mechanics and Physics of Solids*, Vol. 17, 1969, pp. 201-217.
[7] McClintock, F. A., *Transactions*, American Society of Mechanical Engineers, *Journal of Applied Mechanics*, Dec. 1958, pp. 582-588.
[8] Rice, J. R. in *Fracture*, H. A. Liebowitz, Ed., Vol. 2, Academic Press, New York, 1968.
[9] Rice, J. R. and Johnson, M. A. in *Inelastic Behavior of Solids*, M. F. Kanninen, W. F. Adler, A. R. Rosenfield, and R. I. Jaffee, Eds., McGraw-Hill, New York, 1970, pp. 641-672.
[10] McMeeking, R. M., "Finite Deformation Analysis of Crack Tip Opening in Elastic-Plastic Materials and Implications for Fracture Initiation," Technical Report C00-3084/44, Division of Engineering, Brown University, Providence, R. I., May 1976.
[11] Ritchie, R. O., Knott, J. F., and Rice, J. R., *Journal of the Mechanics and Physics of Solids*, Vol. 21, 1973, pp. 395-410.
[12] d'Escatha, Y. and Devaux, J. C., *Transactions*, 4th Structural Mechanics in Reactor Technology Conference, San Francisco, Calif., Vol. G, Paper G2/4, Aug. 1977.
[13] Mackenzie, A. C., Hancock, J. W., and Brown, D. K., *Engineering Fracture Mechanics*, Vol. 9, 1977, pp. 167-188.
[14] Tracey, D. M., *Engineering Fracture Mechanics*, Vol. 3, 1971, pp. 301-315.
[15] Needleman, A., *Transactions*, American Society of Mechanical Engineers, *Journal of Applied Mechanics*, Dec. 1972, pp. 964-970.
[16] Andersson, H., *Journal of the Mechanics and Physics of Solids*, Vol. 21, 1973, pp. 337-356.

[17] Kobayashi, A. S., Chiu, S. T., and Beeuwkes, R., *Engineering Fracture Mechanics*, Vol. 5, 1973, pp. 293–305.
[18] Kfouri, A. and Miller, K. J., "Separation Energy Rates in Elastic-Plastic Fracture Mechanics," CUED/C. MAT/TR 18, Department of Engineering, University of Cambridge, Cambridge, U.K., Dec. 1974.
[19] Rousselier, G., "Croissance Subcritique de Fissure et Critéres de Rupture: Une Approche Numérique," *Transactions*, 4th International Conference on Fracture, Waterloo, Canada, June 1977.
[20] Wilkinson, J. P. D., Hahn, G. T., and Smith, R. E. E., *Transactions*, 4th Structural Mechanics in Reactor Technology Conference, San Francisco, U.S.A., Vol. G, Paper G2/2, Aug. 1977.

Experimental Test Techniques and Fracture Toughness Data

P. C. Paris,[1] *H. Tada,*[1] *H. Ernst,*[1] *and A. Zahoor*[1]

Initial Experimental Investigation of Tearing Instability Theory

REFERENCE: Paris, P. C., Tada, H., Ernst, H., and Zahoor, A., "**Initial Experimental Investigation of Tearing Instability Theory,**" *Elastic-Plastic Fracture, ASTM STP 668,* J. D. Landes, J. A. Begley, and G. A. Clarke, Eds., American Society for Testing and Materials, 1979, pp. 251-265.

ABSTRACT: An initial experimental investigation was conducted to confirm the theory of tearing instability developed in previous work. A simple testing program was selected which employed 3-point bend specimens with various crack size to specimen depth ratios and an additional spring bar to easily adjust effective specimen span (or loading system compliance).
All stable-unstable behaviors observed in the tests are in good agreement with those predicted by the theory. Thus the present study demonstrates the appropriateness of the tearing instability analysis, presenting guidelines for its further development.

KEY WORDS: tearing instability, crack stability, experimental fracture mechanics, J-integral, 3-point bend tests, crack propagation

In previous work the authors have developed an analysis of tearing instability[2] for application to cracking instability predictions based on J-integral R-curve representation of material characteristics. In this previous work, emphasis was placed on applications to fully plastic cracked ligament and plane-strain conditions. However, it was noted that the theory is quite general and is applicable from small-scale yielding through the fully plastic range for plane-strain through plane-stress conditions with work hardening, etc. For a fuller understanding, one is referred to the previous report.[2]

Since application of this new cracking instability theory seems relevant for certain fully plastic plane-strain situations of practical interest (such as reactor vessels at operating temperatures with low upper-shelf Charpy materials), it is of immediate and special importance to experimentally verify the analysis for those situations where no other instability analysis has as yet

---

[1] Professor of mechanics, senior research associate, and graduate research assistants, respectively, Washington University, St. Louis.
[2] Paris, P., Tada, H., Zahoor, A., and Ernst, H., this publication, pp. 5-36.

been formulated. For that reason, a simple testing program was developed which could be performed quickly using a material which was pedigreed and would exhibit fully plastic plane-strain behavior (in the J-integral tearing sense) in reasonably sized specimens.

Therefore, the objective herein is to present results of a testing program which clearly demonstrates the appropriateness of the tearing instability analysis and which illustrates its broad potential for future application, as well as presenting guidelines for its further development.

Testing Program

The material selected was nickel-chromium-molybdenum-vanadium (NiCrMoV) rotor steel supplied by Westinghouse Research. This material was previously subjected to extensive testing by Westinghouse; for example, as reported by Logsdon.[3] The material has flow properties and J-integral R-curve properties (J_{Ic} and dJ/da for tearing) for temperatures well above the transition temperature, which make it quite convenient for crack instability tests. It is very convenient to be able to select reasonable test specimen proportions requiring moderate test loads and the usual instrumentation while being able to change from stable to unstable results from test to test due to simple modifications to the test variables.

The test specimen configuration was selected to be a 3-point bend specimen of a full span, L, of 8 in. with a specimen depth, W, of 1 in. and thickness, B, of ½ in. Specimens were notched and fatigue precracked to various crack size to specimen depth ratios, a/W. A schematic diagram of the test configuration is shown in Fig. 1.

In the tests, stability was affected mainly by varying the a/W of the test specimen or the effective (or equivalent) elastic span of the test specimen, or both. The method of adjusting the effective elastic span was by inserting in the test arrangement an elastic spring bar of adjustable span for a variable spring constant. Analysis details for this arrangement will be presented subsequently.

The testing arrangement as shown in Fig. 1 also permitted measuring load, from a load cell, and displacement, from the ram displacement, in a standard MTS servo-hydraulic testing machine to produce a load displacement record for analysis of the stability of the situation. An important feature of the arrangement was the ability to remove various components individually (spring bar, test specimen, and appropriate rollers) in order to make direct elastic compliance calibrations of the various components of the test arrangement, including the test machine itself. These compliance

---

[3] Logsdon, W. A. in *Mechanics of Crack Growth, ASTM STP 590*, American Society for Testing and Materials, 1976, pp. 43–61.

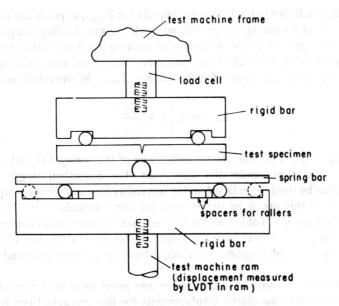

FIG. 1—*Testing arrangement.*

calibrations will be seen to be an important feature of the analysis of the test results.

Tearing Instability Analysis of the Fully Plastic 3-Point Bend Specimen Test

The formula for instability of a 3-point bend specimen where the remaining uncracked ligament, b (or $W - a$), is fully plastic was given in the earlier paper (footnote 2). It is

$$T_{\text{mat}} = \frac{dJ}{da} \times \frac{E}{\sigma_0^2} \leq \frac{2b^2 L}{W^3} - \frac{\theta_c E}{\sigma_0} = T_{\text{applied}} \qquad (1)$$

That is to say, if the right-hand side, T_{applied}, exceeds the left hand side, T_{mat}, instability will occur when the uncracked ligament becomes fully plastic *and* J_{Ic} is exceeded so that tearing begins. This formula assumes a rigid testing machine (fixed displacements) and rigid test fixtures. On the other hand, the driving energy or force for instability comes from the elastic unloading of the bend specimen as the limit load diminishes due to crack extension. Thus T_{applied} contains these influences through specimen proportions L, b, and W, but in addition it is affected by the bend angle, θ_c, of the uncracked ligament section of the test specimen. This is explained further in the earlier paper (footnote 2) (and its appendices).

Now L as a factor in the foregoing formula for T_{applied} appears due to elastic compliance of the test specimen with no crack present. Adding compliance to the test arrangement by using in addition a spring bar is equivalent to adding compliance to the uncracked test specimen. Therefore, analysis leads to an equivalent (increased) length for the test bar, L_{equiv}, by the relationship

$$L_{\text{equiv}} = L\left[1 + \frac{\delta_{SB}}{\delta_{TB}}\right] \quad (2)$$

where δ_{SB} and δ_{TB} are the elastic deflections of the spring bar and test bar (with no crack) are under the same load. The equivalent length, L_{equiv}, should then be used to replace L in the instability formula. From compliance calibrations, this could be determined for any particular test. The compliances are given in Table 1 and it is noted that the compliance of the testing machine and fixtures will have a negligible effect on results (when compared with the much smaller spring constants of both the test specimen and spring bars).

In addition, θ_c could be analyzed for any point on a load displacement record by subtracting elastic displacements for the uncracked test bar and spring bar from compliance calibration information and thus obtaining the displacement due to the crack alone, δ_{crack}. This includes both the elastic and plastic deformations of the uncracked ligament sections. Then θ_c is obtained directly from geometry

$$\theta_c = \frac{4\delta_c}{L} \quad (3)$$

Thus all factors in T_{applied} can be obtained directly from specimen dimensions, compliance calibrations, and a given point on the load displacement record.

TABLE 1—*Instability test component stiffnesses.*

Component	Size, in.	Spring Constant, lb/in.
A. test machine and fixtures		320 000
B. test machine fixtures and spring bar	$L_{SB} = 8$	30 047
	$L_{SB} = 11$	13 035
	$L_{SB} = 12$	10 000
	$L_{SB} = 14$	6 472
	$L_{SB} = 15\frac{3}{4}$	4 638
C. test bars ($L = 8$ in.)	no crack	100 000
	$a/W = 0.4$	90 000
	$a/W = 0.5$	50 397
	$a/W = 0.6$	36 570
	$a/W = 0.7$	19 217

On the other hand, the so-called "tearing modulus," T_{mat}, can be evaluated in a quite independent manner. Indeed, T_{mat} depends only on material properties and thus should be the same throughout the testing program (except for a small effect of temperature). It depends only on the tearing slope, dJ/da, of the J-integral R-curve (a plot of J versus crack length change, Δa), the elastic modulus, E, and the flow stress, σ_0, of the material. (For purposes herein, the flow stress was taken as the average of the yield and ultimate strengths of the material.) Also, though T_{mat} therefore could be measured separately, it could also be evaluated from each test in the following manner.

During any test, J can be evaluated from the usual Rice et al pure bending analysis.[4] It is

$$J = \frac{2}{bB} \times \text{area} \quad (4)$$

where the "area" is the area under the load versus displacement (due to the crack) record up to any point at which J is desired. Changes in J, that is, in ΔJ, can be computed[5] from differences in results or

$$\Delta J = \frac{2}{bB} \times \Delta(\text{area}) \quad (5)$$

On the other hand, changes in crack length are less easy to evaluate directly. However, if test proportions are selected so that limit load is reached prior to beginning of crack extension (that is, the J_{Ic} point), then crack extension can be evaluated from the reduction in limit load due to crack growth. The analysis is as follows. The limit load is

$$P_L = \frac{4M_L}{L} = \frac{4}{L} A b^2 B \sigma_0 \quad (6)$$

where A is a constant. Differentiating for crack extensions ($da = -db$) gives

$$dP_L = \frac{4}{L} A B \sigma_0 (-2b \, da) \quad (7)$$

Dividing Eq 7 by Eq 6 and rearranging leads to

$$\Delta a = -\frac{b}{2} \frac{\Delta P_L}{P_L} \quad (8)$$

---

[4] Rice, J. R., Paris, P. C. and Merkle, J. G. in *Progress in Flaw Growth and Fracture Toughness Testing, ASTM STP 536,* American Society for Testing and Materials, 1973, pp. 231-245.

[5] The method here neglects correction terms, consistent with methods in the literature (see Paris et al (footnote 2), Appendix I).

Thus, upon reaching limit load, Δa can be evaluated from one point to the next simply from the ligament size, b, the change in limit load, ΔP_L, and the limit load itself, P_L.

Hence, T_{mat} can be determined, by the methods described in the foregoing, as

$$T_{mat} = \frac{\Delta J}{\Delta a} \frac{E}{\sigma_0^2} \tag{9}$$

This method was used in the present testing program, that is, based on Eqs 5, 8, 9, and their implied assumptions. Now in Eqs 5 and 8 the remaining ligament, b, appears in such a manner that in T_{mat}, Eq 9, upon substitution of Eqs 5 and 8, it is squared. In all computations, the original ligament size was used, which implies that the analysis is limited to small crack extensions compared with the ligament size.

Moreover, the J-integral method itself becomes suspect if substantial crack extension occurs, so that restriction to small crack extension was already implied.

Results of the Testing Program

The testing program is enumerated on Table 2. Test specimens were machined from a single piece of material which was originally cut from a rotor forging as a blank for a 4-T compact specimen of ASTM-A471 steel (that is, NiCrMoV from Westinghouse Research as mentioned earlier). The cracking plane and direction were as intended in the 4-T blank so that results can be compared [see Paris et al (footnote 2) and Logsdon (footnote 3)].

Specimens are numbered according to a positive location system. First, Slabs A, B, and C were removed from the blanks, starting with the side opposite to the notch location in the 4-T blank. They were then sliced into Specimens 1 through 6 for each slab, dividing the thickness dimension of the 4-T blank. Therefore, each test specimen is numbered with a letter and number identifying its location in the 4-T blank. Although care was taken in identifying with location, no effect of location was necessarily anticipated, nor was any found in test results to be described further here.

The test bars were machined with notches 0.2 in. deep and fatigue precracked to give crack depths, a, from 0.313 to 0.713 in. (the same as a/W values since $W = 1$ in.). Precracking loads were applied in the same testing fixtures but with the spring bar removed to give tension-tension loading of the notched side with maximum fatigue loads less than half the maximum (limit) load in each stability test.

Test temperatures were chosen to be 130 and 230°C, so that each is at least 100°C above the transition temperature of the material to avoid cleavage and

TABLE 2—Tests and results.

Test No.	a or a/W	Test Temperature °C	$L_{\text{spring bar}}$, in.	L_{equiv}, in.	T_{mat}	T_{applied}	Physical Stability	Apparent J_{Ic}, in.-lb/in.^{2}
A-1	0.513	130	8.0	12.9	36.2	4.6	stable	700.0
A-2	0.516	130	8.0	12.9	40.8	5.8	stable	840.0
A-3	0.313	130	8.0	(did not reach P_{limit})			stable	...
A-4	0.411	130	8.0	(did not reach P_{limit})			stable	...
A-5	...	...	...	(broken in precracking)			...	...
A-6	0.411	130	12.0	34.0	34.5	41.5	unstable	720.0
B-1	0.427	230	12.0	34.0	35.5	51.5	unstable	600.0
B-2	0.518	130	11.0	27.1	33.0	18.7	stable	800.0
B-3	0.510	130	14.0	51.6	32.0	45.4	unstable	750.0
B-4	0.513	130	12.0	34.0	33.7	29.4	stable	725.0
B-5	0.510	130	13.0	42.1	34.9	33.2	unstable	850.0
B-6	0.510	130	0	8.0	41.0	−2.2	stable	650.0
C-1	0.502	230	13.0	42.1	36.5	35.6	(marginal)	720.0
C-2	0.516	230	14.0	51.6	38.5	45.2	unstable	750.0
C-3	0.505	230	12.0	34.0	34.6	26.6	stable	750.0
C-4	0.713	130	15.75	76.0	34.1	19.7	stable	600.0
C-5	0.620	130	15.75	76.0	40.8	40.1	unstable	600.0
C-6	0.706	130	13.0	42.1	35.5	7.5	stable	840.0

T_{mat} (avg) = 36.1

Other dimensions of the 3-point bend specimens:

depth, $W = 1$ in.
thickness, $B = \frac{1}{2}$ in.
span, $L = 8$ in.

thereby examine upper-shelf tearing instability behavior as intended here. Occasionally at 130 °C some minor pop-in was noted, which was entirely absent at 230 °C. However, tearing instability results of these tests show no substantial difference for the two temperatures, as may be noted later.

Table 2 gives the length of spring used in each test and the equivalent span of the test specimen, as determined from Eq 2 and compliance calibration information in Table 1.

The equivalent span and other test information from load displacement records were then used to determine independently T_{mat} and $T_{applied}$ as described in the preceding section. Thus the columns for T_{mat} and $T_{applied}$ in Table 2 may be compared to determine according to the theory if unstable behavior is expected, that is, if $T_{mat} \leq T_{applied}$. These results can be compared with the column of Table 2 marked "stability," where the actual physical state of stability observed in each test is listed. Such a comparison is shown in Fig. 2. For each test a point is plotted from its T_{mat} versus $T_{applied}$ values. Lying on one side of the 45-deg theory line is a theoretical prediction of stable or unstable behavior as noted. On the other hand, actual physical behavior is noted for each point as solid points, stable and open points,

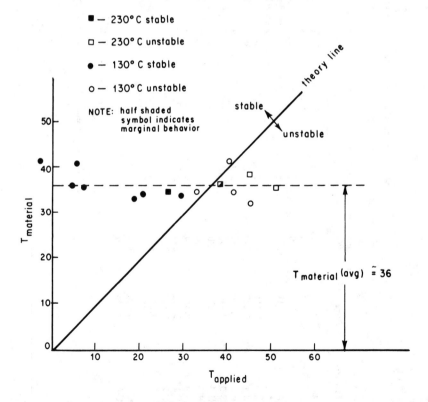

FIG. 2—*Stability test results.*

unstable (and one-half shaded points as marginal). It is noted that the agreement between physical behavior and theory is very good! Only two points exhibiting adjudged unstable physical behavior lie on the stable side, but very close to the theory line. Moreover, the data in Fig. 2 show that T_{mat} is reasonably constant for a wide variety of variables (temperature and also a/W, L_{equiv}, etc. as the effective $T_{applied}$). Thus this is felt to be a strong verification of the theory for a single type of specimen—the bend test specimen.

Further Discussion of Load Displacement Records and Physical Instability

Physical instability behavior can be clarified and better understood by a more specific discussion of load-displacement records from the testing program. Furthermore, it is of special interest to make these observations for a sequence of tests, where temperature and test specimen dimensions including a/W are held constant and only the elastic compliance of the testing system is varied by changing only the spring bar length. Referring to Table 2 for a set of tests, where the temperature is 130°C and a/W is about 0.5 (that is, 0.510 to 0.518), it is seen that Tests A-2 ($L_{SB} = 8.0$), B-2 ($L_{SB} = 11.0$), B-4 ($L_{SB} = 12.0$), B-5 ($L_{SB} = 13.0$), and B-3 ($L_{SB} = 14.0$) comprise such a set of tests. Figs. 3-7 show load displacement records from these particular tests in the same respective order.

Now in each of these tests, tearing starts at the beginning of the descending load portion where the record departs from the maximum load level in the test. The beginning of tearing is denoted also as the J_{Ic} point, and values for comparison are listed in Table 2 (it is noted that they are reasonably constant). Tearing stability or instability then depends on the character of the descending part of the load displacement records beyond this point of beginning of tearing.

As noted from Figs. 3-7, as elastic compliance, that is, spring bar length, is added, the descending portion of the records becomes steeper. Since the testing system is in displacement control as measured by a linear variable differential transformer (LVDT) in the hydraulic ram, when sufficient elastic compliance is added to cause the descending load displacement record to be vertical, unstable behavior ensues. Noting the sequence of behaviors with added compliance in Figs. 3-7, it is judged that Figs. 6 and 7 show substantially unstable behavior and that Fig. 5 shows substantially stable behavior (though a short segment of vertical record occurs just after maximum load).

Moreover, when observing the test, unstable behavior as just described is occasioned by a sudden increase in crack size and deformation of the test specimen, whereas with stable behavior no sudden deformations occur. It is by these observations and analysis of test records, such as Figs. 3-7, that the "physical instability" was judged and recorded in the column so entitled in Table 2.

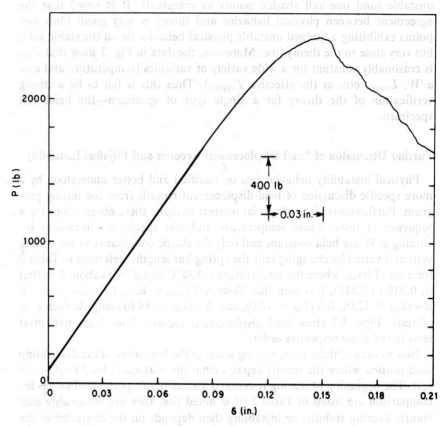

FIG. 3—*Load-displacement record for Test A-2, with* $L_{SB} = (8 \text{ in.})$, *resulting in stable behavior. (Displacement zero-point displaced slightly here and on Figs. 4–7.)*

It is relevant to observe here and to emphasize that as described in the foregoing and in the more general analysis (footnote 2), tearing instability is a system behavior involving not only the local fracture characteristics of a test specimen (or structural component) but also the elastic compliance of the overall test specimen and loading system. Perhaps cleavage fracture instability is a local material instability phenomenon, but it is clear here that tearing instability is a general system type of instability. Moreover, this testing program, where, in Table 2 and Fig. 2, stability is judged by theory, comparing T_{mat} and $T_{applied}$, and judged separately by physical instability behavior, clearly verifies the general approach taken by the theory of tearing instability (footnote 2) and especially that tearing instability is not treatable as a local phenomenon (in the K_{Ic} local critical field intensity sense).

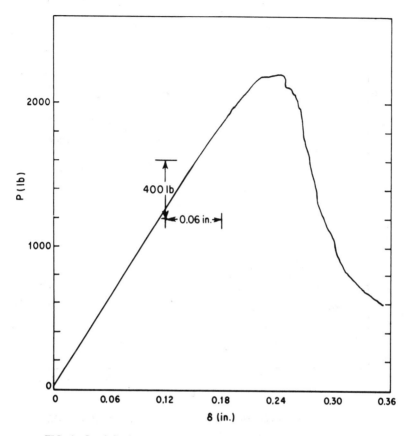

FIG. 4—*Load-displacement record for Test B-2, $L_{SB} = 11$ in., stable.*

Some Additional Comments on Testing Results

In the preceding discussion, verification of the concepts of the tearing instability theory was emphasized as a system phenomenon, where transition from stable to unstable behavior is caused by varying the system compliance, that is, the spring bar length. In addition, other transitions caused by varying local test specimen dimensions are observed from the data in the test program.

For example, with other conditions identical, a switch from unstable behavior is observed in Tests C-6 and B-5 due to a change in a/W from 0.706 to 0.510, respectively; see Table 2. Since $b = W - a$, the implied effect is noted in Eq 1 as an increase in b, which increases $T_{applied}$, causing the transition to instability (even though total compliance increases).

Two other pairs of specimens, B-4 and A-6, and also C-3 and B-1, each exhibits a switch from stable to unstable behavior due to a change in a/W from

262 ELASTIC-PLASTIC FRACTURE

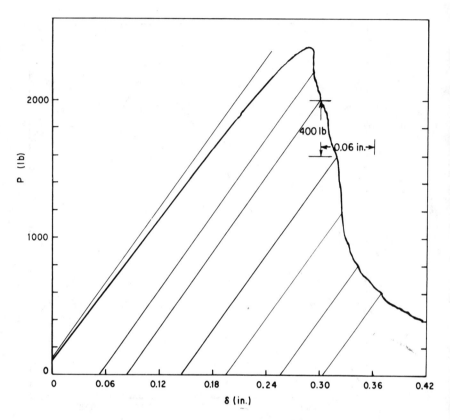

FIG. 5—*Load-displacement record for Test B-4. $L_{SB} = 12$ in., stable.*

about 0.5 to 0.4. Both of these pairs were tested with spring bar lengths of 12 in., but one pair was at 130°C and the other at 230°C. Thus it was shown that the switching of behavior from stable to unstable was not appreciably affected by this temperature change, even though switching was induced by a relatively small change in a/W. Again, these results were expected from the theory, Eq 1, and the fact that temperature changes on the upper shelf only weakly affect J-integral R-curve tearing behavior (footnote 3) (specifically, dJ/da), flow stress, σ_0, and modulus, E. Therefore the results in Table 2 verify the theory even more strongly than the conclusions drawn in the previous section herein.

Additional Experimentation which Seems Relevant for Future Work

Although verification of the theory seems strong here, it would be interesting to go beyond the limited scope of this initial testing program to explore effects of additional variables such as the following.

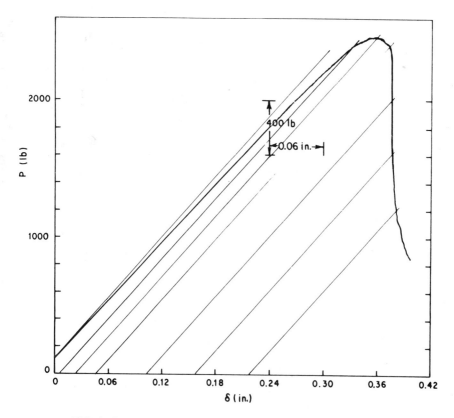

FIG. 6—*Load-displacement record for Test B-5, $L_{SB} = 13$ in., unstable.*

1. Test specimen thickness, that is, thicker specimens to verify that plane strain was in fact fully achieved and thinner to explore the effects of plane stress [especially in relation to J-integral plane-strain size criteria, for example, size $\geq (25$ or $50) J/\sigma_0]$.

2. Other specimen size effects such as proportionately scaling up dimensions toward linear-elastic fracture mechanics behavior (that is, toward large-scale structural component behavior).

3. A wider range of temperature variation to include large changes in the upper shelf and their effects, as well as including temperatures down into the transition range to observe effects of partial and greater amounts of cleavage behavior.

4. Exploring the effects of material changes both through heat treatment and other material processing (such as perhaps including Charpy upper-shelf level changes as occur for irradiation damage, etc.), as well as other types of materials which are vastly different, such as aluminum alloys (where cleavage is nonexistent).

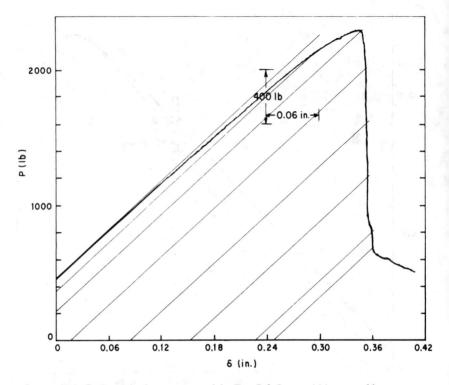

FIG. 7—*Load-displacement record for Test B-3, $L_{SB} = 14$ in., unstable.*

5. Testing of other specimen configurations, such as tension-type specimens instead of bending specimens, to assure relevance of results between different types of configurations (from laboratory specimens to structural components).

Therefore, though this testing program has produced very positive results, much is left to be explored.

Conclusions

1. Tearing instability in three-point bend tests was shown to occur under fully plastic plane-strain conditions.
2. The tests demonstrated the systems aspects of tearing instability by transition from stable to unstable behavior through changes in loading system compliance.
3. The tests demonstrated the effect of local cracked section geometry on tearing instability by transition from stable to unstable behavior through changes in a/W (that is, remaining ligament size effect).
4. A temperature change of 100 °C (within upper-shelf Charpy behavior

range) was shown to not affect tearing instability behavior appreciably for the material tested, ASTM-A471 (NiCrMoV) rotor steel.

5. All test behavior patterns observed tended to support the theory of "instability of the tearing mode of elastic plastic crack growth" as developed in the earlier work (footnote 2) and its approach to the phenomenon.

6. The testing program described herein was intentionally limited in scope in order to develop rapid results and thus has left many aspects of tearing instability to be explored.

Acknowledgments

The support of this testing program at Washington University's Materials Research Laboratory by the U.S. Nuclear Regulatory Commission (NRC) is gratefully acknowledged. The interest and encouragement of NRC personnel, especially the late Mr. E. K. Lynn, and W. Hazelton, and R. Gamble, was a prime factor in this work. Moreover, the timely provision of the material tested by the Westinghouse Research Fracture Mechanics Group under E. T. Wessel aided in an essential way to proper test planning without time-consuming pretesting of material. The assistance of N. Nguyen in performing the test is also acknowledged with thanks. The program was also aided by the consulting assistance of Professors J. W. Hutchinson and J. R. Rice.

J. D. Landes,[1] *H. Walker,*[1] *and G. A. Clarke*[1]

Evaluation of Estimation Procedures Used in J-Integral Testing

REFERENCE: Landes, J. D., Walker, H., and Clarke, G. A., "**Evaluation of Estimation Procedures Used in J-Integral Testing,**" *Elastic-Plastic Fracture, ASTM STP 668*, J. D. Landes, J. A. Begley, and G. A. Clarke, Eds., American Society for Testing and Materials, 1979, pp. 266-287.

ABSTRACT: Estimation techniques for the calculation of J have enabled the development of simpler data reduction methods for multiple specimen J-integral tests and also prompted the development of single-specimen tests. This report describes an experimental program conducted to evaluate the accuracy of these estimation techniques. Comparisons between the values of J as calculated by the energy rate definition and those calculated by the estimation techniques for compact toughness, three-point bend, and center-cracked tension specimens are made.

KEY WORDS: elastic, methods, estimates, plastic, techniques, experiments, fracture, crack propagation

The J-integral as proposed by Rice [1][2] is becoming widely used as a parameter to characterize the fracture toughness of metals when the amount of plasticity in the specimen or structure excludes the use of linear elastic fracture mechanics parameters [2,3]. With the development of the fracture toughness methodology based on J, various methods for experimentally determining J from a load versus load point displacement test record have been proposed [2-8]. The goal of these proposed methods is to simplify the experimental procedure and the data analysis in a J_{Ic} fracture toughness test. Some of these methods were formally exact representations of the J energy line integral value and others were approximations which were exact only for limiting cases.

The first method proposed was the energy rate interpretation of the J line integral. This method proposed by Begley and Landes [2] is formally exact when the material behavior conforms to a deformation plasticity description.

[1] Fellow engineer, associate engineer, and senior engineer, respectively, Westinghouse Electric Corporation, Research and Development Center, Pittsburgh, Pa. 15235.
[2] The italic numbers in brackets refer to the list of references appended to this paper.

The energy rate interpretation of J is expressed by [4]

$$J = -\frac{1}{B}\frac{dU}{da}\bigg|_{v=\text{const}} \qquad (1)$$

where

U = energy under the load-displacement curve,
a = crack length,
v = displacement of the applied force, and
B = specimen thickness.

The experimental approach for using Eq 1 to determine J is shown schematically in Fig. 1. Load-displacement curves were generated for identical specimens with varying crack lengths and a procedure was followed to evaluate the change in energy at a fixed displacement with change in crack length, Fig. 1.

This method for determining J had some major disadvantages, the main one being that several specimens, 5 to 10, had to be tested to simply develop the calibration of J versus displacement. These specimens were not necessarily sufficient to provide any information about the fracture toughness of the material. A major step in the development of the test method occurred

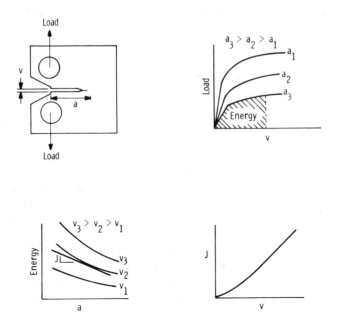

FIG. 1—*Energy rate determination of* J.

when the beginning portion of the crack growth resistance curve was used to determine the J_{Ic} fracture toughness [9]. This development was aided by the work of Rice et al [6] which proposed methods for estimating J as a function of displacement for a single specimen. This work was originally based on the analysis of a deeply cracked bend bar where the value of J is a function of the work done on the cracked body

$$J = \frac{2U_c}{Bb} \qquad (2)$$

where U_c is the work done in loading due to the introduction of a crack and b is the uncracked ligament. Additional estimation formulas were developed for other specimen geometries [6].

These estimation formulas for determining J from a single specimen were instrumental in the development of the current procedures for measuring J_{Ic} where a crack growth resistance curve generated either from several specimens or a single specimen is used to determine J_{Ic} at the point of the initiation of crack growth. Along with the development of the test method was the development of some controversy about the exact form and the use of these estimation formulas. Questions concerning the details of the use of expressions like the one in Eq 2 were asked such as: How deep must the crack be in a deeply cracked bend bar? Should U_T, total energy, be used rather than U_c? Should some modification be made to account for the small tension component in a compact toughness specimen? Over what ranges of crack length and to what accuracy do these estimations work in an actual experimental evaluation? In an attempt to resolve some of the questions, alternative methods for approximating J have been suggested [7,8,10].

The work reported in this paper was undertaken to answer these questions from an experimental basis. Three specimen types that are most frequently used for a J_{Ic} test were evaluated: the compact toughness specimen, the three-point bend bar, and the center-cracked tension specimen. Calibration curves were generated for each of these specimen types over a range of crack lengths. Specimens were tested in monotonic loading with a small radius blunt notch so that no crack extension could occur and a deformation plasticity description of the flow behavior could be closely approximated.

The standard measure of J was taken as the energy rate formulation expressed by Eq 1 which is formally correct and deviates from the energy line integral definition of J only to the degree that the material flow properties vary from specimen to specimen. The various estimation formulations of J as presented by Rice et al [6] were evaluated relative to the energy rate formulation. Several of the alternative methods suggested in answer to some of the questions posed in the foregoing were also evaluated. The results cover a wide range of specimen conditions and provide an answer from an experimental basis to the question of how accurate are these estimation procedures for determining J_{Ic}.

Experimental Procedure

The material used in this work was a high-strength HY130 steel whose properties are given in Table 1. Three specimen types were used for the investigation: the compact toughness specimen, the three-point bend bar, and the center-cracked tension specimen. For each specimen configuration, 10 specimens were machined to the general dimensions shown in Fig. 2 with various crack length to width ratios. The specimens were machined with a blunt notch of radius 1.02 mm (0.04 in.). This was done to eliminate any crack growth during the test. In order to determine that crack extension did not take place, the specimens were heat tinted at the conclusion of each test and finally broken open in liquid nitrogen. The specimens were then checked for stable crack growth by examining the fracture surface for oxidation. At no time during these experiments did any stable crack growth take place.

The range of crack length to width ratios tested was $a/w = 0.4$ to $a/w = 0.85$ for the three-point bend and compact toughness specimens, and $2a/w = 0.4$ to $2a/w = 0.85$ for the center-cracked tension specimens. Load versus load point displacement records for each specimen were taken to prescribed displacement values. The area was then measured (by use of a planimeter) in increments of .51 mm (0.02 in.) of displacement. A plot of the area under the load displacement curve versus total crack length was made at various displacement values. Calculations of the J-integral were then made by the more exact energy rate definition of J [4]. Estimations of the values of J were made by methods described in the following section.

TABLE 1—*Mechanical and chemical properties of HY130 steel used in investigation.*

Mechanical Properties								
0.2% yield strength				974.9 MN/m^2 (141.4 ksi)				
Ultimate strength				1041.1 MN/m^2 (151.0 ksi)				
Elongation				20.0% ...				
Reduction in area				66.5% ...				
Room temperature Charpy energy				107.4 J (79 ft·lb)				
Chemical Composition, weight %								
C	Mn	P	S	Si	Ni	Cr	Mo	V	Al
0.12	0.79	0.004	0.005	0.35	4.96	0.57	0.41	0.057	0.059

270 ELASTIC-PLASTIC FRACTURE

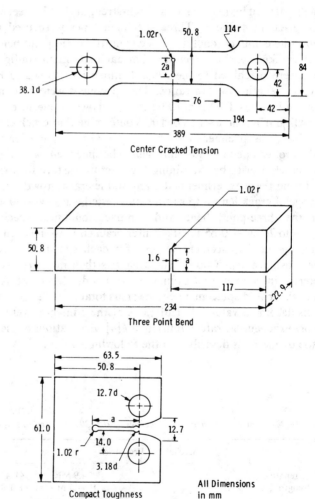

FIG. 2—*Blunt notch toughness specimens, 22.86 mm (0.90 in.) thick.*

Analysis

The estimation procedures for each specimen type are presented in this section. In the interest of clarity, the analysis is presented in three separate subsections, covering each specimen type.

Compact Toughness Specimen

Until recently, the estimation procedure in most common use was the one developed by Rice at al [6]

$$J = \frac{2U_c}{Bb} \qquad (3)$$

where U_c is the total energy, U_T, in the specimen minus the energy, U_{nc}, that would normally exist in the specimen if the specimen did not have a crack. In practice, the energy due to the no-crack situation is negligible in a compact specimen. Therefore the term U_c in Eq 3 can be replaced by the term U_T, which can simply be calculated from the area under the load versus load point displacement curve. The terms B and b in Eq 3 are the specimen thickness and the remaining ligament, respectively. Equation 3 can be rewritten into the form most often used to estimate the value of J for bend-type specimens

$$J = \frac{2A}{Bb} \qquad (4)$$

where A is the area under the load versus load point displacement curve, as shown in Fig. 3. In the development of this equation, Rice assumed that the specimen was in pure bending or, at least, that the contribution due to tension was negligible. However, an analysis by Merkle and Corten [7] showed that the tensile contribution could indeed cause a significant error in the value of J as estimated by Eq 4. The amount of correction to Eq 4 necessary to account for the tension component is not only a function of the crack length to width ratio but it is also a function of the total load and displacement value. The proposed J-integral estimation equation by Merkle and Corten has

$$J = G_1 + \frac{2}{b}\frac{(1+\alpha)}{(1+\alpha^2)}\int_{v_p=0}^{v_p}\left(\frac{P}{B}\right)dv_p$$

$$+ \frac{2}{b}\frac{\alpha(1-2\alpha-\alpha^2)}{(1+\alpha^2)^2}\int_{P/B=0}^{P/B} v_p\, d\left(\frac{P}{B}\right) \qquad (5)$$

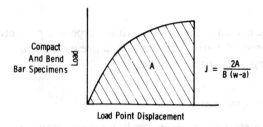

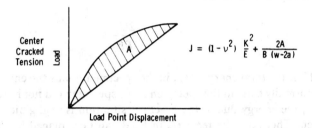

FIG. 3—*Description of the graphical evaluation of J from load versus load point displacement records.*

where

$$\alpha = 2\sqrt{(a/b)^2 + (a/b) + \tfrac{1}{2}} - 2(a/b + \tfrac{1}{2}) \qquad (6)$$

G_1 is the elastic strain energy release rate per unit crack extension and v_p is the plastic displacement value.

Merkle and Corten have shown that, for $a/w > 0.5$, Eq 5 can be replaced by one that contains total displacements [7] leading to the more readily usable form [11]

$$J = \left(\frac{(1+\alpha)}{(1+\alpha^2)} - \frac{\alpha(1-2\alpha-\alpha^2)}{(1+\alpha^2)^2} \right) \frac{2A}{Bb} + \frac{\alpha(1-2\alpha-\alpha^2)}{(1+\alpha^2)^2} \frac{2Pv}{Bb} \qquad (7)$$

A further simplification can be made if it is assumed that the complementary energy is much smaller in magnitude than the total energy, resulting in the following expression

$$J = \left(\frac{1+\alpha}{1+\alpha^2} \right) \frac{2A}{Bb} \qquad (8)$$

A comparison of these three forms of the Merkle-Corten equations can be seen in Table 2. An additional method of estimating J, as proposed by Mc-

Cabe [12] was evaluated. This method uses the secant offset technique to calculate an effective crack length. This effective crack length along with the effective modulus derived from the loading line is used to calculate an elastic-plastic strain energy release rate G_1 from

$$G_1 = \frac{\hat{K}^2}{\hat{E}} \approx J_1 \qquad (9)$$

where $\hat{K}$ is the effective stress intensity and $\hat{E}$ the effective modulus.

Bend Bar Specimens

A number of authors [8,10] have reported that problems exist when using Eq 4 for bend bar specimens.

Srawley [10] shows that there is a considerable difference when using the total energy as compared with the energy due to the presence of a crack. To further illustrate this point, the estimation Eqs 3 and 4 were compared with the strain energy release rate G_1 in the elastic regime. In order to demonstrate this comparison, the constant 2 in Eq 3 is replaced by the variable β. For a one-to-one comparison between J and G in the elastic region, β will be 2.0.

The values of β for three-point bend specimens with a span to width ratio of 4 and for crack length to width ratios between 0.6 and 0.9 are listed in Table 3 and are also plotted in Fig. 4, along with the curves originally presented by Srawley [10]. These results show that by using the total energy, U_T, the value of β varies between 2.002 and 2.035, whereas by using the energy due to the presence of a crack, U_c, the value of β varies between 2.03 and 2.638 for a/w's between 0.6 and 0.9.

Center-Cracked Tension Specimens

The J versus Δa R-curve generated from center-cracked tension specimens are, in many cases different from those generated by compact specimens and bend bar specimens. While the J_{1c} point appears to be exactly the same, the slope of the J versus Δa R-curve is greater for the center-cracked tension specimen than for the bend-type specimens. One of the initial concerns was the possibility that the value of J for large-scale plasticity was not being estimated correctly by the equation developed by Rice et al [6]. This method for estimating J used the sum of the linear elastic energy release rate, and an estimate of the value of J in the plastic regime. The equation for the estimation of J for center-cracked tension specimens has the form

$$J = \frac{(1-\nu^2)K^2}{E} + \frac{2A}{B(w-2a)} \qquad (10)$$

274 ELASTIC-PLASTIC FRACTURE

TABLE 2—*Comparison of J estimation techniques and J as calculated by energy rate definition for compact toughness specimens.*

a/w		Inches of Deflection[a]						
		0.020	0.040	0.060	0.080	0.100	0.120	0.140
0.4	J_1	523	1595	2996	...	...	...	...
	J_2	366	1351	2680	...	...	...	...
	J_3	423	1562	3099	...	...	...	...
	J_4	503	1791	3432	...	...	...	...
	J_5	507	1736	3287	...	...	...	...
0.45	J_1	396	1401	2680	...	...	...	...
	J_2	314	1202	2407	...	...	...	...
	J_3	360	1376	2755	...	...	...	...
	J_4	372	1511	2966	...	...	...	...
	J_5	362	1342	2647	...	...	...	...
0.5	J_1	322	1219	2375	5045	...	...	...
	J_2	277	1060	2143	3357	...	...	...
	J_3	313	1200	2427	3801	...	...	...
	J_4	350	1310	2596	4000	...	...	...
	J_5	308	1256	2462	3728	...	...	...
0.55	J_1	274	1050	2082	3216	4430	...	...
	J_2	242	925	1866	2972	4091	...	...
	J_3	271	1035	2089	3327	4580	...	...
	J_4	292	1106	2203	3457	4725	...	...
	J_5	296	1081	2132	3299	4523	...	...
0.6	J_1	235	892	1800	2823	3858	...	...
	J_2	209	799	1637	2590	3569	...	...
	J_3	231	884	1811	2865	3948	...	...
	J_4	245	936	1891	2959	4052	...	...
	J_5	247	919	1829	2879	3957	...	...
0.65	J_1	200	748	1528	2420	3302	4331	5216
	J_2	177	678	1395	2210	3055	3928	4800
	J_3	193	741	1524	2415	3338	4292	5245
	J_4	205	777	1584	2490	3416	4375	5330
	J_5	216	802	1604	2544	3420	4467	5669
0.7	J_1	163	615	1270	2025	2762	3571	4373
	J_2	146	565	1162	1823	2555	3267	4002
	J_3	157	609	1254	1967	2757	3526	4319
	J_4	152	585	1201	1898	2635	3360	4125
	J_5	166	628	1279	2022	2828	3718	4613
0.75	J_1	127	492	1023	1624	2239	...	...
	J_2	117	457	933	1477	2064	...	...
	J_3	126	487	994	1574	2199	...	...
	J_4	131	508	1030	1619	2248	...	...
	J_5	132	502	1009	1598	2261	...	...

TABLE 2—Continued.

a/w		\multicolumn{7}{c}{Inches of Deflection[a]}
		0.020	0.040	0.060	0.080	0.100	0.120	0.140
0.8	J_1	90	381	790	1228	1733	2175	2642
	J_2	92	353	717	1135	1590	2038	2492
	J_3	97	371	754	1194	1673	2144	2622
	J_4	103	389	784	1231	1714	2188	2667
	J_5	102	383	780	1234	1731	2275	2865
0.85	J_1	57	278	569	840	1250	1549	1810
	J_2	66	252	505	828	1127	1481	1814
	J_3	67	254	509	834	1135	1491	1822
	J_4	69	262	524	860	1170	1538	1884
	J_5	70	266	542	871	1271	1604	1996

$$J_1 = -\frac{1}{B}\frac{dU}{da} \quad J_2 = \frac{2A}{Bb} \quad J_3 = \left(\frac{1+\alpha}{1+\alpha^2}\right)\frac{2A}{Bb}$$

$$J_4 = G_1 + \frac{2}{b}\frac{(1+\alpha)}{(1+\alpha^2)}\int_0^{v_P} P/B\, dV_P + \frac{2}{b}\alpha\,\frac{(1-2\alpha-\alpha^2)}{(1+\alpha^2)^2}\int_0^{P/B} v_P\, d\left(\frac{P}{B}\right)$$

$$J_5 = \frac{\hat{K}_1^2}{\hat{E}}$$

[a] 1 in. = 25.4 mm.

where A is the area between the load versus load displacement curve and the secant offset line to the displacement of interest. This area is shown in Fig. 3.

Results

The load versus load point displacement curves generated for the compact toughness, bend bar, and center-cracked tension specimens are shown in Figs. 5 through 7, respectively. After calculating the area under each curve at various displacement values, the energy values for each area are plotted against its corresponding displacement value. The value of J as defined by the energy rate definition is then calculated by the technique as shown in Fig. 1. The J versus displacement curves for the compact toughness specimens are shown as solid lines in Fig. 8 at various a/w ratios. The estimated values of J determined by Eqs 4 and 8 are also shown in Fig. 8. It can be seen that the estimated values from Eq 8 are in extremely good agreement with the more exact definition of J, whereas Eq 4 considerably underestimates the value of J by Eq 1.

TABLE 3—*Values of β for three-point bend specimen J estimation equation, using total energy, U_T, and energy due to presence of crack, U_c, only.*

a/w	β (Using U_c)	β (Using U_T)
0.6	2.638	2.035
0.7	2.315	2.019
0.8	2.129	2.008
0.9	2.030	2.002

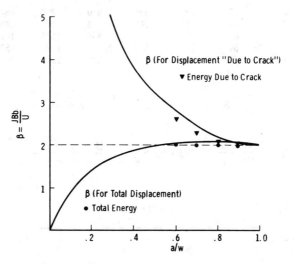

FIG. 4—*Values of the nondimensional coefficient β as used in the form $J = \beta U/Bb$ within the elastic range for a three-point bend specimen with $S/W = 4$. (The solid lines are from Ref 10 and the points calculated from elastic compliances.)*

The initial J estimation equation by Rice et al [6], the Merkle-Corten equation [7], the proposed variation on the Merkle-Corten equation, Eq 8, and the secant offset method by McCabe are all compared in Fig. 9 in a nondimensional form with the value of J as calculated by the energy rate definition. It can be seen that Eq 8 most closely approximates the value of J derived by Eq 1. It is quite fortuitous that Eq 8 is the simplest of the methods used to account for the tension component in the area estimation procedures. The values of J by the various estimation methods are compiled and presented in Table 2 along with J as calculated by Eq 1.

As discussed earlier, the total energy, U_T, appears to be more appropriate for the estimation of J for three-point bend specimens. A comparison of the estimated value of J by Eq 4 and the energy rate definition of J is shown in Fig. 10 with the more exact value of J being represented by the solid lines. It can be seen that the estimated value of J is extremely close to the value of J as

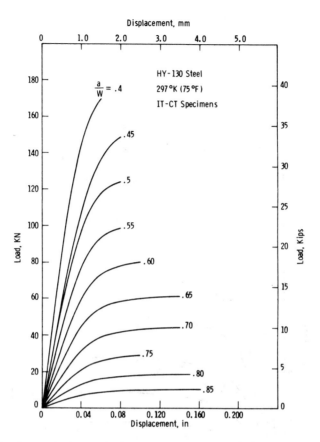

FIG. 5—*Load versus load point displacement curves for compact specimens at various* a/W *ratios.*

calculated by Eq 1 throughout the range of a/w ratios between 0.4 and 0.8. Slight differences should be expected due to the possibility of inaccuracies in curve fitting the energy versus crack length values used to calculate the value of dU/da. The values of J by the estimation Eq 4 and the energy rate definition of J are presented in Table 4.

The comparison between the value of J by the energy rate definition for center-cracked tension specimens and the estimation Eq 10 is shown in Fig. 11. It can be seen that at larger values of $2a/w$ the estimated value of J is somewhat lower than the value of J as calculated by Eq 1. This may well be due to the inaccuracy of curve fitting the values of the energy between the load displacement curve and the offset secant curve, versus crack length at larger values of crack length.

The slope of the energy versus crack length at large crack lengths is much higher than at smaller crack lengths. Even small inaccuracies in curve fitting

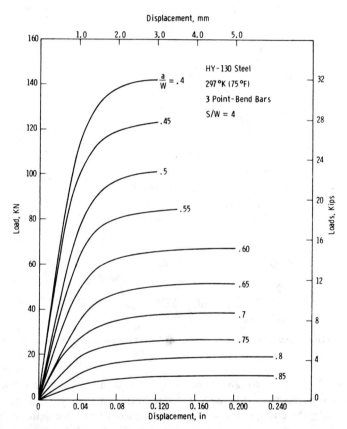

FIG. 6—*Load versus load point displacement curves for three-point bend specimens with S/W = 4 at various a/W ratios.*

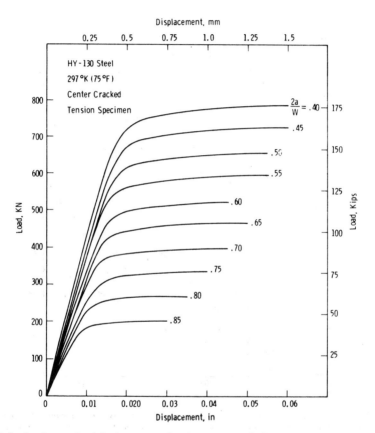

FIG. 7—*Load versus load displacement points for center-cracked tension specimens at various 2a/W ratios.*

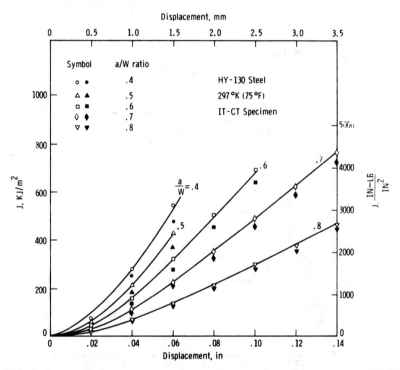

FIG. 8—*J versus displacement curves showing the comparison of the energy rate definition of J (solid lines) and the estimated values of J by Eq 4 (solid points) and Eq 8 (open points) for compact toughness specimens.*

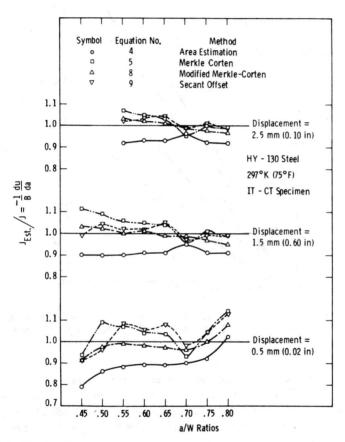

FIG. 9—*Showing the comparison of the various estimates of J with the energy rate definition of J at given displacement values.*

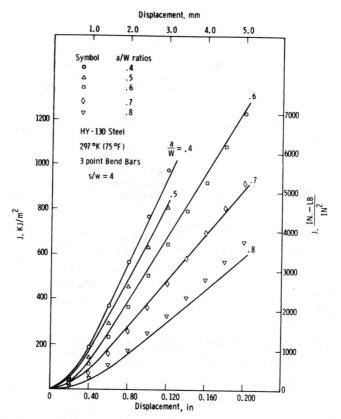

FIG. 10—*J versus displacement curves showing comparison of the energy rate definition of J (solid lines) and the estimated value of J from Eq 4, for three-point bend bar specimens.*

at these higher values of crack lengths can have substantial effect on the value of dU/da. With this in mind, the comparison of the estimated value of J with the energy rate definition of J appears to be quite good. The estimated values of J can be seen in tabular form in Table 5 along with the more exact value of J as calculated by Eq 1.

Discussion

The results from these studies answer most of the questions posed relative to J estimation methods and illustrate, to within experimental limitations, how well these approximations work. Each specimen will be discussed separately.

The compact toughness specimen involved the greatest number of methods for developing estimation of J. From these results it is clear that a simple

TABLE 4—Comparison of the J estimation technique and J as calculated by the energy rate definition for three-point bend specimens S/W = 4.

a/w		\multicolumn{10}{c}{Inches of Deflection[a]}									
		0.020	0.040	0.060	0.080	0.100	0.120	0.140	0.160	0.180	0.200
0.4	J_1	283	1176	2111	3314	4310	5340	...	...	...	...
	J_2	284	1067	2104	3233	4399	5579	...	...	...	...
0.45	J_1	289	1084	2007	3076	4058	5049	...	...	...	...
	J_2	272	1009	1984	3051	4146	5259	...	...	...	...
0.5	J_1	272	984	1852	2817	3767	4708	...	...	...	...
	J_2	206	809	1659	2610	3598	4612	...	...	...	...
0.55	J_1	243	876	1687	2539	3437	4317	5492	...	...	...
	J_2	193	740	1520	2389	3298	4221	5158	...	...	...
0.6	J_1	206	759	1497	2224	3069	3878	4748	5418	6418	7257
	J_2	167	646	1326	2087	2880	3694	4518	5351	6190	7033
0.65	J_1	164	636	1282	1928	2664	3390	4036	4682	5420	6137
	J_2	129	500	1060	1719	2405	3116	3838	4568	5306	6044
0.7	J_1	126	504	1047	1596	2223	2854	3357	3953	4517	5122
	J_2	102	420	906	1473	2075	2698	3333	3975	4621	5268
0.75	J_1	90	369	790	1248	1749	2274	2713	3233	3696	4198
	J_2	92	379	793	1268	1771	2290	2820	3353	3890	4430
0.8	J_1	60	230	516	887	1244	1656	2106	2524	2943	3348
	J_2	67	277	610	999	1427	1871	2329	2796	3268	3751
0.85	J_1	38	91	229	519	718	1004	1536	1830	2242	2554
	J_2	59	215	455	753	1077	1402	1708	2095	2453	2819

$$J_1 = -\frac{1}{B}\frac{dU}{da} \qquad J_2 = \frac{2A}{Bb}$$

[a] 1 in. = 25.4 mm.

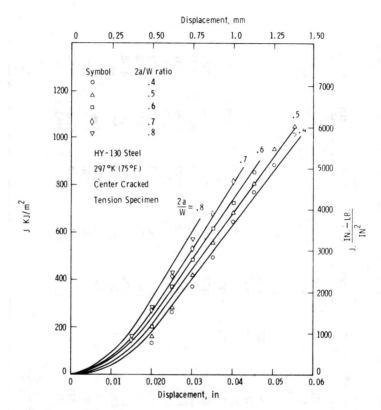

FIG. 11—*J versus displacement curves showing the comparison between the energy rate definition of J (solid lines) and the estimated value of J from Eq 10 for the center-cracked tension specimen.*

bending solution as expressed by Eq 3 provides the least accurate of the approximations. To determine *J* more accurately, some modification must be made to account for tension. The modification proposed by Merkle and Corten [7] provides a better approximation; however, a simplification of the Merkle-Corten approach in Eq 8 provides the best approximation to *J*. The procedure suggested by McCabe also provides a reasonable estimation of *J*.

The estimation formula for the bend bar in Eq 2 works best only for very deeply cracked specimens. For specimens cracked in the range normally tested, $0.5 < a/w < 0.75$, the total energy of loading should be used rather than simply the energy due to a crack. This modification has been demonstrated to work only for a span to specimen width ratio, S/W, of 4.

The estimation formula for the center-cracked tension specimen is also accurate. Any difference that occurs in the development of the crack growth resistance curve between this specimen and bend-type specimens comes from sources other than this formula.

TABLE 5—Comparison of the J estimation technique and J as calculated by the energy rate definition for center-crack tension specimens.

a/w		0.15	0.02	0.025	0.030	0.035	0.040	0.045	0.050	0.055
0.4	J_1	...	1073	1821	2430	2968	3960	4521	4791	5294
	J_2	...	764	1527	2123	2841	3709	4412	5082	5807
0.45	J_1	...	1119	1874	2490	3076	4025	4654	4980	5553
	J_2	...	797	1514	2223	3010	3733	4547	5329	6044
0.5	J_1	...	1172	1934	2560	3200	4101	4826	5225	5880
	J_2	...	928	1608	2403	3183	3922	4638	5458	6039
0.55	J_1	...	1233	2002	2641	3344	4192	5047	5546	6298
	J_2	...	1083	1777	2523	3371	4112	4931	5761	6625
0.6	J_1	...	1307	2080	2738	3513	4301	5339	...	...
	J_2	...	1156	2122	2759	3513	4143	4893	...	...
0.65	J_1	...	1396	2172	2855	3716	4435	5732	...	...
	J_2	...	1226	1995	2734	3517	4299	5069	...	...
0.7	J_1	881	1507	2284	3002	3967	4604	...	...	...
	J_2	849	1564	2391	3054	3895	4693	...	...	...
0.75	J_1	704	1652	2422	3193	4288	4826	...	...	...
	J_2	676	1390	2174	2903	3780	4567	...	...	...
0.8	J_1	958	1849	2602	3454	...	...	...	...	...
	J_2	933	1641	2457	3264	...	...	...	...	...
0.85	J_1	1139	2143	2853	...	...	...	...	...	...
	J_2	1118	1899	2661	...	...	...	...	...	...

[a] Inches of Deflection

$$J_1 = -\frac{1}{B}\frac{dU}{da}$$

$$J_2 = (1 - v^2)\frac{K^2}{E} + \frac{2A}{B(W - 2a)}$$

in. = 25.4 mm.

These results provide a sound basis for the continued use of J estimation formulas for experimental evaluation of J_{Ic}. However, they are only valid when a deformation model of the plasticity behavior of the material is approximated by the test. Additional work should concentrate on developing methods for experimentally approximating J for such cases as large amounts of crack growth or periodic unloading during the test.

Conclusions

The following conclusions can be made from the comparisons of the estimated values of J and the values of J as determined from Eq 1.

1. It is necessary to account for the tension component when estimating the value of J for compact specimens.

2. The area estimation technique for compact specimens which approximates the energy rate definition of J most accurately in this investigation was a variation of the Merkle-Corten technique given by Eq 8.

3. The total energy, U_T, should be used when estimating the value of J for three-point bend specimens.

4. The value of β should be equal to 2 for three-point bend specimens when the span to width ratio is 4.

5. The estimation values of J for center-cracked panel specimens appear to closely approximate the values calculated by the energy rate definition of J.

Acknowledgments

This study was undertaken as a result of questions raised after examining the results of the Cooperative Test Program by members of the ASTM E24.01.09 Task Group on elastic-plastic fracture. The material used in this study was the same material as used in the Cooperative Test Program, which was generously supplied by the United States Steel Corp. Acknowledgment is also made of the care taken in the testing portion of this program by P. J. Barsotti, F. X. Gradich, and R. B. Hewlett of the Structural Behavior of Materials Department of Westinghouse R&D Center. The work of W. H. Pryle and Donna Gongaware, of the same department, is also appreciated for the design and procurement of the specimens and the manuscript typing, respectively.

References

[1] Rice, J. R., *Journal of Applied Mechanics, Transactions,* American Society of Mechanical Engineers, June 1968, pp. 379-386.
[2] Begley, J. A. and Landes, J. D. in *Fracture Toughness, ASTM STP 514*, American Society for Testing and Materials, 1972, pp. 1-20.
[3] Landes, J. D. and Begley, J. A. in *Developments in Fracture Mechanics Test Methods Standardization, ASTM STP 632*, W. R. Brown, Jr. and J. G. Kaufman, Eds., American Society for Testing and Materials, 1977, pp. 57-81.

[4] Rice, J. R. in *Fracture*, Vol. 2, H. Liebowitz, Ed., Academic Press, New York, 1968, pp. 191-311.
[5] Bucci, R. J., Paris, P. C., Landes, J. D. and Rice, J. R. in *Fracture Toughness, ASTM STP 514*, American Society for Testing and Materials, 1972, pp. 40-69.
[6] Rice, J. A., Paris, P. C. and Merkle, J. G. in *Progress in Flaw Growth and Fracture Toughness Testing, ASTM STP 536*, American Society for Testing and Materials, 1973, pp. 231-245.
[7] Merkle, J. G. and Corten, H. T., "A J Integral Analysis for the Compact Specimen, Considering Axial Force as Well as Bending Effects," ASME Paper No. 74-PVP-33, American Society of Mechanical Engineers, 1974.
[8] Sumpter, J. D. G. and Turner, C. E. in *Cracks and Fracture, ASTM STP 601*, American Society for Testing and Materials, 1976, pp. 3-18.
[9] Landes, J. D. and Begley, J. A. in *Fracture Analysis, ASTM STP 560*, American Society for Testing and Materials, 1973, pp. 170-186.
[10] Srawley, J. E., *International Journal of Fracture*, Vol. 12, No. 3, 1976, pp. 470-474.
[11] Embley, G. T., Knolls Atomic Power Laboratory, private communication, 1976.
[12] McCabe, D. E. and Landes, J. D., this publication, pp. 288-305.

D. E. McCabe[1] and J. D. Landes[1]

An Evaluation of Elastic-Plastic Methods Applied to Crack Growth Resistance Measurements

REFERENCE: McCabe, D. E. and Landes, J. D., **"An Evaluation of Elastic-Plastic Methods Applied to Crack Growth Resistance Measurements,"** *Elastic-Plastic Fracture, ASTM STP 668,* J. D. Landes, J. A. Begley, and G. A. Clarke, Eds., American Society for Testing and Materials, 1979, pp. 288–306.

ABSTRACT: Information from tests on blunt notched specimens for J-integral calibration by conventional $J = -1/B \; \partial U/\partial a$ analysis is used to evaluate the significance of plastic zone adjustment to physical crack length in crack growth resistance, K_R, calculations. Secants are drawn to load versus displacement test records to determine plastic zone adjusted crack lengths. Tests on three specimen geometries [compact specimen (CS), single-edge notched bend (SENB), and center-notched tension (CNT)] and on two materials (HY130 steel and 2024 aluminum) have shown that this procedure develops K_R values that are equivalent to J. This demonstration opens possibilities that J can now be applied to cases where there is subcritical crack growth such as in R-curve work, K_{Iscc}, and possibly in creep cracking studies. Also, this provides a simplified method for computing J experimentally on complex geometries for which elastic K_I solutions are available.

Alternative J computational procedures are compared. These include J by a Ramberg-Osgood work-hardening law fit to load-displacement records, and J by area approximation methods. The Ramberg-Osgood modeling appeared to work reasonably well in tests on compact specimens but was found to be unreliable on SENB and CNT specimens and therefore is not recommended. With no stable crack propagation, the area approximation procedures for J determination produce reasonably accurate estimates of J as might have been anticipated from past experience. Tests on the compact specimen geometry required a Merkle-Corten correction procedure which worked well on large crack aspect ratios, where $a/w \geq 0.5$, but tended to overcorrect for short cracks, giving nonconservative results.

KEY WORDS: fracture (materials), J-integral, elastic, compliance, cracks, toughness, deformation, plastic, crack propagation

The use of fracture mechanics on structural materials is presently a reasonably well-established and practical practice, so long as the use is

[1]Senior engineer and fellow engineer, respectively, Structural Behavior of Materials, Westinghouse R&D Center, Pittsburgh, Pa. 15235.

restricted to the lower shelf or lower transition temperature range [1,2].[2] Because of the difficulty in applying fracture mechanics to upper shelf toughness evaluation, where there is extensive plasticity and slow-stable crack growth to contend with, the problem has been handled in an interim manner by defining fracture toughness in terms of conservative values. Presently the toughness is defined using either J_{Ic} or crack-opening displacement (COD) procedures where attention is concentrated on the onset of slow-stable crack growth [3,4]. The conservatism contained within such an approach is satisfying from a safety standpoint, but this more often than not results in critical flaw size and stress level predictions that are not indicative of the true (upper shelf) in-service performance. The true load-carrying capability of flawed structures can be considerably underestimated, depending upon geometry and the crack growth resistance capability of the materials used. Overdesign in materials selection can be as serious an engineering problem as underdesign. Inexpensive alloys may be ruled out for applications in which they would ordinarily be perfectly suitable. Thus the refinement of our upper shelf instability prediction capability can prove to be a very worthwhile objective.

The tool by which upper shelf instability predictions can be made is available in the form of R-curve analysis. However, elastic-plastic methods have not been developed for handling slow-stable crack growth and a means for extending these methods is needed. In addition, instability criteria under large-scale plasticity conditions are presently under review. In all cases, R-curve development is accepted as an expression of fundamental material behavior, and modeling for instability prediction is presently subject to interpretation. One approach would be to incorporate all elastic-plastic effects in the R-curve and then use the conventional match-up procedure with elastic crack drive curves to predict instability [5]. A second possibility is to consider a fresh approach through stable tear modeling as suggested by Paris et al [6]. All considerations, however, depend upon having a valid elastic-plastic R-curve to work with.

Several investigators have suggested that K_I calculated with a plastic zone correction to crack size (effective crack size) produces K_R values that are equivalent to J [7,8]. This has been proposed for limited cases where the plastic zones are relatively small in comparison to the overall size of a surrounding elastic stress field. The experimental approach, using compliance for determining this "effective" or plastic-zone corrected crack size has been well established and demonstrated to be satisfactory for tests on ultra-high-strength sheet materials. Here, the plastic zone is usually small in comparison to overall crack size and any errors involved, either in principle or in practice, tend to be minor. Therefore, it is not entirely clear to some that the equivalency between J and K_R has been adequately dem-

---

[2]The italic numbers in brackets refer to the list of references appended to this paper.

onstrated. The crux of the contention is that G_1 or K_1 become inaccurate field parameters under any appreciable crack tip plasticity conditions and that only J is a supportable computational approach. It is with this argument in mind that the present project was undertaken, namely, to explore the significance of effective crack size when used in K_R calculations. If proved viable, this approach can then be justified for use under elastic-plastic conditions where slow-stable crack growth intercedes at some intermediate point in the crack growth resistance development for the material.

The principal technique proposed in this investigation is to calculate K_R using compliance determined effective crack sizes in the linear-elastic expressions for K_1, and to compare these values with J obtained by conventional means. For displacement levels where there is no slow-stable crack growth, these comparisons give a one-to-one evaluation of the extent of plasticity that can be handled by plastic zone corrected K_R. The comparison is made for three specimen types: compact specimen (CS), single-edge notched bend (SENB), and center notched tension (CNT). Much of the raw data for this analysis were obtained from a companion project entitled "Evaluation of Estimation Procedures used in J-Integral Testing" [9], where the principal objective was to evaluate area approximation methods for determining J. An additional study of value included in this investigation is to compute J by other available or alternative means so that we can compare and evaluate the reliability of these methods as well.

Computational Methods

A number of computational methods which are in varied stages of development and acceptance are available for calculating the J-integral. This section is devoted to a somewhat simplified presentation of the techniques tried. Most specimens were prepared with instrumentation designed to provide the full range of data needed to employ the various computational approaches.

J-Integral by K_R

Again, the principal point of interest is to test the validity of compliance-determined effective crack length as an elastic-plastic methodology. The procedure for effective crack size determination is outlined in Fig. 1, and the crack size so determined is used in the K_1 expression in place of the actual crack size. The J-integral is estimated using the following expression

$$J = \alpha K_R^2/E$$

where

α = a constraint factor varying between 1 (plane stress) and 0.9 (plane strain),

K_R = crack growth resistance using effective crack size, and
E = effective elastic modulus.

The effective elastic modulus is determined from the initial linear slope of the test record. With initial crack size and theoretical compliance known, the apparent elastic modulus behavior of the specimen can be determined. This modulus will inherently contain crack tip constraint effects, thereby justifying the use of $\alpha = 1$ in conversion to J throughout the experiment.

J-Integral, Begley-Landes Method (B-L)

This experimental procedure for the determination of J-integral involves the use of blunt notched specimens of identical geometry but with varying initial crack sizes. The testing sequence, first outlined by Begley and Landes [10], is used where the procedure is aimed at directly satisfying the working definition $J = -1/B \; \partial U/\partial a$ (see Fig. 2). Since this is the most direct method for determining J_R, it is regarded herein as the benchmark of comparison for other computational procedures. Total energies up to fixed displacement levels are determined from areas under load-displacement records. This energy, normalized for material thickness, U/B, is plotted against crack size. For a selected crack size, the family of slopes plotted against displacement level constitutes the J-integral calibration curve for that crack size. Calibration curves can be generated for any chosen crack size within the range of a given family of U/B versus a curves. This procedure is best performed using blunt notched specimens where slow-stable crack extension is suppressed.

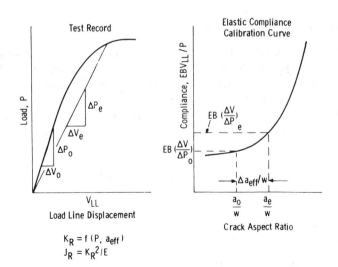

FIG. 1—*Illustration of secant technique for determining effective crack size.*

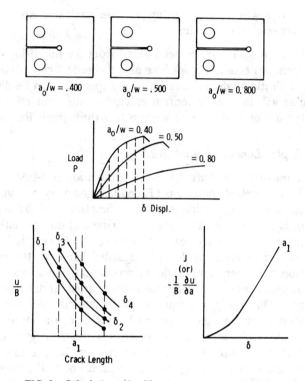

FIG. 2—*Calculation of J calibration curve by* $-1/B\ \partial U/\partial a$.

J-Integral—Area Approximation

An alternative method for determining J_R is the area approximation procedure, as suggested by Rice [11], which is more commonly used but is perhaps subject to acceptably small errors (see Fig. 3). Here all necessary information is obtained from the test record of one specimen. The approximation is to determine the energy input into the specimen, U, from the area under the load-displacement record. This can be converted to J_R by

$$J_R = 2/Bb \int_0^\theta M d\theta = 2U/Bb$$

where

M = bending moment,
θ = bend angle,
B = specimen thickness, and
b = original uncracked ligament size.

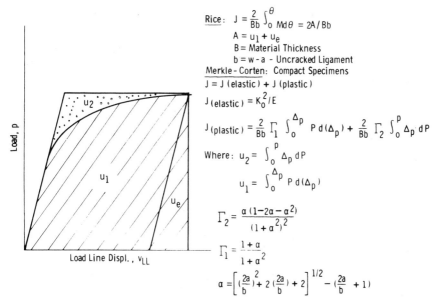

FIG. 3—*J-integral determination by area methods.*

The development is based upon the assumption that the specimen is in pure bending and therefore can be regarded as being a good approximation only for deeply notched bend specimens. This same expression has been used on compact specimens, but an unaccounted-for superimposed tension component tends to make the pure bend model conservative, especially so for crack aspect ratios, a/w, less than 0.5. Recently, Merkle and Corten [12] have suggested a modified computational procedure for the compact specimen (CS), which is also outlined in Fig. 3. Here energy distributions are broken down into elastic and plastic contributions. The plastic contribution is shown as a function of crack aspect ratio, a/w, which accounts for superimposed tension, and typically J_R is increased on the order of 10 to 20 percent over values predicted by the pure bend expression. This development can be simplified considerably from that shown by combining terms such that J (elastic) is approximated from other measurements (P and V_{LL}) and with the Γ_1 and Γ_2 terms reduced to tabular form.

For center-cracked panels, Rice et al [11] have set up an area under the curve method of approximately the same form as the Merkle-Corten treatment for compact specimens. Elastic contribution to J is treated independently, and the J (plastic) area under the load-displacement record corresponds to that between the load-displacement trace and a secant drawn to the test record. To use J (plastic) $= 2A/Bb$, where A is the area just described and B the material thickness, the b dimension corresponds to the sum of the two ligaments on either side of the central crack.

J-Integral—Ramberg-Osgood (R-O)

Another elastic-plastic approach available is to estimate the J-integral by characterizing load-displacement records using the Ramberg-Osgood work-hardening law

$$V = (V/P)_o P + \kappa (V/P)_o^n P^n$$

where

V = load-line displacement,
P = applied load,
$(V/P)_o$ = initial load-line linear elastic compliance slope,
κ = work-hardening coefficient, and
n = work-hardening exponent.

Figure 4 shows the development for calculation of J from the foregoing expression. This development is basically similar to the more rigorous Begley-Landes approach in that an attempt is made to define J in terms of $-1/B \, \partial U/\partial a$. The hazard present in the R-O development, however, is that the work-hardening constants κ and n are determined from one test record and these may not necessarily correctly define the trend in load-displacement records for changing initial crack size.

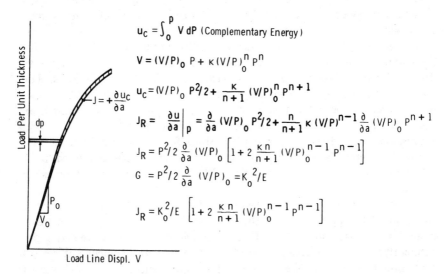

FIG. 4—*Computation of J from Ramberg-Osgood work-hardening expression.*

It may be recognized that the expression for J_R is essentially the same as that developed by Eftis and Liebowitz for plasticity corrected G where the term in brackets corresponds to $\tilde{C}$ [13]. In experiments by the subject investigators, on specimens having different initial flaw sizes, values of n were shown to be variable, but this was not considered to be objectionable. From the J viewpoint, this represents a significant oversight.

Experimental Program

The specimens used were blunt notched so that slow stable crack growth was suppressed and all nonlinear effects observed on test records were due to developing plasticity. J-integral calibration curves were developed according to the Begley-Landes (B-L) procedure on three specimen geometries of 23-mm-thick (0.9 in.) HY130 steel, [1T compact, 5.08-cm-wide (2 in.) CNT, and 5.08-cm-wide (2 in.) by 20.32-cm-long (8 in.) three-point bend specimens]. Variability of material was provided by an aluminum alloy, 2024-T3 of 6.3-mm (0.25 in.) thickness, tested in the CS configuration. Specimen dimensions are reported in Table 1. The crack aspect ratios denoted in the next to last column in Table 1 are for test records that were analyzed to compare the various J procedures. A significantly larger population of specimens was used to develop the benchmark values of J by the B-L method [10].

Results and Discussion

The results of J calculations on the blunt notched 1T compact specimens of HY130 and 2T compact specimens of 2024 Al-T3 are shown in Figs. 5 through 9. Five computational methods for determining J are compared. The lines designated real J are best-fit curves through the open-square data points which represent the B-L method and are regarded here as the benchmark values. The lower fit curves represent elastic computation for G (or elastic J) which is uncorrected for plasticity. Values of nominal stress at the crack tip, corresponding to the various levels of displacement where J-values were calculated, are shown in parentheses. Limit load corresponding to a nominal crack tip stress level of 1.62 times material ultimate strength [14] is indicated by vertical arrows.

The general observation that can be made here is that all methods predicted J with reasonable accuracy, even with extreme plasticity adjustments ranging up to 100 percent of elastic values.

The J calculations derived from the Ramberg-Osgood work hardening law tended toward slightly higher values of J. In order to fit load-displacement records for different initial crack aspect ratios, the work-hardening constants had to be varied appreciably as Table 2 shows. This suggests that κ and n have no significance with regard to the material flow properties,

TABLE 1—*J-integral calibration (comparison of computational procedures).*

Material	Thickness	Specimen Type	Size, in.			Crack Aspect Ratios	Notch Root Radius
			Width	Length			
HY130	0.9	CS[a]	2	...		0.4, 0.5, 0.6	0.05
HY130	0.9	CNT[b]	2	6		0.4, 0.45	0.05
HY130	0.9	SENB[c]	2	8		0.4, 0.5	0.05
2024 Al-T3	0.25	CS	4⅛	...		0.451, 0.589	0.05

Material Properties

Material	Yield Strength, ksi	Tensile Strength, ksi	Percent Elongation, 2 in.
HY130	141.4	151	20 (1-in. gage length)
2024 Al-T3	51	68	18 (2-in. gage length)

[a] CS = compact specimen.
[b] CNT = center-notched specimen.
[c] SENB = single-edge notched bend specimen.
Metric conversion factors:
1 in. = 2.54 cm.
1 ksi = 6.8948 MPa.

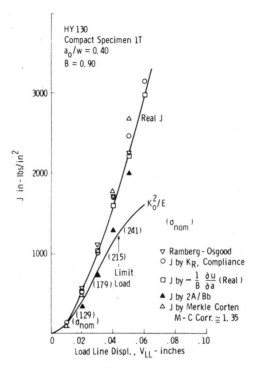

FIG. 5—*J-integral calibration curve for blunt-notched 1T compact specimen; a/w = 0.4.*

and, because of this, it is suggested that the R-O expression has no fundamental significance in the context of being a viable index of fracture toughness. The expression appears only to be a convenient method for curve fitting test records.

Values of J predicted from compliance-determined K_R, according to Fig. 1, are shown as open-circle data points. Again, the elastic moduli, E, used in conversion to J by K_R^2/E, are the "effective" values indicated by the initial elastic slope of the test records. This is done because all computational procedures for J used herein are dependent upon initial slope, and the use of an effective modulus is the best way to compare the computational methods on an equal basis.

These calculations of J from compliance-adjusted K_R tended to be the most consistent in comparison with the benchmark J curve. This not only tends to support the suggestion that plastic zone correction to K_I is equivalent to the computation of J, for small plastic zones embedded in dominant elastic stress fields, but the present data have carried the suggestion well beyond this limitation into large-scale plasticity. Therefore, these claims now appear to be supportable at extensive plastic strain levels on the basis of experimental evidence.

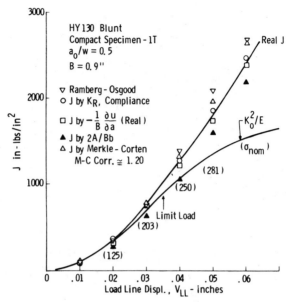

FIG. 6—*J-integral calibration curve for blunt-notched 1T compact specimen; a/w = 0.5.*

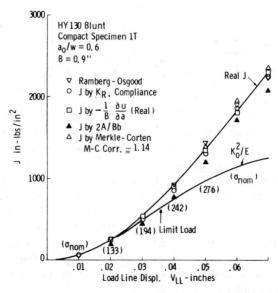

FIG. 7—*J-integral calibration curve for blunt-notched 1T compact specimen; a/w = 0.6.*

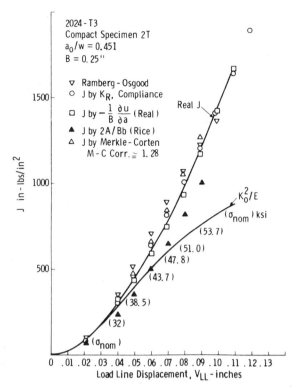

FIG. 8—*J-integral calibration curve for blunt-notched 2T compact specimen; a/w = 0.451.*

The determination of J by area approximation methods was first calculated in terms of the Rice approximation [*11*], denoted by the closed triangles. Although the Merkle-Corten method [*12*] is handled as a separate and distinct calculation, according to Fig. 3, it can also be treated as a correction procedure to the Rice pure bend approximation method. Here we find that correction magnitudes tended to be constant over a range of displacement levels for a given initial crack size. The magnitude of the correction is dominated by initial crack aspect ratio, and Figs. 5-7 show the correction to be nominally 1.14 for $a_o/w = 0.6$, 1.20 for $a_o/w = 0.5$, and 1.35 for $a_o/w = 0.4$. From Figs. 5-9, it is concluded, therefore, that the Rice approximation procedure will always yield conservative estimates of J in compact specimens and that the M-C method improves the J estimate to compare favorably with real J. It is noted, however, that for short crack aspect ratios, the M-C correction tends to overcorrect and J estimates will be on the nonconservative side.

Figures 10 and 11 compare J determinations in the center notched tension specimens, CNT, listed in Table 1, for initial crack aspect ratios,

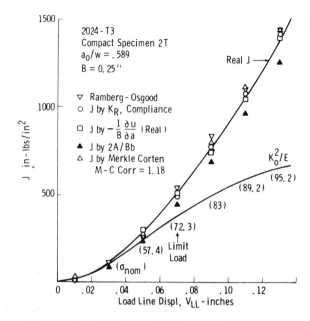

FIG. 9—*J-integral calibration curve for blunt-notched 2T compact specimen;* $a/w = 0.589$.

$2 a_o/w$, of 0.40 and 0.45. Again real J is represented by the curve which is a best fit to the B-L method, open-square data points. The lower curves are again the elastic values of G, and in these cases are almost an order of magnitude less than real J. Levels of net section stress are shown and theoretical limit load levels are indicated by vertical arrows.

The J-integral derived from the Ramberg-Osgood fit did not work satisfactorily in the CNT cases. The load-displacement records were almost elastic-perfectly plastic in nature and, because of this, n-values were of the order of 30 and κ was of the order of 10^{50}. These values are highly unrealistic for the modeling of plasticity effects, and this evidently proved to be the principal cause of the breakdown of J prediction by R-O.

A good comparison between J_R from K_R and real J is shown for stress levels well beyond limit load. This comparison tended to break down at

TABLE 2—*Ramberg-Osgood work-hardening constants.*

Material	a_0/w	κ	n
HY130	0.4	108.4	2.866
HY130	0.5	187.9×10^4	5.751
HY130	0.6	805.4×10^4	6.376
2024-T3	0.451	131.8	3.075
2024-T3	0.589	260.6×10^2	5.019

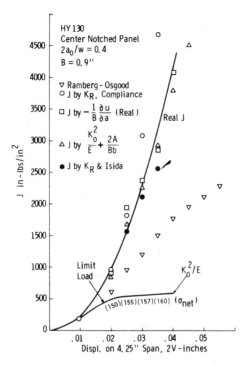

FIG. 10—*J-integral calibration curve for blunt-notched center-cracked panels; 5.08 cm (2 in.) wide,* $2a/w = 0.4$.

large displacements, however. Examination of the data output for these calculations and for specimens with other initial crack aspect ratios, not shown, indicated that this deviation developed when $2a$ (effective) became greater than 80 percent of the total cross-sectional area. Further, it was noted that the rapid increase in K_R was apparently not aggravated by compliance-indicated crack size nor by gross stress level. The sharp deviation appears to be due to the secant expression, $(\sec \pi a/w)^{1/2}$, adjustment to K_1 for finite specimen width. It was interesting to note that if the polynomial approximation to the Isidas' expression for the CNT configuration is used (not supposed to be valid beyond $2a/w = 0.7$), the fit to real J is conservative for crack aspect ratios greater than 0.8. These data are indicated by closed circles in Figs. 10 and 11. The modified area approximation method suggested by Rice [11] for the CNT configuration, indicated by open triangles, gave satisfactory values at all levels of displacement, accurate to within 15 percent of real J [10]. This indicates that even at displacements well beyond theoretical limit load, comparable estimates of the fracture resistance of the material are possible using J. The calculation of K_R from a-effective, on the other hand, becomes extremely sensitive

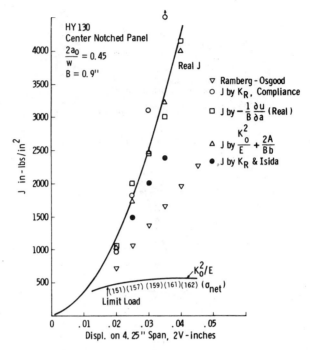

FIG. 11—*J-integral calibration curve for blunt-notched center-cracked panels;* w = 5.08 cm (2 in.) *wide,* 2a/w = 0.45.

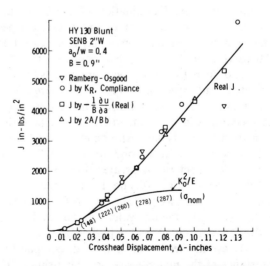

FIG. 12—*J-integral calibration curve for blunt-notched SENB specimen;* w = 5.08 cm (2 in.), a/w = 0.4.

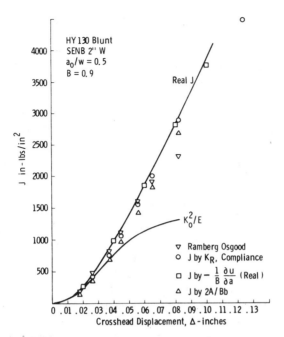

FIG. 13—*J-integral calibration curve for blunt-notched SENB specimen; w = 5.08 cm (2 in.), a/w = 0.5.*

under such conditions and therefore tends to be unreliable, especially when crack aspect ratios are greater than 0.8.

Figures 12 and 13 compare J calibration developed for blunt notched SENB specimens. Two example crack aspect ratios of 0.4 and 0.5 are shown. In these cases there was some difficulty in developing good Ramberg-Osgood fits to the test records, and the deviation in predicted J perhaps reflects this problem. Again, J calculated from plastic zone adjusted K_R values corresponds well with real J over the full range of displacements. Since these specimens have negligible tensile component, the Rice pure bend area approximation method compared quite favorably with real J.

Summary of Computational Methods

Tests on blunt notched specimens for three specimen geometries and for two materials were helpful in exploring the strengths and weaknesses of alternative J-integral computational methods. The principal categories are (1) the Ramberg-Osgood modeling with work-hardening constants from which J can be estimated, (2) area approximation methods, and (3) J obtained from compliance-corrected K_R values.

The Ramberg-Osgood work-hardening law can be used to fit load-displacement records and the determination of J using κ and n works reasonably well so long as the test record does not approach elastic-perfectly plastic behavior as had developed in the present CNT tests. However, each variation in initial crack size required the development of unique work-hardening constants. Because of this, there can be no solid rationale developed to justify the use of this approach. In order to make the R-O method work more effectively, it would be necessary to find the best fit of κ and n for a family of test curves generated from specimens of varied initial crack sizes. For all the difficulty that this would involve, it would be more expedient to determine J directly using the Begley-Landes procedure.

Area approximation methods of computing J proved to develop good estimates of real J. In the case of three-point bend and centrally notched specimens, the expressions suggested by Rice et al [11] worked satisfactorily. For compact specimens, the Rice expression, developed for pure bending, was substantially improved through tensile component adjustments suggested by Merkle and Corten [12]. These M-C corrections tended to be fixed over a range of displacements for a given initial crack aspect ratio. In the present work, the adjustment varied between 35 and 14 percent for initial crack aspect ratios, a_o/w, varying from 0.4 to 0.6, respectively. It was generally observed here that the best comparisons to real J were obtained with the larger initial crack aspect ratios. For short a/w, the corrected values tended to overestimate J.

The compliance technique of drawing secants to test records for predicting effective crack size, and the substitution of these values in elastic stress intensity expressions for crack length, develops K_R values that are essentially equivalent to J. The procedure was tested on two materials and on three specimen geometries. The comparison to real J was generally the best for all computational procedures tried. The only exception to this was for center-notched panels where a-effective was extended to more than 80 percent of the section width and applied stress was well beyond calculated limit load. The compliance concept was tested over a considerable variability in test record shape, a comparison of which is shown in Fig. 14. The point at which the CNT values of J from K_R started to deviate significantly from real J is indicated by the vertical arrow.

Conclusions

1. Tests on blunt notched compact specimens, center-notched panels, and single-edge notched bend specimens were used to evaluate K_R (calculated from compliance-indicated crack size) as an indicator of elastic-plastic toughness. Comparisons were made between these results and J determined directly using the Begley-Landes procedure. It was demonstrated that there

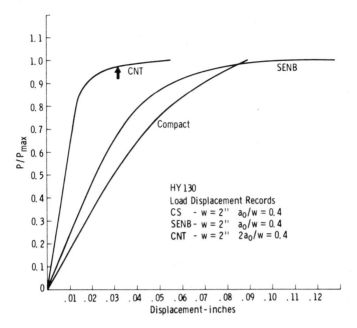

FIG. 14—*Load-displacement records for three specimen types.*

is equivalency between K_R and J and that the equivalence is retained up to theoretical limit load on all specimen geometries. Although this demonstration was made in the absence of subcritical crack growth, its main utility would be to handle cases where subcritical crack growth develops such as R-curves, K_{Iscc}, and possibly in creep studies. Also the concept could be used to analyze for J experimentally and diminish the need for J analysis of geometries where J solutions do not presently exist.

2. Area approximation methods for determining J were tested and were confirmed as being reasonably accurate estimates of real J. For compact specimens, the Merkle-Corten computational procedure was applied and was found to yield better estimates of real J than the pure bend model. For initial crack aspect ratios, a_o/w less than 0.5, the M-C method tended to slightly overcorrect, resulting in nonconservative estimates of J.

3. The Ramberg-Osgood work-hardening expression was applied to the calculation of J and was found to be not completely reliable as a computational procedure. The technique appeared to work reasonably well on compact specimens, but ran into difficulty in CNT and SENB tests. The probable reason for this is that the R-O work-hardening constants, κ and n, in this type of application are values of convenience, lacking fundamental significance with regard to material properties.

References

[1] Barsom, J. M. and Rolfe, S. T., *Journal of Engineering Fracture Mechanics*, Vol. 2, 1971, p. 341.
[2] Shoemaker, A. K. and Rolfe, S. T., *Engineering Fracture Mechanics*, Vol. 2, 1971, pp. 319-339.
[3] J. D. Landes and J. A. Begley in *Fracture Toughness, ASTM STP 514*, American Society for Testing and Materials, 1972, pp. 24-39.
[4] British Standards D.D. 19, "Methods for Crack Opening Displacement (COD) Testing," 1972.
[5] *Fracture Toughness Evaluation by R-Curve Methods, ASTM STP 527*, American Society for Testing and Materials, 1973.
[6] Paris, P. C., Tada, H., Zahoor, A., Ernst, H., this publication, pp. 5-36.
[7] Irwin, G. R. and Paris, P. C., "Elastic-Plastic Crack Tip Characterization in Relation to R-Curves," Plenary Paper for ICF-4, Fourth International Conference on Fracture, Waterloo, Ont., Canada, June 1977.
[8] Turner, C. E. and Sumpter, J. D. G., *International Journal of Fracture Mechanics*, Vol. 12, No. 6, Dec. 1976.
[9] Landes, J. D., Walker, H., and Clarke, G. A., this publication, pp. 266-287.
[10] Begley, J. A. and Landes, J. D. in *Fracture Toughness, ASTM STP 514*, American Society for Testing and Materials, 1972, pp. 1-20.
[11] Rice, J. R., Paris, P. C., and Merkle, J. G. in *Progress in Flaw Growth and Fracture Toughness Testing, ASTM STP 536*, American Society for Testing and Materials, 1973, pp. 231-245.
[12] Merkle, J. G. and Corten, H. T., *Transactions*, American Society of Mechanical Engineers, *Journal of Pressure Vessel Technology*, Vol. 96, No. 4, Nov. 1974, pp. 286-292.
[13] Liebowitz, H. and Eftis, J., *Engineering Fracture Mechanics*, Vol. 3, No. 3, Oct. 1971, p. 267.
[14] Newman, J. C. Jr. in *Properties Related to Fracture Toughness, ASTM STP 605*, American Society for Testing and Materials, 1976, pp. 104-123.

M. G. Dawes[1]

Elastic-Plastic Fracture Toughness Based on the COD and *J*-Contour Integral Concepts

REFERENCE: Dawes, M. G., "**Elastic-Plastic Fracture Toughness Based on the COD and *J*-Contour Integral Concepts,**" *Elastic-Plastic Fracture, ASTM STP 668,* J. D. Landes, J. A. Begley, and G. A. Clarke, Eds., American Society for Testing and Materials, 1979, pp. 307-333.

ABSTRACT: The paper reviews the definition, fracture characterizing roles, and measurement of critical COD and *J*-values. It is proposed that COD should be defined as the opening displacement at the original crack tip position. This definition avoids much of the ambiguity of previous definitions based on the crack tip profile and the elastic-plastic interface. Attention is drawn to a fundamental problem which limits the general application of the *J*-contour integral concept to elastic-plastic descriptions of the crack tip environment when cracks occur in overmatching yield strength weld regions. A comparison of recent three-point single-edge notch bend (SENB) testing techniques, based on the standard instrumentation used in K_{Ic} tests, shows there is a close mathematical link between the estimated values of COD and *J*.

Experimental data, obtained over a wide range of temperatures, are used to demonstrate how the critical values of COD and *J* for unstable fracture are affected by variations in specimen geometry. Also, it is shown that measurements of J_{Ic} may lead to overestimates of K_{Ic} in materials having yield strengths less than approximately 700 N/mm^2.

KEY WORDS: mechanical properties, fracture tests, crack initiation, toughness, crack opening displacement, *J*-integral, elastic-plastic cracking (fracturing), fracture properties, structural steels, crack propagation

Nomenclature

- a Half length of through-thickness crack, or depth of surface crack
- B Section and specimen thickness
- ∂u_i Work term for *J*
- E' E for plane stress or $E/(1 - \nu^2)$ for plane strain
- e Strain
- e_{ij} Strain tensor

[1] Principal research engineer, The Welding Institute, Cambridge, England.

G Crack extension force
J J-contour integral
K Elastic stress intensity factor
m Plastic stress intensification factor
P Load
P_L Limit load
P^e Potential energy
q Load point displacement
q_p Plastic component of q
r_p Rotational factor after net section yielding in bend test
T_i Work term for J
U Work done
U_e Elastic strain energy
U_p Work done in plastic deformation
V Clip gage notch mouth opening in bend test
V_p Plastic component of V
W Width of single-edge notched bend specimen or half width of center-cracked tension specimen
w Strain energy density $= \int_0^e \sigma_{ij} de_{ij}$
z Knife edge thickness
Γ Path contour for J
δ Crack tip opening displacement
η_e, η_p Constants for elastic and plastic work terms, respectively
ν Poisson's ratio
σ_{flow} $(\sigma_Y + \sigma_U)/2$
σ_{ij} Stress tensor
σ_U Tensile strength
σ_Y Uniaxial yield strength

The requirement for a yielding fracture mechanics (YFM) approach in design is exemplified by elastically stressed structures that contain precracked regions in which the in-plane and antiplane dimensions, combined with local stresses, result in unstable quasi-brittle fracture extension, that is, the initiation of unstable fracture after the attainment of sufficient plasticity in the crack tip region to invalidate a description of resistance to crack extension in terms of the plane-strain stress intensity factor, K_{Ic}. Attempts to match the crack tip region plasticity and constraint in these situations commonly involve post net section yielding fracture behavior in small laboratory specimens. This paper is concerned with two parameters which attempt to characterize fracture toughness under such conditions. These are the critical crack tip opening displacement (COD) [1],[2] and a critical value based on the J-contour integral [2].

---

[2] The italic numbers in brackets refer to the list of references appended to this paper.

The following sections review the definition, fracture characterizing roles, and measurement of critical COD and J-values. The results of recent experiments are also presented. Emphasis is given to three-point single-edge notch bend (SENB) tests and specimen geometries that may be used to assess the fracture toughness associated with both deeply buried cracks and shallow cracks in weld metals and weld heat-affected zones (HAZs).

Crack Tip Opening Displacements

Approximately 17 years have passed since Wells [1] suggested COD as a parameter which might be used to describe the capacity of material near a crack tip to deform before crack extension. Numerous encouraging investigations in different countries have followed this suggestion, and in the United Kingdom a simple COD design curve approach [3-7] has evolved which has found widespread application to welded structural steels [8,9]. Nevertheless, doubts are still expressed regarding both the physical occurrence and the definition of COD. These aspects have been the subject of an extensive review [10], which led to the following observations.

Physical Evidence of COD

The literature shows that the profiles of deformed cracks are dominated by such factors as the orientation of the notch relative to the microstructure of the material, the type of microstructure, the grain size, and nonmetallic inclusions. Figure 1 illustrates some typical crack tip profiles [11], starting with an undeformed 0.15-mm-wide notch, Fig. 1a, and successive opening displacements in Fig. 1b-d. The extension ahead of the original crack tip, indicated by x in Fig. 1c, is approximately equal to the region which is generally referred to as the 'stretch' zone. This has been correlated with both K_{Ic} values [12] and COD values [13-15].

Figure 1e and f show examples of COD [16,17] on a scale an order of magnitude smaller than the examples referred to in the foregoing. At this scale the crack tip profiles are dominated by microstructural features such as grain boundaries and nonmetallic inclusions.

Analytical Models of COD

In the Wells small-scale yielding model [1] and the Dugdale strip yielding-based models [3,18,19], there is an implicit assumption that the COD occurs at both the original crack tip and the elastic-plastic boundary. However, nothing is implied regarding radial displacements ahead of the crack tip. It is perhaps for the latter reason rather than from definite experimental evidence that researchers became preoccupied with the search for a square 'nose' in the crack tip profile, ostensibly located at the crack tip.

310 ELASTIC-PLASTIC FRACTURE

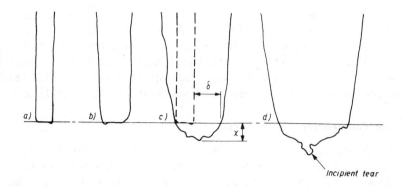

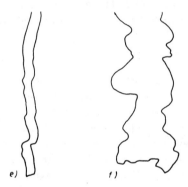

FIG. 1—*Examples of COD.* (a) *to* (d): *0.15-mm-wide sawcut* ×33 [11]; (e) *fatigue crack* ×330 [16]; *and* (f) *ductile tear* ×158 [17].

By considering shear strains in relation to a Prandtl field, Rice and Johnson [20] predicted that the crack tip would be deformed by radial displacements, such that for both small-scale and large-scale yielding the displacement tangential to the original crack tip would be approximately double the displacement ahead of this position. A similar amount of stretching of the deformed crack ahead of the original crack tip position is predicted by Pelloux's [21] alternating shear model. Thus, both Rice and Johnson's [20] and Pelloux's [21] models show approximate agreement with the physical evidence of deformed crack tip profiles.

Numerical Estimates of COD

Because of the complexity of analytical solutions for elastic-plastic materials, finite-element analyses have been used extensively in investigations of crack tip profiles. However, the early investigations were preoccupied with the search for a clearly defined nose in the crack tip profile. Unfortunately, for tension situations and work-hardening materials the computed crack tip profiles are generally more rounded. In these circumstances, therefore, there was a problem of defining a near-tip COD. In response to work by Srawley, Swedlow, and Roberts [22], Wells and Burdekin [23] suggested that the COD should be redefined as the displacement at the elastic-plastic boundary. While this definition is reasonable for small-scale yielding conditions, it is not acceptable for more extensive yielding in materials that work harden, since in these cases the elastic-plastic boundary may move back a significant distance along the flanks of the crack. When this happens, the COD at the elastic-plastic boundary is dependent on crack length, and the COD is not, therefore, a one-parameter description of the near crack tip environment. Boyle [24] has suggested a number of alternative methods of defining COD from a rounded crack profile. Although these definitions may be justified in relation to constant strain triangle finite-element analyses, which have a single fixed node at the crack tip, they appear to be unnecessary for those analyses [25,26] which use special element designs to model the crack tip deformations more closely. According to Rice [26], it is essential to use sophisticated crack tip elements to obtain satisfactory modeling of such deformations.

Fortunately, from the viewpoint of crack tip deformations, the foregoing and other limitations of present finite-element analyses [27] are less important in bending situations. This is because the notch flanks tend to remain straight during bending, and the intersections of the tangents to the notch flanks and crack tip give a COD which is negligibly smaller than the COD a small distance behind the crack tip, that is, δ in Fig. 1c.

Definition of COD

The foregoing considerations suggest that the Mode I COD can be defined as the displacement at the original crack tip position, namely, the tip of the fatigue precrack in a COD test specimen or a natural crack in a structure. This definition recognizes the formation of a stretch zone ahead of the original crack tip and avoids most of the problems associated with earlier definitions based on the deformed crack tip profile and the elastic-plastic boundary. Also, by defining the original crack tip as the reference position, consistency is maintained with both experimental measurements of COD and the early analytical models [1,3,18-21].

The J-Contour Integral Concept

Since rigorous analytical stress analysis solutions for cracked bodies in 'real' (elastic-plastic) materials have proved too difficult, analysts have gained an insight into real material behavior by assuming materials which have linear and nonlinear *elastic* behaviors. These studies have led several workers [2,28,29] to derive path-independent line integrals which may be used to obtain an approximate description of the crack tip environment prior to fracture. The first of these integrals to gain prominence in engineering fracture studies was that due to Rice [2]. He defined the path independent J-contour integral as

$$J = \int_\Gamma \left(w dy - T_i \frac{\partial u_i}{\partial x} ds \right) \quad (1)$$

which, for both linear elastic and nonlinear elastic material, was shown to be equal to the potential energy release rate per unit thickness, that is

$$J = -\frac{dP^e}{da} \quad (2)$$

Furthermore, when limited to linear elastic behavior and small-scale yielding, Eq 2 reduces to

$$J = G = \frac{K^2}{E'} \quad (3)$$

The important implication of path independence is that measurements of J remote from the crack tip can be used to describe conditions near the crack tip. Also, if there is a singularity of stress or strain near the crack tip, a critical value of J can be used as a fracture characterizing parameter.

For incremental plasticity, which is the behavior more appropriate to real materials, path independence has not been proved analytically. However, elastic-plastic finite-element computations have shown J to be virtually path independent except for contours very close to the crack tip [24,30,31], which is the region where finite-element analyses are least accurate.

Unfortunately, as demonstrated by Sumpter and Turner [32], when applied to a real material, the physical meaning of J as an energy release rate, Eq 2, is lost, since in either deformation or incremental plasticity the energy term is no longer potentially available for propagating a crack. This is because a proportion of the energy has been dissipated in plastic deformation. For an elastic-plastic material, therefore, Eqs 1 and 2 must be interpreted as

$$J = -\frac{1}{B} \times \frac{dU}{da} \quad (4)$$

which represents the rate with respect to crack length of elastic and plastic work done. In YFM, therefore, the use of Eq 4 relies not on energy balance arguments, but instead on path independence and the degree to which the value of J is related to a singularity of stress or strain in the crack tip region.

In the absence of complete solutions for cracks in elastic-plastic materials, present assessments of J as a fracture characterizing parameter depend almost entirely on experimental fracture studies.

Relationships Between COD and J

Wells's [1] original small-scale yielding estimate of COD [8] was in the form

$$\delta = \frac{G}{m\sigma_Y} = \frac{K^2}{m\sigma_Y E'} \tag{5}$$

It follows from Eqs 3 and 5, therefore, that

$$J = m\sigma_Y \delta \tag{6}$$

This relationship has been investigated for both small- and large-scale yielding conditions. A review [10] of available analytical, numerical, and experimental investigations shows that the factor m is generally between approximately 1.0 and 2.0. However, discrepancies between the theoretical and experimental values of m raise a number of questions regarding definitions, methods of measurement, and also the relevance of discrete values of σ_Y. These questions will be returned to at a number of points in the following sections.

COD and J as Fracture Characterizing Parameters

Since values of δ are directly proportional to values of J (Eq 6), it is helpful to discuss the critical values of these parameters together. However, it is first necessary to distinguish between the critical δ/J values for the onset of crack extension by unstable fracture and the critical values for the onset of stable crack growth. As mentioned in the introduction to this paper, in the context of YFM, unstable fracture refers to a quasi-brittle fracture, which in the common structural steels is usually associated with a significant proportion of cleavage crack growth. On the other hand, stable crack growth is usually associated with a coalescence of the voids that form around inclusions ahead of the crack tip. Unstable fracture in the ductile/brittle transition temperature range may occur either before or after the onset of stable crack growth.

In subsequent discussions, the critical δ/J values for unstable and stable crack extension will be symbolized by δ_c/J_c and δ_i/J_i, respectively. Furthermore, δ_c/J_c will be used exclusively to indicate fracture with a rising load.

Unstable Fracture: δ_c and J_c

There is much evidence in the literature to show that resistance to quasi-brittle fracture may be affected in a given material by temperature, strain rate, and section thickness. Similarly, it is well known that resistance to fracture decreases with increasing notch acuity. Several major investigations have demonstrated that the aforementioned factors have a similar effect on δ_c values [3,11,33–36] and J_c values [37,38].

Recent investigations have also indicated that δ_c values are dependent on loading type and the in-plane dimensions of specimens [33,34,39–42]. This dependency is such that for tension plates containing through-thickness cracks and geometries having $a/W < \approx 0.5$, the minimum δ_c values occur when a is approximately one half of B. For $a \ll B$, δ_c may be an order of magnitude higher than the minimum value, whereas for $a \gg B$ the δ_c values are generally no more than approximately double the minimum value. From the viewpoint of simple laboratory tests for δ_c, it is significant that the minimum δ_c values for through-thickness cracks in tension plates are slightly underestimated by the preferred three-point SENB geometries for K_{Ic} testing, that is, the ASTM Test for Plane-Strain Fracture Toughness of Metallic Materials (E 399-74) and British Standard Methods for Plane Strain Fracture Toughness (K_{Ic}) Testing (BS 5447-1977). The experimental evidence also indicates that full section thickness three-point SENB specimens will underestimate the δ_c values for surface notches in tension plates provided that the bend specimen crack length and width match the crack depth and section thickness for the tension plate. Ideally, however, the laboratory specimen should be designed to match the plastic constraint in the structural part of interest. This aspect of δ_c testing, and also J_c testing, is especially relevant to the design of weld HAZ [43] and weld metal [44] fracture toughness specimens.

Initiation of Stable Crack Growth: δ_i and J_i

Before fracture mechanics approaches were used to assess resistance to brittle fracture in structural steels, little, if any, consideration was taken of small amounts of stable crack growth before unstable fracture. In fact, the presence of a ductile 'thumbnail' on the fracture surface at the crack tip was often interpreted as an indication of ductility. However, the definition of the critical event as that corresponding to the first detectable extension of a crack has focused attention on δ_i and J_i values when these occur before unstable fracture. While there is some evidence to show that it can be very conservative to use values of δ_i and J_i in design [9], this aspect remains controversial and awaits the further development of theoretical R-curve relationships between, for example, bend test values of δ_i/J_i, δ_c/J_c, and unstable fracture in structural situations.

The fact remains, however, that the δ_i and J_i values come nearest to being 'material properties.' For example, many studies of δ_i [13,15,45-51] and J_i [50,52-56] have shown that under sufficiently 'plane-strain' conditions these values are independent of geometry and loading type. For steels, at least, a sufficient degree of plane strain for a constant J_i is generally ensured when [54]

$$a, B \text{ and } W - a \geq 25 \frac{J_i}{\sigma_{\text{flow}}} \quad (7)$$

Estimates of K_{Ic} *from* J_c *or* J_i

In the United States much interest has been expressed regarding the prediction of K_{Ic} from critical J-values which have been obtained from considerably smaller specimens than those required by the ASTM E 399-74 and BS 5447. For example, when the requirements of Eq 7 are met, the values of J_i or J_c prior to stable crack growth are termed J_{Ic} values, and are used to predict K_{Ic} from Eq 3, for example

$$K_{Ic} = (J_{Ic} E')^{1/2} \quad (8)$$

A similar approach to the foregoing was used by Robinson and Tetelman [49] to estimate K_{Ic} values from critical COD values. These approaches seem reasonable when the same micromodes of fracture initiation can be guaranteed in both the small specimen and the much larger valid K_{Ic} ASTM E 399-74 specimen. In fact, under the latter conditions, a J_i-value that meets the requirements of Eq 7 may underestimate a valid K_{Ic} when this is based on up to 2 percent crack growth, which is permitted by ASTM E 399-74.

Unfortunately, for those conditions when a standard K_{Ic} test gives a valid result following unstable fracture with no prior stable crack growth, values of J_{Ic} from small generally yielded specimens may overestimate K_{Ic} via Eq 8, depending on the ductile/brittle transition temperature behavior for the materials and designs of specimen. In Fig. 2, for example, at temperatures below the ASTM E 399-74 plane strain fracture initiation mode change (that is, from cleavage to microvoid coalescence), a J_{Ic}-value for a specimen of thickness B might considerably overestimate the J_c value for an ASTM E 399-74 valid K_{Ic} test specimen at the same temperature. Several investigations [10,57,58] have drawn attention to this behavior, which is featured in the experimental results. However, before describing these

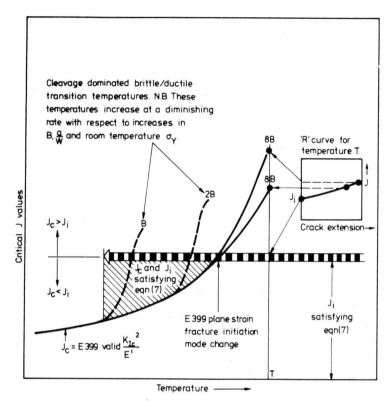

FIG. 2—*Schematic diagram of the relationships between* J_c, J_i *section thickness* (B), *and temperature.*

and other recent experimental studies, it is helpful to describe the test methods.

Three-Point SENB Tests

The emphasis on bend tests stems from the experience that these allow the maximum economy in weld zone material, machining, and fatigue cracking costs, and also the maximum versatility in specimen design and testing equipment. The versatility in specimen design is considered especially important in regard to YFM assessments of cracks in discrete regions of welds [43,44], as mentioned earlier.

Unless stated otherwise, the tests refer to full section thickness (B) square specimens, or $B \times 2B$ section specimens complying with ASTM E 399-74 and BS 5447.

COD Tests

Since standard K_{Ic} tests involve measurements of notch mouth opening displacement, correlations between this displacement and COD offered the prospect of a unified test method which could be interpreted in terms of COD or K_{Ic} depending on the toughness of the material being tested. Consequently this approach has been pursued by many investigators and has been justified to a large extent by experiments [11,33,39,49], finite-element analyses [33,59], and theoretical considerations [59,60]. The latter investigations generally support the British Standards Institution (BSI) Draft for Development, DD19, which was the basis for the COD tests described later. However, it should be noted that a draft BSI standard for COD testing is in the final stages of preparation. This new document specifies that the COD should be calculated using the following relationship [10]

$$\delta = \frac{K^2}{2\sigma_Y E'} + \frac{V_p}{1 + 2.5\left(\frac{a+z}{W-a}\right)} \quad (9)$$

which gives similar estimates to the former more complex relationships in DD19, due to Wells [60].

In Eq 9 the value of K is calculated from the load at the point of interest on the load versus notch mouth opening record.

J-Tests

The estimation procedures in these tests were based on the areas under single load versus displacement diagrams that were either directly or indirectly related to the work done.

A general consideration of different specimen geometries led Sumpter [31] to suggest the following relationship which can be used for any geometry for which the elastic compliance and limit load are known

$$J = \frac{\eta_e U_e}{B(W-a)} + \frac{\eta_p U_p}{B(W-a)} = \frac{K^2}{E'} + \frac{\eta_p U_p}{B(W-a)} \quad (10)$$

Using this relationship, the total energy under the experimental load versus point displacement record is divided into elastic and plastic components as shown in Fig. 3. The value of η_e is constant for a particular geometry and loading type, and can be easily derived from the elastic compliance and stress intensity factor [62]. The second term in Eq 10, and therefore, η_p, may be obtained by treating plastic work done, U_p, as that correspond-

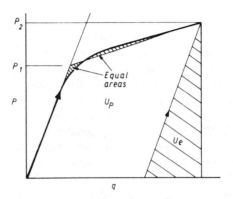

FIG. 3—*Schematic load versus load point displacement diagram.*

ing to the deformation of a rigid plastic body, that is, $U_p = q_p P_L$, where q_p is the plastic component of load point displacement and P_L is the limit load. Hence, from Eq 4

$$\frac{\eta_p U_p}{B(W-a)} = -\frac{1}{B} \times \frac{d(P_L q_p)}{da}$$

For three-point SENB specimens having $a/W > 0.15$, it can be shown that $\eta_p = 2.0$. Also, when these specimens are tested over a span of $4W$, and have $0.45 \leq a/W \leq 0.65$, Eq 10 reduces to

$$J = \frac{2(U_e + U_p)}{B(W-a)} = \frac{2U}{B(W-a)}$$

which is the deep notch form used in the ASTM studies [54].

There are two problems concerning the measurement of load point displacements in SENB tests. The first is the difficulty of separating the true load point displacement of the test specimen from the displacements under the loading points and in the testing system [49]. An investigation of these displacements led the author [10] to develop the equipment shown in Fig. 4. With this equipment the vertical displacement of the notch mouth is measured *relative* to the top surface of a 'comparison' bar. The bar rests on pins which are attached to the specimen at the ends of the loading span. The initial contact points between the comparison bar and the pins are on the neutral axis of the specimen. It was shown [10] that the vertical displacement of the notch mouth represents q to an accuracy of better than ± 2 percent, provided that the total angle of bend is less than 8 deg, which is approximately the maximum value of interest in fracture initiation tests.

FIG. 4—*Equipment for simultaneous measurements of* q, *using a linear transducer, and* V, *using linear clip gage.*

The second problem in measurements of q concerns SENB tests on shallow-notched specimens, especially when the notches are located in weld zones. As shown schematically in Fig. 5, situations can arise where plastic hinges form in the base metal adjacent to notches in an overmatching yield strength weld zone. In these instances the load point displacement and, therefore, the estimated J, will not represent the effective value of J in the notched region, which may involve a relatively small component of work done in plastic deformation. Herein lies a fundamental difficulty in applying the path independent J-contour integral concept to elastic-plastic deformations in varying yield strength regions of welded structures. From the viewpoint of a laboratory test, however, the effective value of J can be estimated by considering contours around the crack, but within the welded region, for example, by estimating J from measurements of the crack mouth opening displacement. This can be done using the following relationships (compare Eq 10) proposed by Sumpter and Turner [61]

$$J = \frac{K^2}{E'} + \frac{2P_L}{B(W-a)} \left[\frac{WV_p}{a + z + r_p(W-a)} \right] \quad (11)$$

where $r_p = 0.4$ for SENB specimens having $a/W > 0.45$, and 0.45 for $a/W < 0.45$. Equation 11 has the added advantage that it can be used with the standard instrumentation for K_{Ic} tests.

The only difficulty with Eq 11 concerns the definition of the limit load, P_L, on the load versus clip gage displacement record. For the experimental work which follows, therefore, Eq 11 was modified to give

$$J = \frac{K^2}{E'} + \frac{P_1 + P_2}{B(W-a)} \left[\frac{WV_p}{a + z + r_p(W-a)} \right] \qquad (12)$$

where P_1 and P_2 are determined as shown in Fig. 3.

Finally it is worth noting the similarity between estimates of δ and J using Eqs 9 and 11, respectively. For instance, taking $P_L = [1.5 \sigma_Y B(W - a)^2]/4W$ it can be shown that the equations predict $J \leq 2\sigma_Y \delta$, that is, $m \leq 2.0$ (Eq 6). Also, it may be observed that the substitution of $(P_1 + P_2)/2$ for P_L (Eq 12) will generally result in larger post net section yield values of m in materials having low ratios of yield to tensile strength.

Experimental Studies

Materials

The Ducol W 30 Grade B and BS 4360 Grade 50C steels used had the chemical analyses and basic material properties summarized in Table 1. Both materials were supplied in the form of 25-mm-thick normalized plate.

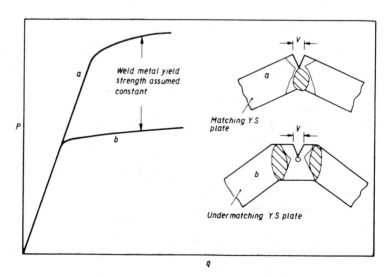

FIG. 5—*Schematic load versus load point displacement behavior in weldment specimens containing shallow notches.*

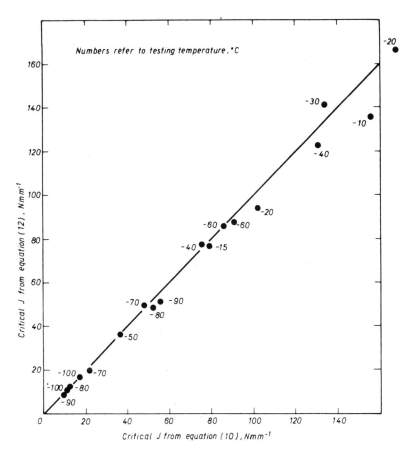

FIG. 6—*Comparison of J-estimates from Eqs 12 and 10: Ducol W30 steel plate.*

Comparison of Eqs 10 and 12

Although Eq 12 provides the most direct and simple method for obtaining J-values from a P versus V record, some doubt must surround the use of a constant rotational factor, r_p, for different materials [33]. It was of interest, therefore, to use the equipment shown in Fig. 4 to obtain a comparison of the J-estimates from Eqs 10 and 12, that is, from simultaneous measurements of q and V. This was done using a series of full section thickness (25 mm) Ducol W 30 SENB specimens having $a/W = 0.5$, $W/B = 2.0$, and $S/W = 4$. As shown in Fig. 6, there was excellent agreement between the values of J_c estimated from Eqs 10 and 12 for temperatures ranging from -100 to $-10°C$. It may be noted that all the test results in Fig. 6 refer to unstable fractures without prior stable crack growth.

While the foregoing results are encouraging, it may be necessary to check the relationships between q and V in different materials and different SENB geometries. For example, in the case of shallow-notched, overmatching-yield-strength weld metal specimens (Fig. 5), it would be helpful to check the relationships using plain material or all-welded metal specimens having similar geometries to the ones of interest, for example, by making simultaneous measurements of the plastic components of q and V using the equipment in Fig. 4.

Measurements of Critical COD and J-Values

These tests were carried out to compare the effects of geometry and temperature on critical values of COD and J for both the onset of stable crack growth and unstable cleavage fracture.

Full thickness (25 mm) three-point SENB specimens were extracted from the BS 4360 Grade 50C steel plate (Table 1) and prepared with through-thickness notches. Tests were then carried out using the ASTM E 399-74 instrumentation and the results were interpreted using the Wells [60] DD19 COD relationship (which gives values similar to Eq 9) and the slightly modified Sumpter and Turner [61] J-relationship, Eq 12.

Figures 7 and 8 show that the ductile/brittle transition temperature ranges were generally raised by increasing the a/W ratios from 0.2 to 0.5, and increasing the notch acuity from a 0.075-mm-radius notch to a fatigue crack. This behavior was true for unstable quasi-cleavage fractures occurring both before and after the onset of stable crack growth. The approximate lengths of stable crack growth in the fatigue precracked specimens are shown in Fig. 9a and b, which are in the form of 'R curves'.

Figure 10 illustrates the near linear relationships between COD and J over a wide range of temperatures. The data in Fig. 10 were combined with more detailed tensile strength data (Fig. 11) to give the values of m (Eq 6) in Fig. 12. This shows that, generally, $m < 2.0$ over a wide range of temperatures, as predicted earlier.

Since there is a close mathematical link between the present COD and J estimation procedures, the δ_c and J_c must be equally useful as fracture characterizing parameters, at least for the three-point SENB specimen geometries examined. This conclusion was confirmed by a variance ratio analysis, which showed that the nondimensional standard deviations of the total δ_c and total J_c values for each geometry in Fig. 7 were equally insensitive to geometry.

Experimental Estimates of K_{Ic} from J_{Ic}

Figure 13 shows the results of a partially completed program of work on a BS 4360 Grade 50D steel. This material has a chemical analysis and

TABLE 1—*Chemical analyses and mechanical properties of plain plate materials.*

Material	Element, weight %										
	C	S	P	Si	Mn	Ni	Cr	Mo	V	Cu	Nb
Ducol W30	0.13	0.013	0.02	0.37	1.23	0.82	0.5	0.25	0.08	0.10	0.044
BS 4360 50C	0.15	0.018	0.031	0.37	1.17	0.06	0.07	0.01	<0.01	0.14	0.028

	Yield Strength, N/mm^2	Tensile Strength, N/mm^2	Elongation, %	Reduction of Area, %
Ducol W30	448	596	30	74
	454	594	31	73
	451	600	30	73
BS 4360 50C	362	529	31	68
	359	526	40	67

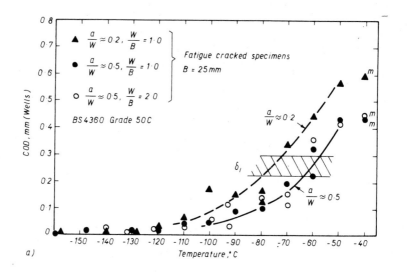
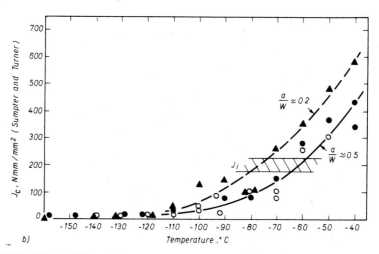

FIG. 7—*Critical COD and J versus temperature for fatigue cracked specimens in BS 4360 Grade 50C steel.*

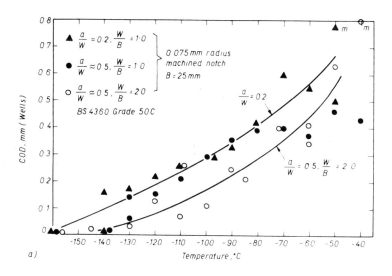
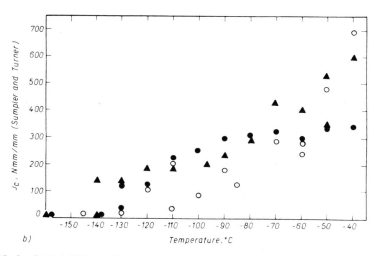

FIG. 8—*Critical COD and J versus temperature for machined notch specimens in BS 4360 Grade 50C steel.*

326 ELASTIC-PLASTIC FRACTURE

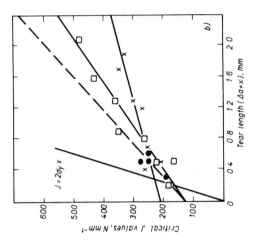

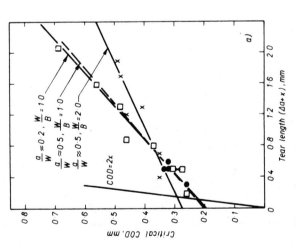

FIG. 9—*Crack growth resistance curves for fatigue-cracked BS 4360 Grade 50C steel, (a) based on COD and (b) based on J.*

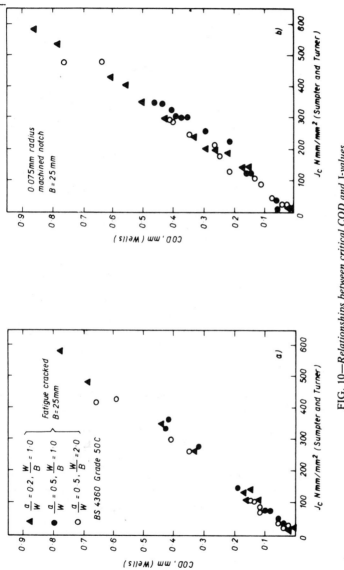

FIG. 10—*Relationships between critical COD and J-values.*

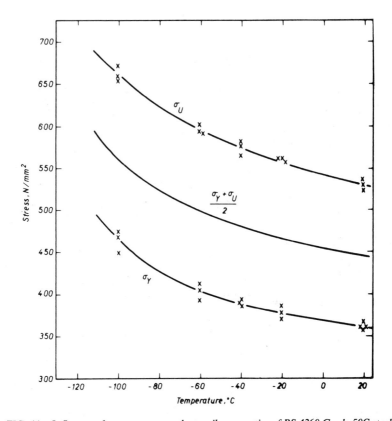

FIG. 11—*Influence of temperature on the tensile properties of BS 4360 Grade 50C steel.*

mechanical properties similar to those summarized for the Grade 50C material in Table 1. In these tests the small (10 mm) specimens were extracted from near the center of the 100-mm-thick plates.

Figure 13 shows the valid J_{Ic} results (according to Eq 7) obtained for specimens in each thickness. It can be seen from Fig. 13 that valid J_{Ic} values [54] can give large overestimates of valid K_{Ic} values when applied to materials which show significant shifts in the ductile/brittle transition temperature range with variations in section thickness or in-plane dimensions.

Conclusions

1. Since a deformed crack tip stretches beyond the original crack tip position, it is proposed that the crack tip COD should be defined as the displacement at the original crack tip position. This definition avoids much

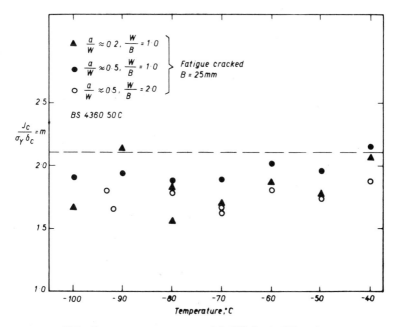

FIG. 12—m versus temperature, BS 4360 Grade 50C steel.

of the ambiguity of previous COD definitions, which were based on the crack tip profile and the elastic-plastic interface on the flanks of the crack.

2. When cracks are sited within weld regions, significant local variations in yield strength cause a fundamental problem in the application of the J-contour integral concept to an elastic-plastic material. For example, when cracks exist in an overmatching yield strength region, estimates of J derived from measurements on contours outside the weld region may grossly overestimate the plastic work component of J within the weld region.

3. Full section thickness three-point SENB tests carried out at the temperatures, strain rates, and section thicknesses of interest that give invalid K_{Ic} results may be interpreted in terms of critical COD and J-values. It can be shown that there is a close mathematical link between the COD and J-values obtained with these interpretations. Also, when using the latter interpretations the experimental results indicated that COD and J are equally useful as fracture characterizing parameters.

4. The critical values of COD and J for unstable cleavage fracture in common structural steels are affected by section thickness and in-plane geometry. It is important, therefore, that laboratory tests be designed to match or overmatch the plastic constraint in the structural situations of interest.

5. Values of J_{Ic} from small laboratory tests may give large overestimates

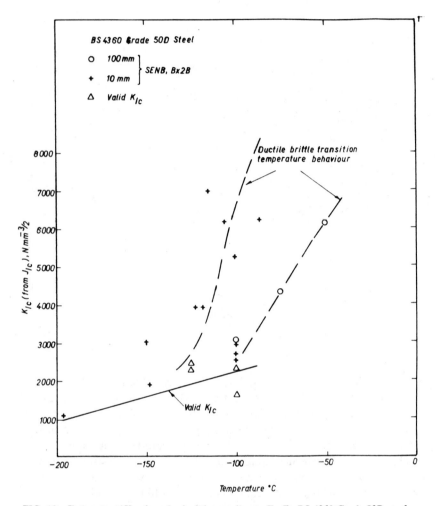

FIG. 13—*Estimates of K_{Ic} from J_{Ic} (valid according to Eq 7): BS 4360 Grade 50D steel.*

of valid K_{Ic} values in common structural steels having yield strengths less than approximately 700 N/mm².

Acknowledgments

The help and encouragement of the author's colleagues is acknowledged, and especial thanks are given to Mr. B. A. Wakefield and the staff of The Welding Institute brittle fracture laboratory. The author also appreciates the generosity of his colleague Dr. H. G. Pisarski for making available the preliminary test results in Fig. 13.

References

[1] Wells, A. A. in *Proceedings*, Crack Propagation Symposium, Cranfield, England, Vol. 1, Paper B4, 1961.
[2] Rice, J. R., *Journal of Applied Mechanics*, June 1968, p. 379.
[3] Burdekin, F. M. and Stone, D. E. W. *Journal of Strain Analysis*, Vol. I, No. 2, 1966, p. 144.
[4] Harrison, J. D., Burdekin, F. M., and Young, J. G., "A Proposed Acceptance Standard for Weld Defects Based upon Suitability for Service," 2nd Conference on the Significance of Defects in Welded Structures, The Welding Institute, London, England 1968.
[5] Burdekin, F. M. and Dawes, M. G. in *Proceedings*, Institution of Mechanical Engineers Conference, London, England, May 1971, pp. 28–37.
[6] Dawes, M. G., *Welding Journal Research Supplement*, Vol. 53, 1974, p. 369s.
[7] Dawes, M. G., and Kamath, M. S. in *Proceedings*, Conference on the Tolerance of Flaws in Pressurised Components, Institution of Mechanical Engineers, London, England May 16–18, 1978, pp. 27–42.
[8] Draft British Standards Rules for the Derivation of Acceptance Levels for Defects in Fusion Welded Joints, BSI WEE/37, Document 75/77081 D.C., British Standards Institute, Feb. 1976.
[9] Harrison, J. D., Dawes, M. G., Archer, G. L., and Kamath, M. S., this publication, pp. 606–631.
[10] Dawes, M. G., "The Application of Fracture Mechanics to Brittle Fracture in Steel Weld Metals," Ph.D. thesis (Council for National Academic Awards), The Welding Institute, London, England, Dec. 1976.
[11] Nichols, R. W., Burdekin, F. M., Cowan, A., Elliott, D., and Ingham, T. in *Proceedings*, Conference on Practical Fracture Mechanics for Structural Steel, Risley, England April 1969, UKAEA/Chapman and Hall, Risley, 1969.
[12] Wessel, E. T. in *Practical Fracture Mechanics for Structural Steels*. UKAEA/Chapman and Hall, Risley, England, 1969, p. H1.
[13] Green, G., Smith, R. F., and Knott, J. F. in *Proceedings*, Conference on Mechanics and Mechanisms of Crack Growth, Churchill College, Cambridge, England, Paper 5, April 1973.
[14] Broek, D. in *Proceedings*, Third International Conference on Fracture, Munich, Germany, April 1973, Vol. 4, p. III 422.
[15] Fields, B. A. and Miller, K. J., "A Study of COD and Crack Initiation by a Replication Technique," Cambridge University Engineering Department Report CUED/C-Mat/TR 17, Cambridge, England, Oct. 1974.
[16] De Morton, M. E. in *Proceedings*, WI/ASM Conference on Dynamic Fracture Toughness, The Welding Institute, London, England, July 1976.
[17] Smith, R. F. in *Proceedings*, Institution of Metallurgists Spring Meeting, Newcastle, England, Paper 4, March 1973, p. 28.
[18] Goodier, J. N. and Field, F. A. in *Fracture of Solids*, Drucker and Gilman, Eds., Interscience, New York, 1963, p. 103.
[19] Hahn, G. T. and Rosenfield, A. R., *Acta Metallurgica*, Vol. 13, No. 3, 1965, p. 293.
[20] Rice, J. R. and Johnson, M. A. in *Inelastic Behavior of Solids*, McGraw-Hill, New York, 1970, p. 641.
[21] Pelloux, R. M. N., *Engineering Fracture Mechanics*, Vol. 1, No. 4, 1970, p. 697.
[22] Srawley, J. E., Swedlow, J., and Roberts, E., *International Journal of Fracture Mechanics*, Vol. 6, 1970, p. 441.
[23] Wells, A. A. and Burdekin, F. M., *International Journal of Fracture Mechanics*, Vol. 7, 1971, p. 242.
[24] Boyle, E. F., "The Calculation of Elastic and Plastic Crack Extension Forces," Ph.D. thesis, Queen's University, Belfast, U.K., 1972.
[25] Levy, N., Marcel, P. V., Ostergren, W. J., and Rice, J. R., *International Journal of Fracture Mechanics*, Vol. 7, No. 2, 1971, p. 143.
[26] Rice, J. R., *International Journal of Fracture Mechanics*, Vol. 9, 1973, p. 313.

[27] Turner, C. E. in *Proceedings*, Conference on the Mechanics and Physics of Fracture, Churchill College, Cambridge, England, Paper 2, Jan. 1975, pp. 2-1-2-16.
[28] Eshelby, J. D. in *Solid State Physics*, Vol. 3, Academic Press, New York, 1956, p. 79.
[29] Cherepanov, G. P., *International Journal of Solids and Structures*, Vol. 4, 1968, p. 811.
[30] Hayes, D. J., "Some Applications of Elastic-Plastic Analysis to Fracture Mechanics," Ph.D. thesis, University of London, London, England, 1970.
[31] Sumpter, J. D. G., "Elastic Plastic Fracture Analysis and Design Using the Finite Element Method," Ph.D. thesis, University of London, London, England, Dec. 1973.
[32] Sumpter, J. D. G., and Turner, C. E., *International Journal of Fracture Mechanics*, Vol. 9, No. 3, 1973, p. 320.
[33] Kanazawa, T., Machida, S., Hagiwara, Y., and Kobayashi, J., "Study on Evaluation of Fracture Toughness of Structural Steels Using COD Bend Test," International Institute of Welding, Document X-702-73, June 1973.
[34] Serensen, S. V., Girenko, V. S., Deinega, V. A., and Kir'yan, V. I., *Automatic Welding*, Vol. 28, No. 2, 1975, p. 1.
[35] Burdekin, F. M., Dawes, M. G., Archer, G. L., Bonomo, F., and Egan, G. R., *British Welding Journal*, Vol. 15, 1968, p. 590.
[36] Priest, A. H. and May, M. J., "Strain Rate Transitions in Structural Steels," British Iron and Steel Research Association Report MG/C/95/69, 1969.
[37] Egan, G. R., "Application of Yielding Fracture Mechanics to Design of Welded Structures," Ph.D. thesis, London University, London, England, 1972.
[38] Kanazawa, T., Machida, S., Onozuka, M., and Kaneda, S. A., "Preliminary Study on the J-Integral Fracture Criterion," International Institute of Welding, Document X-779-75, May 1975.
[39] Ingham, T., Egan, G. R., Elliott, D., and Harrison, T. C. in *Proceedings*, Institution of Mechanical Engineers Conference on Practical Application of Fracture Mechanics to Pressure Vessel Technology, London, England, Paper C54/71, May 1971.
[40] Knott, J. F., *Materials Science and Engineering*, Vol. 7, No. 1, 1971, p. 1.
[41] Veerman, C. C. and Muller, T., "The Effect of Notch Depth and Pre-Strain on the COD at Fracture," 2nd Progress Report of the European Community Research Program 6210-55/0/50, Nov. 1972.
[42] Kanazawa, T., Machida, S., Momota, S., and Hagiwara, Y. A. in *Proceedings*, 2nd International Conference on Fracture, Brighton, England, Paper 1, 1969, p. 1.
[43] Dolby, R. E., and Archer, G. L. in *Proceedings*, Institution of Mechanical Engineers Conference, London, May 1971, pp. 190-199.
[44] Dawes, M. G., *Welding Journal*, Vol. 55, No. 12, Dec. 1976, p. 1052.
[45] Harrison, T. C. and Fearnehough, G. D., *International Journal of Fracture Mechanics*, Vol. 5, 1969, p. 348.
[46] Fearnehough, G. D., Lees, G. M., Lowes, J. M., and Weiner, R. T. in *Proceedings*, Institution of Mechanical Engineers Conference on Practical Application of Fracture Mechanics to Pressure Vessel Technology, London, England, Paper C33/71, May 1971.
[47] Smith, R. F. and Knott, J. F. in *Proceedings*, Institution of Mechanical Engineers Conference on Practical Application of Fracture Mechanics to Pressure Vessel Technology, London, England, Paper C9/71, May 1971.
[48] Chipperfield, C. G., Knott, J. F., and Smith, R. F. in *Proceedings*, 3rd International Conference on Fracture, Munich, Germany, Vol. 2, April 1973, pp. 1-233.
[49] Robinson, J. N. and Tetelman, A. S. in *Fracture Toughness and Slow-Stable Cracking*, ASTM STP 559, American Society for Testing and Materials, 1974, pp. 139-158.
[50] Robinson, J. N., *International Journal of Fracture Mechanics*, Vol. 12, 1976, p. 723.
[51] Green, G. and Knott, J. F. in *Proceedings*, ASME Conference on Micromechanical Modelling of Flow and Fracture, Troy, New York, American Society of Mechanical Engineers, June 1975.
[52] Begley, J. A. and Landes, J. D. in *Fracture Toughness, ASTM STP 514*, American Society for Testing and Materials, 1972, pp. 1-20.
[53] Landes, J. D., and Begley, J. A. in *Fracture Toughness, ASTM STP 514*, American Society for Testing and Materials, 1972, pp. 24-39.
[54] Landes, J. D. and Begley, J. A. in *Fracture Analysis, ASTM STP 560*, American Society for Testing and Materials, 1974, pp. 170-186.

[55] Clarke, G. A., Andrews, W. R., Paris, P. C., and Schmidt, D. W. in *Mechanics of Crack Growth, ASTM STP 590,* American Society for Testing and Materials, 1976, pp. 27-42.
[56] Garwood, S. J., "The Measurement of COD and *J* at the Initiation of Crack Growth and Their Interpretation," Imperial College, London, England, July 1976.
[57] Sumpter, J. D. G., *Metal Science,* Oct. 1976, p. 354.
[58] Milne, I., "The Ductile-Brittle Transition and Fracture Toughness of Ferritic Steels," to be published.
[59] Hayes, D. J. and Turner, C. E., *International Journal of Fracture Mechanics,* Vol. 10, No. 1, March 1974, p. 17.
[60] Wells, A. A. in *Proceedings,* Canadian Congress of Applied Mechanics, Calgary, Alta., Canada, 1971, pp. 59-77.
[61] Sumpter, J. D. G. and Turner, C. E. in *Cracks and Fracture, ASTM STP 601,* American Society for Testing and Materials, 1976, pp. 3-18.
[62] Turner, C. E., *Materials Science and Engineering,* Vol. 11, 1973, p. 275.

J. Royer,[1] J. M. Tissot,[1] A. Pelissier-Tanon,[2] P. Le Poac,[3] and D. Miannay[3]

J-Integral Determinations and Analyses for Small Test Specimens and Their Usefulness for Estimating Fracture Toughness

REFERENCE: Royer, J., Tissot, J. M., Pelissier-Tanon, A., Le Poac, P., and Miannay, D., "**J-Integral Determinations and Analyses for Small Test Specimens and Their Usefulness for Estimating Fracture Toughness,**" *Elastic-Plastic Fracture, ASTM STP 668,* J. D. Landes, J. A. Begley, and G. A. Clarke, Eds., American Society for Testing and Materials, 1979, pp. 334–357.

ABSTRACT: General estimation procedures for the J-integral determination are reviewed for the three-point bend and the compact tension specimens. Tests were made using 10 CD 9-10 steel. Experimental results are presented and are in partial agreement with analytical results. The errors due to simplified analysis and experimental procedure are explored. Toughness, as analyzed by the resistance curve technique, is shown to be size dependent for bending and not for tension. Disagreement between the two loading modes, if not fortuitous and due to the steel, suggests that simple strain and stress analyses are not sufficient and that the loading procedure and the T-effect must be taken into account.

KEY WORDS: crack propagation, *J*-contour integral, fracture tests, fracture properties, steels, elastic-plastic fracture, fracture initiation

Nomenclature

J   Energy line integral
V   Pseudo strain energy release rate
G   Elastic strain energy release rate
W   Strain energy density

[1] Professor of mechanics and assistant professor, respectively, Ecole Nationale Supérieure de Mécanique, 1, rue de la Nöe, 44072 Nantes Cédex, France.
[2] Research consultant, Framatome, 77-81, rue du Mans, 92400 Courbevoie, France.
[3] Materials engineer and head, Fracture Mechanics Group, respectively, Commissariat à l'Energie Atomique, service Métallurgie, B. P. No. 511, 75752 Paris-Cédex, France.

U Strain energy or work done in loading a specimen
U^c Complementary strain energy or work
K_I Opening mode stress intensity factor
Y Calibration factor
P Applied load
F Force per unit thickness
v Displacement of the applied load
C Specimen compliance
a Crack length
w, B Specimen width, specimen thickness
S Span in bending
ρ Notch acuity
σ_y Uniaxial yield stress
σu Ultimate tensile strength
E Young's modulus
ν Poisson's ratio

Subscripts
e Value for the elastic behavior
p Value for the plastic behavior
L Value for the fully plastic limit
ϵ Value for plane strain
σ Value for plane stress

To evaluate the toughness of low- or medium-strength metals with specimens not satisfying size requirements of the ASTM Test for Plane-Strain Fracture Toughness of Metallic Materials (E 399-74T) for linear elastic behavior, three concepts are proposed for the elastoplastic range: the J-integral, the crack opening displacement, (COD), and the equivalent-energy concepts. These lead to different experimental and estimation procedures to obtain the adequate parameter and to determine its critical value representing the toughness.

Unfortunately the results obtained by different investigators are not in complete agreement. The controversy centers on the estimation procedure and the possible relationships between the three concepts.

The first part of this paper deals with the calculation or estimation of J from load displacement data of precracked specimens from experimental and theoretical points of view and the second part with the influence of specimen size, geometry, and mode of loading on the value of toughness.

Materials and Experimental Procedure

One 110-mm-thick rolled plate of 10 CD 9-10 steel with two heat treatments was tested. Chemical composition, heat treatments, and tension test results are given in Table 1.

TABLE 1—*Properties of materials tested parallel to the rolling direction.*

\multicolumn{9}{c}{Chemical Composition, percent by weight}								
C	P	S	Si	Mn	Cr	Mo	Al	Cu	Sn
0.125	0.012	0.017	0.260	0.425	2.35	0.99	0.012	0.14	0.022

Material No.	Heat Treatment
1	normalized 940°C, 1 h, water quenched; tempered 700°C, 1 h
2	normalized 940°C, 1 h, water quenched; tempered 700°C, 1 h; stress relieved 675°C, 1 h

Room Temperature Tensile Properties ($\dot\epsilon^0 = 1.11\ 10^{-4}\ s^{-1}$)

Material No.	Yield Strength 0.2% Offset, MPa σ_y 0.2	Tensile Strength, MPa σ_u	Elongation, %	Parameters ($\sigma = \sigma_0 + K\epsilon_p^n$)		
				σ_0	K	n
1	520	650	22.6	502.9	51.4	0.707
2	500	625	23.8	483.4	48.9	0.704

Young's modulus: $E = 206\ 000$ MPa.
Poisson's ratio: $\nu = 0.3$.

Two specimen geometries were machined at midthickness in the longitudinal (LT) direction with respect to rolling: the three-point bend (TPB) and the compact tension (CT) specimens with dimensions as given in Fig. 1. Specimens were precracked at different crack lengths, a, by fatigue according to the ASTM specifications or drilled with a hole of different radius, ρ, at the notch end. For short cracks in bend specimens, precracking was done before machining. Crack length was measured at seven equally distant points, excluding the specimen faces, on the crack surface.

Specimens were tested with a constant crosshead velocity of 0.005 mm/s^{-1}. Displacement of the load point, v, was measured with a clip gage located on two knife edges, one fixed at the end of the upper roll and the other fixed on a sliding bar supported by the ends of the lower rolls for bending and between the two center knife edges localized on the loading line in the CT specimen. Load, P, was measured with the load cell. Electrical monitoring was also performed by the d-c drop potential method with the configuration as shown in Fig. 1 [1].[4]

---

[4]The italic numbers in brackets refer to the list of references appended to this paper.

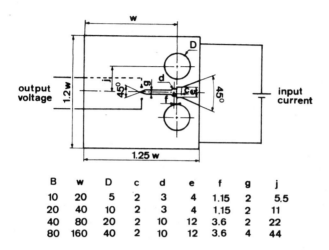

B	w	D	c	d	e	f	g	j
10	20	5	2	3	4	1.15	2	5.5
20	40	10	2	3	4	1.15	2	11
40	80	20	2	10	12	3.6	2	22
80	160	40	2	10	12	3.6	4	44

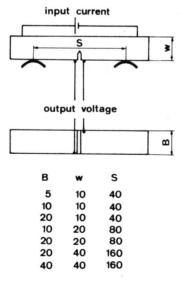

B	w	S
5	10	40
10	10	40
20	10	40
10	20	80
20	20	80
20	40	160
40	40	160

FIG. 1—*CT and TPB specimens*.

Theoretical Background

J is defined for two-dimensional problems as the line integral

$$J = \int_\Gamma \left(W\, dy - T\, \frac{\partial u}{\partial x}\, ds \right)$$

where

W = strain energy density,
T = traction vector on path Γ, and
u = displacement vector.

Its properties are largely reviewed in the literature [2-7] and will not be reported here. For an elastic material, J is representative of the potential energy variation with respect to crack length. Begley and Landes [4] proposed to extend this interpretation in the nonlinear range and J, or V as denoted here, is given by

$$V = -\left(\frac{\partial U}{\partial a}\right)_v = \int_0^v \left(-\frac{\partial F}{\partial a}\right)_v dv$$

or

$$V = \left(\frac{\partial U}{\partial a}\right)_F = \int_0^F \left(\frac{\partial v}{\partial a}\right)_F dF \qquad (1)$$

with U the work of the applied force F per unit length of the crack front. Thus V can be evaluated experimentally from the load displacement curves for identical specimens of differing crack lengths. This method has been used here at constant value of displacement.

The displacement v may be separated into two parts

$$v = v_\text{no crack} + v_\text{crack}, \text{ or } v = v_\text{elastic} + v_\text{plastic}$$

It follows that

$$V = V_\text{crack} \text{ and } V = V_\text{elastic} + V_\text{plastic} = G + V_\text{plastic} \qquad (2)$$

G being the elastic strain energy release rate given in different equivalent forms

1. $$G = \frac{K_I^2}{E'}$$

with $E' = E/(1 - v^2)$ in plane strain (subscript ϵ) and $E' = E$ in plane stress (subscript σ), with K_1 the stress intensity factor $= (F/\sqrt{w})\, Y(a/w)$, w being the width of the specimen.

2.
$$G = \frac{1}{2} F^2 \frac{\partial C}{\partial a} \qquad (3)$$

with C the compliance v/F, generally tabulated as Ev/F or calculated as

$$C = C(0) + C(a) = C(0) + \frac{2}{E'} \int_0^a \frac{K^2}{F^2}\, da \text{ or } C = C(0)$$
$$+ \frac{1}{FE'} \int_0^a \frac{\partial K^2}{\partial F}\, da \qquad (4)$$

3.
$$G = -\left(\frac{\partial U_e}{\partial a}\right)_v = -\frac{\partial \frac{1}{2}\frac{v^2}{C}}{\partial a} =$$
$$+ \frac{C'}{C}\left(1 - \frac{a}{w}\right) \frac{U_e}{(w-a)} = \eta_e \frac{U_e}{(w-a)} \qquad (5)$$

the prime denoting differentiation with respect to a/w and with U_e the elastic strain energy.

4.
$$G = \frac{2Y^2}{E'C}\left(1 - \frac{a}{w}\right) \frac{U_e}{(w-a)} = \eta_e \frac{U_e}{(w-a)} \qquad (6)$$

The functions and values used in this paper are given in the Appendix and tabulated in Tables 2-4 with $v = 0.3$.

Analyses have been developed for calculating the value of V from a single load displacement curve. They are based on three assumptions:

1. The actual load displacement curves are approximated by the two limiting cases, purely elastic and rigid plastic behaviors [5,6]. That is

$$V = -\left(\frac{\partial(U_{eM} + U_p)}{\partial a}\right)_v = +\frac{C'}{C}\frac{U_{eM}}{w}$$
$$+ \frac{F'_L}{F_L}\frac{U_p}{w} = \eta_e \frac{U_e}{w-a} + \eta_p \frac{U_p}{w-a} \qquad (7)$$

with U_{eM} the maximum elastic energy when $v = v_L = CF_L$, $P_L = BF_L$ being the limit load, and with $U_p = F_L(v - v_L)$, the plastic energy. η_p values are given in Tables 4 and 5.

TABLE 2—*Values of EC for the TPB specimen with S/W = 4.*

Reference	0	0.1	0.2	0.3	0.4	$\frac{a}{w}$ 0.5	0.6	0.7	0.8	0.9
Eq 14	19.12	...	...	...	...	...	...	...	...	...
Eq 15	20.07	...	...	...	...	...	...	...	...	...
Eq 16										
plane strain	0	1.11	4.13	9.41	18.26	33.74	63.22	128.19	315.85	1342.99
plane stress	0	1.22	4.54	10.34	20.06	37.07	69.48	140.86	347.08	1475.81
Present results										
(5 × 10 × 40)	26.5	27.6	30.7	36.1	41.4	57.3	83.6	143.5	...	...
(20 × 10 × 40)	25.4	26.9	29.0	34.4	41.4	58.5	83.8	142.3	...	...
(20 × 20 × 80)	26.8	29.6	32.3	35.1	45.5	62.1	86.0	151.8	...	...

TABLE 3—Values of EC for the standard CT specimen.

Reference	0	0.2	0.3	0.4	$\frac{a}{w}$ 0.5	0.6	0.7	0.8
Eq 22	2.2							
Eq 23								
plane strain	0	3.52	7.98	15.50	28.37	52.52	106.68	272.66
plane stress	0	3.87	8.76	17.03	31.18	57.71	117.23	299.63
Ref [16]								
plane strain	...	7.8	12.98	20.78	33.64	57.62	114.24	280.8
plane stress	...	8.57	14.26	22.84	36.97	63.32	125.54	308.57
Present results								
$\rho = 0$	...	...	12.4	18.6	30.6	52.6	106.0	255.0
$\rho = 0.5$	...	...	14.4	22.3	34.3	61.4	119.1	287.4
$\rho = 2$	...	...	15.0	23.6	38.8	64.1	133.8	329.5

TABLE 4—Values of η_e and η_p for the TPB specimen with $S/w = 4$.

η	Reference	$\frac{a}{w}$								
		0.1	0.2	0.3	0.4	0.5	0.6	0.7	0.8	0.9
η_e	Eqs 16 + 15									
	plane strain	0.89	1.33	1.60	1.79	1.92	1.99	2.02	2.02	2.03
	plane stress	0.97	1.43	1.70	1.87	1.98	2.03	2.04	2.03	2.03
	Eq 18									
	plane strain	1.08	1.50	2.00	2.00	2.00	2.00	2.00	2.00	2.00
η_p	Eq 19									
	plane strain	1.04	1.70	2.00	2.00	2.00	2.00	2.00	2.00	2.00
	Eq 17									
	plane stress	2.00	2.00	2.00	2.00	2.00	2.00	2.00	2.00	2.00
η_e	Present results									
	(5 × 10 × 40)	0.75	1.15	1.43	1.82	1.98	2.18	2.29	...	...
	(20 × 10 × 40)	0.77	1.22	1.50	1.82	1.94	2.17	2.31	...	...
	(20 × 20 × 80)	0.70	1.09	1.47	1.66	1.83	2.12	2.16	...	...

TABLE 5—*Values of* η_e, η_p, *and* η_c *for the standard CT specimen.*

η	Reference	0.1	0.2	0.3	0.4	0.5	0.6	0.7	0.8	0.9
η_e	Eq 21 + Ref *16*	...	3.78	3.42	3.05	2.76	2.58	2.44	2.41	...
	Eqs 22 + 23									
	plane strain	4.20	4.42	3.84	3.21	2.75	2.47	2.32	2.25	2.15
	plane stress	4.48	4.59	3.93	3.26	2.77	2.48	2.33	2.25	2.15
	Eq 23									
	plane stress	...	...	5.05	3.74	2.99	2.59	2.38	...	...
η_p	Eq 24									
	plane strain	2.63	2.59	2.53	2.46	2.38	2.29	2.21	2.14	2.07
	Eq 25									
	plane stress	2.46	2.44	2.40	2.35	2.29	2.23	2.17	2.11	2.05
η_c	Ref *7*									
	plane stress	...	...	0.19	0.21	0.20	0.17	0.14	0.11	...
η_e	Present results									
	$\rho = 0$	...	...	3.57	3.42	3.05	2.83	2.63	2.66	...
	$\rho = 0.5$	...	...	3.07	2.85	2.72	2.43	2.34	2.36	...
	$\rho = 2$	...	...	2.95	2.69	2.41	2.33	2.08	2.06	...

2. The plastic displacement due to the crack is a function of the ratio of the applied load to the fully plastic limit load [7], from which

$$V = \eta_e \frac{U_e}{w-a} + \eta_p \frac{U_p}{w-a} + \eta_c \frac{U_p{}^c}{w-a} \tag{8}$$

with $U_p{}^c$ the plastic complementary work of the applied load. η_c is null for pure bending and its low values are given in Table 5 for the CT specimen.

3. The initial elastic behavior of the cracked body may be corrected for plasticity by considering an equivalent elastic or effective crack size up to the limit load [5]. This leads to an evaluation of the V versus v relationship. Thus the first two assumptions give relations between V and areas easily measured on the records. Such relations moreover, are more practical when $\eta_e = \eta_p$ or when one contribution is greater than the others. Our first objective has been to verify experimentally the adequacy of such an approach.

The J-integral has been advocated to represent strain and stress fields near the crack tip in the plastic range [2-4], so that the condition for onset of growth from a precrack can be phrased in terms of J. As shown by Larsson and Carlsson [9], however, in the small scale yielding regime, plastic zone sizes are different for the same elastic limit condition, suggesting that damage will vary with geometry. This phenomenon is probably attributable to the effects of in-plane biaxiality as suggested by Rice [10], biaxiality being negative in tension testing and positive in bend testing, but not influencing greatly the J-value. For stably growing cracks, no similar characterizing parameter has yet been identified. Begley and Landes [11], however, proposed to determine a critical value, $J_{\rm Ic}$, from a resistance curve. Such a technique is studied in the second part of this paper.

Results for the Estimation Procedure

For the two geometries, analysis has been made with the same procedure as outlined in Fig. 2. For a geometry and size, areas under load displacement record at given displacement v less than the displacement corresponding to any encountered maximum load have been plotted as the normalized term U/Bw^2 versus a/w. The following function was fitted to the data by the method of least squares so that the deviation $\Sigma'(U - U_{\rm adj})^2/(U)^2$ is minimum, $U_{\rm adj}$ being the adjusted value

$$\frac{U}{Bw^2} = \left(1 - \frac{a}{w}\right)^2 \left[\alpha_0 + 2\alpha_0 \frac{a}{w} \right.$$

$$\left. + \alpha_2 \left(\frac{a}{w}\right)^2 + \alpha_3 \left(\frac{a}{w}\right)^3 + \alpha_4 \left(\frac{a}{w}\right)^4 \right] \tag{9}$$

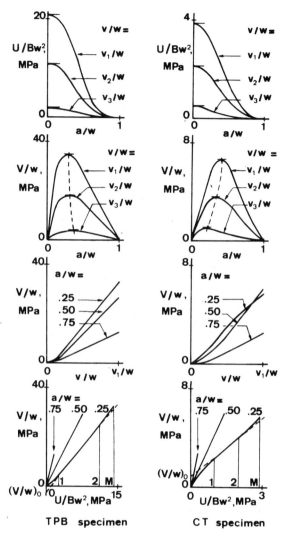

FIG. 2—*Schematic of the analysis.*

This polynomial gives values of V and U null for $a = w$ and V null for $a = 0$, which is in agreement with the physical significance of these parameters. The degree of this function has been chosen over the range up to 12 to give the minimum deviation. Then this function is derived with respect to a/w to give V/w versus a/w at given displacement. Finally V/w is plotted versus v/w or $(U/Bw^2)_{adj}$ and the best fit to the data is found.

Three-Point Bend Specimen

Seventeen (5 × 10 × 40), twenty (20 × 10 × 40), and five (20 × 20 × 80) precracked specimens of the first material with different relative crack lengths over the range 0 to 0.77, 0 to 0.79, and 0 to 0.75, respectively, were loaded up to a relative displacement $v/w = 0.128$, 0.128, and 0.104, respectively. The observed elastic compliance for $v/w \leq 0.010$ is reported in Table 2. For plain bar, the actual compliance is greater than the theoretical, emphasizing perhaps the indentation effects of rolls or revealing the inaccuracy of the equation; the crack contribution is relatively independent of size and in good agreement with theoretical values. Corresponding values of η_e calculated with Eq 6 with $E' = E$ are reported in Table 4 and are slightly greater than the generally accepted value of 2 for $a/w > 0.5$.

Due to the appearance of the curves of Fig. 2, the method outlined in the introduction leads to a decomposition of the V/w versus U/Bw^2 relationship into three ranges. The first one is defined between $U/Bw^2 = 0$ and $U/Bw^2 = (U/Bw^2)_1$ by an initial slope at the origin, giving the presumed value η_e and a final slope at $(V/w)_1$, $(U/Bw^2)_1$, giving the presumed value η_p; $(V/w)_1$, $(U/Bw^2)_1$, and η_p are the values obtained from the determination for the second range. The polynomial of order three in U/Bw^2 fulfilling these requirements is given as

$$\frac{V}{w} = \frac{\eta_e}{1 - \frac{a}{w}} \frac{U}{Bw^2} + \left[3\left(\frac{V}{w}\right)_1\right.$$

$$\left. - \frac{\left(\frac{U}{Bw^2}\right)_1 (\eta_p + 2\eta_e)}{\left(1 - \frac{a}{w}\right)}\right] \left[\frac{\frac{U}{Bw^2}}{\left(\frac{U}{Bw^2}\right)_1}\right]^2 \qquad (10)$$

$$+ \left[\frac{\left(\frac{U}{Bw^2}\right)_1 (\eta_e + \eta_p)}{1 - \frac{a}{w}} - 2\left(\frac{V}{w}\right)_1\right] \left[\frac{\frac{U}{Bw^2}}{\left(\frac{U}{Bw^2}\right)_1}\right]^3$$

The second one is linear between the limits $(U/Bw^2)_1$ and $(U/Bw^2)_2$ such that

$$\frac{V}{w} = \frac{\eta_p}{\left(1 - \frac{a}{w}\right)} \frac{U}{Bw^2} + \left(\frac{V}{w}\right)_0 \qquad (11)$$

with $(V/w)_0$ being the intercept with the V/w axis.

Beyond this, no relationship is proposed and we just report the maximum values $(V/w)_M$ and $(U/Bw^2)_M$ attained.

Values of the variables with a fit within 2 percent are given in Table 6. We note that the linear relationship is sufficient for $a/w \geq 0.6$, coefficients are relatively independent of geometry, and the mean value is 1.86. This value is lower than the values given in Table 4. However, we have seen in a similar treatment that two is obtained if the parameter V is plotted versus U_{crack} instead of U.

From a practical point of view, we note that the V/w versus a/w relationship at given displacement shows a maximum localized near $a/w = 0.3$ and going toward increasing a/w with decreasing v/w. Therefore, when several specimens are used to determine toughness, it seems judicious to choose a/w near this range, if some scatter at precracking is to be assumed, because V/w varies slowly with a/w. Moreover, our results show that an efficient and easy initial calibration may be obtained with as few specimens as five, an uncracked specimen included, which is proved by the (20 × 20 × 80) specimen testing for which the results are in the scatterband.[5]

Some limitation may subsist after this study, however, because slow crack growth always occurred during testing.

Compact Tension Specimen

To alleviate this last restriction, testing was made with single size specimens of the second material, CT 20 mm, but with three notch radii: $\rho = 0$ obtained by fatigue precracking, $\rho = 0.5$ mm, and $\rho = 2$ mm. Several relative crack lengths were investigated: 13 ranging from 0.255 to 0.853 for $\rho = 0$, 12 ranging from 0.250 to 0.803 for $\rho = 0.5$ mm, and 12 ranging from 0.250 to 0.801 for $\rho = 2$ mm. Slow crack growth was observed only for $\rho = 0$. Plain specimens were not considered due to a lack of definition. In this kind of test, indentation by the pins has no direct effect on measured displacement, but for short cracks, plastic flow occurring through pin holes, as noted in the Appendix, obliged us to reject our tests with lower crack length than mentioned in the foregoing. Experimental compliances are given in Table 3 and the corresponding η_e in Table 5; compliance for $v/w \leq 0.003$ increases with increasing radius in agreement with theory [1]; and when testing was stopped for the first limit load encountered for $\rho = 0$, that is, for $v/w \leq 0.03$, limit loads for other radii were not yet observed and must increase with increasing radius. Better agreement with theory is noted for the medium radius; the improvement on $\rho = 0$ may be due to a straight crack front.

In view of the appearance of the curves in Fig. 2, two analytical treatments were tried. The first one, called the "polynomial function," is the same as ap-

---

[5] Actually, four specimens are sufficient to determine the four coefficients of the polynomial but with five specimens deviation is tested.

TABLE 6—*Data for the V/w versus U/Bw^2 relation for the TPB specimen.*

Specimen	$\frac{a}{w}$	η_e	η_p	$(V/w)_0$, MPa	$(U/Bw^2)_1$, MPa	$(U/Bw^2)_2$, MPa	$(U/Bw^2)_M$, MPa	$(V/w)_M$, MPa
(5 × 10 × 40)	0.1	0.64	1.22	−0.75	3.03	10.63	19.17	26.89
	0.2	0.96	1.73	−1.00	2.62	10.11	15.97	35.01
	0.3	1.19	1.91	−1.01	2.91	8.89	12.48	33.81
	0.4	1.36	1.90	−0.52	1.09	7.34	9.32	29.18
	0.5	1.57	1.84	−0.22	0.80	6.65	6.65	24.22
	0.6	1.80	1.80	0	0	4.45	4.45	19.96
	0.7	1.83	1.83	0	0	2.64	2.64	16.11
	0.8	1.90	1.90	0	0	1.24	1.24	11.81
	0.9	1.98	1.98	0	0	0.32	0.32	6.37
(20 × 10 × 40)	0.1	0.58	1.05	−0.60	2.07	13.33	21.36	25.14
	0.2	1.03	1.57	−0.73	1.83	10.14	18.20	36.13
	0.3	1.31	1.83	−0.66	1.52	8.12	14.43	38.11
	0.4	1.51	1.95	−0.50	1.18	7.64	10.75	34.98
	0.5	1.64	1.96	−0.30	0.86	5.38	7.51	29.52
	0.6	1.93	1.93	0	0	4.86	4.86	23.46
	0.7	1.89	1.89	0	0	2.81	2.81	17.69
	0.8	1.88	1.88	0	0	1.31	1.31	12.30
	0.9	1.91	1.91	0	0	0.35	0.35	6.76
(20 × 20 × 80)	0.1	0.47	1.04	−0.88	3.15	9.65	15.31	17.57
	0.2	0.96	1.56	−0.91	2.78	8.33	13.09	25.49
	0.3	1.34	1.83	−0.69	2.28	7.63	10.42	27.05
	0.4	1.58	1.93	−0.38	1.14	6.44	7.80	24.90
	0.5	1.68	1.93	−0.21	0.82	5.50	5.50	21.04
	0.6	1.86	1.86	0	0	3.61	3.61	16.79
	0.7	1.81	1.81	0	0	2.13	2.13	12.84
	0.8	1.80	1.80	0	0	1.03	1.03	9.24
	0.9	1.86	1.86	0	0	0.29	0.29	5.37

plied to bend testing; the second, called "the exponential function," was intended to cover a larger range, and the function is given by

$$\frac{V}{w} = \frac{\eta_p}{1 - \frac{a}{w}} \frac{U}{Bw^2} + \left(\frac{V}{w}\right)_0 + \left\{ \left[\frac{\eta_e - \eta_p}{1 - \frac{a}{w}} \right. \right.$$

$$\left. \left. - \left(\frac{V}{w}\right)_0 r \right] \frac{U}{Bw^2} - \left(\frac{V}{w}\right)_0 \right\} \exp - r \frac{U}{Bw^2} \quad (12)$$

with r taking the value

$$r = \frac{\eta_e - \eta_p}{\left(1 - \frac{a}{w}\right)\left(\frac{V}{w}\right)_0} - \frac{1}{\left(\frac{V}{w}\right)_1}$$

to provide a good fit; the curve goes through the experimental point $(U/Bw^2)_1$, $(V/w)_1$.

Values of the variables with a fit within 2 percent are given in Table 7. We see that the linear relationship is sufficient for $a/w \geq 0.7$ for $\rho = 0$ and for $a/w \geq 0.6$ for the two other radii, and that $\eta_e = \eta_p$ is relatively radius independent with a decreasing value with increasing crack length, but this value is lower than the theoretical ones reported in Table 5. Below this limit, some decreasing trend is observed which is in disagreement with theory. No clear explanation has been found, though limited backward plasticity may be the reason. Moreover, we may conclude that slow crack growth does not affect these results very much in the interpolation range.

From a practical point of view, we note that the V/w versus a/w relationship at a given displacement as obtained in our treatment shows a maximum at about $a/w = 0.4$ which is displaced by v/w in the reverse direction of bending. Thus more accuracy is to be assumed when testing over this range. Moreover, we have observed that the Merkle and Corten treatment [8] leads to calibration curves displaced toward higher V/w values.

Results for the Fracture Criterion

The previous calibration results have been used to obtain the crack growth resistance curves with some extension to other specimen sizes. The double bend technique [11] has been applied with heat tinting or with brittle fracture at $-196\,°C$ to measure the ductile crack extension Δa. Partial results are shown in Figs. 3 and 4, with the extreme point corresponding to the maximum load. No curve could be easily fit to data points.

For the effect of specimen size, we see that what is partially attributed to the stretch zone development is at the left of the $V = (\sigma_y + \sigma_u) \Delta a$ straight

TABLE 7—Data for the V/w versus U/Bw^2 relation for the CT 20 specimen.

Specimen	Method	$\frac{a}{w}$	η_e	η_p	$(V/w)_0$, MPa	$(U/Bw^2)_1$, MPa	$(U/Bw^2)_2$, MPa	$(U/Bw^2)_M$, MPa	$(V/w)_M$, MPa
$\rho = 0$	1[a]	0.2	2.56	0.84	1.27	1.27	2.01	3.33	5.08
	1	0.3	2.82	1.41	0.80	0.99	2.07	2.74	6.45
	1	0.4	2.76	1.84	0.38	0.88	2.07	2.07	6.73
	2[b]	0.5	2.49	2.07	0.16	0.48	1.42	1.42	6.07
	2	0.6	2.26	2.18	0.05	0.29	0.88	0.88	4.82
	2	0.7	2.17	2.17	0	0	0.47	0.47	3.37
	2	0.8	2.06	2.06	0	0	0.20	0.20	2.04
	2	0.9	1.96	1.96	0	0	0.05	0.05	0.97
$\rho = 0.5$	1	0.2	2.62	0.75	1.53	1.48	3.29	3.29	4.62
	1	0.3	2.83	1.36	0.93	1.17	2.74	2.74	6.23
	1,2	0.4	2.79	1.82	0.40	0.86	2.08	2.08	6.72
	2	0.5	2.50	2.12	0.11	0.47	1.43	1.43	6.16
	2	0.6	2.25	2.25	0	0	0.87	0.87	4.90
	2	0.7	2.22	2.22	0	0	0.46	0.46	3.37
	2	0.8	2.10	2.10	0	0	0.19	0.19	1.99
	2	0.9	1.96	1.96	0	0	0.05	0.05	0.93
$\rho = 2$	1	0.2	2.68	0.67	1.65	1.64	3.12	3.37	4.52
	1	0.3	2.85	1.35	0.96	1.10	2.82	2.82	6.42
	1,2	0.4	2.77	1.87	0.40	0.65	2.13	2.13	7.05
	2	0.5	2.52	2.19	0.09	0.43	1.45	1.45	6.45
	2	0.6	2.32	2.32	0	0	0.87	0.87	5.04
	2	0.7	2.25	2.25	0	0	0.45	0.45	3.37
	2	0.8	2.06	2.06	0	0	0.19	0.19	1.94
	2	0.9	1.91	1.91	0	0	0.05	0.05	0.92

[a] Polynomial function.
[b] Exponential function.

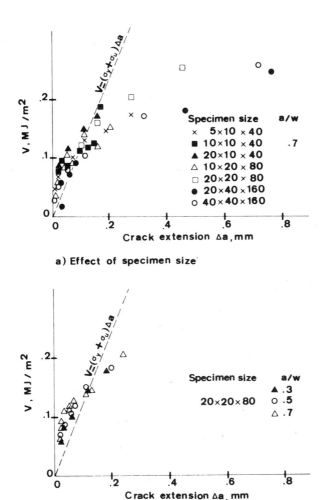

FIG. 3—*Resistance curves of V versus Δa for TPB specimen and Material 1.*

line. No true initial extension can be pointed out, because very early, all along the fatigue crack front, several small ductile tunnels appear as observed by microfractography, and because when the deviation point from the second straight line on the drop potential versus displacement record occurs in the CT specimens, propagation has taken place all along the crack length. However, when fibrous fracture concerns all the front, some trends can be deduced in spite of the scatter of the data points: for the CT specimen the resistance is independent of size; for the TPB specimen, for similar geometries, resistance and resistance gradient decrease with increasing size;

352 ELASTIC-PLASTIC FRACTURE

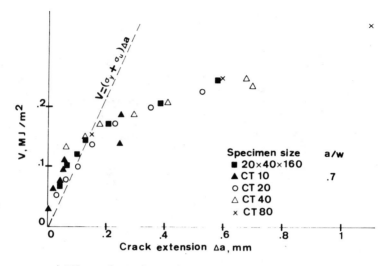

a) Effect of specimen size

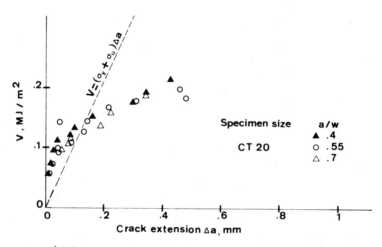

b) Effect of crack size

FIG. 4—*Resistance curves of* V *versus* Δa *for CT specimen and Material 2.*

for a geometry where thickness is only varying, resistance and resistance gradient increase with increasing size. No explanation is apparent. However, these observations may be put together with the extension mode: in all geometries, onset of crack growth occurs at midthickness; then, in CT specimens this initial "fibrous thumbnail" develops to spread uniformly all along the fatigue front; in TPB specimens, this thumbnail develops forward by tunneling when thickness is low and stops and two other thumbnails develop on both sides when thickness is high. The two extreme modes may be relevant to plane-stress and plane-strain conditions.

For the effect of initial crack length, nothing is noted.

For the effect of geometry, according to our limited data and with the use of $\eta = 1.86$ and 2.2 in the relation $V = \eta U/B \ (w - a)$ for $a/w = 0.7$, for the two geometries, nothing is noted.

From a practical point of view, in our state of knowledge, it seems very difficult to define a critical value to characterize toughness, though it should be possible to define a critical Δa value for which all values would be the same. However, crack growth resistance may represent toughness. In this respect, CT specimens appear more attractive, but their behavior may be fortuitous here and due to the material under investigation. But it seems reasonable to consider toughness as geometry and size dependent. Increasing size would allow resistance over a larger range to be obtained.

Conclusion

V/w versus U/Bw^2 relationships have been established for the TPB and CT specimens. Due to normalizing, they are to cover all specimen and crack sizes. From them it is shown, with a limitation due to the uniqueness of the material under study, that toughness is independent of crack length and loading mode, but depends on size.

Acknowledgment

This investigation was made possible by a research grant from the Délégation Générale à la Recherche Scientifique et Technique. The support of our respective laboratories is gratefully acknowledged.

APPENDIX

Three-Point Bend Specimen

Elastic Behavior

Srawley [12]:

$$K_1 = \frac{P}{B\sqrt{w}} \times \qquad (13)$$

$$\left\{ \frac{3 \left(\frac{S}{w}\right) \left(\frac{a}{w}\right)^{1/2} \left[1.99 - \frac{a}{w}\left(1 - \frac{a}{w}\right)\left(2.15 - 3.93 \frac{a}{w} + 2.7 \left(\frac{a}{w}\right)^2\right)\right]}{2\left(1 + 2\frac{a}{w}\right)\left(1 - \frac{a}{w}\right)^{3/2}} \right\}$$

for $0 \leq a/w \leq 1$
Elastic beam theory:

$$C(0) = \frac{1}{4E} \frac{S^3}{w^3} \left[1 + \frac{12}{5} \left(\frac{w}{S}\right)^2 (1 + \nu) \right] \quad (14)$$

Bucci et al [5]:

$$C(0) = \frac{0.24}{E} \frac{S^3}{w^3} \left[1.04 + 3.28 \left(\frac{w}{S}\right)^2 (1 + \nu) \right] \quad (15)$$

Srawley [12] (from):

$$C(a) = \frac{2S^2}{E'w^2} \left[-19.37 \frac{a}{w} + 8.72 \left(\frac{a}{w}\right)^2 - 6.10 \left(\frac{a}{w}\right)^3 \right.$$

$$+ 2.98 \left(\frac{a}{w}\right)^4 - 0.82 \left(\frac{a}{w}\right)^5 + 13.54 \ln\left(1 + 2\frac{a}{w}\right)$$

$$- 2.26 \ln\left(1 - \frac{a}{w}\right) - 10.39 \frac{\frac{a}{w}}{1 + 2\frac{a}{w}} - 0.57 \frac{\frac{a}{w}}{1 - \frac{a}{w}}$$

$$\left. + 0.49 \frac{\frac{a}{w}\left(2 - \frac{a}{w}\right)}{\left(1 - \frac{a}{w}\right)^2} \right] \quad (16)$$

Fully Plastic Behavior

In plane strain:

$$P_{L\epsilon} = \beta f \sigma_f \frac{B}{S} (w - a)^2 \quad (17)$$

where

$\beta = 1$ for a Tresca material,
$2/\sqrt{3} =$ for a Von Mises material,
$\sigma_f =$ uniaxial tensile flow stress $= (\sigma_y + \sigma_u)/2$,
$\sigma_y =$ uniaxial yield stress, and
$\sigma_u =$ ultimate tensile strength.

From Ewing [13]:

$f = 1.261$ for $\frac{a}{w} \geq 0.290$ (Charpy notch)   0.296 (sharp notch)

$f = 1$ for $\frac{a}{w} = 0$

From Chell and Spink [14]:

$$f = 1.261 - 2.72 \left(0.31 - \frac{a}{w}\right)^2 \quad \text{for } 0 \leq \frac{a}{w} \leq 0.31$$

with the assumption of f being a continuous and differentiable function between the Ewing' limits from which V is continuous—from adjustment to Knott's results [15] in $w/(w - a)$ for Charpy notch

$$f = 1.26 - 1.56 \left(\frac{w}{w - a} - 1.408\right)^2 \text{ for } 0 \le \frac{a}{w} \le 0.29 \quad (18)$$

and with the other condition of $V = 0$ for $a/w = 0$

$$f = 1.261 - 3.8314 (0.261 - a/w)^2 \quad (19)$$

In plane stress:

$$P_{L_\sigma} = \frac{2\beta}{1 + \beta} \sigma_f \frac{B}{S} (w - a)^2 \quad (20)$$

for $a/w \ge 0.02$ for a Von Mises material and $a/w \ge 0$ for a Tresca material.

Compact Tension Specimen

Elastic Behavior

Srawley [12]:

$$K_1 = \frac{P}{B\sqrt{w}} \frac{\left[0.886 + 4.64 \frac{a}{w} - 13.32 \left(\frac{a}{w}\right)^2 + 14.72 \left(\frac{a}{w}\right)^3 - 5.6 \left(\frac{a}{w}\right)^4\right]}{\left(1 - \frac{a}{w}\right)^{3/2}}$$

(21)

for $0.2 \le a/w \le 1$

Adams and Munro [16]:

$$C(0) = \frac{2.2}{E} \quad (22)$$

Srawley [12] (from):

$$C(a) = \frac{1}{E} \left[119.11 \frac{a}{w} + 75.04 \left(\frac{a}{w}\right)^2 - 92.83 \left(\frac{a}{w}\right)^3 \right.$$

$$- 69.00 \left(\frac{a}{w}\right)^4 + 27.01 \left(\frac{a}{w}\right)^5 + 32.76 \left(\frac{a}{w}\right)^6$$

$$- 15.62 \left(\frac{a}{w}\right)^7 - 7.84 \left(\frac{a}{w}\right)^8 + 134.84 \ln\left(1 - \frac{a}{w}\right)$$

$$\left. + 15.82 \frac{\frac{a}{w}\left(2 - \frac{a}{w}\right)}{\left(1 - \frac{a}{w}\right)^2} - 9.64 \frac{\frac{a}{w}}{1 - \frac{a}{w}} \right]$$

(23)

with the restriction of $Y(a/w)$ not defined for $(a/w) < 0.2$.
Newman [17]: tabulated values of EC.

Fully Plastic Behavior

Ewing and Richards [18] (lower-bound solution):

$$P_L = \beta \sigma_f w B \left[\sqrt{2.7 + 4.59 \left(\frac{a}{w}\right)^2} - \left(1 + 1.7 \frac{a}{w}\right) \right] \qquad (24)$$

$$P_L = \sigma_f w B \left[\sqrt{(1 + \beta)\left(1 + \beta \left(\frac{a}{w}\right)^2\right)} - \left(1 + \beta \frac{a}{w}\right) \right] \qquad (25)$$

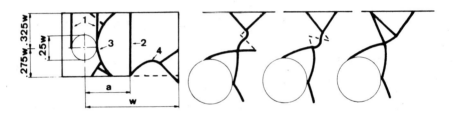

1. $F_L = .325\ w^2 k$
2. $F_L = .3\ w^2 k$
3. $F_L = .18\ w^2 k$
4. EWING and RICHARDS [17]

(k = shear yield stress)

Partial enlarged view of solution 3.

FIG. 5—*Schematic of slip line fields in the CT specimen.*

For the geometry under consideration, these solutions are valid above a limit which is given for plane strain by the upper-bound theorem as $a/w \simeq 0.27$, or with more refined slip line fields as shown in Fig. 5 as $a/w \simeq 0.25$. Therefore, due to the strain hardening of the material, flow may take place through the pinhole for values of a/w less than, say, 0.35.

References

[1] Ritchie, R. O., Garrett, G. G., and Knott, J. F., *International Journal of Fracture Mechanics*, Vol. 7, 1971, pp. 462-467.
[2] Rice, J. R. in *Fracture*, H. Liebowitz, Ed., Vol. 2, Academic Press, New York, 1968, pp. 191-311.
[3] McClintock, F. A. in *Fracture*, H. Liebowitz, Ed., Vol. 3, Academic Press, New York, 1971, pp. 47-225.
[4] Begley, J. A. and Landes, J. D. in *Fracture Toughness*, *ASTM STP 514*, American Society for Testing and Materials, 1972, pp. 1-23 and pp. 24-39.
[5] Bucci, R. J., Paris, P. C., Landes, J. D. and Rice, J. R. in *Fracture Toughness*, *ASTM STP 514*, American Society for Testing and Materials, 1972, pp. 40-69.
[6] Rice, J. R., Paris, P. C., and Merkle, J. G. in *Progress in Flaw Growth and Fracture Toughness Testing*, *ASTM STP 536*, American Society for Testing and Materials, 1973, pp. 231-245.

[7] Miannay, D. and Pelissier-Tanon, A., *Mécanique Matériaux Electricité*, No. 328-329, 1977, pp. 29-40.
[8] Merkle, J. G. and Corten, H. T., *Journal of Pressure Vessel Technology*, No. 11, 1974, pp. 286-292.
[9] Larsson, S. G. and Carlsson, A. F., *Journal of the Mechanics and Physics of Solids*, Vol. 21, 1973, pp. 263-277.
[10] Rice, J. R., *Journal of the Mechanics and Physics of Solids*, Vol. 22, No. 1, 1974, pp. 17-26.
[11] Landes, J. D. and Begley, J. A. in *Fracture Analysis, ASTM STP 560*, American Society for Testing and Materials, 1974, pp. 170-186.
[12] Srawley, J. E., *International Journal of Fracture Mechanics*, Vol. 12, No. 3, 1976, pp. 475-476.
[13] Ewing, D. J. F., *Journal of the Mechanics and Physics of Solids*, Vol. 16, 1968, pp. 205-213.
[14] Chell, G. G. and Spink, G. M., *Engineering Fracture Mechanics*, Vol. 9, 1977, pp. 101-121.
[15] Knott, J. F. in *Fracture 1969, Proceedings of the Second International Conference on Fracture*, Chapman and Hall Ltd., London, 1969, pp. 205-218.
[16] Adams, N. J. I. and Munro, H. G., *Engineering Fracture Mechanics*, Vol. 16, 1974, pp. 119-132.
[17] Newman, J. C., Jr. in *Fracture Analysis, ASTM STP 560*, American Society for Testing and Materials, 1974, pp. 105-121.
[18] Ewing, D. J. F. and Richards, C. E., *Journal of the Mechanics and Physics of Solids*, Vol. 22, 1974, pp. 27-36.

I. Milne[1] *and G. G. Chell*[1]

Effect of Size on the *J* Fracture Criterion

REFERENCE: Milne, I. and Chell, G. G., "**Effect of Size on the *J* Fracture Criterion,**" *Elastic-Plastic Fracture, ASTM STP 668*, J. D. Landes, J. A. Begley, and G. A. Clarke, Eds., American Society for Testing and Materials, 1979, pp. 358-377.

ABSTRACT: Experimental evidence showing the size dependence of J_{Ic} in ferritic steels is presented. This behavior is described in terms of a relatively simple mechanistic model for cleavage fracture. It is shown that the mechanisms of cleavage fracture and ductile slow crack growth are always in competition and this leads to the behavior frequently encountered in fracture tests where initially ductile crack extension leads to fast brittle failure. The size effect is also discussed in terms of a shift in the ductile brittle transition temperature. The implications on failure assessment procedures are mentioned.

KEY WORDS: fracture criterion, size effect, cleavage failure, slow crack growth, *J*-integral, fracture toughness, ferritic steels, ductile brittle transition (toughness), failure assessment, elastic-plastic, crack propagation

Although in the small-scale yielding regime failure can be characterized a one-parameter criterion, such as the fracture toughness K_{Ic}, it is not clear if this is applicable after appreciable yielding. Experimental evidence based upon evaluation of the *J*-integral at failure suggests that under some circumstances this may be the case [1].[2] Alternatively, there is an increasing amount of evidence demonstrating situations in which this is not so, and a geometry effect is apparent. In this paper we briefly review the latter evidence and investigate some of its implications on fracture toughness testing and service assessments, confining ourselves to ferritic steels. Existing concepts and models of fracture are used to provide a description of fracture behavior consistent with the observed size dependence of the *J* failure criterion. The model predictions are related to fractographic features and the occurrence of slow crack growth prior to fast failure.

---

[1]Research officer and Fracture Mechanics Project leader, respectively, Materials Division, Central Electricity Research Laboratories, Kelvin Avenue, Leatherhead, Surrey, U.K.
[2]The italic numbers in brackets refer to the list of references appended to this paper.

Failure Criterion

The failure criterion is based on the attainment of a critical value, J_{Ic}, of the parameter J [1,2]. Consistent with the usual experimental method of determining J [2], we interpret J_{Ic} as characterizing the maxima in the total energy. This interpretation is similar to that proposed by Griffith for brittle failure. J_{Ic} therefore represents a critical force which is exerted on the crack and plastic zone at failure. It should be noted that the energy released by the propagation of the crack an increment Δa is not $-J\Delta a$ [1,3].

The problem of relating this macroscopic parameter J to the metallurgical mechanisms of failure is still unresolved. The assumption that J characterizes the crack tip environment [1], although supported by some theoretical evidence [4], is still not proven in the case of real materials. Thus while it may have credence for monotonically loaded cracked bodies at constant temperature, it is certainly not true in general [5].

To aid comparison with fracture toughness, we define a plastic stress intensity factor, K_p, which is equal to a critical value K_{IJ} at failure and is related to J through the equation

$$K_p = \sqrt{\frac{EJ}{(1-\nu^2)}} \qquad (1)$$

where E is Young's modulus and ν Poisson's ratio. Putting $J = J_{Ic}$ in Eq 1 provides the necessary relationship between K_{IJ} and J_{Ic}. In circumstances where plastic deformation is negligible, K_p is, of course, equal to the stress intensity factor calculated linear elastically. It thus follows that, provided J_{Ic} is a valid fracture criterion, K_{IJ} is numerically equal to K_{Ic} since $J_{Ic} = (1 - \nu^2)K_{Ic}^2/E$.

Experimental Evidence Showing a Geometry Dependence of J_{Ic}

Much of the work in elastic-plastic fracture mechanics has been aimed at establishing the equivalence of K_{IJ} and K_{Ic}. This has been successful within limits and the current argument tends to center around the definition of these limits [2]. This tends to beg the question, however, since there obviously is a geometry dependence to J_{Ic}. Moreover, size limits cannot be imposed on real structures. In particular, where tests have been performed without regard to geometry limitations, large variations in J_{Ic}, and hence K_{IJ}, have been observed as a function of crack size a, a/w (where w is the specimen width), and specimen size. Table 1 lists much of the recent work performed in this area. Clearly the effect is not confined to one type of steel or one specimen geometry. This effect is also dramatically illustrated in Fig. 1, where the results for the single-edge notched tension (SENT) tests listed in Table 1 are plotted in terms of K_{IJ}/K_{Ic} against the crack

360 ELASTIC-PLASTIC FRACTURE

TABLE 1—*Published work where a size dependence of toughness in ferritic steels has been noted.*

Reference	Material	Testing Geometry
[6]	maraging steel	three-point bend
[7]	high-temperature bolting steel and a rotor steel	three-point bend
[8]	pressure vessel steel	three-point bend
[9]	high-temperature bolting steel	single edge notched tension
[10]	high-temperature bolting steel and rotor steel	single edge notched tension
[11]	A533B pressure vessel steel	compact tension

length a, normalized for convenience by K_{IJ}^2/σ_u^2, σ_u being the ultimate tensile strength. The values of K_{IJ} were obtained using a load displacement curve fitting technique which enables J to be determined for a given specimen using only the recorded data for that specimen [10,30]. The increase in K_{IJ}

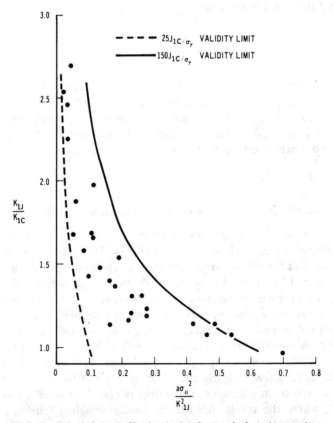

FIG. 1—*Effect of size on K_{IJ} for single-edge notched tension specimens.*

occurs not only as a/w decreases, but also as the specimen size decreases although this is not apparent from the figure.

It is interesting to note that if

$$a = 25 J_{Ic}/\sigma_Y \,(=25(1 - \nu^2)K_{Ic}^2/E\sigma_Y)$$

the minimum value required for a valid J_{Ic} (K_{Ic}) value [32], then the curve

$$a\sigma_u^2/K_{IJ}^2 \simeq 0.1 \,(K_{Ic}/K_{IJ})^2$$

in Fig. 1 (dotted line) is the validity limit below which size effects should become apparent. The data in Fig. 1 thus meet the proposed validity requirement for a J_{Ic} analysis, indicating that this proposal is inadequate. A limit of 150 J_{Ic}/σ_Y, full line in Fig. 1, would be more satisfactory [10].

Fractography

The behavior illustrated in Fig. 1 could be attributed to several causes: slow crack growth prior to fast fracture, a change in the mode of failure, or the inability of the analytical technique to correctly predict the failure criterion. This latter point can be countered by the similarity of the predictions using different methods of J analyses [10] and by the fact that in at least two instances [6,8] the failure criterion calculated without any plasticity correction still exceeded K_{Ic}. Thus we are left with having to explain the phenomenon in terms of mechanistic or fractographic behavior.

To investigate these features, the fracture surfaces of four of the steels listed in Table 1 were examined in detail using a high-resolution Camebax scanning electron microscope. The four steels, whose composition and mechanical properties are listed in Table 2, were

(A) BS 1501 271A: A fine grained, pearlitic pressure vessel steel which had been tested in three-point bend at 130°C [8]. The fracture toughness of this steel had previously been measured at between 56 and 69 MNm$^{-3/2}$ at $-30°$C, yet small specimens produced K_{IJ} values as high as 270 MNm$^{-3/2}$.

(B) A medium-strength quenched-and-tempered bainitic steel which had been tested in three-point bend between $-70°$C and room temperature [12]. This steel suffered from bands of inclusions, mainly of manganese sulphide and titanium nitride, oriented in the rolling plane. Loss of linearity in the load displacement curves, where this occurred, was primarily due to the slow stable crack growth. Since the initiation point was not known, K_{IJ} was not determined.

(C) A quenched-and-tempered bainitic rotor forging steel which had been tested using SENT specimens at room temperature [10,13]. Results shown in Fig. 1 are for this steel.

(D) A high-temperature bolting steel heat treated to give a bainitic

TABLE 2—Analysis and mechanical properties of the steels.

Steel	Type	σ_Y, MNm^{-2}	σ_u, MNm^{-2}	Analysis, %								
				C	Si	Mn	S	P	Ni	Cr	Mo	V
A	BS1501 271A	550	720	0.17	0.28	1.43	0.004	0.007	0.43	0.65	0.33	0.06
B	bainitic steel	730	794	0.13	0.47	0.88	0.017	0.012	0.05	0.62	0.28	<0.01
C	IP rotor forging	570	743	0.40	0.19	0.80	0.01	0.01	0.02	0.73	0.64	0.27
D	bolting steel	565	710	0.23	0.36	0.59	0.018	0.009	0.20	1.07	1.00	0.64

structure [10]. Tests were performed on SENT specimens and the results are also shown in Fig. 1.

In each case the area studied was confined to that region immediately ahead of the fatigued starter crack.

Despite the differences in the four steels studied, the fracture surfaces exhibited similar features which could be categorized in the following way.

1. Fully brittle, as in Fig. 2. Here immediately adjoining the stretch zone the fracture surface was made up almost entirely of cleavage facets, often with microcracks and sharp stepped features associated with it. There was no evidence that changes in orientation from one cleavage facet to another involved any substantial amount of ductile fracture. These fractures were associated with the lower temperatures of Steel B and the lower K_{IJ} values in the other steels ($K_{IJ}/K_{Ic} < 1.3$).

2. Brittle, but with isolated ductile regions, as in Fig. 3. Here, although the fracture surface was predominantly brittle, as in Fig. 2, regions of ductile fracture, generally no larger than a grain, occurred randomly distributed along the tip of the stretch zone. The dimpled features of these areas were always much finer than the main features associated with ductile slow crack growth (compare Figs. 3 and 4), were similar to that observed in the shear lip regions, Fig. 5, but were never associated with inclusions. It should be emphasized that the ductile areas were always isolated from each other and should not be confused with ductile slow crack growth. They were, however, present in those specimens which exhibited high K_{IJ} values ($K_{IJ}/K_{Ic} > 1.3$) and in particular although not exclusively where the crack lengths were short. There were less of these regions of ductile fracture in areas of the fracture surface remote from the fatigue starter crack.

3. Ductile slow crack growth as in Fig. 4. These regions were observed only for the shorter cracked specimens of the smallest size of Steels A and D, and for the higher temperatures of Steel B. In this latter instance some tests failed entirely in the slow ductile mode, while others started in the ductile mode but changed eventually to fast brittle fracture. Often in these cases there was a sharp transition between the ductile and the brittle regions, yet some brittle fracture could be observed well within the slow crack growth areas and some ductile dimpling within the predominantly brittle region. The ductile areas contained features on two different scales: (1) voids 15 to 25 μm in size which generally contained inclusions or other nonmetallic particles and (2) fine dimples less than $\sim 2 \mu$m in size.

The dimpled regions tended to link one void with another. The individual dimples were comparable in scale to similar features observed in the ductile regions in 2 of the foregoing, and also to the ductile dimples observed in the shear lip regions (Fig. 5).

The voids, on the other hand, apart from containing inclusions, had a similar appearance to the stretch zones developed in all of the specimens,

364 ELASTIC-PLASTIC FRACTURE

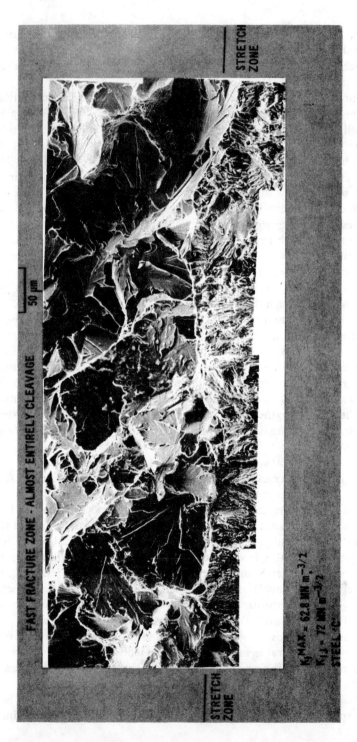

FIG. 2—*An example of brittle fracture.*

these being characteristic of stretch zones typically found in steels of this nature [14]. This similarity between the stretch zones and the voids suggests that the voids are really a quasi-continuous extension of the stretch zone.

Ductile tearing of this nature, that is, void growth generally around inclusions linked by bridges of ductile dimples, can occur either at plastic collapse or during slow stable crack growth. The latter is crack tip controlled rather than controlled by the dimensions of the uncracked ligament, and as such, although the microprocesses appear to be the same, the mechanical description of failure is different from plastic collapse. As mentioned previously, this slow crack growth is not just a special case of ductile fracture but is a mode of propagation in its own right [12].

This should be compared with the general observation that, where the fracture process was fast, a high measure of fine ductile dimple fracture was observed on the surfaces of specimens where K_{IJ} was much higher than K_{Ic}; for example, compare Fig. 3a and 3b.

Descriptions of Fracture Behavior Based upon Simple Mechanistic Models

Cleavage Fracture

Recently a model of cleavage fracture from sharp cracks has been proposed based upon the postulate that fracture will occur when the stress normal to the crack plane exceeds the cleavage stress over a characteristic distance ahead of the crack tip [15]. This distance is associated with some microstructural feature such as the grain size or carbide spacing. The model is appealing since it explains an apparent anomaly in the comparative magnitudes of Charpy energies and fracture toughness values of two steels [16], as well as the increase in toughness with temperature due to changes in yield stress [15,31].

In applying this model to the size effect described in the foregoing and to relate J to the failure mechanism, we assume that the stress field ahead of the crack can be characterized by J, even in the large-scale yielding regime. Hence, contrary to general belief, when the failure mechanism is taken into account, a size dependence of J is qualitatively predictable.

If $\bar{\sigma}$ represents a flow stress and X_0 the characteristic distance over which the normal stress σ_{YY} must exceed the cleavage fracture stress σ_f, then the point of failure, F, for a given specimen and material, is shown schematically in Fig. 6 for plane-strain conditions. The quantity X, a measure of the distance ahead of the crack, has been normalized by $EJ/\bar{\sigma}^2$. In a smaller specimen of identical material loaded to the same value of J, the stress field directly ahead of the crack should be the same as before. However, let us assume that in the smaller specimen failure occurs after large-scale plasticity such that some through-thickness deformation occurs with

FIG. 3a—*Isolated ductile regions in a predominantly brittle fracture surface.*

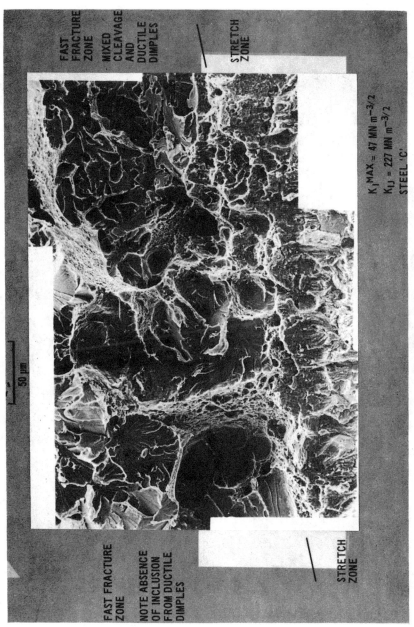

FIG. 3b—Continued.

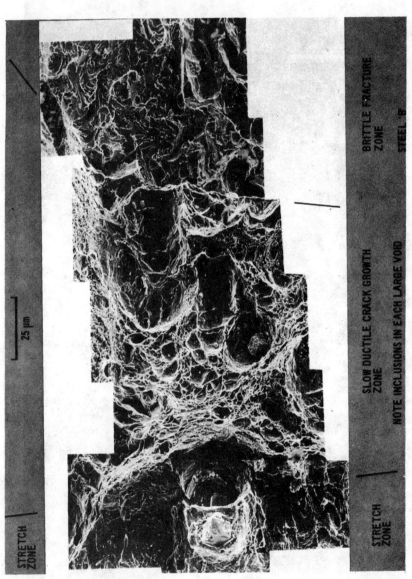

FIG. 4—*Ductile slow crack growth.*

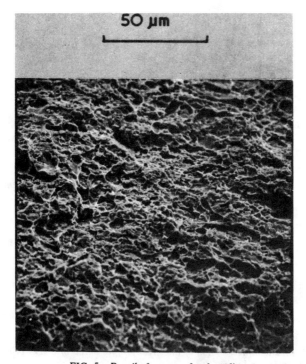

FIG. 5—*Ductile features of a shear lip.*

a resulting loss of stress triaxiality. The new stress field, although still characterized by the same value of J, will now fall below the previous level near the crack tip, and thus, when $J = J_{Ic}$, failure will not occur because the critical stress condition for cleavage is not satisfied (see Fig. 6). To satisfy this condition, extra load must be added so that J at failure becomes greater than J_{Ic}. This results in a value of K_{IJ} which exceeds K_{Ic}. Since the effect is likely to be most pronounced in failures occurring after general yielding, the extra load needed to fracture the specimen will result in an even greater loss of stress elevation. Thus in this regime the conditions necessary to attain cleavage are in direct competition with the consequences of trying to attain it. At some stage, cleavage will not be attainable and another mode of failure will take over.

After general yielding there is some loss of constraint. A measure of the plastic constraint, R, in the specimen is given by the ratio of plane-strain to plane-stress collapse loads. If the observed increase in K_{IJ} is a consequence of loss of stress elevation, then it should be less pronounced in the more highly constrained geometries. The predictions of slip-line field theory for both three-point bend [18] and SENT geometries [19] indicate increasing values of R with increasing ratio a/w. (In the case of SENT this

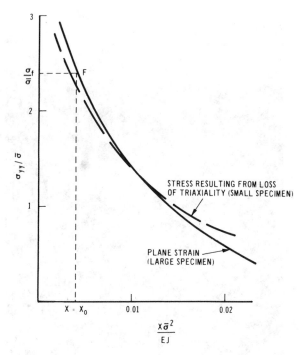

FIG. 6—*Schematic representation of the stress profiles ahead of cracks in large and small specimens with the same J-value.*

peaks at about $a/w = 0.5$ and decreases again.) This is in agreement with the observed increase in K_{IJ} as crack length a is decreased in a specimen of given size.

There are effects due to crack tip blunting in addition to the complications inherent in trying to achieve the critical stress over the critical distance where plasticity is large. Blunting creates a localized area of plane stress ahead of the crack and a resulting intense strain region [17] which will amplify the effects discussed in the foregoing. Indeed the ductile appearance of the stretch zone is a direct surface manifestation of this straining. Furthermore, if ρ is the root radius of a notch, the maximum stress elevation ahead of it will depend on a/ρ and the effect of blunting will be most pronounced for small crack sizes. Since the increasing values of K_{IJ} are consistent with increasing stretch zone size and, in general, since ρ will be a function of the stretch zone size, the effect of the blunting is to make the attainment of the critical stress more difficult.

The increase in the amount of ductility on the fracture surfaces of specimens with increasing K_{IJ} values clearly reflects the increasing amount of strain occurring ahead of the crack as triaxiality is lost. The rise in K_{IJ} values as size is decreased (see Fig. 1) is also consistent with the sudden increased sensitivity of J to the applied load after general yielding is reached (Fig. 7a). After general yielding, the macroscopic deformation of the specimen becomes progressively more consistent with plane-stress rather than plane-strain conditions [20] and large displacements occur (Fig. 7b). In this region small changes in load can result in large changes in both the value of J and displacements which are consistent with the sudden increase in K_{IJ}.

Ductile Slow Crack Growth[3]

As previously mentioned, at some stage the critical stress criterion cannot be satisfied because the "maximum achievable stress levels are essentially limited, even with continuous strain hardening" [17]. This is "suggestive of abrupt toughness transitions," although, unlike Ref 17, we refer to changes in the value of the failure parameter at constant temperature, and not to changes due to increases in temperature or loading rate. Eventually a transition in failure mode occurs, cleavage fracture cannot be initiated, and the crack advances instead in a stable manner.

The initiation of this slow crack growth arises due to the formation of voids ahead of the crack tip, and their subsequent linking by local necking of the remaining material [17]. Thus the crack can be thought to advance in a series of jumps [26] as each void becomes linked with its neighbor.

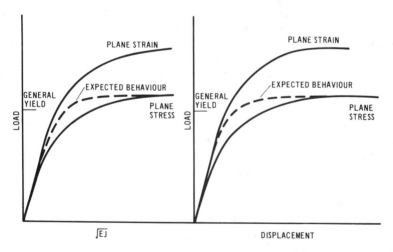

FIG. 7—*Schematic representation showing sudden increase in* J *and displacement as a function of load after general yielding.*

---

[3]To the authors' knowledge there is no evidence of slow crack growth by brittle cleavage, although a quasi-static fast fracture mode is possible in specimens where the stress intensity factor falls with increasing crack length, or where crack front geometry changes result in a lower stress intensity.

The crack tip displacement required to do this is inversely related to the spacing of the metallurgical features, inclusions, precipitates, and grain boundary triple points, around which the voids nucleate. Thus the microstructure will have an important influence over the incidence of slow crack growth, and we would expect steels containing a high density of inclusions, or spheroidals carbides, to more readily exhibit slow crack growth prior to fast fracture than those steels without any obvious nuclei where holes can nucleate.

For voids to initiate and grow around inclusions, there must be a large amount of plastic strain resulting from loss of stress triaxiality. Nevertheless, voids will not grow without a certain amount of stress elevation [17,23]. This is borne out by observations on the fracture surfaces of specimens where slow crack growth occurs. The ductile stable crack is held up at the surfaces, where triaxiality is a minimum, and advances most rapidly in the center of the specimen in a thumbnail geometry. In the surface regions of the specimen where there is no stress elevation, ductile shear lips develop without the presence of large voids around inclusions. These shear lips are similar to those developed on surfaces which have failed by fast fracture, and contain fine ductile dimples. Moreover, specimens which have been side grooved to artificially increase constraint [24], and which fail by slow crack growth, do so with a uniformly straight crack front rather than a thumbnail. These considerations lead to a mechanism for slow crack growth based upon the postulate that ductile crack propagation will occur when a critical plastic strain (which will depend on the state of stress) is exceeded over a characteristic distance associated with the microstructure (for example, inclusion spacing) [17,21,22].

Frequently after some amount of slow crack growth the crack propagates in a brittle manner. This can result from several causes, not least the heterogeneity of the material. Regardless of this, as the geometry of the crack front becomes more convex, through-thickness deformation diminishes and triaxiality increases local to the advancing crack tip. It has also been suggested that the stress level is raised further by crack tip sharpening as the crack propagates [25]. The overall effect is to make cleavage more favorable so that a change from ductile slow crack growth to brittle fast fracture can occur.

In summary, both slow crack growth and cleavage are favored by a large degree of triaxiality, but slow crack growth also needs large plastic strains to induce void growth. For slow crack growth to be preferred, these conditions must be satisfied before the stress ahead of the crack can be elevated above σ_f over the critical distance for cleavage [31]. The mode of failure which predominates depends upon microstructural features (for example, carbide and inclusion spacings) as well as mechanical effects such as loss of through-thickness constraint and crack tip blunting. In a sense, therefore, the fast cleavage and slow ductile modes of crack propagation are always in competition, even at temperatures below the ductile brittle transition.

Geometry Dependence of K_U and the Ductile Brittle Transition

We assume that the plane-strain fracture toughness, K_{Ic}, is, as postulated, a genuine material parameter which is measurable only on large enough (valid) specimens. Brittle cleavage fracture is then expected in the lower-shelf region; fast ductile fracture, which is distinct from slow crack growth and has not so far been discussed here, is expected in the upper-shelf region. These two regions are linked by the transition region, which is a mixed fracture mode.

The observed increases in K_U as a function of decreasing a, a/w or specimen size, and associated with extra ductility in the fracture surface, have an analogy in the effects of temperature on toughness in the ductile brittle transition region. This transition, in body centered cubic alloys, and the associated notch sensitivity was first explained by Orowan [27]. It is a consequence of the temperature dependence of the yield or flow stress in these materials, and the temperature independence of the cleavage stress. Increasing geometric constraint, by notching a specimen or by increasing its size, raises the stress triaxiality and increases the ductile brittle transition temperature, creating the notch or size sensitivity. There is no apparent reason why these concepts should not extend to specimens containing sharp cracks. Indeed it is of interest to note that in a fracture analysis of invalid compact tension test data on A533B steel, it was concluded that there were indications of a size effect on the ductile brittle transition temperature [11].

Using models of cleavage failure it should be possible to quantify the effects of size, at a given temperature, T, in terms of a shift in the ductile brittle transition temperature. For example, the model proposed in Ref 15 allows changes in toughness resulting from variations in yield stress and stress elevation to be taken into account. If the stress field ahead of a crack is known, and the effects of crack tip blunting are suitably simulated, then the model enables the effects of size to be expressed in terms of a change in yield stress. This can be directly related to a temperature shift through the yield stress-temperature curve.

Thus continuum analyses based upon only macroscopic variables (for example J) cannot predict K_{Ic} from an invalid test if the size, geometry, and material of the specimen is such that K_U is greater than K_{IC} at the testing temperature. For K_{Ic} to be obtained from such a test, the mechanism for failure must be defined. Indeed the attainment of a critical value of J, although macroscopically necessary, is not, on the microscale, always a sufficient condition to initiate fast fracture. Hence, if $K_{Ic}(T)$ represents the temperature dependence of the toughness, we can write

$$K_{IJ}(T) = K_{Ic}(T + \Delta T)$$

where ΔT is the shift in the transition curve resulting from size effects.

In the transition region the difference between K_{IJ} and K_{Ic} depends very much upon the magnitude of this shift in the transition curve. For a given shift, a steel with a sharp transition will show a greater effect than one with a gradual transition. The lower end of the transition is always gradual, so tests performed in this region may exhibit only a small increase in K_{IJ} over K_{Ic}, which may be contained within the experimental scatter band.

The foregoing description has excluded the possibility of slow crack growth, which complicates the problem even further and makes a general description difficult. It also leads to questions concerning the relationship of failure parameters determined at the initiation of slow crack growth to the same failure parameters calculated at the onset of fast fracture. These questions have not, as yet, been satisfactorily resolved.

Failure Assessments

Although much of the previous discussion has revolved around the microprocesses of fracture, these have to be represented in some mechanical way (that is, via macroscopic variables) before they can be used in an assessment. Figure 8 represents how the order of events (from Path 1 to Path 5) leading to failure of a cracked body can change as the triaxiality ahead of the crack is reduced. This also shows that, excluding fast ductile failure, there are only two mechanical descriptions of failure, brittle fracture and plastic collapse [28]. The question to be answered in a failure assessment is which of the alternative paths to failure will be followed by the structure, and how can some measure of control be introduced into each of these paths? Once the path to failure has been established, the difficult task of obtaining relevant materials parameters must then be faced.

Current assessments are based upon initiation data. Thus if the two failure limits can be reconciled, and there are procedural manuals now available for doing this [29], the problem can be handled in principle. However, for tough materials, the size limitations of test specimens will cause them to fail on any of the paths from 2 to 5. The previous discussion and the data used therein indicate that there is a risk that specimens failing along Path 2 may produce K_{IJ} in excess of K_{Ic}, especially in the ductile brittle transition region. This can lead to an overestimate of the defect tolerance of a structure. Moreover, for those specimens which follow Path 3, slow crack growth initiation occurs before the specimen reaches its full load bearing capacity. Here it is generally thought that there is a likelihood of underestimating the defect tolerance of the structure. This is not always the case since K_{Ic} is definable as the minimum possible value for K_{IJ} at the relevant

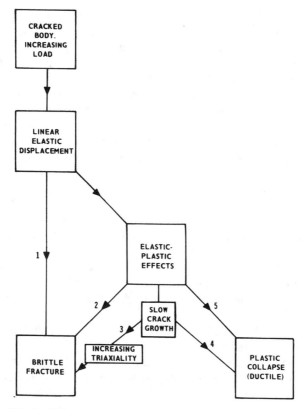

FIG. 8—*Different failure paths which a cracked body can follow.*

temperature. It follows therefore that the ductile initiation value for K_{IJ} can also exceed K_{Ic}. This argument is equally applicable to Path 4, of course, since the only difference between Paths 3 and 4 is that sufficient triaxiality for cleavage cannot be generated in the latter instance. Thus a simple mechanical representation of the failure conditions of a body is not sufficient to accurately predict the load-carrying capacity of a structure where failure is likely to be by any route other than Paths 1 and 5.

There are two things needed to cover Routes 2 to 4 in a way that avoids undue pessimism.

1. It is necessary to accurately assess the triaxiality local to the crack tip, taking into account the changing shape of a growing crack and the location of the crack front.

2. A way must be discovered to relate how this triaxiality determines which of the microprocesses, cleavage or void growth, is favored.

If the triaxiality in a structure cannot be developed to the level where cleavage is initiated, assessment can be based upon the amount of growth necessary to reach plastic collapse under the applied loads. Alternatively, it can be based upon the geometric changes required of the crack to invoke the triaxiality necessary for cleavage. Clearly, despite the current activity in helping to understand the mechanics of slow crack growth, we are a long way from developing the techniques for solving this problem.

In the meantime the problems discussed in the foregoing are best handled by designing to initiation. In a slow crack growth situation, this is highly pessimistic since it does not use the full load bearing capacity of the structure. However, it should be recognized that there is a chance that the initiation data obtained from small specimens could exceed K_{Ic}.

Conclusions

1. There is a distinct possibility for ferritic steels that fracture toughness values obtained by elastic-plastic analyses of invalid-sized specimens can exceed K_{Ic}.
2. This behavior can be qualitatively described using relatively simple mechanistic models of fracture, and results from a loss of stress triaxiality ahead of the crack due to loss of through thickness constraint and crack tip blunting.
3. This size effect can be represented in terms of a shift in the ductile brittle transition temperature.
4. In general the attainment of a critical value of J, although a macroscopic requirement for fast fracture, is not necessarily a sufficient one on the microscopic level.

Acknowledgments

The authors wish to thank Drs. V. Vitek and I. L. Mogford for their comments on the manuscript.

This work was performed at the Central Electricity Research Laboratories and is published by permission of the Central Electricity Generating Board.

References

[1] Landes, J. D. and Begley, J. A. in *Fracture Toughness, ASTM STP 514*, American Society for Testing and Materials, 1972, pp. 24–39.
[2] Begley, J. A. and Landes, J. D. in *Fracture Toughness, ASTM STP 514*, American Society for Testing and Materials, 1972, pp. 1–23.
[3] Vitek, V. and Chell, G. G., *Materials Science and Engineering*, Vol. 27, 1977, p. 209.
[4] McClintock, F. A., in *Fracture*, Vol. 3, H. Liebowitz, Ed., Academic Press, New York, 1971, p. 47.
[5] Chell, G. G. and Vitek, V., *International Journal of Fracture Mechanics*, Vol. 13, 1977, p. 882.
[6] Brown, W. F. and Srawley, J. E., *Plane Strain Crack Toughness Testing, ASTM STP 410*, American Society for Testing and Materials, 1966, p. 16.
[7] Chell, G. G. and Spink, G. M., *Engineering Fracture Mechanics*, Vol. 9, 1977, p. 101.

[8] Milne, I. and Worthington, P. J., *Materials Science and Engineering*, Vol. 26, 1976, p. 585.
[9] Chell, G. G. and Davidson, A., *Materials Science and Engineering*, Vol. 24, 1976, p. 45.
[10] Chell, G. G. and Gates, R. S., *International Journal of Fracture Mechanics*, Vol. 14, 1978, p. 233.
[11] Sumpter, J. D. G., Metal Science, Vol. 10, 1976, p. 354.
[12] Milne, I., *Materials Science and Engineering*, Vol. 30, 1977, p. 241.
[13] Batte, D. A., Blackburn, W. S., Elsender, A., Hellen, T. K., Jackson, A. D., and Poynton, W. A., to be published.
[14] Elliott, D., "The Practical Implications of Fracture Mechanisms," Spring Meeting, Institute of Metallurgists, Newcastle-upon-Tyne, U.K., 1973, p. 21.
[15] Ritchie, R. O., Knott, J. F., and Rice, J. R., *Journal of the Mechanics and Physics of Solids*, Vol. 21, 1973, p. 395.
[16] Ritchie, R. O., Francis, B., and Server, W. L., *Metallurgical Transactions*, Series A, Vol. 7A, 1976, p. 831.
[17] Rice, J. R. and Johnson, M. A. in *Inelastic Behavior of Solids*, M. F. Kanninen et al, Eds., McGraw-Hill, New York, 1970, p. 641.
[18] Ewing, D. J. F., Ph.D. thesis, Cambridge University, Cambridge, England, 1969.
[19] Ewing, D. J. F. and Richards, C. E., *Journal of the Mechanics and Physics of Solids*, Vol. 22, 1974, p. 27.
[20] Andersson, H., *Journal of the Mechanics and Physics of Solids*, Vol. 20, 1972, p. 33.
[21] McClintock, F. A., *International Journal of Fracture Mechanics*, Vol. 4, 1968, p. 101.
[22] MacKenzie, A. C., Hancock, J. W., and Brown, D. K., *Engineering Fracture Mechanics*, Vol. 9, 1977, p. 167.
[23] McClintock, F. A., *Journal of Mechanics*, Vol. 35, 1968, p. 363.
[24] Green, G. and Knott, J. F., *Metals Technology*, Vol. 2, 1975, p. 422.
[25] Hancock, J. W. and Cowling, M. J. in *Fracture, 1977, 4th International Conference on Fracture*, D. M. R. Taplin, Ed., University of Waterloo, Waterloo, Ont., Canada, Vol. 2, 1977.
[26] Clayton, J. Q. and Knott, J. F., *Metal Science*, Vol. 10, 1976, p. 63.
[27] Orowan, E., *Reports of Progress in Physics*, Vol. 12, 1948, p. 185.
[28] Dowling, A. R. and Townley, C. H. A., *International Journal of Pressure Vessel Piping*, Vol. 3, 1975, p. 77.
[29] Harrison, R. P., Loosemore, K., and Milne, I., Report No. R/H/6-Revision 1, Central Electricity Generating Board, 1977.
[30] Chell, G. G. and Milne, I., *Materials Science and Engineering*, Vol. 22, 1976, p. 249.
[31] Rawal, S. P. and Gurland, J., *Metallurgical Transactions*, Series A, Vol. 8A, 1977, p. 691.
[32] Landes, J. D. and Begley, J. A. in *Fracture Analysis, ASTM STP 560*, American Society for Testing and Materials, 1973, pp. 170–186.

C. Berger,[1] H. P. Keller,[2] and D. Munz[2]

Determination of Fracture Toughness with Linear-Elastic and Elastic-Plastic Methods

REFERENCE: Berger, C., Keller, H. P., and Munz, D., "**Determination of Fracture Toughness with Linear-Elastic and Elastic-Plastic Methods,**" *Elastic-Plastic Fracture, ASTM STP 668,* J. D. Landes, J. A. Begley, and G. A. Clarke, Eds., American Society for Testing and Materials, 1979, pp. 378–405.

ABSTRACT: Fracture toughness was determined for two nickel-chromium-molybdenum steels with linear-elastic and elastic-plastic methods. From the evaluation of linear-elastic/ideal plastic load-displacement curves and from the experimental results it follows that the relation of Merkle and Corten should be applied for J-integral determination for compact specimens. The extrapolation method for the determination of a critical J-value implies some problems. As an alternative it is proposed to determine J at a fixed distance from the blunting line. A comparison between stress intensity factors calculated from J and with linear-elastic methods shows that linear-elastic fracture mechanics can be applied to much smaller specimens than given by the ASTM Test for Plane-Strain Fracture Toughness of Metallic Materials (E 399-74). The equivalent-energy method and measurement of crack tip opening displacement with different clip gages at different distances from the crack tip yield stress intensity factors in agreement with the J-integral method.

KEY WORDS: crack propagation, fracture tests, steel, J-integral, linear-elastic fracture mechanics, crack opening displacement

Nomenclature

a Crack length
Δa_{max} Crack extension at maximum load
Δa_{st} Crack extension due to crack blunting
b Ligament width
B Specimen thickness
COD   Crack tip opening displacement
COD_{Io} Plane-strain crack tip opening displacement at the onset of crack extension

---
[1] Research engineer, Kraftwerk-Union AG, Muelheim, Germany.
[2] Research engineer and division head, respectively, Deutsche Forschungs- und Versuchsanstalt für Luft- und Raumfahrt, Cologne, Germany.

E Young's modulus
F_L Limit load for ideal plastic behavior
G Strain energy release rate
$\tilde{G}$ J-integral calculated according to Eftis and Liebowitz
J J-integral according to Eq 2
J_{Io} Plane-strain J-integral at the onset of crack extension
$J_1, J_2, J_3, J_4, J_{41}, J_{42}$ J-integral according to approximate Eqs 3–8
K Stress intensity factor calculated with Eq 10
K^* Stress intensity factor calculated with Eq 10 and plasticity correction
K_{Io} Plane-strain stress intensity factor at the onset of crack extension
K_J Stress intensity factor calculated from J with Eq 32
K_Q Stress intensity factor determined with 5 percent secant method
K_{equ} Stress intensity factor calculated according to equivalent energy method
U Deformation energy
U_o Deformation energy at onset of crack extension
U_{nocr} Deformation energy of a specimen without a crack
U_{el} Elastic component of deformation energy
W Specimen width
Y Function of a/W; see Eq 10
α Size factor; see Eq 31
β Size factor for ligament width; see Eq 33
γ Function of a/W; see Eq 12
δ Load point displacement
δ_{cr} Load point displacement due to the crack
δ_{nocr} Load point displacement of a specimen without a crack
σ_y Yield strength
σ_{fl} Mean flow stress
ν Poisson constant (0.33)
ψ Function of a/W; see Eq 9

For the determination of fracture toughness K_{Ic} with subsized specimens, elastic-plastic evaluation methods are used. The most important methods are known under the terms J-integral, method of Eftis and Liebowitz, crack tip opening displacement (COD), and equivalent energy. For all methods a critical load of the load-displacement curve for the evaluation of K_{Ic} has to be determined. For specimens with large plastic zone sizes, the secant method of the ASTM Test for Plane-Strain Fracture Toughness of Metallic Materials (E 399-74) for linear-elastic fracture mechanics evaluation cannot be applied directly. Also, the maximum load, sometimes used during the development of elastic-plastic fracture mechanics, is not suitable, because very often crack extension begins before the maximum load is reached and then the maximum load depends on the specimen geometry in a complicated manner. Therefore, in all elastic-plastic evaluation methods the attempt is now made to use the load F_o at the onset of crack extension. Determination of this load, however, is the crucial point in all elastic-plastic methods.

There are two prerequisites for an elastic-plastic procedure for the determination of plain-strain fracture toughness. The critical value, for example, J_{Io} or COD_{Io}, has to be independent of specimen size in a broader range than K_{Io}. Furthermore, there must exist an unequivocal relation between the elastic-plastic parameter, for example, J or COD, and the stress intensity factor K in the linear-elastic region.

The most promising method seems to be the J-integral procedure. After the first experimental investigation by Begley and Landes [1,2][3] with the pseudo-compliance method, requiring specimens with different crack lengths, approximate methods for the determination of J from one load-displacement curve were sought.

In this paper results are presented for two alloy steels. A comparison between the different evaluation procedures is made; the problem of determining the onset of crack extension is discussed, and the effect of specimen size on the critical values is shown.

The J-integral

The J-integral evaluation is based on the relation

$$J = -\frac{1}{B}\frac{dU}{da} \qquad (1)$$

Begley and Landes [1,2] used specimens of different crack length and calculated J with the equation

$$J = \frac{1}{B}\frac{U(a + \Delta a) - U(a)}{\Delta a} \qquad (2)$$

In Eq 1, dU is the change of the deformation energy during crack propagation da, whereas in Eq 2 the difference of the deformation energy of two specimens with crack length a and $a + \Delta a$ is determined. By finite-element calculations it was shown that J, according to Eq 2, leads to higher values than according to Eq 1 [3]. The pseudo-compliance method of Begley and Landes requires a large number of specimens. Therefore, different approximate equations were developed to determine J from one load-displacement curve. For compact specimens these equations are as follows:

[3] The italic numbers in brackets refer to the list of references appended to this paper.

Rice, Paris, and Merkle [4]:

$$J_1 = \frac{2}{Bb}(U - U_{nocr}) \quad (3)$$

Landes and Begley [5]:

$$J_2 = \frac{2}{Bb} U \quad (4)$$

Kanazawa et al [6]:

$$J_3 = \frac{1}{B}\left[\left(\frac{2}{b} - \frac{\psi}{W}\right)U + \left(\frac{\psi}{W} - \frac{1}{b}\right)F\delta - \frac{\psi}{W}U_{nocr}\right] \quad (5)$$

Merkle and Corten [7]:

$$J_4 = G + J_{pl} = G + \frac{2}{Bb}[D_1 \int F d\delta_{pl} + D_2 \int \delta_{pl} dF] \quad (6)$$

$$= G + \frac{2}{Bb}[(D_1 - D_2)U + D_2 F\delta - (D_1 + D_2)U_{el}] \quad (6a)$$

Merkle and Corten simplified [7]:

$$J_{41} = \frac{2}{Bb}[D_1 \int F d\delta_{cr} + D_2 \int \delta_{cr} dF] \quad (7)$$

$$= \frac{2}{Bb}[(D_1 - D_2)U + D_2 F\delta - (D_1 + D_2)U_{nocr}] \quad (7a)$$

Merkle and Corten simplified, replacing the displacement due to the crack by the total displacement δ:

$$J_{42} = \frac{2}{Bb}[D_1 \int F d\delta + D_2 \int \delta dF] \quad (8)$$

$$= \frac{2}{Bb}[(D_1 - D_2)U + D_2 F\delta] \quad (8a)$$

In these equations there is

$$\psi = \frac{Y^2}{\int Y^2 da/W} \quad (9)$$

where $Y(a/W)$ is given by the linear-elastic relation

$$K = \frac{F}{B\sqrt{W}} \times Y(a/W) \qquad (10)$$

$$D_1 = \frac{1 + \gamma}{1 + \gamma^2}, \quad D_2 = \gamma \frac{1 - 2\gamma - \gamma^2}{(1 + \gamma^2)^2} \qquad (11)$$

$$\gamma = \frac{\sqrt{2}[1 + (a/W)^2]^{1/2} - (1 + a/W)}{1 - a/W} \qquad (12)$$

By means of idealized load-displacement curves it is possible to calculate J with the different approximate equations and to compare the results with J from Eq 1.

1. For linear-elastic behavior there is

$$\frac{\delta}{F} = \frac{\delta_{cr} + \delta_{nocr}}{F} = \frac{2(1 - \nu^2)}{EB} \left(\int Y^2 \, da/W + A\right) \qquad (13)$$

$$U = \frac{F^2 (1 - \nu^2)}{EB} \left(A + \int Y^2 \, da/W\right) \qquad (14)$$

$$U_{nocr} = \frac{F^2 (1 - \nu^2)}{EB} \times A \qquad (15)$$

$$J = -\frac{1}{B} \frac{dU}{da} = \frac{F^2 (1 - \nu^2)}{EB^2 W} Y^2 \qquad (16)$$

For calculation of the different J_i according to Eqs 3–8 and of J according to Eq 16, Y was calculated using the equation of Srawley [8]. For the total displacement, values of Gross [9], given in Table 1, were used. These values were calculated assuming a more realistic load distribution than earlier calculations by Roberts [10]. The displacement of a specimen without a crack is dependent on the load distribution and on the gage length of the extensometer. In this investigation, A in Eq 13 was obtained by comparing the values of Gross for the total displacement and $\int Y^2 \, da/W$, using the equation of Srawley [8]. An average value of $A = 1.92$ was obtained (see Table 1). In Fig. 1 the ratios J_i/J are plotted against a/W. These ratios are independent of the materials properties. The calculations of Kanazawa et al (J_3) and of Merkle and Corten (J_4) lead to correct J-values. The simplified equation of Merkle and Corten, based on the total displacement (J_{42}), leads to correct values for $a/W = 0.6$, whereas $J_2 = 2U/Bb$ is about 17 percent below the correct value.

2. For linear-elastic/ideal plastic behavior, a linear $F - \delta$ curve is assumed until a limit load F_L is reached, which is given according to Merkle and Corten [7] by

$$F_L = \sigma_y B(W - a)\gamma \qquad (17)$$

where γ is given by Eq 12.
The displacement where F_L is reached is

$$\delta_L = \frac{2\sigma_y}{E'}(W - a)\gamma(\int Y^2 \, da/W + A) \qquad (18)$$

Calculation of J, J_1, J_2, J_3, J_4, and J_{42} leads to

$$J_1/J = \frac{\delta - \frac{1}{2}\delta_L - \frac{1}{E'}\sigma_y W(1 - a/W)\gamma A}{D_1(\delta - \delta_L) + \frac{1}{2E'}\sigma_y W(1 - a/W)^2 \gamma Y^2} \qquad (19)$$

$$J_2/J = \frac{\delta - \frac{1}{2}\delta_L}{D_1(\delta - \delta_L) + \frac{1}{2E'}\sigma_y W(1 - a/W)^2 \gamma Y^2} \qquad (20)$$

$$J_3/J = \frac{\delta - \delta_L + \frac{1}{E'}\sigma_y W(1 - a/W)^2\gamma Y^2}{2D_1(\delta - \delta_L) + \frac{1}{E'}\sigma_s W(1 - a/W)^2 \gamma Y^2} \qquad (21)$$

$$J_4/J = 1 \qquad (22)$$

$$J_{42}/J = \frac{\delta - \delta_L(D_1 - D_2)/2D_1}{\delta - \delta_L + \frac{1}{2E'D_1}\sigma_y W(1 - a/W)^2 Y^2 \gamma} \qquad (23)$$

For $\delta \to \infty$, J_{42}/J approaches 1, whereas J_1/J and J_2/J approach $1/D_1$ and J_3/J approaches $1/2D_1$. In Fig. 2 the different ratios are plotted against J for $a/W = 0.6$, $\sigma_y = 500$ N/mm^2, $E = 2 \times 10^5$ N/mm^2, $\nu = 0.33$, $W = 40$ mm, and $A = 1.23$.

From these calculations the following conclusions can be drawn:
1. $J_2 = 2U/Bb$ underestimates J also for $a/W = 0.6$.
2. The equation of Kanazawa et al, which is correct for elastic behavior, underestimates J in the elastic-plastic region.

TABLE 1—*Displacements for compact specimens; see Eq 13.*

a/W	$A + \int Y^2 \, da/W^a$	$\int Y^2 \, da/W^b$	A
0.3	6.20	4.38	1.82
0.4	10.49	8.51	1.98
0.5	17.57	15.58	1.99
0.6	30.75	28.85	1.90
0.7	60.49	58.60	1.89

[a] Gross [9].
[b] Y from Srawley [8].

3. The equation of Merkle and Corten is a good approximation of J. For $a/W = 0.6$ the simplified equation leads to correct J-values.

The Method of Eftis and Liebowitz

Eftis and Liebowitz [11,12] discussed fracture behavior for nonlinear load-displacement curves. On the basis of energy considerations they introduced a fracture criterion $\tilde{G}$ for stable crack propagation. Using Eq 1 or 2 for J-integral definition, $\tilde{G}$ is then identical with J for the onset of crack extension.

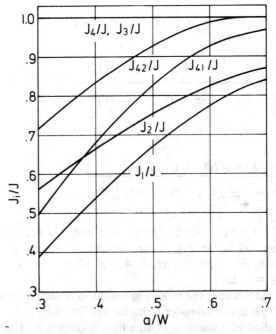

FIG. 1—J_i/J *for linear-elastic behavior for compact specimens.*

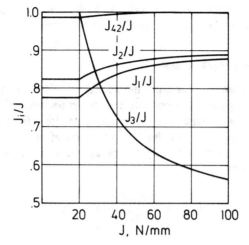

FIG. 2—J_i/J for linear-elastic/ideal plastic behavior for compact specimens.

Besides the more fundamental considerations, Eftis and Liebowitz proposed an evaluation procedure for $\tilde{G}$ or J, respectively. In this procedure the load-displacement curve is described by

$$\delta = F/M + k (F/M)^n \tag{24}$$

with the two parameters n and k, leading to

$$\tilde{G} = J = G \left[1 + \frac{2nk}{n+1} \left(\frac{F}{M}\right)^{n-1} \right] \tag{25}$$

n and k can be determined from two points, (δ_1, F_1) and (δ_2, F_2), on the load-displacement curve according to

$$n = \frac{\lg \dfrac{\delta_2 - F_2/M}{\delta_1 - F_1/M}}{\lg F_2/F_1} \tag{26}$$

$$k = (\delta_1 - F_1/M) \left(\frac{M}{F}\right)^n \tag{27}$$

The validity of Eq 24 has to be proved. The crack length dependence of the nonlinear component of δ in this equation is given by the term $(F/M)^n$, where M is the slope of the linear part of the load-displacement curve. There is some arbitrariness in the choice of $(F/M)^n$. It can be shown that different results for J are obtained, if the nonlinear component of δ in Eq 24 is replaced for instance by $k \times F^n \times M^{-2n}$.

The Equivalent-Energy Concept

The equivalent-energy concept, developed by Witt and Mager [13,14], is an essential empirical method for the determination of K_{Ic} with small specimens. Witt and Mager used the maximum load for the K_{Ic}-determination. Later on, however, the load F_o at the onset of crack extension was used [15,16]. The stress intensity factor at the onset of crack extension K_{equ} is calculated by

$$K_{equ} = \frac{F_1 \sqrt{U_o}}{B \sqrt{W U_1}} Y(a/W) \qquad (28)$$

where

F_1 = arbitrary load within the linear region of the F-δ-curve,
U_1 = corresponding deformation energy, and
U_o = deformation energy at F_o.

Begley and Landes [17] have shown that the equivalent energy concept and the J-integral concept do not give identical results. For linear-elastic/ideal plastic material behavior the two methods can be compared. The comparison is made in the form of the ratio $p = J/(K^2_{equ}/E')$. This ratio is equal to one if both methods coincide. For $\delta < \delta_L$, that is, in the elastic region, there is $p = 1$. For $\delta > \delta_L$, p deviates from 1 and reaches a boundary value for $\delta \to \infty$. For compact specimens this boundary value is given by

$$p_\infty = J/(K^2_{equ}/E') = \frac{\{[1 + (a/W)^2]^{1/2} + \sqrt{2}\} (\int Y^2 da/W + A)}{Y^2(1 - a/W)[1 + (a/W)^2]^{1/2}} \qquad (29)$$

In Fig. 3, p_∞ is plotted against a/W. It can be seen that the equivalent-energy method leads to higher values than the J-integral method. For $a/W = 0.6$ the difference is about 10 percent.

Crack Tip Opening Displacement

In most experimental investigations an attempt is made to calculate COD from crack mouth displacement. Different relations between COD and crack mouth displacement have been published, leading to different COD-values [15,18-21]. Therefore it is useful to measure COD at different distances from the crack tip and to find COD by extrapolation to the crack tip. By displacement measurements and finite-element calculations

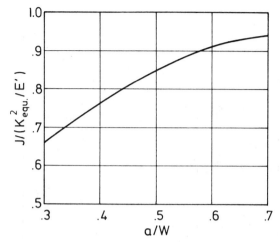

FIG. 3—$J/(K^2_{equ}/E')$ versus a/W for linear-elastic/ideal plastic behavior ($\delta \to \infty$).

it could be shown that for compact specimens there exists a linear relation between COD and distance from the crack tip only up to the region of the pinholes [22].

The relation between COD and K is given by

$$\text{COD} = C \frac{K^2 (1 - \nu^2)}{E \times \sigma_y} \qquad (30)$$

The constant C is not exactly known. Finite-element calculations resulted in $C = 0.5$ for three-point bend specimens [23,24]. Experimental investigations have shown that C depends on the material. Robinson and Tetelman [25] found $C = 1$ for different materials. Later on, different values of C between 0.38 and 1 were determined [26-28].

Thus there are two uncertainties in the determination of K_{Ic} from crack opening displacement: first, the determination of COD, and second, the relation between COD and K.

Experimental Procedure

Two nickel-chromium-molybdenum steels were investigated. The chemical composition, heat treatment, and mechanical properties are given in Table 2. Steel 1 was available as a turbine disk (outer diameter 2925 mm, inner diameter 885 mm, thickness 670 mm). Steel 2 was supplied as bars of dimensions 450 by 250 by 100 mm. From both materials compact specimens of different sizes, given in Tables 3 and 4, were machined. The larger specimens had a W/B ratio of about two. The specimens with $B =$

TABLE 2—Composition and mechanical properties of the investigated steels.

Composition

	C	Si	Mn	P	S	Sn	Cr	Mo	Ni	V
Steel 1	0.28	0.19	0.26	0.006	0.007	0.005	1.55	0.35	3.47	0.09
Steel 2	0.32	0.28	0.41	0.016	0.014	...	1.68	0.43	4.20	...

Thermal Treatment

| Steel 1 | 12 h, 850°C, quenched in boiling water, 29 h 600°C |
| Steel 2 | normalized |

Mechanical Properties

	σ_y, N/mm^2	σ_u, N/mm^2	Elongation, %	Reduction of Area, %	E, N/mm^2	NDTT,[a] °C	FATT,[b] °C
Steel 1	850	980	18	60	200 600	−40	−35
Steel 2	497	728	...	...	207 200	...	...

[a] Nondestructive testing temperature.
[b] Fracture appearance transition temperature.

TABLE 3—J in N/mm for Steel 1.

B, mm	W, mm	J_2	J_3	J_4	J_{42}	$\tilde{G}$	J_2	J_3	J_4	J_{42}	$\tilde{G}$
			$\Delta a = \Delta a_{st}$						Extrapolation		
100	200	91	102	97	108	84	173	204	210	205	185
50	100	66	80	73	79	71	135	150	150	159	145
25	50	80	95	79	93	83	131	139	133	154	138
14	28	84	93	77	99	91	...	...	...	...	...
5	50	70	77	74	81	72	...	...	...	...	...

390 ELASTIC-PLASTIC FRACTURE

TABLE 4—*J in N/mm and K in MNm$^{-3/2}$ at the kneepoint of the J-Δa-curve for Steel 2.*

B, mm	W, mm	J_1	J_2	J_3	J_4	J_{41}	J_{42}	$\tilde{G}$	K_{J4}	K_{equ}	K	K^*
90	200	151	155	157	162	176	181	155	194	192	169	194
49	100	148	150	132	162	170	173	167	194	192	141	171
24	50	148	150	124	169	169	171	174	198	199	114	148
9	50	168	170	127	189	192	194	181	210	209	106	129
5	50	239	240	165	258	271	271	197	245	225	103	125

5 mm and $B = 9$ mm had a width of 50 mm. The specimens were precracked by fatigue up to a crack length of $a/W = 0.6$. The specimens were loaded to different load levels and the load-load point displacement curves recorded. For some specimens with $B = 50$ mm and $B = 100$ mm, crack opening displacement was measured with four clip gages at different distances from the crack tip. For specimens with $B = 25$ mm, displacement could be measured only with three clip gages. The different displacements measured were extrapolated to the crack tip for the determination of COD as shown in Fig. 4. Crack extensions were measured with a scanning electron microscope (SEM) on the fracture surface after unloading and fatigue cracking of the specimens. At least 10 measurements were made at equal distances along the crack front and an average crack extension was calculated. The stretched zone width was included in the crack extension. For some specimens the width of the stretched zone was measured also at different points along the crack front.

Some results for Steel 2 have already been published [29]. In this paper, some additional results are presented.

Results and Discussion

Determination of critical J-values

At first for all specimens, J_4 (Merkle/Corten relation) was calculated and plotted against crack extension.

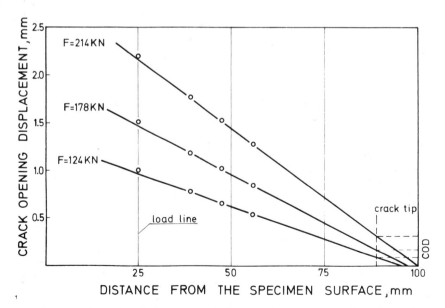

FIG. 4—*Crack opening displacement for different loads versus distance from the specimen surface for a specimen with* $B = 50$ mm *of Steel 1.*

Results for Steel 1 are shown in Figs. 5 and 6. In Fig. 5 the crack extensions at maximum load Δa_{max} are marked for the smaller specimens. It can be seen that some points are included in the figures where the specimens are loaded beyond maximum load. Up to a crack extension of about 0.3 mm, J increases considerably. At larger crack extensions the slope of the J-Δa-curve is smaller. From Fig. 6 it can be seen that the J-Δa-curves intersect the blunting line $J = 2\sigma_{fl} \times \Delta a$ at a crack extension between 40 and 50 μm. It was assumed [5] that the crack extension up to $\Delta a = J/2\sigma_{fl}$ is due to the blunting of the crack tip, which can be seen on the fracture surface as a stretched zone Δa_{st}. This assumption could not be confirmed by fracture surface observations in the SEM. For all specimen sizes, the stretched zone was measured at different points along the crack front. An example of the stretched zone between fatigue crack and static fracture is shown in Fig. 7a. The average values of Δa_{st} are plotted against specimen thickness in Fig. 8. For the larger specimens a stretched zone width of about 28 μm was found, and for the smaller specimens a lower value of about 20 μm was measured. These stretched-zone values are lower than the values determined by the intersection of the J-Δa-curve with the blunting line.

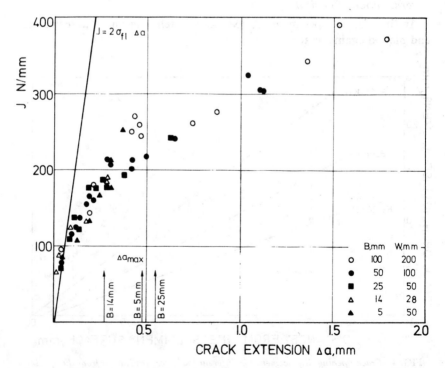

FIG. 5—J-Δa-*curve for Steel 1.*

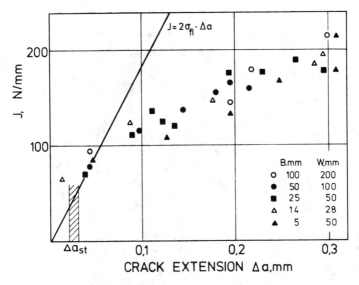

FIG. 6—*Initial part of* J-Δa-*curve for Steel 1.*

For Steel 2, crack extensions were measured only up to about 0.3 mm. As an example, a J_4-Δa-curve for specimens with $B = 9$ mm, $W = 50$ mm is shown in Fig. 9. Two straight lines were drawn through the points. Originally it was assumed that crack extension begins at the intersection of the two straight lines [29]. Detailed investigations with the SEM, however, have shown that crack extension occurs also below the intersection point. In Fig. 7b the extension of the stretched zone is shown. In Fig. 8 it can be seen that almost the same values Δa_{st} were observed as for Steel 1. Again a decrease with decreasing thickness occurred.

From these results a generalized J-Δa-curve can be drawn (see Fig. 10). It is supposed that crack blunting begins if the maximum load during fatigue precracking is exceeded. The corresponding J is called $J_{f\max}$. Between $J_{f\max}$ and J_o crack blunting occurs, leading to a stretched zone Δa_{st} on the fracture surface. For the investigated steels, J increases considerably at the beginning of stable crack extension. Then there exists a transition region or—as shown in Fig. 9—a kneepoint. The intersection of the "blunting line" $J = 2\sigma_{fl} \Delta a$ with the J-Δa-curve can occur in the steep region (at Point 2 in Fig. 10a for Steel 1) or in the flat region (at Point 2 in Fig. 10b for Steel 2).

During the development of the J-integral method it was suggested that J_o at the onset of stable crack extension should be determined to predict K_{Ic} for large structures. For materials with a steeply rising crack growth resistance curve or with a large amount of crack blunting before the onset of crack extension, an exact determination of J_o can be very difficult. It is

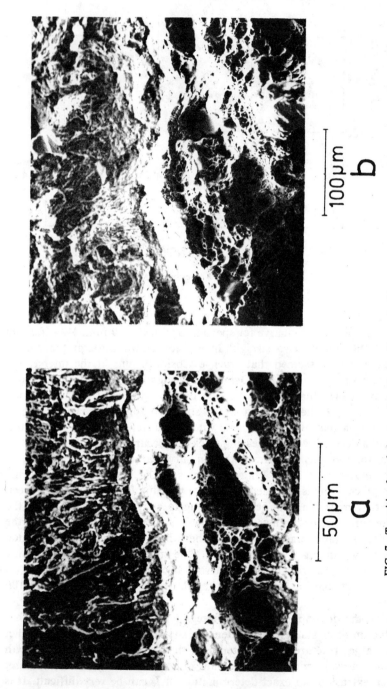

FIG. 7—*Transition from fatigue crack to slow stable crack with stretched zone: (a) Steel 1; (b) Steel 2.*

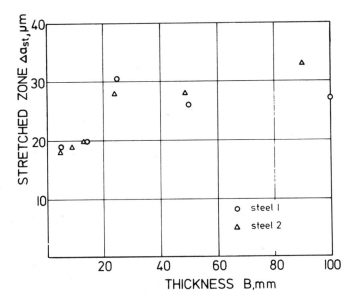

FIG. 8—*Stretched zone versus specimen thickness.*

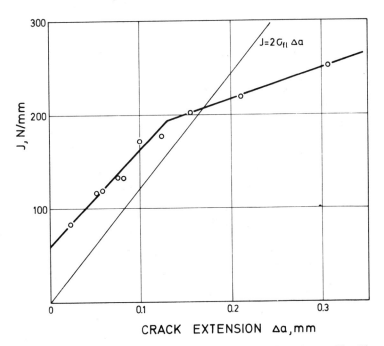

FIG. 9—*J-Δa-curve for specimens with* B = *9 mm and* W = *50 mm of Steel 2.*

396 ELASTIC-PLASTIC FRACTURE

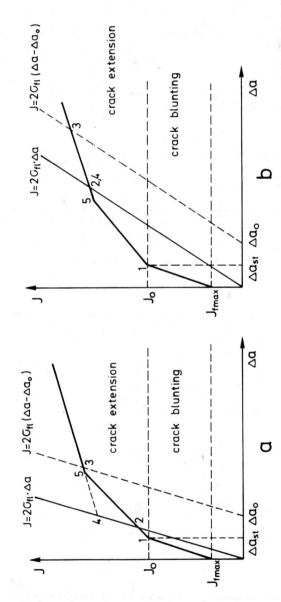

FIG. 10—*Generalized J-Δa-curve: (a) generalized from the behavior of Steel 1; (b) generalized from the behavior of Steel 2.*

also a subject of discussion whether it is desirable to determine J_o for a steeply rising crack growth resistance curve or, better, a J-value after some crack extension. In linear-elastic fracture mechanics, a "K_{Ic}" is determined also after some crack extension. The problem is to find a procedure which is simple and leads to unequivocal values independent of specimen size.

The extrapolation method proposed by Landes and Begley [17] is an attempt to find such a J-value. With regard to the test results for the two steels, this method has some critical points:

1. For small specimens the maximum load can occur at small crack extensions. Then the extrapolation method can be applied only if measuring points beyond maximum load are included. For instance, for specimens with $B = 14$ mm of Steel 1, the maximum load occurred at $\Delta a = 275$ μm. Therefore the range of Δa is too small to determine a straight line (see Fig. 11). It is possible to calculate J-integral for specimens loaded beyond maximum load. It has to be proved, however, if the measured J-value is affected by the unloading of the whole specimen.

2. If the transition region from the steep to the flat region of the J-Δa-curve occurs beyond the blunting line, then the intersection with the blunting line can depend strongly on the number of points below the transition region.

For the foregoing reasons, some modifications of this evaluation procedure should be considered. One possibility is to define the critical J at a fixed deviation from the blunting line—for instance, at the intersection of the J-Δa-curve with the straight line parallel to the blunting line at a distance of 0.1 mm (see Fig. 10).

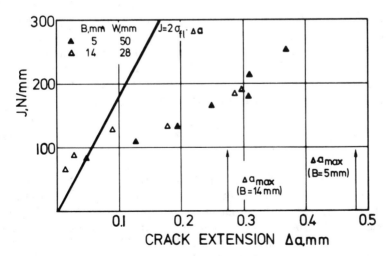

FIG. 11—*J-Δa-curve for specimens with* B = 5 *mm,* W = 50 *mm and* B = 14 *mm,* W = 28 *mm for Steel 1.*

The possible critical J-values, mentioned in the foregoing, are marked in Fig. 10:
1. at $\Delta a = \Delta a_{st}$ (Point 1),
2. at the intersection of the J-Δa-curve with the blunting line (Point 2),
3. at the intersection of the J-Δa-curve with a parallel line at a distance Δa_o to the blunting line (Point 3),
4. at the intersection of the blunting line with a straight line through the points, neglecting all points within the steep part of the J-Δa-curve (Point 4), and
5. at the kneepoint of the J-Δa-curve (Point 5).

Effect of Specimen Size on the Critical J-Values

The effect of specimen size on the J-Δa-curve can be seen from Fig. 6. For crack extensions up to 0.3 mm there is no effect of the specimen size for the proportionally sized specimens. The points for the specimens with $B = 5$ mm, $W = 50$ mm are at the lower bound of the scatter band. In the range between 0.3 and 0.5 mm the points for the specimens with $B = 5$ mm and $B = 100$ mm are above the points of the specimens with $B = 25$ mm and $B = 50$ mm (see Fig. 5).

Some of the different J-values, mentioned before, are plotted against specimen thickness for Steel 1 in Fig. 12. The values according to 1, 2, and 3 of the foregoing are almost independent of thickness.

The extrapolation method leads to J-values which increase with increasing thickness.

The results for Steel 2 can be seen from Fig. 13. J at $\Delta a = \Delta a_{st}$ and at the kneepoint of the J-Δa-curves is almost independent of thickness for $B > 9$ mm. J at the intersection with the blunting line, however, is much higher for specimens with $B = 100$ mm than for the smaller specimens. At small thickness in the range of $B = 5$ mm, all J-values increase.

From these results it can be seen that, at small crack extension, J is independent of specimen size above a critical thickness. It was assumed that J increases below a critical thickness given by

$$B = \alpha \times \frac{J}{\sigma_y} \tag{31}$$

with α between 25 and 50 [5,30]. For Steel 1, the J-values tend more to a decrease than to an increase. Possibly the smallest specimen investigated had a thickness larger than that given by Eq 31. From $J = 150$ N/mm (intersection of J-Δa-curve with the parallel to the blunting line) a minimum thickness of $B = 4.4$ mm is calculated with $\alpha = 25$. From $J = 84$ N/mm^2 (onset of crack extension) a minimum thickness of $B = 2.5$ mm is calculated. For Steel 2 for $J = 162$ N/mm (kneepoint of the J-Δa-curve), the minimum

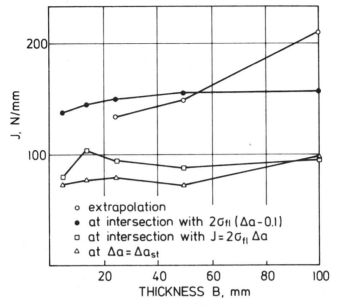

FIG. 12—*Effect of specimen size on different J-values for Steel 1.*

thickness is $B = 8.2$ mm; for $J = 100$ N/mm (onset of crack extension), $B = 5$ mm.

From these calculations it can be concluded that Eq 31 with $\alpha = 25$ is a good approximation to the minimum specimen thickness.

Comparison Between the Different Evaluation Procedures for J

In this investigation all J-values were determined from load-displacement curves of specimens with one crack length. Therefore, J-values according to Eq 2 could not be determined.

The different J-values for both steels are given in Tables 3 and 4. From these tables the following conclusions can be drawn:

1. $J_2 = 2U/Bb$ is smaller than J_4 (correct Merkle/Corten equation).

2. The simplified Merkle/Corten equation, based on the total displacement (J_{42}), leads to slightly higher values than the correct Merkle/Corten equation.

3. The equation of Kanazawa (J_3) is smaller than J_4, especially for small specimens.

4. The evaluation procedure of Eftis and Liebowitz agrees with J_4 within 10 percent.

These results generally are in agreement with the prediction from the idealized load-displacement curves made in the foregoing. Consequently,

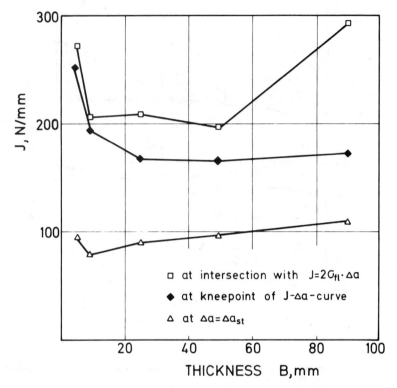

FIG. 13—*Effect of specimen size on different J-values for Steel 2.*

the Merkle/Corten relation for J-integral determination for compact specimens is recommended.

Comparison of J-Integral with Linear-Elastic Determined Stress Intensity Factor

For large specimens, K_J calculated from J with

$$K^2_J = \frac{JE}{(1-\nu^2)} \tag{32}$$

should be identical to K calculated by the linear elastic Eq 10. For smaller specimens there is $K < K_J$, but still $K^* = K_J$, where K^* is calculated by the linear elastic equation, but where crack length a is replaced by $a + (1/6\pi)(K/\sigma_y)^2$. For even smaller specimens there is also $K^* < K_J$. The minimum specimen size, for which $K^* = K_J$, is given by the ligament width criterion

$$W - a = \beta \left(\frac{K}{\sigma_y}\right)^2 \tag{33}$$

For Steel 1, different K-values are plotted against specimen thickness for the proportionally sized specimens in Fig. 14: K_Q (ASTM secant method), K, K^*, and K_{J4} (from J_4) for two critical loads, corresponding to $\Delta a = \Delta a_{st}$ and to the J-value determined with the extrapolation method. From Fig. 14 and the values listed in Table 5, the following conclusions are derived.

1. Comparing K_Q with K at $\Delta a = \Delta a_{st}$ shows that for the specimens with $B = 50$ mm and $B = 100$ mm, crack extension begins below K_Q, and for specimens with $B = 14$ mm and $B = 25$ mm, above K_Q.

2. For the onset of crack extension ($\Delta a = \Delta a_{st}$), K^* and K_{J4} are identical for all specimen sizes. Therefore the critical $W - a$ of Eq 33 is smaller than 11.2 mm (for specimens with $B = 14$, $W = 28$ mm, and $a/W = 0.6$), leading to $\beta < 0.40$.

3. For the critical point, determined with the extrapolation method, K_{J4} and K^* agree also for the smallest specimens, for which the extrapolation method could be applied ($B = 25$ mm, $W = 50$ mm, and $a/W =$

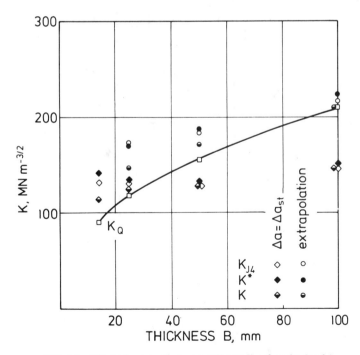

FIG. 14—*Effect of specimen size on different K-values for Steel 1.*

TABLE 5—K in $MNm^{-3/2}$ for Steel 1.

B, mm	W, mm	K_Q	K_{J4}	K_{equ}	K_{COD}	K	K^*	K_{J4}	K_{equ}	K_{COD}	K	K^*
					$\Delta a = \Delta a_{st}$					Extrapolation		
100	200	211	148	148	140	146	151	217	218	211	209	224
50	100	156	128	128	133	127	134	184	184	178	172	188
25	50	119	133	131	137	125	137	173	170	177	147	169
14	28	90	132	137	...	120	141	...	...	...	...	...
5	50	90	129	128	...	114	123	...	...	...	...	...

0.6). The minimum ligament width therefore is smaller than 20 mm, leading to $\beta < 0.41$.

For Steel 2, $\beta = 0.4$ was found in an earlier investigation [29].

From these results it can be concluded that linear elastic fracture mechanics can be applied to much smaller specimens than given by ASTM Method E 399-74 if the plasticity correction is used for the K-calculation.

Other Elastic-Plastic Methods for K_{Ic} Determination

The K-values determined from COD for Steel 1 are given in Table 5. K_{COD} was calculated from COD by means of Eq 30 with $C = 1$. There is an excellent agreement between K_{J4} and K_{COD} for Steel 1. Therefore it can be concluded that, for the investigated steel, $C = 1$.

The stress intensity factors K_{equ} determined with the equivalent-energy method are given in Tables 4 and 5. In accordance with the predictions from the idealized load-displacement curves, there is very good agreement between K_J and K_{equ}.

Conclusions

From the evaluation of idealized load-displacement curves, especially linear-elastic/ideal-plastic behavior, and from experiments on two nickel-chromium-molybdenum steels, the following conclusions can be drawn for compact specimens.

1. From the different equations for J-integral determination from one load-displacement curve, the relation of Merkle and Corten even in its simplified form yields the best results.

2. Crack extension begins below the J-value determined with the extrapolation method. This method can be applied only for small specimens, if data points beyond maximum load are included. As an alternative it is proposed to determine J at a fixed distance from the blunting line.

3. A comparison between stress intensity factors calculated from J and linear-elastic including the plasticity correction shows that linear elastic fracture mechanics can be applied to much smaller specimens than given by ASTM Method E 399-74.

4. The equivalent-energy method agrees fairly well with the J-integral method, if the evaluation is made at the same load.

5. Crack tip opening displacement can be determined using different clip gages at different distances from the crack tip. K_{COD} calculated from COD with $C = 1$ in Eq 30 agrees with K_J calculated from J.

Acknowledgment

We wish to thank J. Eschweiler and F. Vahle for their help during the performance of the tests. The financial support of the Deutsche Forschungsgemeinschaft is gratefully acknowledged.

References

[1] Begley, J. A. and Landes, J. D. in *Fracture Toughness, ASTM STP 514*, American Society for Testing and Materials, 1972, pp. 1-20.
[2] Landes, J. D. and Begley, J. A. in *Fracture Toughness, ASTM STP 514*, American Society for Testing and Materials, pp. 24-39.
[3] Boyle, E. F., "The Calculation of Elastic and Plastic Crack Extension Forces," Ph.D. Thesis, Queen's University, Belfast, U.K., 1972.
[4] Rice, J. R., Paris, P. C., and Merkle, J. G. in *Progress in Flaw Growth and Fracture Toughness Testing, ASTM STP 536*, American Society for Testing and Materials, 1973, pp. 231-245.
[5] Landes, J. D. and Begley, J. A. in *Fracture Analysis, ASTM STP 560*, American Society for Testing and Materials, 1974, pp. 170-186.
[6] Kanazawa, T., Machida, D., Onozuka, M., and Kaned, S., "A Preliminary Study on the J-Integral Fracture Criterion," Report No. IIW-779-75, University of Tokyo, Tokyo, Japan, 1975.
[7] Merkle, J. R. and Corten, H. T., *Journal of Pressure Vessel Technology, Transactions, American Society of Mechanical Engineers*, Vol. 96, 1974, pp. 286-292.
[8] Srawley, J. E., *Engineering Fracture Mechanics*, Vol. 12, 1976, pp. 475-476.
[9] Gross, B., unpublished results.
[10] Roberts, E., *Materials Research & Standards*, Vol. 9, 1969, p. 27.
[11] Liebowitz, H. and Eftis, J., *Engineering Fracture Mechanics*, Vol. 3, 1971, pp 267-281.
[12] Eftis, J., Jones, D. L., and Liebowitz, H., *Engineering Fracture Mechanics*, Vol. 7, 1975, pp. 491-503.
[13] Witt, F. J. and Mager, T. R., "A Procedure for Determining Bounding Values on Fracture Toughness K_{Ic} at any Temperature, Report ORNL-TM 3894, Oak Ridge National Laboratory, 1972.
[14] Witt, F. J. and Mager, T. R., *Nuclear Engineering and Design*, Vol. 17, 1971, pp. 91-102.
[15] Robinson, J. N. and Tetelman, A. S., "Comparison of Various Methods of Measuring K_{Ic} on Small Precracked Bend Specimens that Fracture After General Yield," Technical Report No. 13, School of Engineering and Applied Science, University of California, Los Angeles, Calif.
[16] Schieferstein, U., Berger, C., Czeschik, H., and Wiemann, W. in *Berichtsband der 8. Sitzung des Arbeitskreises Bruchvorgänge, Deutscher Verband für Materialprüfung*, 1976, pp. 50-57.
[17] Begley, J. A. and Landes, J. D. in *Progress in Flaw Growth and Fracture Toughness Testing, ASTM STP 536*, American Society for Testing and Materials, 1973, pp. 246-263.
[18] "Methods for Crack Opening Displacement (COD) Testing," Draft for Development 19, British Standards Institution, 1972.
[19] Barr, R. R., Elliott, D., Terry, P., and Walker, E. T., *Journal of the Welding Institute*, Vol. 7, 1975, pp. 604-610.
[20] Hollstein, T., Blauel, J. G. and Urich, B., "Zur Beurteilung von Rissen bei Elasto-Plastischem Werkstoffverhalten," Report of Institut für Festkörpermechanik der Fraunhofer-Gesellschaft, Freiburg, Germany, 1976.
[21] Schmidtmann, E., Ruf, P. and Theissen, A., *Materialprüfung*, Vol. 16, 1974, pp. 343-348.
[22] Berger, C. and Friedel, H., unpublished results.
[23] Levy, N., Marcal, P. V., Ostergren, W. J., and Rice, J. R., *International Journal of Fracture Mechanics*, Vol. 7, 1971, pp. 143-150.
[24] Hayes, D. J. and Turner, C. E., *International Journal of Fracture Mechanics*, Vol. 10, 1974, pp. 17-32.

[25] Robinson, J. N. and Tetelman, A. S. in *Fracture Toughness and Slow-Stable Cracking, ASTM STP 559,* American Society for Testing and Materials, 1974, pp. 139-158.
[26] Robinson, J. N., *International Journal of Fracture Mechanics,* Vol. 12, 1976, pp. 723-737.
[27] Hollstein, T. and Blauel, J. G., *International Journal of Fracture Mechanics,* Vol. 13, 1977, pp. 385-390.
[28] Chipperfield, G. G., *International Journal of Fracture Mechanics,* Vol. 12, 1976, pp. 873-886.
[29] Keller, H. P. and Munz, D. in *Flaw Growth and Fracture, ASTM STP 631,* American Society for Testing and Materials, 1977, pp. 217-231.
[30] Paris, P. C. in *Fracture Toughness, ASTM STP 514,* American Society for Testing and Materials, 1973, pp. 21-22.

D. Munz[1]

Minimum Specimen Size for the Application of Linear-Elastic Fracture Mechanics

REFERENCE: Munz, D., "**Minimum Specimen Size for the Application of Linear-Elastic Fracture Mechanics,**" *Elastic-Plastic Fracture, ASTM STP 668,* J. D. Landes, J. A. Begley, and G. A. Clarke, Eds., American Society for Testing and Materials, 1979, pp. 406–425.

ABSTRACT: The minimum thickness and the minimum ligament width for the determination of plane-strain fracture toughness with linear-elastic methods can be considerably smaller than given by the ASTM Test for Plane-Strain Fracture Toughness of Metallic Materials (E 399-74). For the ligament width, the factor 2.5 in the size requirement equation can be replaced at least by 1, possibly by 0.4. The size dependence of K_Q determined with the 5 percent secant method is due to plasticity at the crack tip and to the existence of a rising plane-strain crack growth resistance curve. With a variable secant, adjusted to the specimen width, it is possible to determine size-independent fracture toughness values.

KEY WORDS: fracture properties, crack propagation, toughness, tests, aluminium alloys

Nomenclature

a Crack length
a^* Effective crack length
Δa Crack extension
Δa_{KQ} Crack extension at K_Q
Δa_c Crack extension at K_{Ic} (onset of unstable crack extension)
B Specimen thickness
B_c Minimum thickness according to ASTM method E 399-74
$B_{pl.st.}$ Minimum thickness of a specimen with plane-strain region in the center
B_{LE} Minimum thickness of a proportional-sized specimen, for which linear-elastic fracture mechanics can be applied

---

[1] Division head, Deutsche Forschungs- und Versuchsanstalt für Luft- und Raumfahrt, Cologne, FR Germany.

C Constant in $K_{Qpl} = C\sigma_y\sqrt{W}$
COD   Crack tip opening displacement
E Young's modulus
F_o Load at the onset of crack extension
G Strain energy release rate
G^* Strain energy release rate calculated with a^*
J_{Io} Plane-strain J-integral at the onset of crack extension
K Stress intensity factor
K^* Stress intensity factor calculated with a^*
K_o Stress intensity factor at the onset of crack extension
K_{Io} Stress intensity factor at the onset of crack extension under plane strain
K_Q Stress intensity factor at the 5 percent secant intersection
K_{Qpl} K_Q, if $K_o > K_Q$
$K_{0.1}$ Stress intensity factor at 0.1-mm crack extension
K_s Stress intensity factor, determined with a variable secant according to Eq 20
m_o Slope of the linear part of the F-v-curve
m Slope of the secant
r_{pl} Radius of plastic zone
v Crack mouth displacement
v_{el} Elastic component of v
v_{pl} Plastic component of v
v_{cr} Component of v due to crack extension
Δv $= v_{pl} + v_{cr}$
W Specimen width
W_o Specimen width for which $K_o = K_Q$
$(W - a)_{LE}$ Minimum ligament width for which linear-elastic fracture mechanics can be applied
$(W - a)_c$ Minimum ligament width according to ASTM Method E 399-74
α Constant in minimum thickness relation, Eq 7
β_1 Constant in minimum thickness relation, Eq 3
β_2 Constant in minimum ligament width relation, Eq 4
σ_y Yield strength
ν Poisson's ratio
ω Plastic zone size

According to ASTM Method E 399-74, the size requirements for the determination of plane-strain fracture toughness K_{Ic} are given by

$$B > B_c = 2.5 \left(\frac{K_Q}{\sigma_y}\right)^2 \quad (1)$$

$$(W - a) > (W - a)_c = 2.5 \left(\frac{K_Q}{\sigma_y}\right)^2 \tag{2}$$

It is assumed that the same size factor of $\beta = 2.5$ for thickness B and ligament width $W - a$ has to be used. There are, however, different requirements for thickness and width. The critical thickness—called $B_{\text{pl.st.}}$—is given by the requirement of a sufficient amount of plane strain along the crack front in the center of the specimen. From J-integral investigations it is known that the factor 2.5 in Eq 1 can be reduced considerably [1,2].[2] The critical ligament width—called $(W - a)_{\text{LE}}$—is given by the requirement of a sufficiently small plastic zone size ω. Munz et al [2,3] have shown also that the factor 2.5 in Eq 2 can be reduced. For these reasons the size requirements are written in the following form:

$$B > B_{\text{pl.st.}} = \beta_1 \left(\frac{K_Q}{\sigma_y}\right)^2 \tag{3}$$

$$(W - a) > (W - a)_{\text{LE}} = \beta_2 \left(\frac{K_Q}{\sigma_y}\right)^2 \tag{4}$$

If these requirements are fulfilled it should be possible to determine a size-independent fracture toughness K_{Ic}. However, if the 5 percent secant method of the ASTM Test for Plane-Strain Fracture Toughness of Metallic Materials (E 399-74) is used, K_{Ic} can depend on the ligament width, also for $W - a > (W - a)_{\text{LE}}$. There are two reasons for the effect of ligament width on K_{Ic}: the existence of a rising plane-strain crack growth resistance curve and the plastic deformation at the crack tip. After discussing the size limits for plane strain and for linear elastic fracture mechanics, it will be shown in this paper that size-independent "K_{Ic}"-values can be determined using a variable secant, adjusted to the specimen size.

Stable and Unstable Crack Extension

There are two possible definitions of fracture toughness K_{Ic}. It can be defined as the critical stress intensity factor at the onset of unstable crack extension under plane-strain conditions. In ASTM Method E 399-74, "K_{Ic} is based on the lowest load at which significant measurable extension of the crack occurs," that is, at the onset of stable crack extension. During the development of the standard method for K_{Ic} determination, it was assumed that for thick enough specimens, failing under plane-strain conditions, no stable crack growth but immediate unstable crack extension occurs. From theoretical [4] and experimental [3,5-7] investigations, however, it is known

---

[2] The italic numbers in brackets refer to the list of references appended to this paper.

that under plane-strain conditions stable crack growth can also occur. In this case, plane-strain stable crack growth can be characterized by a rising K_I, Δa-curve (see Fig. 1). K_{Io} at the onset of stable crack extension is independent of specimen size, if Eqs 3 and 4 are fulfilled. K_{Ic} at the onset of unstable crack extension depends on the crack length or specimen width, respectively [8]. With the 5 percent secant method a "fracture toughness" K_Q at a crack extension of 2 percent or less is determined. Therefore K_Q increases with increasing specimen width or crack length, respectively. It is a point of discussion which value of K along the K_I, Δa-curve should be used to characterize the fracture behavior. Should it be K_{Io} or a value near K_{Io}, or should the characterization be done with K_{Ic} or another K-value in the upper range of the K, Δa-curve?

Minimum Thickness for Plane Strain

No exact solution exists of the three-dimensional stress distribution within the plastic zone in the near crack tip region. Of special interest is the stress σ_z in the direction of the crack front and the minimum specimen thickness $B_{\text{pl.st.}}$ for which a plane-strain region exists in the center of the specimen. From experimental investigations, some approximate values of $B_{\text{pl.st.}}$ or β_1 in Eq 1, respectively, can be determined.

Vosikowsky [9] found in steels that the thickness constraint at the crack tip begins to collapse when the size of the plastic zone at the surface (plane stress) is equal to the specimen thickness. With

$$\omega_{\text{plane stress}} = \frac{1}{\pi}\left(\frac{K}{\sigma_y}\right)^2 \tag{5}$$

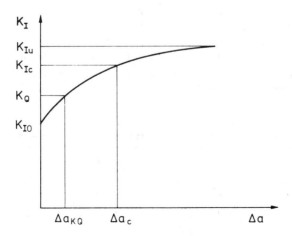

FIG. 1—*Plane-strain K_I-Δa-curve.*

$\beta_1 = 0.32$ is calculated. Robinson and Tetelman [10] measured the transversal strain parallel to a notch and found

$$\beta_1 = 25 \text{ COD} = \frac{25(1-\nu^2)\sigma_y}{E} \tag{6}$$

For an aluminium alloy with $\sigma_y = 500$ N/mm² there is $\beta_1 = 0.18$; for a steel with $\sigma_y = 1000$ N/mm² and $E = 2 \times 10^5$ N/mm² there is $\beta_1 = 0.11$.

The minimum thickness for plane strain can also be found indirectly by measuring the stress intensity factor at the onset of crack extension K_o with specimens of different thickness but constant width. It can be assumed that K_o deviates from K_{Io} for $B < B_{pl.st.}$. It is also possible to find $B_{pl.st.}$ from the thickness-dependence of the J-integral at the onset of crack extension J_o. J_o should deviate from the J_{Io} also at $B_{pl.st.}$. The minimum thickness in J-integral investigations for the determination of J_{Io} is given in the form

$$B_{pl.st.} = \alpha \times \frac{J_{Io}}{\sigma_y} = \alpha \frac{K_{Io}^2 (1-\nu^2)}{E\sigma_y} \tag{7}$$

or

$$\beta_1 = \frac{\alpha(1-\nu^2)\sigma_y}{E} \tag{8}$$

α-values of 25 and 50 were proposed [1,11]. For $\alpha = 25$, Eq 8 is identical to Eq 6.

From all these considerations it can be concluded that the minimum specimen thickness for the determination of plane-strain fracture toughness can be given by Eq 8 with $\alpha = 50$ or less.

A comparison for the minimum thickness between B_c according to Eq 1 and $B_{pl.st.}$ according to Eq 8 with $\alpha = 25$ and $\alpha = 50$ is given in Table 1 for some materials. It can be seen that the minimum thickness of ASTM Method E 399-74 can be reduced considerably. (The ratio $B_{LE}/B_{pl.st.}$ in Table 1 is explained later on.)

TABLE 1—*Comparison of minimum thickness* B_c, $B_{pl.st.}$ *and* B_{LE}.

Material	E, N/mm²	σ_y, N/mm²	$B_c/B_{pl.st.}$		$B_{LE}/B_{pl.st.}$	
			$\alpha = 25$	$\alpha = 50$	$\alpha = 25$	$\alpha = 50$
Al-alloy	7×10^4	400	20	10	3.1	1.6
Steel	2×10^5	400	56	28	9.0	4.5
		800	28	14	4.5	2.2
Ti-alloys	$1,2 \times 10^5$	1000	13	7	2.1	1.1

Minimum Ligament Width

The minimum ligament width for the application of linear elastic fracture mechanics in terms of the factor β_2 in Eq 4 can be obtained by comparison of J-integral and strain energy release rate G. Such a comparison can be made by finite-element calculations. The results can be plotted in the form of J/G-σ/σ_y-curves (Fig. 2). At a critical G, the J-curve deviates from the G-curve. The deviation occurs at considerably higher values, if strain energy release rate is calculated using Irwin's plasticity correction with an effective crack length

$$a^* = a + r_{pl} = a + \frac{1}{6\pi}\left(\frac{K}{\sigma_y}\right)^2 \tag{9}$$

Stress intensity factor and strain energy release rate calculated with a^* are designated K^* and G^*. For sufficiently large specimens or sufficiently low loads, $K^* = K$ and $G^* = G$.

For three-point bend specimens with $a/W = 0.5$, the calculations of Hayes [12] for a non-work-hardening material showed that G^* was in agreement with J up to $F/BW\sigma_y = 0.10$, leading to $\beta_2 = 0.45$. Markström and Carlsson [13] calculated J and G for compact specimens for linear elastic/ideal plastic behavior. From these results, $\beta_2 = 0.42$ is obtained.

Another method to find out the minimum ligament width is the determination of the onset of crack extension for specimens with different ligament width. In Fig. 3, K_o calculated from F_o and crack length a and K^*_o calculated from F_o and a^* are plotted schematically against $W - a$. Below a critical ligament width $(W - a)_{LE}$ K^*_o is lower than $K_{Io} = K^*_{Io}$.

Munz and coworkers [2,3,6,14] have determined K^*_o for some materials with specimens of different size. In Table 2 the minimum ligament width and β_2 are given. For some materials, only an upper limit for $(W - a)_{LE}$ can be given, because it was possible to determine K_{Io} also with the smallest

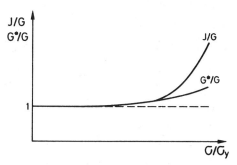

FIG. 2—J/G and J/G^* versus σ/σ_y.

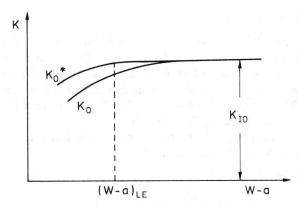

FIG. 3—*Stress intensity factor K_o and K^*_o at the onset of crack extension versus ligament width.*

specimen. From Table 2 it can be seen that in any case $\beta_2 = 2.5$ can be replaced by $\beta = 1.0$. From the results for the steels, $\beta_2 = 0.4$ is suggested.

For proportionally sized specimens with $W/B = 2$, the minimum specimen size for the determination of plane-strain fracture toughness is given by the minimum ligament width $(W - a)_{LE}$. For even smaller specimens, elastic-plastic methods, such as J-integral, have to be applied. The lower size limit for these methods is given by the minimum thickness $B_{pl.st.}$. The range of specimen size, where only elastic-plastic methods can be applied for fracture toughness determination in terms of thickness, is given by the ratio $B_{LE}/B_{pl.st.}$. For proportionally sized specimens with $W/B = 2$ and $a/W = 0.5$, $B_{LE} = (W - a)_{LE}$ and

$$B_{LE}/B_{pl.st.} = \frac{\beta_2}{\beta_1}$$

For some materials this ratio is given in Table 1 for $\beta_2 = 0.4$ and β_1

TABLE 2—*Minimum ligament width $(W - a)_{LE}$ for the determination of fracture toughness for different materials.*

Material	σ_y, N/mm^2	K_{Io}, MN m$^{-3/2}$	$(W - a)_c$, mm	$(W - a)_{LE}$, mm	β_2	Ref
Ti-6Al-4V	910	48	7.0	< 3	<1.1	[6]
7475-T7351	426	40	22.0	< 6.25	<0.7	[3]
7475, annealed	326	44	45.5	<20	<1.1	[2]
NiCrMo Steel I	497	188	358	56	0.39	[2]
NiCrMo Steel II	850	141[a]	70	<11.2	<0.41	[14]
	850	188[b]	121	<20	<0.41	[14]

[a] Onset of crack extension.
[b] Determined according to ASTM Task Group Elastic Plastic Fracture.

according to Eq 8 with $\alpha = 25$ and $\alpha = 50$. From this table it can be seen that there is only a small range of specimen size where elastic-plastic methods have to be applied for the determination of plane-strain fracture toughness.

Plastic Component of COD

The plastic deformation at the crack tip leads to a deviation from the linear-elastic load-displacement curve. For small specimens the deviations until the 5 percent secant intersection at K_Q can be due only to the plastic deformation. The plastic component of crack opening displacement v_{pl} can be calculated by finite-element methods. A simple estimation was made by Irwin, assuming linear-elasticity with the crack tip in the center of the plastic zone. The elastic component of COD is given by

$$v_{el} = \frac{F}{EB} f(a/W) \tag{10}$$

According to Irwin the plastic component then is given by

$$v_{pl} = \frac{dv_{el}}{da} \times r_{pl} = \frac{F}{EBW} \frac{df(a/W)}{da/W} \times \frac{1}{6\pi} \left(\frac{K}{\sigma_y}\right)^2 \tag{11}$$

and

$$\frac{v_{pl}}{v_{el}} = \frac{f'}{6\pi \, W \times f} \left(\frac{K}{\sigma_y}\right)^2 \tag{12}$$

For three-point bend specimens with $a/W = 0.5$, $f'/f = 5$ and therefore

$$\frac{v_{pl}}{v_{el}} = \frac{0.265}{W} \left(\frac{K}{\sigma_y}\right)^2 \tag{13}$$

For the 5 percent secant method, $\Delta v/v_{el} = 0.0526$. If the onset of crack extension occurs at $K_{Io} > K_Q$, then at K_Q the total deviation from the linear F, v-curve Δv is identical to v_{pl} and

$$K_Q = K_{Qpl} = C \sqrt{W \times \sigma_y} \tag{14}$$

with $C = 0.45$.

Finite-element calculations lead to somewhat different results [15-17]. The C-values obtained are listed in Table 3. An average value of 0.72 was found.

TABLE 3—C in $K_{Qpl} = C\sqrt{W}\sigma_y$, calculated by finite-element method.

C	Work Hardening	Ref
0.73	low	Brown and Srawley [22]
0.67	low	Markström [23]
0.78[a]	high	Markström [23]
0.69[b]	no	Hayes and Turner [24]

[a] Extrapolated value.
[b] From Fig. 7 of Ref 24.

C can also be obtained from experimental investigations from measured K_{Qpl} and σ_y. Results for aluminum alloys and a low alloy steel are given in Table 4. With two exceptions, the values are between 0.50 and 0.60. From all results, an average value of $C = 0.55$ is obtained. This experimental value is between the value from finite-element calculations and the estimation of Irwin. From results of Griffis and Yoder [18] for an aluminum alloy, $C = 0.50$ can be obtained. It is not clear why the experimental C-values are below the calculated ones.

Specimen Size Effect on K_Q

Within the application range of linear-elastic fracture mechanics, K_Q, determined with the secant method, can be dependent on specimen size. In different investigations it was found that K_Q is almost independent of thickness but can increase considerably with increasing width [3,6,19]. Two effects can be responsible for the increase of K_Q with increasing width: the plastic deformation at the crack tip and the rising crack growth resistance curve. The increase of K_Q with increasing width is shown schematically in Fig. 4. The maximum possible K_Q is K_{Qpl}, if $K_{Io} > K_Q$. For small specimens, $K_Q = K_{Qpl}$. At a critical width W_o, crack extension starts at K_Q. For a level K, Δa-curve there is no further increase of K_Q. For a rising K, Δa-curve, K_Q increases according to slope of the K, Δa-curve. The stable crack extension Δa_{KQ} at K_Q increases with increasing width, beginning at W_o.

Figure 5 shows results obtained with three-point bend specimens of the aluminum alloy 7475-T7351. K_Q is plotted against width and a considerable increase can be seen. With the electrical potential method for the onset of crack extension, $K_{Io} = 40$ MNm$^{-3/2}$ was found. In Fig. 5 also, the K_{Qpl}-W-relation according to Eq 14 with $C = 0.55$ is plotted. For the specimens with $W = 12.5$ mm and $W = 25$ mm, crack extension occurs at $K_{Io} > K_Q$ and therefore $K_Q = K_{Qpl}$. The increase of K_Q from $W = 50$ mm to $W = 100$ mm is due to the rising K, Δa-curve.

With the electrical potential method it was also possible to determine $K^*_{0.1}$ at a crack extension of 0.1 mm. The calculation was done with the

TABLE 4—C in $K_{Qpl} = C \sqrt{W} \sigma_y$ for different materials.

Material	Specimen Type	Specimen Orientation	a/W	σ_y N/mm²	W, mm	B, mm	K_Q MNm$^{-3/2}$	C
Al-alloy 7475-T7351	3-point bend	L-T	0.5	426	25	12.5	35.2	0.52
					12.5	12.5	28.7	0.60
					25	25	39.2	0.58
	3-point bend			436	25	12.5	36.4	0.53
					12.5	12.5	25.4	0.52
		T-L	0.64	423	12	12	26.9	0.58
		L-S	0.64	436	12	12	25.7	0.54
		S-L	0.64	403	12	12	25.2	0.57
		T-S	0.64	423	12	12	26.5	0.57
		S-T	0.64	403	12	12	23.3	0.53
Al-alloy 7475-T7351 and 15 h, 180°C	compact	L-T	0.6	326	50	25	43.0	0.59
					50	15	41.5	0.57
					50	5	31.9	0.44
Al-alloy 2024 (II)	compact	L-T	0.5	292	40	8	29.0	0.50
Al-alloy 7075 (I)	3-point bend	L-T	0.5	415	12	12	28.2	0.62
		T-L		421	12	12	27.1	0.59
Al-alloy 7050-T73651	3-point bend	L-T	0.5	441	12	12	24.2	0.50
		T-L		428	12	12	25.0	0.53
		S-T		410	12	12	21.8	0.49
Ni-Cr-Mo Steel I	compact	L-T	0.6	497	200	90	125	0.56
					100	49	89	0.57
					50	24	59	0.53
					28	14	43	0.52
					50	9	61	0.55
					50	5	55	0.50

416 ELASTIC-PLASTIC FRACTURE

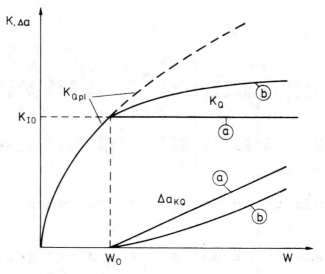

FIG. 4—K_Q (5 percent secant) and crack extension at K_Q versus specimen width for a level (a) and a rising (b) K-Δa-curve.

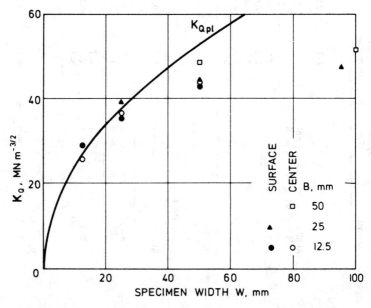

FIG. 5—Effect of specimen width on K_Q for three-point bend specimens of aluminium alloy 7475-T7351.

plasticity correction. From Fig. 6 it can be seen that $K^*_{0.1}$ is independent of width. Thus it is possible to apply linear-elastic fracture mechanics also to the smallest specimens with $W = 12.5$ mm, showing again that the increase of K_Q with increasing W is not due to failing of linear-elastic fracture mechanics.

Variable Secant for the Determination of Size-Independent K-Values

The introduction of the 5 percent secant method was an important step during the development of a standard procedure for fracture toughness determination. This method has two disadvantages. A specimen size effect occurs for some materials, and for small specimens a K-value is determined which is lower than K_o at the onset of stable crack extension. With a variable secant with a slope adjusted to the specimen size, it is possible to determine size-independent K-values.

To determine the relation between the slope of the secant and the crack extension, it is useful to decompose the crack opening displacement v into the three components

$$v = v_{el} + v_{pl} + v_{cr} \tag{15}$$

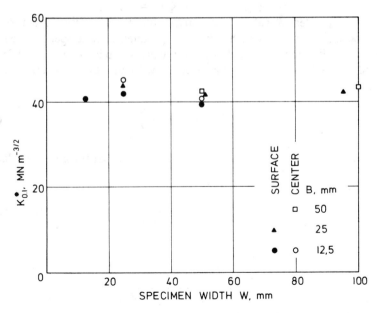

FIG. 6—*Stress intensity factor* $K^*_{0.1}$ *at 0.1-mm crack extension versus specimen width for the aluminium alloy 7475-T7351.*

v_{el} is given by Eq 10. With regard to Eqs 12 and 14 between v_{pl}/v_{el} and $(K/\sigma_y)^2/W$, a linear relation can be assumed

$$v_{pl}/v_{el} = A \frac{1}{W}\left(\frac{K}{\sigma_y}\right)^2 = \frac{0.0526}{C^2}\frac{1}{W}\left(\frac{K}{\sigma_y}\right)^2 \qquad (16)$$

For $C = 0.55$, $A = 0.174$.

The component v_{cr}, due to crack extension, is given by

$$v_{cr} = \frac{dv_{el}}{da} \times \Delta a = \frac{F}{EBW}\frac{df}{da/W} \qquad (17)$$

and

$$\frac{v_{cr}}{v_{el}} = \frac{f'}{f} \times \frac{\Delta a}{W} \qquad (18)$$

For three-point bend specimens with $a/W = 0.5$, $f'/f = 5$ and

$$\frac{\Delta v}{v_{el}} = \frac{v_{pl} + v_{cr}}{v_{el}} = \frac{0.174}{W}\left(\frac{K}{\sigma_y}\right)^2 + 5 \times \frac{\Delta a}{W} \qquad (19)$$

If specimens with different width but identical K, Δa-curves are tested, it can be seen from Eq 19 that the secant intersects the F, v-curve at the same K or Δa, respectively, for $\Delta v/v_{el} \times W = $ const.

Size-independent K-values then can be obtained according to the following procedure. For tests with different specimen sizes a specimen with a width $W = W_s$, to which the 5 percent secant is applied ($\Delta v/v_{el} = 0.0526$), is used as a reference. Then for a specimen with an arbitrary width W there has to be

$$\frac{\Delta v}{v_{el}} = 0.0526 \frac{W_s}{W} \qquad (20)$$

Between $\Delta v/v_{el}$ and the change of the slope of the secant Δm the relation holds

$$\frac{m_o - m}{m_o} = \frac{\Delta m}{m_o} = \frac{1}{v_{el}/\Delta v + 1} \qquad (21)$$

leading for the secant to

$$\frac{\Delta m}{m_o} = \frac{1}{1 + 19 W/W_s} \qquad (22)$$

where m_o is the slope of the linear part of the F, v-curve.

For the three-point bend tests of the aluminum alloy 7475-T7351, for which the K_Q-W-relation was shown in Fig. 5, K-values called K_s were determined with the variable secant. As a reference width $W_S = 50$ mm was used. In Fig. 7, K_s is plotted against width.

Comparing Figs. 5 and 7, it is shown that there is a much smaller increase of K_s than of K_Q with increasing width. This small increase disappears, if the original crack length a is replaced by a corrected crack length $a + 0.5$ mm, leading to K^*_s (see Fig. 8).

Similar results were obtained for the titanium alloy Ti-6Al-4V. As can be seen from Fig. 9 also, a considerable increase of K_Q with width was observed for this alloy [6]. K^*_s, however, obtained for a reference width of 40 mm, is independent of width (Fig. 10).

Fracture Toughness Determination with Subsized Specimens

For a rising plane-strain K, Δa-curve, the question arises which value along the curve should be used as a fracture toughness value. This problem shall not be discussed here. With the variable secant it is possible to determine a K-value at small or larger crack extension, depending on the choice of the reference width.

For small specimens, the slope of the secant has to be low enough to

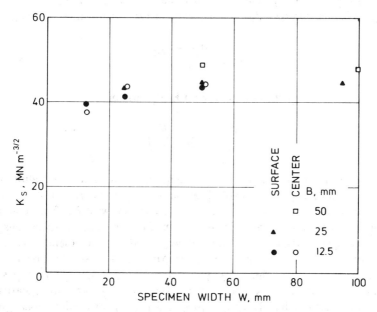

FIG. 7—*Effect of specimen width on* K_s *for the aluminium alloy 7475-T7351.*

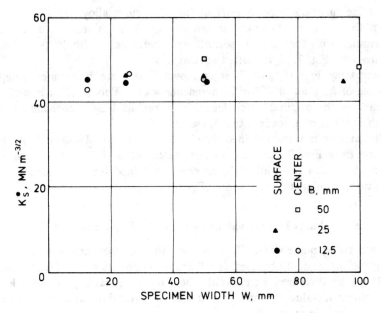

FIG. 8—*Effect of specimen width on K*$_s$ for the aluminium alloy 7475-T7351.*

make sure that the secant intersects the F, v-curve at $K_s > K_{Io}$. The necessary slope can be obtained by means of Eq 19 with $\Delta a = 0$, leading to

$$\frac{\Delta v}{v_{el}} > \frac{0.174}{W}\left(\frac{K}{\sigma_y}\right)^2 \tag{23}$$

Taking into account the scatter of C in Eq 14, use of

$$\frac{\Delta v}{v_{el}} > \frac{0.2}{W}\left(\frac{K}{\sigma_y}\right)^2 \tag{24}$$

is proposed. In addition to the already mentioned results, for some aluminum alloys, K_{Ic} obtained from small three-point bend tests with $B = 12$ mm, $W = 12$ mm and a span of 48 mm was compared with K_{Ic} from compact specimens with $B = 25$ mm, $W = 50$ mm. For the compact specimens the 5 percent secant ($\Delta v/v_{el} = 0.0526$) was used; for the bend specimens according to Eq 20, $\Delta v/v_{el} = 0.22$ or $\Delta m/m_o = 0.18$. The results are given in Table 5. In Fig. 11, K^*_{Ic} for the compact specimens is plotted against K^*_s for the three-point bend specimens. Nearly all measuring points are along a line for which K^*_{Ic} is 5 percent below K^*_s. This small deviation may be due to the effect of specimen type. In some investigations

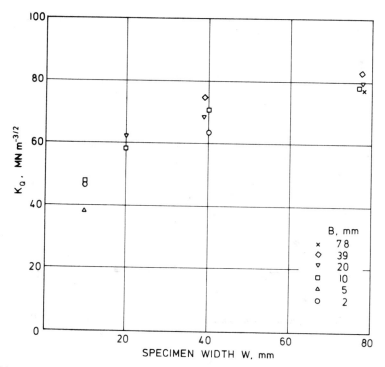

FIG. 9—*Effect of specimen width on K_Q for three-point bend specimens of the titanium alloy Ti-6Al-4V.*

comparing compact and three-point bend specimens, such an effect was observed [20,21]. For some tests the difference was larger than 5 percent.

Conclusions

1. The minimum specimen thickness for the determination of plane-strain fracture toughness, which is due to the requirement for a plane-strain region in the center of the specimen, is much smaller than given in ASTM Method E 399-74. The factor 2.5 of the thickness requirement can be replaced by $\alpha \sigma_y (1 - \nu^2)/E$, with α between 25 and 50.

2. The minimum ligament width for the application of linear-elastic fracture mechanics is also much smaller than given by ASTM Method E 399-74. If K is calculated with Irwin's plasticity correction, the factor 2.5 can be replaced at least by 1, possibly even by 0.4.

3. For proportionally sized specimens with $W/B = 2$, there is only a small range of specimen size where elastic-plastic methods have to be applied for the determination of plane-strain fracture toughness.

4. For specimens for which linear elastic fracture mechanics can be applied, size-dependent K_{Ic}-values can be obtained if the 5 percent secant

422 ELASTIC-PLASTIC FRACTURE

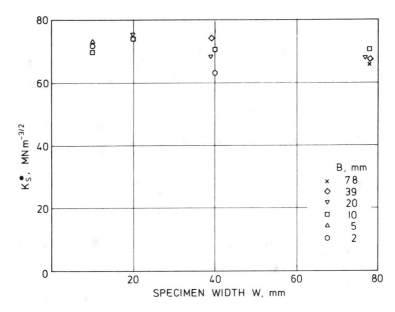

FIG. 10—*Effect of specimen width on* K^*_s *for the titanium alloy Ti-6Al-4V.*

TABLE 5—*Comparison of stress intensity factors (in* $MNm^{-3/2}$*) of 3-point bend and compact specimens of different size.*

Material	Specimen Orientation	3-Point Bend, $W = 12$ mm		Compact, $W = 50$ mm		$2.5\left(\dfrac{K_{Ic}}{\sigma_s}\right)^2$
		K_s	K^*_s	K_{Ic}	K^*_{Ic}	
Al-alloy 7475-T7351	T-L	35.7	42.5	38.7	40.3	22.7
	L-T	39.2	46.5	43.5	45.3	27.0
	S-T	32.1	38.0	34.5	35.9	19.8
Al-alloy 7075 (I)	L-T	39.8	45.8	48.9	50.5	37.0
	T-L	38.0	44.2	40.6	41.9	24.8
Al-alloy 7050	L-T	33.5	38.6	31.8	32.8	13.8
	T-L	36.5	42.3	33.8	34.9	16.6
	S-T	24.6	28.5	26.5	27.4	11.2
Al-alloy 2024 (I)	L-T	21.3	33.1	36.0	37.2	25.6
	T-L	21.3	33.5	34.6	35.7	31.1
	S-T	19.6	28.4	26.6	27.5	21.6
Al-alloy 7075 (II)	L-T	29.1	34.0	32.3	33.4	13.3
	T-L	27.4	31.8	28.9	29.9	11.6
	S-T	21.6	24.8	23.1	23.8	8.6

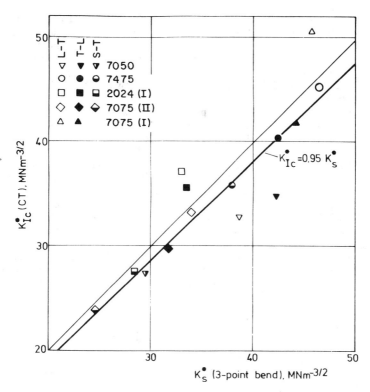

FIG. 11—K^*_{Ic} for compact specimens with $W = 50$ mm versus K^*_s for three-point bend specimens with $W = 12$ mm.

method is applied. The size effect is due to the plastic deformation at the crack tip and to the existence of a rising plane-strain crack growth resistance curve.

5. With a variable secant, adjusted to the specimen width, it is possible to determine size-independent fracture toughness values.

Acknowledgments:

The author thanks J. Eschweiler for performing the tests thoroughly. The financial support of the Deutsche Forschungsgemeinschaft is gratefully acknowledged.

APPENDIX

Materials and Experimental Procedure

The experimental results were obtained for different aluminum alloys, a titanium alloy, and two steels:

1. Aluminum alloy 7475-T7351, plate of thickness 63 mm, fracture toughness and yield strength dependent on the location of the specimens; therefore, specimens from the center of the plate were distinguished from specimens from the surface.
2. Aluminum alloy 7475-T7351, plate of thickness 63 mm; specimens were annealed for 15 h at 180°C.
3. Aluminum alloy 2024-T351 (I), plate of thickness 130 mm.
4. Aluminum alloy 2024 (II), plate of thickness 10 mm.
5. Aluminum alloy 7050-T73651, plate of thickness 100 mm.
6. Aluminum alloy 7075-T7351, plate of thickness 32 mm.
7. Aluminum alloy 7075 (II) (treatment not specified), plate of thickness 63 mm.
8. Titanium alloy Ti-6Al-4V, plate of thickness 82 mm.
9. Nickel-Chromium-Molybdenum Steel I with the composition 0.32C, 4.2Ni, 1.68Cr, 0.43Mo, 0.41Mn, 0.28Si, 0.016P, 0.014S, bars of cross section 100 by 250 mm in normalized condition.
10. Nickel-Chromium-Molybdenum Steel II with the composition 0.28C, 3.47Ni, 1.55Cr, 0.35Mo, 0.26Mn, 0.09V, 0.19Si, 0.006P, 0.007S, 0.005Sn, 12 h 850°C, water-quenched, 29 h 600°C; specimens were cut from a turbine disk.

Compact and three-point bend specimens of different size were cut from the plate and precracked in fatigue. Some of the compact specimens also were used for J-integral evaluation. For these specimens, instead of crack mouth displacement, the displacement at the load line was measured.

References

[1] Landes, J. D. and Begley, J. A. in *Fracture Analysis, ASTM STP 560*, American Society for Testing and Materials, 1974, pp. 170-186.
[2] Keller, H. P. and Munz, D. in *Flaw Growth and Fracture, ASTM STP 631*, American Society for Testing and Materials, 1977, pp. 217-231.
[3] Munz, D., "Fracture Toughness Determination of the Aluminium Alloy 7475-T7351 with Different Specimen Sizes," Deutsche Luft- und Raumfahrt, Forschungsbericht DLR-FB 77-04, 1977.
[4] Rice, J. R. in *Mechanics and Mechanisms of Crack Growth, Proceedings*, British Steel Corp., Cambridge, U.K., 1973, pp. 14-36.
[5] Robinson, J. N., "The Critical Crack-Tip Opening Displacement and Microscopic and Macroscopic Fracture Criteria for Metals," Ph.D. thesis, University of California, Los Angeles, Calif., 1973.
[6] Munz, D., Galda, K. H., and Link, F. in *Mechanics of Crack Growth, ASTM STP 590*, American Society for Testing and Materials, 1976, pp. 219-234.
[7] Green, G. and Knott, J. F. *Journal of the Mechanics and Physics of Solids*, Vol. 23, 1975, pp. 167-183.
[8] Srawley, J. E. and Brown, W. F. in *Fracture Toughness Testing and its Application, ASTM STP 381*, American Society for Testing and Materials, 1965, pp. 133-198.
[9] Vosikowsky, O., *International Journal of Fracture Mechanics*, Vol. 10, 1974, pp. 141-157.
[10] Robinson, J. N. and Tetelman, A. S., *International Journal of Fracture Mechanics*, Vol. 11, 1975, pp. 453-468.
[11] Paris, P. in *Fracture Toughness, ASTM STP 514*, American Society for Testing and Materials, 1972, pp. 21-22.
[12] Hayes, D. J., "Some Applications of Elastic-Plastic Analysis to Fracture Mechanics," Ph.D. thesis, University of London, London, U.K., 1970.
[13] Markström, K. M. and Carlsson, A. J., "FEM-Solutions of Elastic-Plastic Crack Problems—Influence of Element Size and Specimen Geometry," Publication No. 197, Hallfastnetslära, KTH, Stockholm, Sweden, 1973.
[14] Berger, C., Keller, H. P., and Munz, D., this publication, pp. 378-405.

[15] Brown, W. F. and Srawley, J. E. in *Review of Developments in Plane Strain Fracture Toughness Testing, ASTM STP 463*, American Society for Testing and Materials, 1970, pp. 216-248.
[16] Markström, K. M., *Engineering Fracture Mechanics*, Vol. 4, 1972, pp. 593-603.
[17] Hayes, D. J. and Turner, C. E., *International Journal of Fracture Mechanics*, Vol. 10, 1974, pp. 17-28.
[18] Griffis, C. A. and Yoder, G. R., *Transactions, ASME, Journal of Engineering Materials Technology*, American Society of Mechanical Engineers, Vol. 98, 1976, pp. 152-158.
[19] Kaufman, J. G. and Nelson, F. G. in *Fracture Toughness and Slow-Stable Crack Growth, ASTM STP 559*, American Society for Testing and Materials, 1974, pp. 74-85.
[20] Hall, L. R. in *Fracture Toughness Testing at Cryogenic Temperatures, ASTM STP 496*, American Society for Testing and Materials, 1971, pp. 40-60.
[21] Munz, D., unpublished results.
[22] Brown, W. F., Jr. and Srawley, J. E. in *Review of Developments in Plane Strain Fracture Toughness Testing, ASTM STP 463*, American Society for Testing and Materials, 1970, pp. 216-248.
[23] Markström, K. M., *Engineering Fracture Mechanics*, Vol. 4, 1972, pp. 593-603.
[24] Hayes, D. J. and Turner, C. E., *International Journal of Fracture Mechanics*, Vol. 10, 1974, pp. 17-32.

W. R. Andrews[1] and C. F. Shih[1]

Thickness and Side-Groove Effects on J- and δ-Resistance Curves for A533-B Steel at 93°C

REFERENCE: Andrews, W. R. and Shih, C. F., **"Thickness and Side-Groove Effects on J- and δ-Resistance Curves for A533-B Steel at 93°C,"** *Elastic-Plastic Fracture, ASTM STP 668,* J. D. Landes, J. A. Begley, and G. A. Clarke, Eds., American Society for Testing and Materials, 1979, pp. 426–450.

ABSTRACT: A test program was conducted to determine the effects of specimen thickness variations, side grooves, and crack length variations on the deformation and ductile fracture of A533-B, Cl-1 steel at 93°C.
  The crack extensions were estimated using the correlation between elastic compliance and crack length. Crack extensions were also estimated using a correlation among crack-opening displacements, load line displacements ($\delta - V_L$), and crack length. The inferred estimates of crack extension were supplemented by some measurements on heat-tinted fracture surfaces. The results suggest that the observation of thickness or side-groove effects on crack-extension resistance curves is dependent on the method of measuring crack extension.
  The compliance correlation method was less sensitive to crack extension and showed a classical thickness effect: increased crack growth resistance with decreasing thickness, and decreased resistance with the use of side grooves. The $\delta - V_L$ correlation method was more sensitive to crack extension and showed no effect of thickness or of side grooves on crack growth resistance. The presence of side grooves promoted flat fracture and suppressed shear lips. Specimens without side grooves developed large shear lips.

KEY WORDS: ductile fracture, testing, crack extension, side grooves, compact specimen, thickness effects, J-integral, crack-opening displacement, crack propagation

Ductile fracture of low-alloy steel initiating from crack-like defects has been studied in the past, and a number of criteria for predicting initiation and continued stable tearing have been advanced. None of the criteria advanced have found the acceptance that K_{Ic} has found for brittle fracture ASTM Test for Plane-Strain Fracture Toughness of Metallic Materials (E 399-74). The evaluation of these criteria has been made difficult by the frac-

---

[1] Engineer-Mechanics of Materials, Materials and Processes Laboratory and research engineer, Corporate Research and Development, respectively, General Electric Company, Schenectady, New York 12345.

ture process and the difficulty of observing the process through the thickness of the specimen.

A review of the available data, primarily those from the Heavy Section Steel Technology (HSST) Program [1],[2] revealed that both flat fracture and shear lip formation play important roles in the fracture of A533-B steel. Compact specimens as large as 508 mm wide by 254 mm thick produced shear lips on nearly 60 percent of the specimen thickness. In these specimens, the fracture surfaces indicate the sequence of events in the process of ductile tearing. The initiation of fracture occurred at the center of the specimen thickness and proceeded by flat, ductile tearing to form a characteristic thumbnail-shaped crack front. At some critical depth of the crack front, the side ligaments began tearing to form shear lips adjacent to the surface. The width of the shear lips increased as the crack progressed until 60 percent of the specimen thickness fractured by 45-deg shear. This process was observed in practically all specimen sizes.

Thus, the phenomenon of ductile fracture takes on a complex, three-dimensional aspect not found in brittle fracture. The flat fracture near the center thickness develops under nearly plane-strain constraint, whereas the shear lips near the surfaces develop under plane-stress deformation.

The problem is to select fracture criteria which are independent of specimen geometry and size. The objective of a larger program [2,3], of which these tests are a part, is to evaluate plastic fracture criteria beyond small-scale plasticity using results of compact specimen, center-cracked plate specimen, and double-edge notched plate specimen tests. Finite-element calculations based on these specimen geometries were carried out to provide detailed computations of several potential fracture parameters for initiation, stable growth, and instability [4]. The preliminary observations indicate that a fully three-dimensional analytical model is needed to simulate a standard compact specimen when the mode of failure is nonplanar. To avoid the expense of 3-D modeling, incorporating criteria for flat and for shear fracture, the use of side grooves in compact specimens was tried with the objective of simplifying the fracture process to approximate a 2-D, plane-strain model. To accomplish this, the side grooves were expected to suppress to a minimum the formation of shear-lips and to produce a flat-fracture crack which has a straight leading edge through the thickness.

This program investigated the effects of specimen thickness (B), side-groove depth, and initial crack depth (a_0) on compact specimen tests. Two methods for estimating the crack extension (Δa) were used. One method, using the crack-length correlation with elastic compliance [5], measures the average crack depth, whereas the second, using the correlation among crack-tip opening displacement, load-line displacement ($\delta - V_L$), and crack length, measures the crack extension near the center of the specimen. These

---

[2]The italic numbers in brackets refer to the list of references appended to this paper.

tests showed that the crack growth resistance was affected by specimen thickness and side grooves when estimated with the elastic compliance correlation, but these effects were negligible with the $\delta - V_L$ correlation.

Material

The composition of the test materials is given in Table 1, the mechanical properties in Table 2, and the Charpy V-notch impact properties in Table 3. Test Material 1 was from the same heat and heat treatment as was used in the Welding Research Council (WRC) survey on mechanical properties of A533-B steel [6]. The source material was nozzle dropouts from 165-mm-thick (6½ in.) plate. Test Material 2 was A533-B plate designated in this program as EPRI-01-GE-02. This plate was rolled and quenched and tempered as a 203-mm-thick (8 in.) plate.

Test Specimens

The specimen geometry was based on the standard compact specimen (ASTM E399-74) with modifications to permit measurement of the load-line deflection (V_L) and the opening displacement near the crack tip (V_N), Fig. 1, using a linear variable differential transformer (LVDT) and an extension rod across the crack. The varied test specimen dimensions and precrack lengths

TABLE 1—*Ladle analysis of test material, weight percent.*

Heat No.	C	Mn	P	S	Cu	Si	Ni	Cr	Mo	Al
A0999-1	0.22	1.32	0.010	0.014	0.14	0.19	0.60	.09	0.50	0.026
B0256	0.20	1.22	0.011	0.005	...	0.15	0.66	...	0.55	...
	0.19	1.22	0.009	0.016	...	0.15	0.65	...	0.54	...

TABLE 2—*Longitudinal tension test results at 93°C.*

Identification	Material 1 A0999-1				Material 2 B0256
	30319	30321	30322	30XXX	T-Specimen
Tensile Strength, MPa	555	555	552	542	574
Proportional Limit, MPa	368	381	400	423	426
0.02 % yield strength, MPa	409	421	423	439	441
0.2 % yield strength, MPa	421	430	425	436	443
Percent Elongation, 4.57 cm	26.1	26.7	24.4	26.1	26.7
Percent Reduction of Area	70.0	71.8	65.7	72.3	71.6
Fracture stress	1150	1238	1085	1091	1262

TABLE 3—*Charpy impact test results (L-T orientation).*

	Material 1 A0999-1				Material 2 B0256
Identification	30319	30321	30322	30XXX	
40.7 J T.T., °C	−39	−40	−40	−46	< −32
67.8 J T.T., °C	−26	−29	−19	−17	< −32
50 percent FATT, °C	− 4	−18	−18	−12	− 7
0.89 mm L.E.T.T., °C	−29	−23	−32	−32	< −32
Energy at 93°C, J	154	165	190	171	190

NOTES:
 L-T = Longitudinal-transverse (see ASTM E399-74, Fig. 9).
 T.T. = Transition temperature.
 FATT = Fracture appearance transition temperature.
 L.E.T.T. = Lateral expansion transition temperature.

are given in Table 4. Both the nine-point average crack length and center thickness crack length are listed. In subsequent discussion of results, the elastic compliance is referred to the nine-point average crack length and the limit load with the center-thickness, or maximum depth, of the precrack.

Test Procedure

The specimens were loaded in an Instron 1330-kN (300 kip), four-column load frame, using closed-loop position control of the loading ram. The testing machine and grips had a compliance of 2.9 × 10^{-6} mm/N (5.1 × 10^{-7} in./lb). The ram opened the load points of the specimen at about 7.6 mm/h (0.3 in./h). The load-line displacement versus load (P) was recorded on an X-Y plotter. The load-line displacement, the load, the near-tip crack opening displacement, and the ram position were recorded on digital magnetic tape using a Vidar data logger. The transducer signals were scanned sequentially every 3 s at a scan rate of about 30 channels per second. The data were recorded to the nearest millivolt using a ±10 V standard range. The magnetic tape was placed on file on a large, general-purpose computer and the values of δ, Δa, and J_1 (J-integral) were computed using a FORTRAN computer program.

The crack-tip opening displacement was estimated from the measurements of V_L and V_N, assuming the arms of the compact specimens to be rigid (no bending), by extrapolating the two measurements to the center-thickness crack depth.

Two techniques were used for estimating Δa. One method is the unloading compliance method [5] in which the elastic compliance is used to estimate Δa through the use of combined analytical [7] and experimental [8] correlations. The compliances were estimated using a linear least-squares best fit to the P

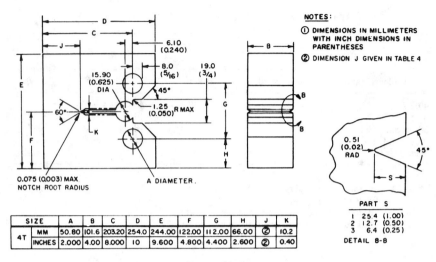

FIG. 1a—*Compact fracture specimen: basic design.*

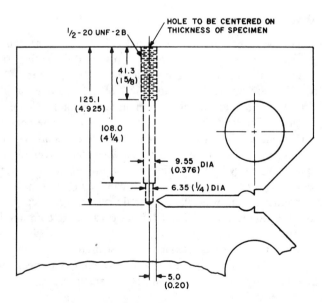

FIG. 1b—*Compact fracture specimen: hole for LVDT displacement measurement V_N.*

TABLE 4—*Test specimen geometry and material properties.*

Nominal Thickness	B, mm	Thickness, in.	Specimen Identification	Test Material	Crack Length, mm 9-Point Avg	Center
1	25.4	1.00	30319-3	1	101.6	113.5
2.5	50.8	2.50	30319-1	1	115.3	117.3
2.5	50.8	2.50	30319-4	1	134.6	136.1
4	101.6	4.00	30322-1	1	122.7	129.3
4	101.6	4.00	30321-1	1	141.0	146.6
4	101.6	4.00(25%SG)[a]	30XXX-1	1	128.3	131.3
4	101.6	4.00(25%SG)	30XXX-2	1	143.5	145.5
4	101.6	4.00(50%SG)	30322-2	1	127.5	128.3
4	101.6	4.00(50%SG)	30321-2	1	141.0	141.5
4	101.6	4.00(25%SG)	T52[b]	2	117.2	117.2
4	101.6	4.00(12.5%SG)	T-71	2	115.8	115.8
4	101.6	4.00(12.5%SG)	T-32	2	125.0	125.0
4	101.6	4.00(12.5%SG)	T-21[b]	2	134.0	134.0
4	101.6	4.00(12.5%SG)	T-31	2	136.0	136.0
4	101.6	4.00(12.5%SG)	T-62	2	144.5	144.5
4	101.6	4.00(12.5%SG)	T-22[b]	2	145.8	145.8
4	101.6	4.00(25%SG)	T-51	2	149.5	149.5
4	101.6	4.00(25%SG)	T-61	2	162.7	162.7
4	101.6	4.00(25%SG)	T-41[b]	2	169.8	169.8

NOTES:
Modulus of elasticity = 200 GPa.
Flow stress, σ_Y = 490 MPa.
[a] SG = side grooved.
[b] Specimen was heat-tinted following test.

versus V_L data, obtained on unloading, ignoring the first one to four data points at each unloading. The calculated crack extension values were corrected for the error in compliance measurement due to the finite deflection of the load line.

The second method is based on the unique relationship between δ and V_L which holds provided no crack extension occurs. When crack extension occurs, the measured excess of δ over the unique value calculated for the original crack length is used to estimate the extent of the crack extension. The derivation of this method is shown in the Appendix. This latter method for estimating crack extension will be referred to as the $\delta - V_L$ method. Four were terminated with a heat-tinting operation which provided calibration points for the $\delta - V_L$ estimate of Δa. (The heat tinting was performed by heating the cracked specimen to about 260°C (500°F) for 4 h, cooling to room temperature, and breaking.) The heat tint points were in good agreement with compliance estimates of Δa, but only the T-41 result is within the range reported here.

The values of J_I were calculated using the Merkle-Corten relationship [9] in the form

$$J_1 = \alpha_1 \frac{2A}{B_N(W - a_0)} + \alpha_2 \frac{2PV_L}{B_N(W - a_0)} \qquad (1)$$

where

$$\alpha_1 = 1.222468 - 0.637295\,(a_0/W) \\ + 0.614937\,(a_0/W)^2 - 0.200797\,(a_0/W)^3 \qquad (2)$$

$$\alpha_2 = -0.006771 + 0.595163\,(a_0/W) \\ - 0.940241\,(a_0/W)^2 = 0.353779\,(a_0/W)^3 \qquad (3)$$

where

A = area under the load deflection curve,
B_N = net thickness,
W = specimen width,
a_0 = initial fatigue-crack depth (nine-point average), and
P = maximum load reached at or prior to the measurement point.

This expression for J_1 was found to be in excellent agreement with J_1 evaluated using finite-element computations along a contour remote from the crack tip for both stationary and growing cracks [5].

A method for obtaining silicone rubber replicas of the crack tip was applied to the specimens not heat tinted. The procedures used are detailed elsewhere [10].

Results and Discussion

The results are presented in two parts. The first part discusses results prior to crack extension and the second focuses on crack initiation and growth results.

Deformation

The deformation of the specimens in terms of load-line deflection versus load is summarized in Fig. 2 using normalized axes. A discussion of the normalizing parameters is found in the Appendix. In Fig. 2, the effect of side grooves on the elastic compliance is made evident by the different slopes in the linear, rising load portion of the curves. The effects of side grooves on the elastic compliance of compact specimens are evaluated in greater detail in Ref 6. The differing limit loads for large plastic deformations are consistent with the transition from plane-strain to plane-stress plastic deformation as the thickness is reduced. The plane-strain and plane-stress limit loads based on slip-line field solutions [4,11] are indicated in Fig. 2. The transition in limit loads from plane-strain to plane-stress levels was observed only when

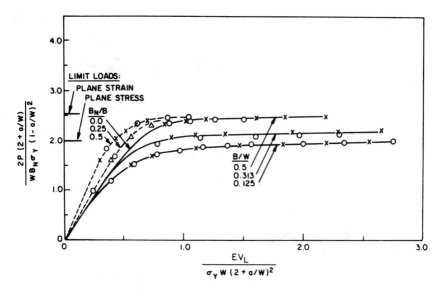

FIG. 2—*Summary of load versus load-line displacement curves, compact specimens, a/W ≥ 0.55.*

the center-thickness crack length was used to calculate notch stress and limit stress. This distinction in crack-length measure was made necessary by the curvature of the crack front in the non-side-grooved specimens (See Table 4).

The normalized values of J_1 with increasing deflection are shown in Fig. 3 for varying specimen thickness. The figure gives evidence that any unique relationship between J_1 and δ deteriorates when there is significant cross slip (out-of-plane slip). This observation is consistent with the transition from plane-strain to plane-stress constraint, and is a direct consequence of the variation of limit loads seen in Fig. 2.

The relationships between J_1 and δ also span a range for varying specimen thicknesses (Fig. 4), and fall in the same order as observed in Fig. 3. It was found that on normalized coordinates for δ versus V_L, a unique relationship exists prior to crack initiation. This unique relationship is developed in the Appendix and forms the basis for a sensitive method for estimating crack extension at the specimen mid-thickness.

Cracking

Fracture Appearance—The fracture surfaces of the various specimens are contrasted in Fig. 5. In every case, the use of side grooves, 12½ percent of the gross thickness or deeper, promoted flat fracture. Some obvious lateral con-

434 ELASTIC-PLASTIC FRACTURE

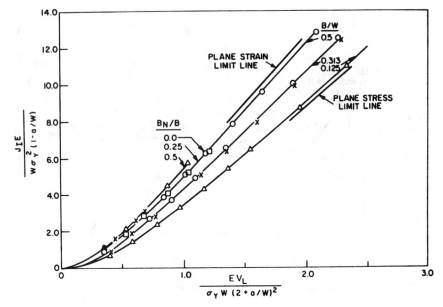

FIG. 3—*Summary of J_I versus load-line displacement curves, compact specimens, a/W $\geq$ 0.55.*

traction of the sides occurred in those specimens side grooved to 12½ percent. To be perfectly clear, percent side groove depth is calculated

$$\text{percent SG} = 100 \left(\frac{B - B_N}{B} \right) \tag{4}$$

where B_N is the net specimen thickness.

When side grooves were not employed, shear lips formed which were of constant absolute size for the several thicknesses. The shear lips formed an increasing proportion of the fracture surface as the thickness decreased and also as distance increased from the precrack-tip in the direction of crack propagation. This latter relation held until the crack approached the back surface to a distance of about three-quarters to half the specimen thickness, at which point the proportion to total thickness of shear lip became constant at about 60 percent. The thinnest specimen ($B/W = 0.125$) developed a full shear fracture after a short transition from the originally flat fatigue precrack. These results are consistent with those of Merkle [1] discussed in the introduction.

Crack Extension Estimates Using Compliance—The *J*-resistance curves developed for Test Material 1 using the unloading compliance estimates of crack extension are shown in Fig. 6. The test results show the effects of side

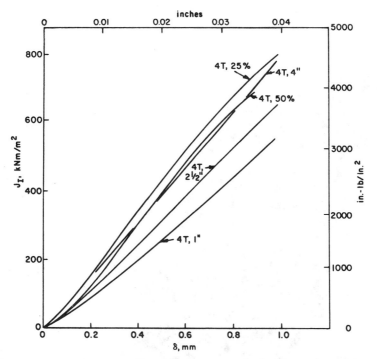

FIG. 4—*Relationships between J_I and δ, compact specimens, A533 Grade B, Class 1, steel; 93°C.*

grooves and thickness. It is seen that the values of J_I for crack initiation, J_c, are dependent upon specimen thickness and on side grooving. The values of J_c increased as the thickness was reduced from the 102-mm (4 in.) side-grooved specimens (largest effective thickness) to 102 mm (4 in.), to 63.5 mm (2.5 in.) and then decreased for the 25.4-mm (1.0 in.) thickness. These results suggest that J_c is dependent on the degree of plane-strain constraint achieved in the fracture process zone.

The slopes of the J_I-resistance curves, Fig. 6, increased with decreasing thickness, reflecting the formation of large shear lips in specimens without side grooves. The increased slope with decreased thickness correlated with the ratio of the estimated plastic zone size to the thickness

$$\beta_c = \frac{JE}{B\sigma_Y^2} \qquad (5)$$

where σ_Y is the average of the ultimate and 0.2 percent yield strengths. This relationship is shown in Fig. 7. A similar, nearly linear, relationship was found by Lake [*12*] for an aluminum alloy.

436 ELASTIC-PLASTIC FRACTURE

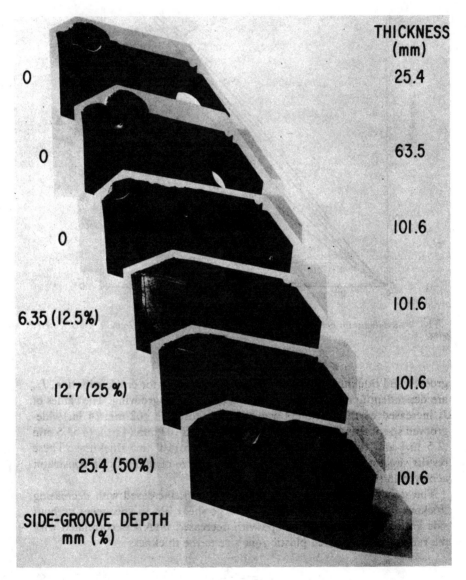

FIG. 5—*Fracture surfaces of 4T compact specimens, A533 Grade B, Class 1 steel tested at 93°C.*

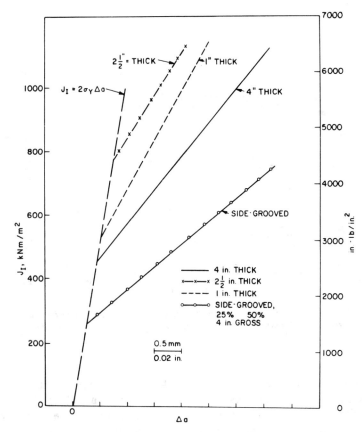

FIG. 6—*Effects of thickness and side grooves on resistance to crack growth, estimated using compliance correlation, 4T compact specimens, A533-B Cl steel (Material 1) tested at 93°C.*

The J_1-resistance curves developed for Material 2 using unloading compliance estimates of Δa are shown in Fig. 8. These tests indicate that initial crack length ($a/w \geq 0.55$) has no effect on the resistance curves.

Crack Extension Estimates Using $\delta - V_L$—The results of the thickness and side-groove effect studies (Material 1) when evaluated using the $\delta - V_L$ estimates of Δa are shown in Fig. 9. Note that there is no effect of thickness on J_c, δ_c, or on the slopes of the resistance curves. The side-mounted gage mentioned in the legends to Fig. 9 was mounted across the crack tip on the side of the specimen, thus representing a slightly different measurement than was made for the other specimens. Comparison of the results in Fig. 9a with those in Fig. 6 indicates that the $\delta - V_L$ method was more sensitive to crack initiation; it seems to be representative of the crack extension at the center of the specimen. The compliance technique, in contrast, represents a through-

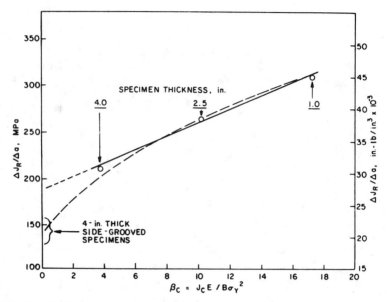

FIG. 7—*Correlation of slopes of J_I-resistance curves, crack growth estimated using compliance correlation, 4T compact specimens, A533-B Cl-1 steel (Material 1) tested at 93°C.*

thickness average crack extension. A further reduction in the data scatter is noted when the data are presented as a δ-resistance curve (Fig. 9b).

Further evidence that supports the validity of the $\delta - V_L$ estimates of Δa was found in silicone rubber crack-tip replicas [10]. The replicas, obtained for the 15 specimens not heat tinted (Table 4) were sectioned and observed in profile, and gave δ values which agreed with those inferred from the two displacement measurements. (See Test Procedure section and Fig. 1.) Furthermore, the critical values of δ for crack initiation were in agreement between estimates from the silicone rubber profile (0.2 to 0.7 mm) and estimates using the $\delta - V_L$ technique (0.3 to 0.5 mm).

The results of evaluating tests of Material 2 side-grooved specimens with varied crack lengths using the $\delta - V_L$ method are shown in Fig. 10a. When compared with Fig. 8, the range of J_c estimates is reduced and the scatter in the crack growth resistance values of J_1 is reduced.

Further reduction of scatter was again noted for these when presented as a δ-resistance curve (Fig. 10b).

The $\delta - V_L$ method for estimating crack extension resulted in J_1- and δ-resistance curves which are independent of thickness and of the presence of side grooves. The method provided a measure of mid-specimen crack extension. This observation is important because the measured limit loads correlated with mid-specimen crack length and not with the nine-point average

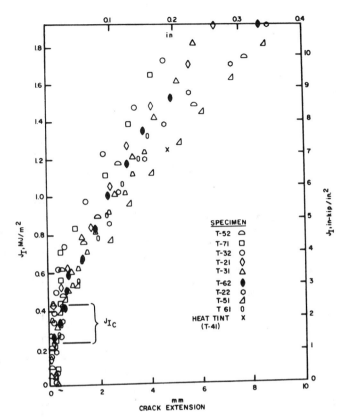

FIG. 8—*Resistance to crack growth estimated using compliance correlation, side-grooved 4T compact specimens, variable crack length, A533-B Cl-1 steel (Material 2) tested at 93°C.*

crack length when the crack fronts were not straight. Like the curves developed using compliance estimates of Δa, the J_1- and δ-resistance curves were independent of initial crack length. The $\delta - V_L$ method resulted in a greater sensitivity to short crack extensions.

Conclusions

1. The load-deflection curves in compact specimens showed variations of limit loads between plane-strain and plane-stress limits as the specimen thickness was reduced from $B/W = 0.5$ to $B/W = 0.125$. A similar and consistent variation was found for the J-integral deflection curves. δ deflection curves were independent of thickness.

2. Side grooves ranging from 12½ percent of gross thickness and deeper successfully suppressed shear-lip formation in A533-B steel at 93°C.

ELASTIC-PLASTIC FRACTURE

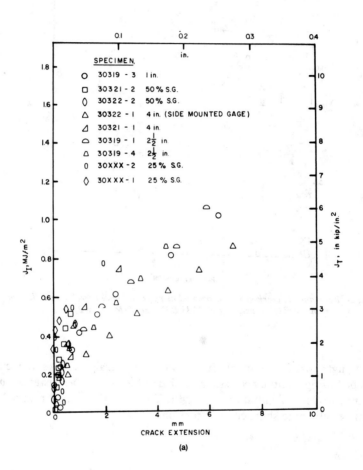

(a)

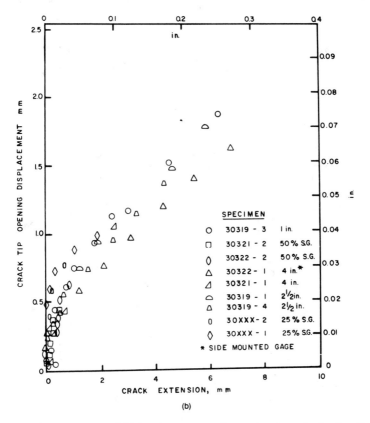

FIG. 9—(a) J_I-resistance and (b) δ-resistance to crack growth estimated using $\delta - V_L$ technique, 4T compact specimens, variable thickness and side grooved, A533-B, Cl-1 steel (Material 1), tested at 93°C.

442 ELASTIC-PLASTIC FRACTURE

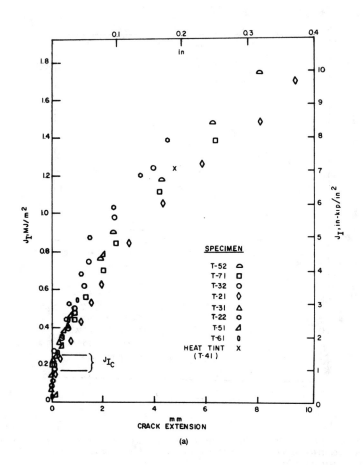

(a)

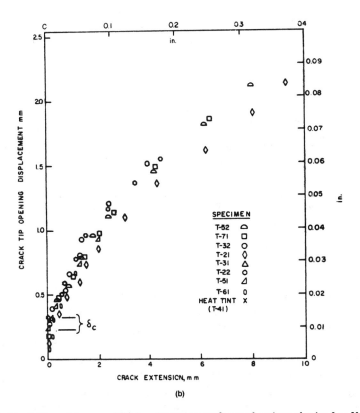

FIG. 10—(a) J_I-resistance and (b) δ-resistance to crack growth estimated using $\delta - V_L$ technique, side grooved, 4T compact specimens, variable crack length, A533-B Cl-1 (Material 2) steel tested at 93°C.

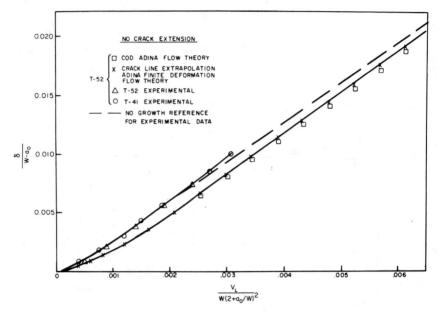

FIG. 11—*Relationship between crack-tip opening displacement (δ) and load-line displacement (V_L) for compact specimens, a/W $\geq$ 0.55, prior to crack growth (A = crack length).*

3. Shear lip dimensions in non-side-grooved (smooth) specimens were relatively independent of specimen thickness.

4. The observation of the effects of specimen thickness and of side-groove depth in the specimen is dependent on the method of estimating the crack extension. The methods used are the elastic compliance technique and the $\delta - V_L$ technique, which used the correlation among crack-tip opening displacement (δ), load line displacement (V_L), and crack extension.

5. Crack extensions estimated using the $\delta - V_L$ technique measure mid-thickness crack extensions and result in J_I- and δ-resistance curves which are independent of thickness in the range studied, are unaffected by the use of side grooves, and are independent of crack length when the crack length to width ratio, a/W, is greater than 0.55.

Acknowledgments

The authors are grateful to D. J. Tinklepaugh and D. F. St. Lawrence for laboratory testing and to S. Yukawa and D. F. Mowbray for technical consultation.

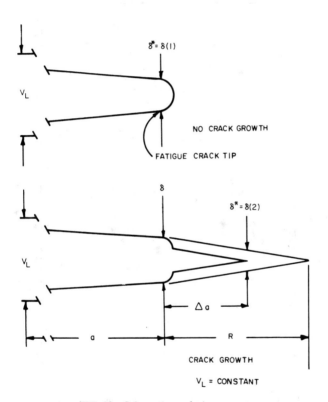

FIG. 12—*Schematic crack-tip geometry.*

446 ELASTIC-PLASTIC FRACTURE

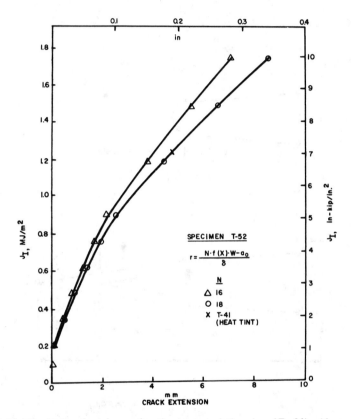

FIG. 13—*Sensitivity of crack-growth estimates to variations in γ (Eq 26), side-grooved 4T compact specimen, T-52; A533-B Cl-1 steel tested at 93°C.*

APPENDIX

Derivation of Basis for Crack Extension Calculation From $\delta - V_L$ Measurements

Srawley and Gross [13] observed for compact specimens that the dimensionless coefficient

$$\frac{K_1 B \sqrt{W}}{P} \times \frac{(1 - a/W)^{3/2}}{(2 + a/W)} = 1.312 = Q \tag{6}$$

is approximately constant for large a/W (that is, $a/W > 0.55$). This observation may be coupled with relations for elastic fracture mechanics

$$K_1^2 = G_1 E' = J_1 E' \tag{7}$$

where $E' = E$ — (plane stress) and $E' = E/(1 - \nu^2)$ — (plane strain) and

$$J_1 = \frac{2A}{B(W-a)} \times f(a/W) \tag{8}$$

where A is the area under the load-displacement curve. For elastic loading

$$A = \int P \, dV_L = \frac{1}{2} P V_L \tag{9}$$

and

K_1 = stress intensity factor,
G_1 = elastic strain-energy release rate,
J_1 = J-integral,
a = crack length,
B = specimen thickness,
P = applied load,
W = specimen width,
V_L = load line displacement,
E = Young's modulus,
ν = Poisson's ratio, and
Q = a constant.

Substituting Eq 8 into Eq 7, we get

$$K_1^2 = \frac{2AE}{B(W-a)} \times f(a/W) \tag{10}$$

The factor

$$f(a/W) \doteq 3/(2 + a/W)$$

in Eqs 8 and 10 is an approximate correction for the tensile loading and for elastic conditions. J_1 so calculated differs from that using the Merkle-Corten correction [9] by 3 to 4 percent for $a/W \geq 0.5$.

Substituting Eq 9 into Eq 10 and this in turn into Eq 6 yields

$$E \frac{V_L}{P} B \times \frac{(1 - a/W)^2}{(2 + a/W)^3} = \frac{Q^2}{3} \quad (11)$$

Introducing the notch stress ratio

$$\frac{\sigma}{\sigma_Y} = \frac{2P(2 + a/W)}{\sigma_Y WB(1 - a/W)^2} = \frac{6}{Q^2} \times \frac{V_L E}{\sigma_Y W(2 + a/W)^2} \quad (12)$$

Equation 12 suggests plotting a load versus load-line deflection curve on coordinates

$$y = \sigma/\sigma_Y = \frac{2P(2 + a/W)}{\sigma_Y WB(1 - a/W)^2} \quad (13)$$

and

$$x = \frac{V_L E}{\sigma_Y W(2 + a/W)^2} \quad (14)$$

which has an elastic slope of $6/Q^2$ or 3.49.

Considering fully plastic deformation of the compact specimen, the limit load [10] for plane-strain constraint is

$$\frac{\sigma_{\lim}}{\sigma_Y} = 2.52 \quad (15)$$

The curve resulting from the combined elastic and plastic deformation, Fig. 1, is independent of crack length for all $a/W \geq 0.55$ and for plane-strain constraint.

A second plot in which J_1 is normalized may be developed by integrating the normalized load-deflection plot. Without going through the step-by-step procedure, the result is

$$\frac{JE}{W\sigma_Y^2(1 - a/W)} = f\left\{\frac{EV_L}{\sigma_Y W(2 + a/W)^2}\right\} \quad (16)$$

Again, similar to the normalized load-deflection curves, the functional relationship is independent of crack length for $a/W \geq 0.55$ as long as no crack growth occurs. See Fig. 2.

In the figure, J_1 is a unique function for a given level of constraint at the crack tip, but different functions exist for differing levels of constraint. The relationship for crack-tip opening displacement (δ) can be examined by noting

$$J_1 = \alpha \times \delta \times \sigma_Y \quad (17)$$

where α is a proportionality constant, and is dependent on the degree of constraint. Substituting Eq 17 into Eq 16

$$\frac{E\delta}{W\sigma_Y(1 - a/W)} = f_2\left\{\frac{EV_L}{\sigma_Y W(2 + a/W)^2}\right\} \quad (18)$$

The empirical data available from the dual-gage estimates of δ and from finite-element calculations show that the relationship given in Eq 18 is unique and independent of a/W and of the degree of constraint as long as no crack extension occurs (Fig. 11).

Simplifying Eq 18 and rearranging

$$\delta = f(V_L) \times \frac{(1 - a/W)}{(2 + a/W)^2} \quad (19)$$

Assuming this relationship is unique for $a/W \geq 0.55$ and $f(V_L)$ is independent of a/W, Eq 19 can be used to estimate crack extension if V_L and δ are known independently. Taking the derivative of Eq 19

$$\frac{\partial(\delta^*)}{\partial(a/W)} = f(V_L) \times \frac{(a/W - 4)}{(2 + a/W)^3} \quad (20)$$

where δ* is measured at the current crack tip (Fig. 12). In Fig. 12

$$\Delta\delta^* = \delta_{(2)} - \delta_{(1)} \quad (21)$$

δ is measured at the original crack tip, thus a correction is needed to relate the virtual value of δ, $\delta_{(2)}$, to the measured value of δ. Referring to Fig. 12, similar triangles give the relationship

$$\frac{\delta}{R} = \frac{\delta_{(2)}}{R - \Delta a} \quad (22)$$

Assuming

$$R = \gamma(\delta) \quad (23)$$

gives

$$\delta = \delta_{(2)} + \frac{\Delta a}{\gamma} \quad (24)$$

Solving Eq 21 for $\delta_{(2)}$ and substituting the result, with $\Delta(\delta^*)$, into Eq 24 gives

$$\delta = \delta_{(1)} + \Delta a \left(\frac{f(V_L)}{W} \times \frac{(a_0/W + 4)}{(2 + a_0/W)^3} + \frac{1}{\gamma} \right) \quad (25)$$

Rearranging and solving for Δa

$$\Delta a = \frac{\delta - f(x) \times (W - a_0)}{\frac{1}{\gamma} + f(x) \frac{(a_0/W - 4)}{(2 + a_0/W)}} \quad (26)$$

where

$$f(x) = f(V_L)/W(2 + a_0/W)^2$$
$$= \delta_{(1)}/(W - a_0) \quad (27)$$

If the value or values of γ are determined, and if the function $f(x)$ is established, the change in crack length, Δa, may be calculated using Eq 26.

The function $f(x)$ was established in part experimentally and in part analytically. Experimentally, the specimen with the deepest crack, T-41, had the largest load-line displacement and the largest value of x (Eq 18) prior to cracking initiation. The $\delta - V_L$ relationship for this specimen was used to cracking initiation. Beyond initiation the relationship was established with the help of finite element calculations for Specimens T-61 and T-52. The calculations for these two specimens established that the $f(x)$ versus x curve (Eq 18) is unique and independent of a/W as long as crack extension does not occur.

The values of γ which gave the best results are given by the relationship

$$\gamma = \frac{(N) \times f(x) \times (W - a_0)}{\delta} \tag{28}$$

where N is a dimensionless constant.

The sensitivity of the slope of the J_1-resistance curve is indicated in Fig. 13. The results of using two values of N for calculating Δa for Specimen T-52 are shown. For reference, a single heat-tint-derived crack extension value is plotted. Note that varying N changes the slope of the crack extension curve, but does not change the critical J_1 at crack initiation.

References

[1] Merkle, J. G., Oak Ridge National Laboratory, private communication.
[2] Wilkinson, J. P. D., Hahn, G. T., and Smith, R. E., "A Program to Study Methods of Plastic Fracture," *Proceedings,* American Society for Metals/American Society for Non-Destructive Testing. Fourth Annual Forum on Prevention of Failure Through Non-Destructive Inspection, Tarpon Springs, Florida, June 15, 1976.
[3] Wilkinson, J. P. D., Hahn, G. T., and Smith, R. E., "Methodology for Plastic Fracture—A Progress Report," *Proceedings,* Fourth International Conference on Structural Mechanics in Reactor Technology, San Francisco, California, Aug. 1977.
[4] Shih, C. F., deLorenzi, H. G., and Andrews, W. R., this publication, pp. 65-120.
[5] Clarke, G. A., Andrews, W. R., Paris, P. C., and Schmidt, D. W. in *Mechanics of Crack Growth, ASTM STP 590,* American Society for Testing and Materials, 1976, pp. 27-42.
[6] Hodge, J. H., "Properties of Heavy Section Nuclear Reactor Steel," Welding Research Council Bulletin No. 217.
[7] Jewett, R. P., *Closed Loop Magazine,* MTS Corp, Vol. 4, No. 3, Summer 1974.
[8] Shih, C. F., deLorenzi, H. G., and Andrews, W. R., *International Journal of Fracture Mechanics,* Vol. 13, 1977, pp. 544-548.
[9] Merkle, J. G., Corten, H. T., *Transactions* ASME, *Journal of Pressure Vessel Technology,* Nov. 1974, pp. 286-292.
[10] Shih, C. F., deLorenzi, H. G., Yukawa, S., Andrews, W. R., van Stone, R. H., and Wilkinson, J. P. D., "Methodology for Plastic Fracture," Contract RP-601-2, Third Quarterly Report, 1 Nov. 1976 to 31 Jan. 1977 for Electric Power Research Institute, Palo Alto, Calif., 16 March 1977.
[11] Green, A. P. and Hundy, B. B., *Journal of the Mechanics and Physics of Solids,* Vol. 4, 1956, pp. 128-144.
[12] Lake, R. L. in *Mechanics of Crack Growth, ASTM STP 590,* American Society for Testing and Materials, 1976, pp. 208-218.
[13] Srawley, J. E. and Gross, B., *Compendium, Engineering Fracture Mechanics,* Vol. 4, 1972, pp. 587-589.

J. A. Joyce[1] *and J. P. Gudas*[2]

Computer Interactive J_{Ic} Testing of Navy Alloys

REFERENCE: Joyce, J. A. and Gudas, J. P., "**Computer Interactive J_{Ic} Testing of Navy Alloys,**" *Elastic-Plastic Fracture, ASTM STP 668*, J. D. Landes, J. A. Begley, and G. A. Clarke, Eds., American Society for Testing and Materials, 1979, pp. 451–468.

ABSTRACT: A computer interactive unloading compliance single-specimen J_{Ic} test procedure has been developed. This procedure utilizes an on-line minicomputer to analyze digitized load-displacement data during testing. Unique values of J_I and crack length are determined from compliance measurement on short unloadings along the load displacement record. The test procedure is presented in detail and analysis procedures are discussed.

Three tasks which demonstrate the validity and utility of the computer interactive test method are discussed. Results for single-specimen and multiple-specimen tests are presented for HY 130, 10Ni steel, 17-4PH steel, Ti-7Al-2Cb-1Ta, and Ti-6Al-4V which show close correspondence between the two methods. Tests on 17-4PH steel compact tension specimens with various thicknesses and crack lengths are summarized and dimensional effects on J_{Ic} and the J-Δa resistance curve slopes are discussed. Finally, tests on HY 130 specimens with various notch root radii demonstrate effects of notch acuity on J_{Ic}.

KEY WORDS: elastic-plastic fracture, toughness testing, single-specimen tests, multiple-specimen tests, computer interactive testing, high-strength steels, titanium alloys, specimen geometry effects, notch acuity, unloading compliance test method, crack propagation

The objective of this effort was to develop a single-specimen test procedure to evaluate the elastic-plastic fracture toughness parameter, J_{Ic}, with the requirement that the method be practical for use in conventional materials testing laboratories. Single-specimen J_{Ic} testing is potentially superior to multiple-specimen procedures for several reasons. In the first place, single-specimen tests define the J_I versus crack growth resistance curve more thoroughly. The determination of a unique J_{Ic} data point from each specimen allows for analysis of material variability and facilitates testing over a range of temperatures and various environments.

[1] Assistant professor of mechanical engineering, U.S. Naval Academy, Annapolis, Md. 21402.
[2] Head, Fatigue and Fracture Branch, David W. Taylor Naval Ship Research and Development Center, Annapolis Laboratory, Annapolis, Md. 21402.

Single-specimen J_{Ic} evaluation methods have been developed by Clarke et al [1][3] utilizing complicated electronic signal amplification which involves considerable data evaluation and calculation after the test is completed. The method described herein involves the analysis of conventional load-displacement signals through the use of readily available electronic instruments and a minicomputer. The method involves digitizing analog load-displacement data, and on-line, real-time computer interactive determination of crack length and J_1. The immediate calculation of crack length allows the test engineer to properly space unloadings, vary test machine speed, or vary unloading length during the test to obtain optimum results from each specimen. Further, the method provides for storage of both digitized load-displacement data and J_1 versus crack extension data for future retrieval and analysis.

J_{Ic} Test Methods

The application of J_1 as an elastic-plastic fracture toughness criterion is based on the nonlinear elastic singularity solution of Hutchinson [2] and Rice and Rosengren [3] which gives the stress, strain, and displacement components near a sharp crack in an incompressible deformation theory plastic material with a uniaxial stress-strain relation

$$\sigma = \bar{\sigma}_1(\epsilon_p)^N \tag{1}$$

as

$$\frac{\sigma_{ij}}{\bar{\sigma}_1} = \left(\frac{J_1}{\bar{\sigma}_1 I_n r}\right)^{N/(N+1)} \tilde{\sigma}_{ij}(\theta)$$

$$\epsilon_{ij} = \left(\frac{J_1}{\bar{\sigma}_1 I_n r}\right)^{1/(N+1)} \tilde{\epsilon}_{ij}(\theta) \tag{2}$$

$$u_{ij} = \left(\frac{J_1}{\bar{\sigma}_1 I_n r}\right)^{N/(N+1)} \tilde{u}_{ij}(\theta)$$

where $\tilde{\sigma}_{ij}(\theta)$, $\tilde{\epsilon}_{ij}(\theta)$, $\tilde{u}_{ij}(\theta)$, and I_n are functions of θ and N, and r and θ define a polar coordinate system about the crack tip. Only J_1 in Eq 2 depends on the applied boundary conditions or specimen geometry and thus J_1 sets the intensity of the stress, strain, and displacement singularity in a manner completely analogous to the stress factor K_1 in the Williams [4] and Irwin [5] elastic solution. Begley and Landes [6] further proposed that a critical value of J_1 exists (which is a material property) such that crack extension occurs when $J_1 > J_{\mathrm{Ic}}$.

---
[3] The italic numbers in brackets refer to the list of references appended to this paper.

For the special case that $N = 1$ in Eq 1, Eq 2 reduces to the Irwin-Williams elastic solution, and for this case Rice [7] has demonstrated that

$$J_1 = \frac{(1 - \nu^2)}{E} K_1^2 \qquad (3)$$

where

E = material modulus of elasticity,
ν = Poisson's ratio, and
K_1 = linear elastic stress intensity factor.

Rice et al [7] has shown that for specimens in which only one length dimension is present, J_1 can be evaluated approximately by

$$J_1 = \frac{2A}{Bb} \qquad (4)$$

where

A = area under the load-load point displacement curve,
b = uncracked ligament, and
B = specimen thickness.

To obtain J_{1c}, J_1 must be evaluated at the point on the load-displacement curve where crack extension initiates. To determine this point of crack initiation, Landes and Begley [8] proposed a multispecimen test procedure which includes the following steps:

1. For each material, temperature, environment, etc., at least four individual specimens are loaded to different crack opening displacement (COD) values and a load-displacement plot is developed for each specimen.
2. Crack extension is then marked by heat tinting or fatigue cracking.
3. Specimens are then pulled apart and crack extension is measured over nine evenly spaced points, including the centerline of the specimen and excluding the points at each surface.
4. J_1-values are then calculated from the load-displacement record using Eq 4.
5. For each material, a plot of J versus crack extension, Δa, is constructed. A least-squares straight line is fit to all data. The critical J_{1c} value is obtained at the intersection of the foregoing line and the crack opening stretch line defined by the relationship

$$J = 2\sigma_{\text{flow}} \Delta a \qquad (5)$$

where σ_{flow} is the average of the material yield strength and ultimate tensile strength, and Δa is the crack extension.

6. Test validity is determined by the following relationship

$$B > \frac{\beta J_{Ic}}{\sigma_{ys}} \qquad (6)$$

where the value β has been suggested as 50 by Paris [10] or 25 to 40 by Landes and Begley [9].

Alternative methods to the multiple-specimen method just outlined have obtained J_{Ic} from a test on a single specimen. For this type of test, a procedure is required which gives an accurate measure of crack extension while the specimen is under load so that the J_1 at initial crack extension can be determined. Clarke et al [1] has used an unloading compliance method to evaluate crack extension. Carlsson and Markström [11] have used electrical impedance and eddy current techniques for this purpose. The description of the computer interactive unloading compliance test method developed in this effort is presented in the next section.

Computer Interactive J_{Ic} Test Method

The schematic of the computer interactive test arrangement is shown in Fig. 1. In this setup, load cell and clip gage signals are conventionally amplified and fed to a scanner which is interfaced to a digital voltmeter through an IEEE Standard 488-1975 interface. The digitized data are then made available to a microprocessor with magnetic tape cartridge and cathode ray tube (CRT) graphics capability. The test arrangements at the U.S. Naval Academy and the David W. Taylor Naval Ship Research and Development Center (DTNSRDC) also employ a peripheral interactive graphics plotter interfaced with the computer. The test results reported herein utilized an Instron Model TTD universal test machine and a Tinius-Olsen test machine, both of which are screw-type displacement-controlled devices. The computer used for this testing was a Tektronix Model 4051.

A single-specimen J_{Ic} test is conducted with this apparatus using the following steps:

1. Load and COD transducers are calibrated using the test machine electronics and a precision micrometer, respectively. This step yields the slope and intercept of a straight-line calibration curve for each data channel. These values are stored in a magnetic tape file for use by the J_{Ic} test program.

2. The clip gage is attached to razorblade knife edges mounted on the load line and the specimen is inserted in the test machine.

3. Initial crack length is determined by loading the specimen between 10 and 50 percent of the expected maximum test load. Between 30 and 50 load-displacement data pairs are gathered and the computer estimates the initial crack length (a/w) using the relation from Saxena and Hudak [12]

$$\frac{a}{w} = 1.000196 - 4.06319Ux + 11.242Ux^2$$

$$-706.043Ux^3 + 464.335Ux^4 - 650.677Ux^5 \qquad (7)$$

where

$$Ux = \frac{1}{\left(\frac{BE'\delta}{P}\right)^{1/2} + 1}, \quad \text{and} \quad E' = \frac{E}{(1-\nu^2)}$$

where

E and ν = specimen elastic modulus and Poisson ratio, respectively,
B = specimen thickness, and
δ/P = load-line compliance.

The specimen compliance, the least-squares correlation, and the crack length estimate are determined and displayed after each unloading and loading on the computer CRT screen. A least-squares correlation of the unloading data to a straight line of 0.9999 or greater and crack length estimates varying by ±0.05 mm are generally obtainable and are required before continuing the test. With the initial crack length estimate completed, the specimen is returned to zero load and the test is begun by starting the test machine and computer data acquisition simultaneously.

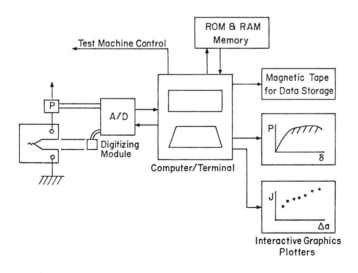

FIG. 1—*Schematic of computer interactive J_{Ic} test apparatus.*

4. Load-displacement data pairs obtained by the computer are plotted on line by the interactive graphics plotter, producing a typical plot as shown in Fig. 2. The area under the load-displacement curve is obtained from each data pair using a trapezoidal quadrature. Unloadings of approximately 10 percent are initiated and spaced at the discretion of the operator. For each unloading, involving 25 to 45 load-displacement pairs, the computer develops the best-fit straight-line slope, which is substituted into Eq 7 to give a crack length estimate. At each unloading, estimated crack length, change in crack length, J_1, and correlation of the regression analysis are determined and printed on the CRT screen. When the desired maximum crack extension is obtained, the test is terminated and the data file is closed.

5. Finally, an additional program is introduced which operates on the J_1-Δa data file to obtain J_{Ic}. This program fits a least-squares straight line to all data with $\Delta a > J/2\sigma_{flow} + 0.05$ mm, then obtains J_{Ic} by solving for the intersection of the foregoing line and the crack opening stretch line.

A critical requirement for a successful test of this type is that an accurate estimate of crack length be made from the data obtained on each unloading. A typical unloading is shown greatly magnified in Fig. 3. The present program eliminates the upper 2 percent of the unloading curve based on maximum load and then uses all other points, both unloading and loading, to obtain a slope or compliance estimate. A slight delay in data gathering for slope calculation is then experienced before data acquisition is resumed. Figure 3

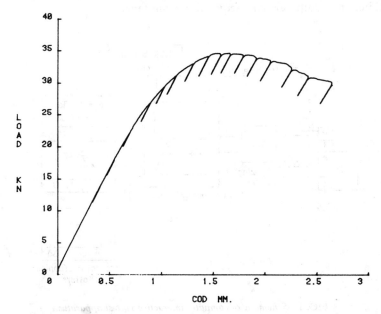

FIG. 2—*Plot of single-specimen load-displacement data for HY 130 steel.*

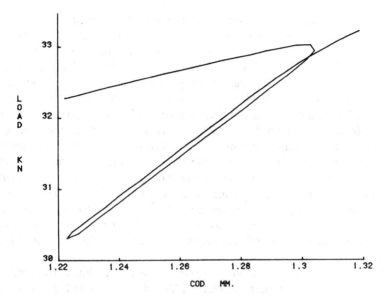

FIG. 3—*Plot of single-specimen unloading detail.*

shows a slight hysteresis between loading and unloading, but nearly identical slopes are obtained. A correlation greater than 0.995 is typical for unloadings involving 25 to 40 data pairs.

Applications of the Computer Interactive J_{Ic} Test Method

The computer interactive J_{Ic} test procedure has been employed in several programs at the U.S. Naval Academy and DTNSRDC. The results of three separate efforts are discussed herein. These include:

1. Evaluation of J_{Ic} of five Navy steels and titanium alloys and comparison with multiple-specimen data.
2. Determination of specimen size limitations on thickness and remaining ligament for 17-4PH steel.
3. Evaluation of effects of notch acuity on J_{Ic} measurement of HY 130 steel.

J_{Ic} Testing of Navy Steels and High-Strength Titanium Alloys

Computer interactive single-specimen tests and conventional multiple-specimen tests were performed on three high-strength steels and two high-strength titanium alloys. The objective of these tests was to assess the ability of the computer interactive test procedure to produce J_{Ic} and J_I-Δa test data which correlated with that produced with the multiple-specimen method. The chemical composition of the test materials included in this effort is described in Table 1 and the mechanical properties are presented in Table 2. The 10Ni steel was provided in the form of 38-mm plate and the Ti-7-2-1 alloy was provided in the form of 102-mm plate. The other materials were in the form of 25-mm plate. Modified compact specimens (ITCT) were used in these tests. In all cases, the crack plane was produced in the T-L orientation. All specimens were fatigue cracked as per the ASTM Test for Plane-Strain Fracture Toughness of Metallic Materials (E 399-74) to a total notch depth of 38 mm ($a/w = 0.75$). During J_{Ic} testing, maximum crosshead speed was 0.25 mm/min. The multiple-specimen test procedure was that by Landes and Begley [9] described earlier, and the computer interactive test procedure described in the preceding section was employed. Specific J_{Ic}-values for these single- and multiple-specimen tests are reported in Table 3. The J_I versus crack extension resistance curves for HY-130, 10Ni steel, 17-4PH steel, Ti-7-2-1, and Ti-6-4 are presented in Figs. 4–8, respectively. Each figure shows the blunting line and least-squares linear regression fit of individual data points.

Analysis of multiple- and single-specimen test data shows that both test methods produce equivalent J_{Ic}-values. The data in Table 3 show that J_{Ic} calculated from single-specimen tests tends to run higher than J_{Ic} from multiple-specimen tests. With the exception of Ti-7-2-1, the single-specimen method consistently produces high J_I-Δa resistance curve slopes. This difference results from the fact that considerable crack front curvature upon crack extension existed with all materials except Ti-7-2-1. When cracks tunnel, the compliance calculated crack length is smaller than the average of nine measured points across the specimen. Final nine-point crack length measurements for the single specimen are shown in Figs. 4–8 and suggest equivalence between single- and multiple-specimen tests when an identical crack length measurement technique is used.

Specimen Size Limitation Analysis

This effort involved using a basic 1T compact specimen configuration with a range of thicknesses and fatigue crack lengths to determine effects of these changes on J_{Ic} and the J_I-Δa resistance curve slope. The 17-4PH steel (σ_{ys} = 895 MPa) used in the test method correlation study was used in this task, but crack planes were placed in the L-T orientation. Tests were carried out at ambient temperatures using the Instron TTD test machine. The computer interactive test procedure described earlier was employed for all tests and in this case the correction for axial force on the ITCT specimen was included in the J_{Ic} calculation [13].

Figure 9 shows the J_I-Δa resistance curves for three specimens with thicknesses of 25.4, 9.5, and 5.0 mm, respectively. For these tests, the initial crack length was identical at $a/w = 0.72$. J_{Ic}-values for these three specimens are presented in Table 4. J_{Ic}-values for specimens with 25.4 and

TABLE 1—Chemical composition of test alloys.

Steels

Material (Code)	C	Mn	P	Si	Ni	Cr	Mo	V	S	Cb	Ta	Cu	Al	Co	Ti
HY 130[a] (FKS)	0.11	0.76	0.005	0.31	5.00	0.42	0.53	0.043	0.004	...	...	0.022	0.021	0.02	0.008
HY 130[c] (FLF)	0.10	0.90	0.011	0.28	5.24	0.56	0.57	0.009	0.009	...	...	0.04	...	...	...
10Ni Steel[a] (EZC)	0.12	0.11	0.007	0.07	10.26	1.99	1.007	...	0.003	...	...	...	...	...	0.002
17-4PH[a,b] ENU	0.037	0.20	0.023	0.57	4.28	15.99	...	...	0.011	0.27	0.01	3.33	...	7.75	...

Titanium Alloys

	C	N	Fe	Al	Cb	Ta	Mo	O	H	V
Ti-7-2-1[a] (FAR)	0.010	0.0057	0.12	6.52	2.22	1.35	...	0.032	0.0032	...
Ti-6-4[a] FGU	0.021	0.012	0.197	6.25	...	...	...	0.15	0.0031	3.95

[a] J_{Ic} testing of Navy steels and high-strength titanium alloys.
[b] Specimen size limitation analysis.
[c] Crack tip acuity effects.

TABLE 2—Mechanical properties of test alloys (transverse orientation).

Alloy (Code)	Heat Treatment	0.2% Yield Strength, MPa	Ultimate Tensile Strength, MPa	Elongation, % in 50 mm	Reduction of Area, %
HY 130[a] (EFI)	830°C, 1.5 h, water quench; 630°C, 1.5 h, water quench	937	978	21	55
HY 130[c] (FLF)	...	972	1034	19	64
10Ni Steel[a] (EZC)	885°C, 1 h, water quench; 815°C, 1 h, water quench; 510°C, 5 h, water quench	1300	1452	19	69
17-4PH[a,b] (ENU)	1900°F, 1 h, air cool; 1100°F, 4 h, air cool	883	972	11	64
Ti-7-2-1[a] (FAR)	1090°C, 1 h, air cool	665	741	10	23
Ti-6-4[a] (FGU)	...	822	934	11	16

[a] J_{Ic} testing of Navy steels and high-strength titanium alloys.
[b] Specimen size limitation analysis.
[c] Crack tip acuity effects.

TABLE 3—*Summary of J_{Ic} test results for Navy steels and high-strength titanium alloys.*

Material	J_{Ic}, kPa·m		Equivalent K_{Ic}, mPa·m$^{1/2}$	
	Single Specimen	Multi-specimen	Single Specimen	Multi-specimen
HY 130	208	186	213	202
10 Ni steel	138	118	174	161
17-4PH steel	112	106	157	153
Ti-7Al-2Cb-1Ta	73	71	97	96
Ti-6Al-4V	48	39	78	71

9.5 mm thickness agree within 1 percent while the specimen which was 5.0 mm thick produced a J_{Ic}-value which was nearly twice that of the thicker specimens. The analyses included in Table 4 based on the Paris [10] thickness requirements for a valid J_{Ic} test show that the thickest specimen is definitely valid, the intermediate thickness is questionable, and the thinnest specimen is not valid. This shows excellent agreement with these single-specimen J_{Ic} test results. Figure 9 also shows that the J_I-Δa points at large crack extension fall below the extrapolated straight-line fit to data for $\Delta a <$ 1.1 mm. Therefore the J_{Ic}-values reported in Table 4 were calculated ex-

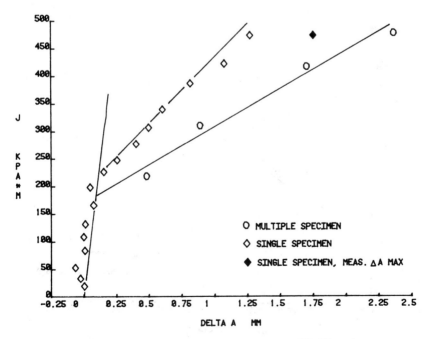

FIG. 4—*Plot of J versus crack extension data for HY 130 steel.*

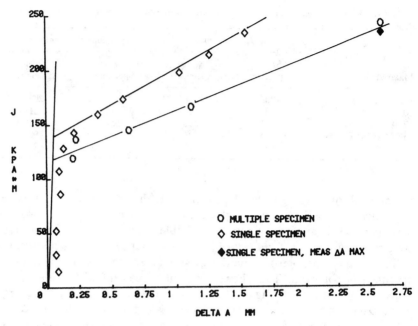

FIG. 5—*Plot of J versus crack extension data for 10Ni steel.*

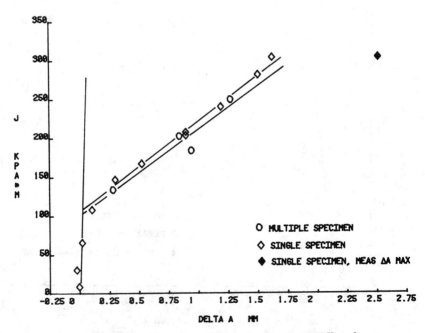

FIG. 6—*Plot of J versus crack extension data for 17-4PH steel.*

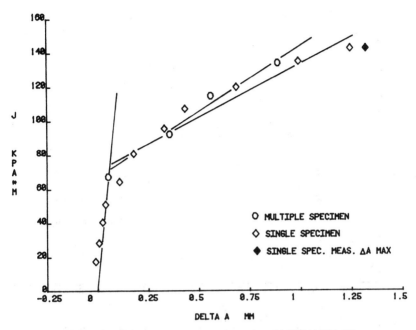

FIG. 7—*Plot of J versus crack extension data for Ti-7Al-2Cb-1Ta.*

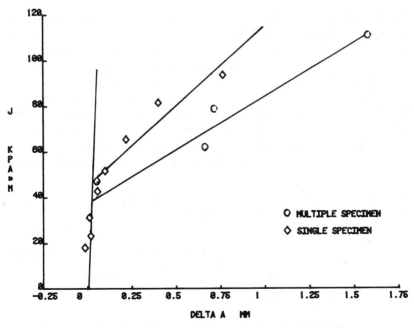

FIG. 8—*Plot of J versus crack extension data for Ti-6Al-4V.*

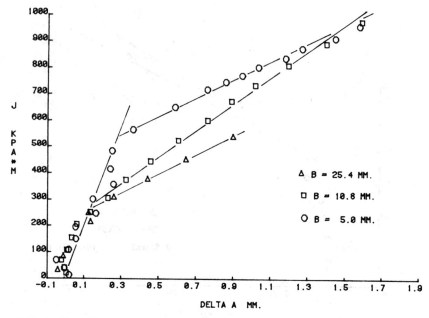

FIG. 9—*Plot of J versus crack extension data for 17-4PH steel as related to specimen thickness.*

cluding data pairs with $\Delta a > 1.1$ mm. Data on specimens of various thickness have shown an increase in the initial J_I-Δa curve slope with decreasing thickness, but subsequent decrease in slope. This is shown clearly with the 5.0-mm specimen and suggests that insufficient thickness can result in nonconservative J_{Ic} measurements for the material investigated.

The second part of this effort involved developing J_I-Δa data with ITCT specimens with crack lengths in the range 0.72 to 0.92 a/w. Figure 10 shows the J_I-Δa resistance curves for the three test specimens and J_I-values are reported in Table 4. J_{Ic}-values for the specimens with $a/w = 0.72$ and 0.80 agree within 3 percent while the resistance curve slopes agree within 10 per-

TABLE 4—*Summary of J_{Ic} test results for 17-4PH steels with various thickness specimens and various crack lengths.*

Specimen	Thickness, mm	Crack Length, a/w	J_{Ic}, kPa·m	25 to 50 J_{Ic}/δ_{ys}, mm	dJ_I/da, kPa·m/mm
1	25.4	0.72	262	7.3 to 14.6	379.5
2	9.5	0.72	264	7.4 to 14.7	551.3
3	5.0	0.72	473	13.1 to 26.4	388.7
4	25.4	0.80	254	7.1 to 14.1	345.0
5	25.4	0.92	300	8.5 to 17.0	745.2

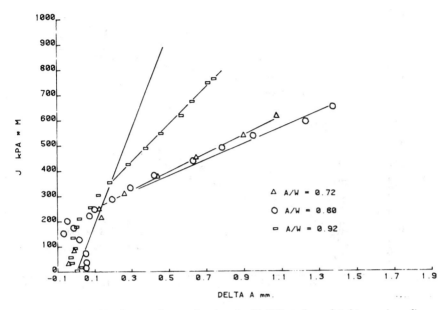

FIG. 10—*Plot of J versus crack extension data for 17-4PH steel as related to specimen ligament length.*

cent. The most deeply cracked specimen ($a/w = 0.92$), however, gives a slightly higher value of J_{Ic} and a much steeper resistance curve slope, indicating that the error in J_{Ic} introduced by subsize ligament effects is nonconservative for the material investigated.

Crack Tip Acuity Effects

The J_I of HY 130 plate was evaluated to determine the effects of notch tip acuity on the J_{crit} measurement. Two specimens were produced conventionally, and four specimens were produced with machined notches to depths in the range $0.7 < a/w < 0.76$ and root radii of 0.051 and 0.076 mm. The two regularly machined specimens were fatigue cracked to depths of 0.74 and 0.79 a/w according to ASTM E 399-74 criteria. All tests were carried out at ambient temperature utilizing the Tinius-Olsen test machine. The single-specimen procedure described earlier was followed and the correction for axial force on the ITCT specimen was included in the J_{Ic} calculation [*13*]. The J_I versus crack extension curves for these tests are presented in Fig. 11 and results are summarized in Table 5. It can be seen that crack tip geometry substantially affects the J_{Ic} measurement. The lowest values were produced with the fatigue crack specimens. Among the specimens with machined notches, the apparent J_{Ic}-values related to the 0.051-mm root radii were below those related to the 0.076-mm root radii. This result is consistent with that

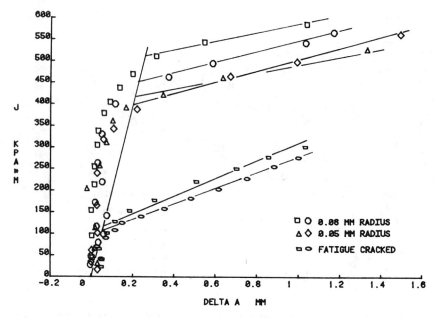

FIG. 11—*Plot of J versus crack extension data for HY 130 steel as related to specimen crack root radius.*

obtained by Begley and Logsdon [*14*] for A471 nickel-chromium-molybdenum-vanadium rotor steel.

It should also be pointed out that the J_{Ic}-values determined from fatigue precracked specimens are substantially lower than those reported for HY 130 in the test method correlation task. The HY 130 plate used for this notch acuity study was not traceable to a particular producer or vintage, precluding the use of these J_{Ic} data to describe the HY 130 system. On the other hand, these data point up the sensitivity of the J_{Ic} measurement and the computer interactive test procedure in evaluating fracture toughness from the simple quality-assurance standpoint.

TABLE 5—*Summary of J_{Ic} test results for HY 130 steel with various notch root radii.*

Specimen	Notch radius, mm	J_{Ic}, kPa·m	dJ_1/da, kPa·m/mm
FLF-2	fatigue cracked	128	186.2
FLF-4	fatigue cracked	103	182.5
FLF-5	0.08	446	123.5
FLF-6	0.08	508	99.5
FLF-7	0.05	407	104.9
FLF-8	0.05	381	119.7

Conclusions

The computer interactive unloading compliance J_{Ic} test method has been shown to produce equivalent J_{Ic}-values for the steels and titanium alloys tested when compared with multiple-specimen data. These single-specimen tests show high J_I-Δa resistance curve slopes when crack tunneling occurs because effective crack length is shorter than that calculated from nine measurements across the thickness.

The computer interactive test method is seen to possess several advantages in comparison with the multiple specimen method. In the first place, the computer interactive method produces more complete and consistent J_I-Δa resistance curve data. The immediate calculation of J_{Ic} and crack extension after each unloading gives the test engineer the capability to space unloadings evenly, to repeat a particular unloading, to change machine speed, etc., so as to obtain optimum results from each specimen. The test method allows for evaluation of material variability and is adaptable for testing at different temperatures and in various environments. Magnetic tape storage of digitized load-displacement data allows for future reanalysis as the J_{Ic} fracture criterion develops. Finally, the fact that a unique J_{Ic} test result is produced from each test suggests that J_{Ic} testing can be carried out on a routine basis as is K_{Ic} testing. These advantages are enhanced by the fact that the test method described herein is readily adaptable to the new generation of computer interactive test machines now being made available.

The computer interactive unloading compliance method was successfully utilized to evaluate effects of specimen thickness, remaining ligament, and notch acuity on J_{Ic} and the shape of the J_I-Δa resistance curve.

Acknowledgment

The authors acknowledge the Naval Sea Systems Command (NAVSEA 03522), the National Science Foundation (Contract ENG76-09623), and the Structures Department of DTNSRDC for supporting various aspects of this research.

References

[1] Clarke, G. A., Andrews, W. K., Paris, P. C., and Schmidt, D. W. in *Mechanics of Crack Growth, ASTM STP 590*, American Society for Testing and Materials, 1976, pp. 27-42.
[2] Hutchinson, J. W., *Journal of Mechanics and Physics of Solids*, Vol. 16, 1968, pp. 13-31.
[3] Rice, J. R. and Rosengren, G. E., *Journal of the Mechanics and Physics of Solids*, Vol. 16, 1968, pp. 1-12.
[4] Williams, M. L., *Journal of Applied Mechanics*, Vol. 24, 1957, pp. 109-114.
[5] Irwin, G. R., *Journal of Applied Mechanics*, Vol. 24, pp. 361-364.
[6] Begley, J. A. and Landes, J. D. in *Fracture Mechanics, ASTM STP 514*, American Society for Testing and Materials, 1972, pp. 1-20.
[7] Rice, J. R., *Journal of Applied Mechanics*, Vol. 35, 1968, pp. 379-386.

[8] Rice, J. R., Paris, P. C., and Merkle, J. G. in *Progress in Flaw Growth and Fracture Toughness Testing, ASTM STP 536*, American Society for Testing and Materials, 1973, pp. 231-245.
[9] Landes, J. D. and Begley, J. A. in *Fracture Analysis, ASTM STP 560*, American Society for Testing and Materials, 1973, pp. 170-186.
[10] Paris, P. C., Discussion to J. A. Begley and J. D. Landes in *Fracture Mechanics, ASTM STP 514*, American Society for Testing and Materials, 1972, pp. 21-22.
[11] Carlsson, A. J. and Markström, K. M. in *Proceedings*, Fourth International Conference on Fracture, Waterloo, Ont., Canada, 1977, pp. 683-691.
[12] Saxena, A. and Hudak, S. J., Jr., "Review and Extension of Compliance Information for Common Crack Growth Specimens," Westinghouse Scientific Paper 77-9E7-AFCGR-P1, Pittsburgh, Pa., 1977.
[13] Merkle, J. G. and Corten, H. T., *Journal of Pressure Vessel Technology, Transactions*, American Society of Mechanical Engineers, Vol. 96, Nov. 1974, pp. 286-292.
[14] Logsdon, W. A. and Begley, J. A., *Engineering Fracture Mechanics*, Vol. 6, 1977, pp. 461-470.

A. D. Wilson[1]

Characterization of Plate Steel Quality Using Various Toughness Measurement Techniques

REFERENCE: Wilson, A. D., "**Characterization of Plate Steel Quality Using Various Toughness Measurement Techniques,**" *Elastic-Plastic Fracture, ASTM STP 668,* J. D. Landes, J. A. Begley, and G. A. Clarke, Eds., American Society for Testing and Materials, 1979, pp. 469–492.

ABSTRACT: The fracture toughness properties of three steels (A516-70, A533B Class 1, and HY-130) are determined in two steel-quality levels (conventional and calcium treated). In addition, a conventional quality A543B Class 1 steel is similarly examined at two locations (quarterline and centerline). Both investigations involved comparing steels with differing inclusion structures. The primary effort was to establish the upper-shelf toughness differences found using the Charpy V-notch impact and dynamic tear tests compared with those found using J-integral determinations of J_{Ic} and K_{Ic}. The J_{Ic} determinations appeared to be more sensitive to changes in inclusion structure than either the Charpy V-notch or dynamic tear tests. This was established by comparing the toughness between quality levels and by measuring the anisotropy of toughness within a steel. Comparisons are made with the Charpy V-notch impact-K_{Ic} upper-shelf correlation values. Comments concerning the suggested graphical method for J_{Ic} determination are also given.

KEY WORDS: crack propagation, fractures (materials), inclusions, steels, plastic properties

The quality of steel plate can be significantly affected by nonmetallic inclusions. The presence of inclusions, such as sulfides and oxides, primarily affects the ductile behavior of the steel. Therefore when the quality of a particular grade of steel is improved by inclusion reduction or modification, the benefits are commonly assessed by conventional testing determinations such as the tensile percent reduction of area or Charpy V-notch (CVN) upper-shelf energy [1-3].[2] In addition, the dynamic tear (DT) test upper-shelf energy can be used to quantify quality enhancement [2,3]. While these measurements give relative indications of improvement, the results cannot

[1] Research engineer, Lukens Steel Company, Coatesville, Pa. 19320.
[2] The italic numbers in brackets refer to the list of references appended to this paper.

be used directly in design. However, through fracture mechanics by determining fatigue crack propagation and fracture toughness properties, these benefits can be directly related to design. Previous work [4] has reported the influence of inclusions on the fatigue crack propagation properties of steels. In this study the effect on the plain-strain fracture toughness, K_{Ic}, will be established.

There is a concern for obtaining the K_{Ic} properties of improved quality steels for a number of reasons. Existing structural designs may already be based on K_{Ic} determinations made on conventional quality steels. Thus, if it is possible that plain-strain conditions may exist in a structure, any design modification to accommodate the improved steel quality would require the actual improved K_{Ic} properties. In addition, it is of interest to determine if existing K_{Ic} correlations with CVN properties are still applicable and to establish whether tensile ductility, CVN, and DT testing provide reliable indications of the enhancement in K_{Ic} of improved quality steels. However, to obtain upper-shelf K_{Ic} values, to the ASTM Test for Plane-Strain Fracture Toughness of Metallic Materials (E 399–74) standards, for common structural steels would require extremely large and expensive test specimens. The J-integral testing approach provides a test method for obtaining K_{Ic} values using reasonable-sized specimens.

The conception [5] and development [6–10] of the J-integral as a fracture criterion and testing technique has been a significant contribution to metallurgists interested in obtaining fracture toughness properties of ductile materials. This is because K_{Ic} values can be determined from the following relationship

$$J_{Ic} = \frac{1 - \nu^2}{E} K_{Ic}^2 \qquad (1)$$

where J_{Ic} is the critical value of J at the point of crack extension and ν and E are Poisson's ratio and the modulus of elasticity, respectively. In steels, for example, J-integral testing has been used to establish the fracture toughness in the transition range and on the upper shelf [11–13] of a number of alloys. Also emphasizing the interest in the J-integral test technique has been the formation of an ASTM task group, originally E24.01.09 and at present E24.08.04, which has developed guidelines for J_{Ic} determinations [14].

In this investigation, a carbon steel (A516-70) and two alloy steels (A533B Class 1 and HY-130) are characterized in two quality levels (conventional and calcium treated). In addition, a conventional quality A543B Class 1 steel is examined at two locations (quarterline and centerline). In both programs the tensile ductility, CVN, and DT upper-shelf energies and J_{Ic} on the upper shelf are established. The primary intent of these evaluations is to determine how the conventional test techniques, particularly the CVN and DT, rate material quality compared with the J_{Ic} measurements.

Experimental Details

Materials

Four steels of a wide range of strength levels were studied. The A516-70 carbon steel has a minimum 0.2 percent offset yield strength (0.2YS) of 262 MPa (38 ksi), the A533B Class 1 low-alloy steel a minimum 0.2YS of 345 MPa (50 ksi), the HY-130 alloy steel a minimum 0.2YS of 896 MPa (130 ksi), and the A543B Class 1 alloy steel a minimum 0.2YS of 552 MPa (80 ksi). The concern for determining the K_{Ic} at the upper shelf for these four steels is due to their primary operating temperature in a number of applications being on the upper shelf. To obtain plain-strain conditions in each of these steels with a K_{Ic} of 165 MPa$\sqrt{m}$ (150 ksi$\sqrt{in}$.) and an average 0.2YS would require a thickness of about 559 mm (22 in.), 305 mm (12 in.), 76 mm (3 in.), and 178 mm (7 in.), respectively, for the A516, A533B, HY-130, and A543 steels. These thicknesses of material have been produced for these steels and thus the concern for K_{Ic} values exists.

The A516, A533B, and HY-130 steels were characterized in two quality levels, namely, steel made by conventional steelmaking practices (CON) and by a calcium treatment (CaT). The mechanical properties and nonmetallic inclusions resulting from these two practices have been reported in detail [1-4]. Briefly, for aluminum-killed, fine-grained steels the CON steels have higher sulfur levels, lower toughness and ductility properties, and anisotropy of these properties due to the presence of two kinds of inclusions. Type II manganese sulfide and galaxies of alumina inclusions lead to this behavior. Figure 1 shows these manganese sulfide inclusions in the CON A533B steel and indicates their elongated and pancaked nature due to their plastic behavior at hot rolling temperatures. The alumina galaxies shown in Fig. 2 for the same steel do not individually deform, but as a group the galaxies are rotated and aligned in a planar fashion due to rolling. Both of these kinds of inclusions lead to the lower level of properties and anisotropy in CON steels.

CaT prevents the formation of both of the foregoing kinds of inclusions by both desulfurization and inclusion shape control. The remaining inclusions in these steels, as shown in Fig. 3 for CaT A533B, are duplex-round compact inclusions. The calcium modification of these inclusions makes them harder at hot-rolling temperatures and thus they do not elongate. CaT steels therefore tend to have better toughness and ductility properties with improved isotropy of these properties, as well as lower sulfur levels.

The chemistries of the steels examined in this part of the study are given in Table 1. In addition to conventional steelmaking techniques, the CON HY-130 material was treated with a ladle flux practice to obtain the rather low sulfur level indicated. However, there is no inclusion shape control in this practice [1,2] and thus there is still anisotropy present.

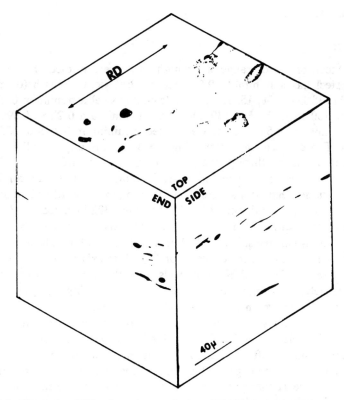

FIG. 1—*Composite of light photomicrographs from CON A533B steel indicating morphology of largest Type II manganese sulfide inclusions.*

The A543 steel studied was of CON quality level. The purpose of this part of the program was to compare the properties of this steel at the quarterline (QL) (quarter thickness) and centerline (CL) (center thickness) locations of the plate. Because of the solidification behavior of large steel ingots, there normally are larger inclusions in both size and number at the CL. This leads to poorer upper-shelf energies and tensile ductilities. The chemistry of this plate is given in Table 1.

The A516 and HY-130 plate steels, which were nominally 51 mm (2 in.) in thickness, were tested at the centerline of the plates. Tension testing was performed in the longitudinal (L) and transverse (T) orientations. CVN, DT, and J_{Ic} testing was performed in the longitudinal (LT) and transverse (TL) orientations. The thicker A533B and A543 plates were tested in all three testing orientations, namely, L, T and through-thickness (S) tensiles and LT, TL and through-thickness (SL) CVN, DT, and J_{Ic} tests. The A533B CON steel was tested at the QL, while the CaT was tested at the CL. As mentioned previously, the A543 tests were performed at the QL and CL.

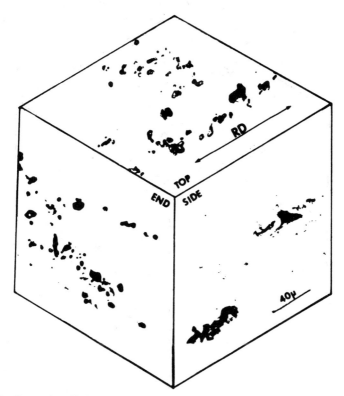

FIG. 2—*Composite of light photomicrographs from CON A533B steel indicating morphology of largest galaxies of alumina inclusions.*

Testing Techniques

The tension, CVN, and DT tests were all performed according to the applicable ASTM specifications, namely: Tension Testing of Metallic Materials (E 8-69); Notched Bar Impact Testing of Metallic Materials (E 23-72); and Test for Dynamic Tear Energy of Metallic Materials (E 604-77). The tensile properties were obtained at room temperature (RT) using two 6.4-mm-diameter (0.252 in.) specimens. The full transition curve was obtained in both CVN and DT testing and the respective upper-shelf energies were obtained by averaging 3 to 5 CVN and 2 to 3 DT results which had 100 percent ductile fracture appearance. The CVN tests used the conventional 10-mm-square (0.394 in.) specimen and the DT tests used the 16-mm-thick (5.8 in.) specimen. The SL-oriented DT tests for the A533B CON material at the QL were performed on specimens with welded-on extensions. No SL-oriented DT tests were performed at the QL of the A543 material.

474 ELASTIC-PLASTIC FRACTURE

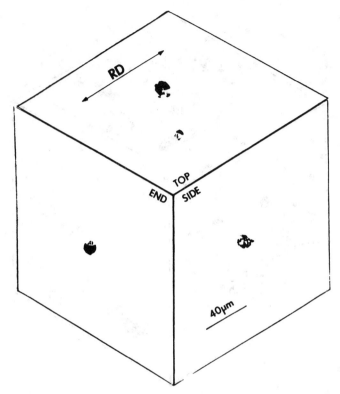

FIG. 3—*Composite of light photomicrographs from CaT A533B steel showing calcium-modified, shape-controlled inclusions.*

The J-integral tests in this study were performed using the single-specimen compliance-unload technique developed by Clarke et al [*15*]. The single-specimen J-integral (SSJ) technique allows obtaining a full J versus crack extension (Δa) resistance (R)-curve, having about 15 to 25 points, using a single specimen. Two specimens were tested for each material at a temperature expected to give upper-shelf behavior. The A543 and HY-130 steels were tested at RT and the A516 and A533B steels were tested at +93°C (+200°F) to assure fully ductile conditions. Those temperatures also coincided with the upper shelves of the CVN and DT tests. The elevated temperature was obtained by wrapping the specimen with resistance heating tapes and insulating with glass wool. The specimen used was the compact design, deeply notched, with provision made for load-line displacement measurements on the specimen. Specimens 25 mm (1 in.) thick were tested in all cases except for the CON A533B and CON HY-130 steels, where 16-mm-thick (5/8 in.) specimens were tested because of limited material availability. The precrack lengths were controlled so that the a/w for most tests was nominally 0.75, where a is the crack length to

TABLE 1—*Chemistry*.

Steel	Heat Treatment[a]	Quality Level[b]	Plate Thickness, mm	Plate Thickness, (in.)	C	S	Mn	P	Cu	Si	Ni	Cr	Mo	V	Al
A516-70	N	CON	51	(2)	0.25	0.022	1.12	0.017	0.29	0.21	0.30	0.15	0.05	0.001	0.034
	N	CaT	51	(2)	0.23	0.003	1.03	0.009	0.12	0.20	0.17	0.07	0.06	0.001	0.041
A533B-1	Q + T + S	CON	169	(6-7/16)	0.18	0.009	1.41	0.008	0.13	0.23	0.48	0.11	0.48	0.001	0.045
	Q + T + S	CaT	203	(8)	0.20	0.003	1.34	0.017	0.12	0.24	0.69	0.16	0.40	0.004	0.069
HY-130	Q + T	CON	51	(2)	0.12	0.006	0.94	0.004	0.13	0.34	4.92	0.58	0.53	0.10	0.047
	Q + T	CaT	57	(2-1/4)	0.08	0.002	0.74	0.006	0.18	0.28	4.98	0.61	0.30	0.11	0.055
A543B-1	Q + T	CON	229	(9)	0.17	0.015	0.25	0.016	0.15	0.22	3.54	1.88	0.43	0.026	0.030

[a] N = normalized; Q + T = quenched and tempered; Q + T + S = quenched, tempered, and stress relieved.
[b] CON = conventional steelmaking practice; CaT = calcium treated.

the centerline of loading and w is the specimen width. The a/w for the CaT A533B tests was 0.60. The loading rate during the tests was 0.953 mm/min (0.0375 in./min).

The J_{Ic} determinations were made from the J versus Δa R-curves following the guidelines of Clarke [14]. Not all specimens met the suggested specimen size requirement of this procedure. The J values used in the R-curves were corrected to account for the tension component in the compact specimen using the Merkle-Corten correction [16]. In addition, a correction to account for the rotation of the compact specimen during testing was developed by Donald [17]. The rotation correction of the Δa values ranged from less than 1 percent for the HY-130 tests to as much as 25 percent for the A516 CaT tests.

Results

The results of the conventional mechanical property tests, tension, CVN, and DT, are presented in Table 2. The items to particularly note are the tensile percent reduction of area (RA) and the CVN and DT upper-shelf energies (CVN USE and DT USE). The improvement in the level of ductility and toughness in the A516, A533B, and HY-130 for the CaT quality is readily apparent. In the A543 steel the QL toughness and ductility levels are better than those at the CL.

Tension tests at $+93\,°C$ ($+200\,°F$) for the A533B CaT steel indicated that the 0.2YS and ultimate tensile strength (UTS) values are about 21 MPa (3.1 ksi) and 36 MPa (5.2 ksi) lower, respectively, than at RT. For A516, the 0.2YS and UTS are reduced by 18 MPa (2.6 ksi) and 35 MPa (5.1 ksi), respectively, at $+93\,°C$ ($+200\,°F$) versus RT. These modifications were made to the tensile strengths used in the later J_{Ic} analyses.

Although the tensile ductility, particularly the percent RA, can be a good measure of steel quality, as shown in Table 2, it is not commonly considered a measure of toughness. The CVN USE and DT USE are measures of notch toughness and thus are often related to fracture toughness values. Therefore, the percent RA values will not be used in the comparisons developed later in the paper, while the CVN USE and DT USE will be used extensively.

The J-integral results are displayed in Figs. 4–7 in the form of the J versus Δa R-curves for the A516, A533B, HY-130, and A543 steels, respectively. An average "blunting line" for all of the data of the particular steel grade is also given for reference purposes. Actual material tensile properties were used in each actual J_{Ic} determination. The "blunting line" is determined from

$$J = 2\,FS\,\Delta a \qquad (2)$$

where FS is the flow stress [$FS = (0.2YS + UTS)/2$]. It can be roughly

TABLE 2—Conventional mechanical properties.

Steel	Orientation	0.2 Yield Strength, ksi		Ultimate Tensile Strength, ksi		Percent Reduction of Area		Charpy V-Notch Upper-Shelf Energy, ft-lb		Dynamic Tear Upper-Shelf Energy, ft-lb	
		CON[a]	CaT[b]	CON	CaT	CON	CaT	CON	CaT	CON	CaT
A516	L(LT)[c]	48.0	51.7	78.5	79.1	66.7	66.3	80	139	710	1025
	T(TL)[d]	46.0	47.4	74.5	78.6	63.2	66.3	54	139	485	1045
A533B	L(LT)	64.4	59.9	85.4	85.3	71.4	72.2	128	177	1160	1420
	T(TL)	63.2	61.7	84.9	85.7	57.5	66.3	70	144	600	1085
	S(SL)[e]	63.7	59.6	80.2	82.7	20.3	38.6	41	117	335	850
HY-130	L(LT)	130.6	134.3	138.6	138.3	70.4	71.7	103	95	870	855
	T(TL)	130.6	132.8	138.3	136.8	63.4	67.3	54	81	435	690
		CL[f]	QL[g]	CL	QL	CL	QL	CL	QL	CL	QL
A543	L(LT)	89.9	96.5	107.7	113.8	61.6	71.9	74	84	755	...
	T(TL)	89.9	96.6	108.1	113.5	37.1	69.0	59	63	465	...
	S(SL)	87.0	96.9	105.3	112.7	26.2	64.7	30	56	335	...

[a] Conventional steelmaking practice.
[b] Calcium treatment.
[c] Longitudinal.
[d] Transverse.
[e] Through-thickness.
[f] Centerline.
[g] Quarterline.

Conversion factors: MPa = ksi × 6.895.
Joules = ft-lb × 1.356.

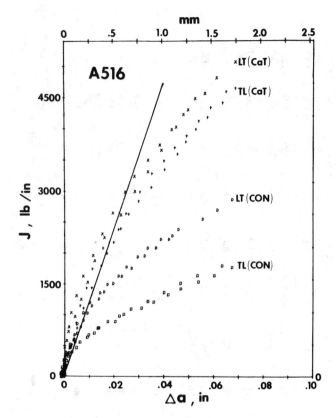

FIG. 4—*J versus Δa R-curves for A516 steels. Line indicates average "blunting line" for these steels; for reference purposes, Eq 2. ($KN/m = lb/in. \times 0.1751$).*

noted by examining Figs. 4–6 that for each steel grade the R-curve is shifted to higher J levels, indicating more toughness for the CaT steels. This is particularly indicated in the TL and SL orientation. Figure 7 shows that the R-curves for the QL A543 steel indicate tougher behavior than at the CL in all testing orientations.

In this investigation the critical value of J is determined by two methods, namely, that J_{Ic} determined using the graphical analysis technique $(J_{Ic})_G$, [14], and that determined at the point of first load drop from the load-displacement curve, $(J_{Ic})_{FLD}$ [15]. In the graphical-analysis method the $(J_{Ic})_G$ was taken at the intersection of the "blunting line" and the visually determined best-fit line through points having the required amount of crack extension according to Ref 14. The $(J_{Ic})_{FLD}$ was established at the point of maximum load, since all of the load-displacement curves were smooth with no discontinuities. These determinations are given in Table 3. Those $(J_{Ic})_G$ values that meet the suggested validity requirements [14],

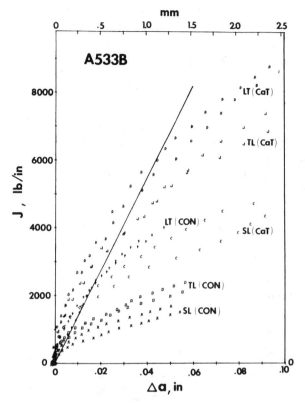

FIG. 5—J versus Δa R-curves for A533B steels. Line indicates average "blunting line" for these steels; for reference purposes, Eq 2. ($KN/m = lb/in. \times 0.1751$).

including that for specimen size, are noted. The specimen size requirement demands that

$$B, b \geq 25J/FS \tag{3}$$

where B is the specimen thickness and b the initial remaining ligament length of the specimen. Those $(J_{Ic})_{FLD}$ points which would also meet this specimen size requirement are also noted in Table 3. All of the specimen tests which did not meet the validity or specimen size requirements failed due to the ligament length, b, being undersized.

The applicable K_{Ic} values were calculated for each of the J_{Ic} determinations, using Eq 1 with a value for ν of 0.3 and a value for E of 207 000 MPa (30×10^6 psi), and are listed in Table 4. Also, the Rolfe-Novak-Barsom (RNB) upper-shelf K_{Ic}-CVN correlation [18,19] was used to determine an

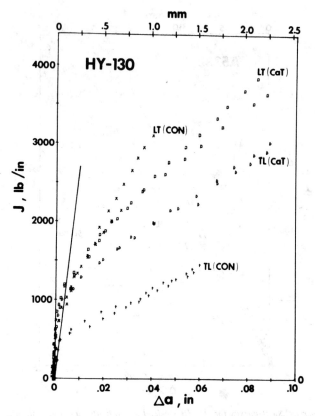

FIG. 6—J versus Δa R-curves for HY-130 steels. Line indicates average "blunting line" for these steels; for reference purposes, Eq 2. (KN/m = lb/in. × 0.1751).

additional K_{Ic} value using the CVN USE values of Table 2. This correlation is

$$\left(\frac{K_{Ic}}{0.2YS}\right)^2 = 5\left(\frac{CVN\ USE}{0.2YS} - 0.05\right) \qquad (4)$$

where 0.2YS is in ksi, CVN USE in ft-lb, and K_{Ic} in ksi $\sqrt{in}$.

Discussion

Rating Steel Toughness

In order to determine how each of the toughness measurement techniques rates the quality of the steels, a normalizing technique was used.

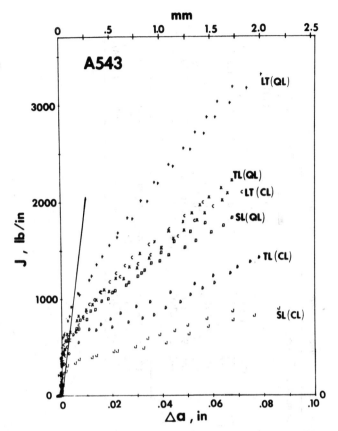

FIG. 7—J versus Δa R-curves for A543 steels. Line indicates average "blunting line" for these steels; for reference purposes, Eq 2. (KN/m = lb/in. × 0.1751).

The normalizing method used determined ratios of toughness levels for each test technique and then compared these ratios between techniques. Two types of ratios were calculated. The first were anisotropy ratios. These quantify the amount of anisotropy of a particular property in a steel by ratioing the property levels in each testing orientation. This is done by dividing the TL or SL value of a property by the LT value. Thus a steel with a great deal of anisotropy such as the CON A533B material would have anisotropy ratios TL/LT and SL/LT of 0.55 and 0.32, respectively, for CVN USE, while the CaT A533B steel ratios would be 0.81 and 0.66, respectively, for the same property. This indicates the more isotropic nature of the CaT steel.

The second type of ratios determined were quality ratios. These quantify the amount of improvement of a particular property by rating the property in a particular orientation of the steel in the poorer quality to that in the

TABLE 3—*Comparison of graphical and first load drop J_{Ic} determinations.*

Steel	Orientation	Conventional Steelmaking Practice, lb/in.		Calcium Treated, lb/in.	
		Graphical (avg)	FLD[a] (avg)	Graphical (avg)	FLD (avg)
A516	LT[b]	1115, 1310 (1213)	1717, 1801 (1759)	2400, 2850 (2625)	3337, 3011 (3174)
	TL[c]	570, 615 (593)[e]	961, 855 (908)[f]	2340, 2100 (2220)	3016, 2711 (2864)
A533B	LT	2525, 2710 (2618)	1770, 1734 (1732)	6540, 5690 (6115)	4770, 4180 (4475)
	TL	965, 1080 (1023)	944, 869 (907)	4480, 4350 (4415)	3839, 3431 (3635)
	SL[d]	625, 525 (575)	623, 507 (565)[f]	2260, 1970 (2115)	2712, 2294 (2503)
HY-130	LT	1120, 972 (1046)	889, 960 (925)[f]	1120, 1320 (1270)[e]	1328, 1277 (1303)[f]
	TL	572, 484 (528)[e]	682, 593 (638)[f]	1108, 1152 (1130)[e]	1265, 1157 (1211)[f]
		Centerline		Quarterline	
A543	LT	728, 784 (756)[e]	881, 833 (857)[f]	976, 992 (984)[e]	1016, 1060 (1038)[f]
	TL	520, 622 (571)[e]	612, 583 (598)[f]	768, 772 (770)[e]	762, 793 (778)[f]
	SL	304, 360 (332)[e]	353, 392 (373)[f]	644, 748 (696)[e]	648, 799 (724)[f]

[a] First load drop.
[b] Longitudinal.
[c] Transverse.
[d] Through thickness.
[e] Valid per graphical technique, Ref 14.
[f] Meets specimen size requirement at point of FLD.
Conversion factor: KN/m = lb/in. × 0.1751.

TABLE 4—K_{Ic} determinations in ksi $\sqrt{in.}$

Steel	Orientation	Conventional Steelmaking Practice[a] (CON)		Calcium Treated[a] (CaT)		(K_{Ic}) RNB[b]	
		$(K_{Ic})_G$ (avg)	$(K_{Ic})_{FLD}$ (avg)	$(K_{Ic})_G$ (avg)	$(K_{Ic})_{FLD}$ (avg)	CON	CaT
A516	LT[c]	192, 208 (200)	238, 244 (241)	281, 307 (294)	332, 315 (324)	132	183
	TL[d]	137, 142 (140)f	178, 168 (173)g	278, 263 (271)	315, 299 (307)	106	175
A533B	LT	289, 299 (294)	242, 239 (241)	465, 433 (449)	397, 371 (384)	196	222
	TL	178, 189 (184)	176, 169 (173)	384, 379 (382)	356, 336 (346)	142	203
	SL[e]	144, 132 (138)	143, 129 (136)g	273, 255 (264)	299, 275 (287)	108	180
HY-130	LT	192, 179 (186)	171, 178 (175)g	201, 209 (205)f	209, 205 (207)f	251	243
	TL	137, 126 (132)f	150, 140 (145)g	191, 195 (193)f	204, 195 (200)f	196	222
		CL		QL		CL	QL
A543	LT	155, 161 (158)f	171, 166 (169)g	179, 181 (180)f	183, 187 (185)f	177	196
	TL	131, 143 (137)f	142, 139 (141)g	159, 160 (160)f	159, 162 (161)f	157	168
	SL	100, 109 (105)f	108, 114 (111)g	146, 157 (152)f	146, 162 (154)f	106	158

[a] Determinations made from J_{Ic}.
[b] Determinations made using Rolfe-Novak-Barsom Charpy V-notch upper-shelf correlation, Refs 18 and 19.
[c] Longitudinal.
[d] Transverse.
[e] Through-thickness.
[f] Valid per graphical technique (G), Ref 14.
[g] Meets specimen size requirement at point of first load drop (FLD).
Conversion factor: MPa$\sqrt{m}$ = ksi$\sqrt{in.}$ × 1.099.

better-quality condition. Therefore, for the A516, A533B, and HY-130 steels, the ratios compare CON and CaT toughness levels, CON/CaT. In the A543 steel, this ratio would be CL/QL. Thus, again using the A533B steel CVN USE for comparison, the ratios would be 0.72, 0.49, and 0.35, respectively, for the LT, TL, and SL orientations. This shows that the largest improvement in CVN USE is obtained by the CaT steel in the SL orientation.

In comparing the ratios for each testing technique it was soon found that the CVN USE and DT USE ratios compared well with each other and were closer to the J_{Ic} ratios than the K_{Ic} ratios. This is plausible since these three parameters are all energy related. Therefore the two J_{Ic} determination ratios, $(J_{Ic})_G$ and $(J_{Ic})_{FLD}$, were compared with each of the other toughness ratios as shown in Fig. 8. The ratios for the K_{Ic} determinations from each J_{Ic} analysis (Eq 1) were compared with the ratios of K_{Ic} values calculated using the RNB correlation (Eq 4) and are also shown in Fig. 8. All six of these graphs also have 45-deg lines for reference purposes.

Generally it can be concluded from these graphs that the ratios obtained from conventional toughness tests are not the same as those obtained by J_{Ic} testing. More specifically, the J_{Ic} test is shown to be more sensitive to changes in inclusion structure than either CVN or DT tests. This is indicated by the general position of the data points to the right of the 45-deg perfect agreement lines in all six plots in Fig. 8. This is shown particularly for the invalid J_{Ic} data points and to a lesser extent for the valid data. For example, in Fig. 8a it is generally shown that when the $(J_{Ic})_G$ ratio is 0.4 the CVN USE ratio is 0.6, demonstrating that the $(J_{Ic})_G$ tends to show more anisotropy in a steel and larger differences between quality levels of a steel than the CVN USE. What this means specifically to these comparisons is that for three of the steels the J_{Ic} determinations demonstrate that CaT results in more improvement in steel quality than would be identified by the CVN or DT tests. For the A543 steel the J_{Ic} determinations indicate a larger difference between CL and QL quality than shown by the conventional tests.

These differences result even though the appearance of the fracture mode in each of the test specimens is the same. The appearance of the manganese sulfide Type II inclusion formations on an SL SSJ fracture of the CON A533B steel is shown in Fig. 9 and the calcium modified inclusions on a SL SSJ fracture of the CaT A533B steel is shown in Fig. 10. Similar fracture appearance has also been noted on tensile and CVN fractures [2] and also on DT fractures.

There are two reasons why the J_{Ic} data are more sensitive to inclusion effects. Most evident is that a sharp crack starter is used in the J_{Ic} test compared with the machined notch in the CVN and the pressed notch in the DT tests. This may therefore indicate a different interaction between the crack or notch and the inclusions in the steel. An additional reason for

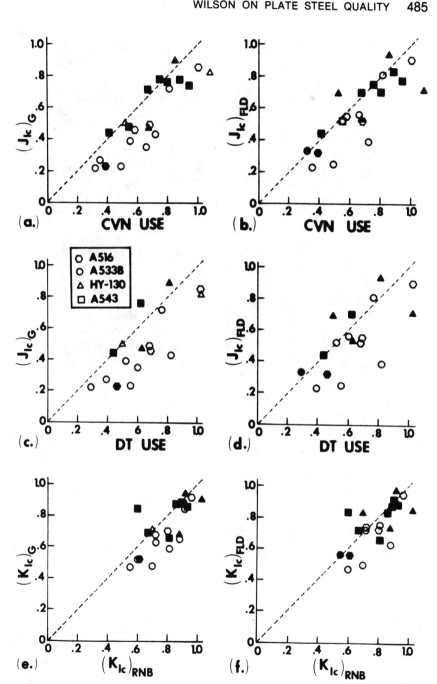

FIG. 8—*Comparison of anisotropy and quality ratios determined from the various toughness measurements. Lines indicate equality between the ratios. Solid points are valid J_{Ic} results.*

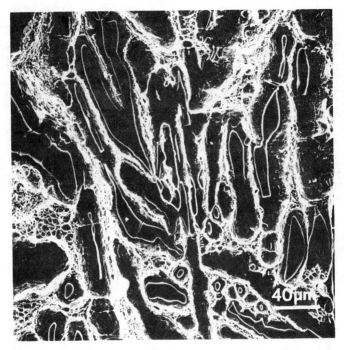

FIG. 9—*Scanning electron microscope fractograph of J-integral specimen in SL orientation of CON A533B steel showing manganese sulfide Type II inclusions.*

the differences just mentioned is that the CVN and DT tests are conducted at impact or high loading rates, while the J_{Ic} tests are performed at a static or low loading rate. This may be the controlling influence since the data points for the lower-strength steels (A516 and A533B), which are more loading rate sensitive, are more prominently shifted to the right in the graphs in Fig. 8. This may indicate that the rate sensitivity of the steel is the reason for the different effect of inclusion structure for the J_{Ic} data. This could only be checked, however, by performing J_{Ic} tests at impact loading rates.

Comparing J_{Ic} Determination Methods

In this paper two methods have been used to determine J_{Ic}, namely, the graphical technique and the first load drop method. Figure 11 presents a plot comparing the K_{Ic} values obtained from the respective J_{Ic} determinations for each of these methods. Also given on the graph are the 45-deg reference line and ± 10 percent lines surrounding this line. This compilation reveals that, generally speaking, using ± 10 percent to account for possible experimental and analytical errors, the $(K_{Ic})_G$ is equal to the $(K_{Ic})_{FLD}$ when

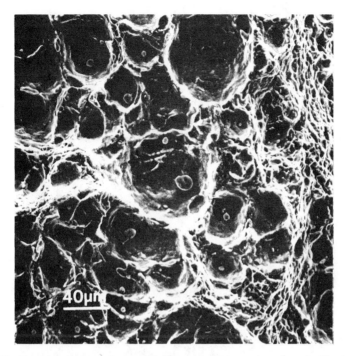

FIG. 10—*Scanning electron microscope fractograph of* J-*integral specimen in SL orientation of CaT A533B steel showing calcium-modified, shape-controlled inclusions.*

the J_{Ic} data being used are valid. When the J_{Ic} data are invalid, that is, not meeting specimen size requirements, there is more of a deviation from this relationship. If this relationship—indicating that the two J_{Ic} determination methods give identical results for valid data—holds for a number of other steels or metals, the J-integral testing and analysis procedure would be simplified significantly. This is because the FLD number can be obtained from a single specimen without the additional instrumentation required by the SSJ technique. The results given in Fig. 11 therefore suggest that other materials should be examined for the presence of this correlation.

If the foregoing relationship exists, it would also appear to allow comment on the specimen size requirement for J_{Ic} determinations [14]. In particular, the two CON HY-130 LT orientation J_{Ic} values and the two CON A533B SL orientation results would appear to be close to validity judging by the fact that their graphical and FLD K_{Ic} values are close or within the scatterband of Fig. 11. Also, on the other hand, the valid A516 data points falling outside the scatterband suggests that possibly these points should not be considered valid. These observations are examined next.

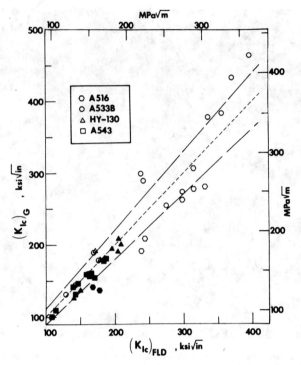

FIG. 11—*Comparison of K_{Ic} values calculated for various J_{Ic} analytical methods, graphical and first load drop. Dashed line represents equality with scatterband around it indicating ± 10 percent. Solid points are valid J_{Ic} results.*

Comments on J_{Ic} Analysis Guidelines

The graphical J_{Ic} analysis method used here is that suggested by the guidelines of Clarke [*14*]. The specimen size requirement in these guidelines is covered by Eq 3. If the factor of 25 in this equation had been 18, the two HY-130 and two A533B values mentioned in the preceding section would be considered valid. However, this would only accentuate the problem that appears to be present for the two valid A516 CON values. This suggests that possibly more weight should be given to the strength level of the steel in Eq 3. This could be done by making the equation similar to the form of the ASTM E 399-74 specimen size requirement, for example

$$B, b \geq \frac{JE}{\lambda (FS)^2} \tag{5}$$

where λ is a number like 15. This equation would invalidate the questioned A516 results while allowing the two HY-130 values to be considered valid; however, the two A533B values would also be invalid.

By way of observation, it was also found that generally the points on the J versus Δa R-curve at low Δa values did not fall on the "blunting line." These points tended at the start to be at zero crack extension and then to come above the blunting line. This has been reported previously also by Clarke et al using the SSJ technique [15]. The lack of correspondence between the data points and the blunting line also appeared to be independent of whether the test turned out to give a valid or invalid J_{Ic}.

An additional remark that can be made on the graphical J_{Ic} procedure is that the points closest to the blunting line should be given more weight in the determination of the line to extrapolate back to the blunting line to obtain the J_{Ic} value. This is shown in Fig. 12 for a CON A533B SL-oriented specimen. If an average line using all of the points were used, a J_{Ic} of 4.85 N/m (850 lb/in.) would have been determined. However, if only the nearest points to the blunting line were considered, the J_{Ic} of 3.54 N/m (620 lb/in.) would be found. The latter result is also closer to the FLD value.

Comment on CVN Upper-Shelf Correlation

The RNB CVN upper-shelf correlation (Eq 4) was found to be initially applicable to steels with 0.2YS greater than 758 MPa (110 ksi) [18,19]. It

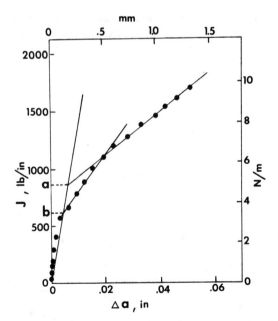

FIG. 12—*J versus Δa R-curve for CON A533B steel in SL orientation. If all data points are used, a J_{Ic} at Point "a" of 4.85 N/m (850 lb-in.) is obtained. If only points at lower Δa's are used, a J_{Ic} at Point "b" of 3.54 N/m (620 lb-in.) is determined.*

has since been found to be useful for rotor forging steel CVN-K_{Ic} correlations with 0.2YS values down to 552 MPa (80 ksi) [20]. It also has been found useful for A533B steels with 0.2YS values as low as 414 MPa (60 ksi) [21]. In addition, Paris has commented that the J-integral concept has made this empirical CVN-K_{Ic} correlation appear more reasonable from a technical standpoint [22]. Figure 13 shows plots of K_{Ic} determined by the J_{Ic} methods plotted versus the K_{Ic} obtained from the RNB correlation. It can be immediately noted that almost all of the *valid* K_{Ic} from J_{Ic} values are lower than those predicted by the correlation. This is explainable since the J_{Ic} determination is made at the point of crack initiation, while the K_{Ic} is determined after an allowable amount of crack extension (5 percent secant offset). Thus the J_{Ic}-determined values should be lower. The *invalid* K_{Ic} from J_{Ic} points, on the other hand, tend to be above the line. This is most likely a result of the large amounts of plasticity involved in these fractures. In addition, the points above the line tend to be from the lower-strength steels, A533B and A516, which would be expected to have a significant effect of loading rate on upper-shelf toughness. Therefore, the RNB correlation would not be expected to hold up, because the K_{Ic} test is performed at a slow loading rate and the CVN at a fast loading rate.

Conclusions

1. The J_{Ic} values appear to be more sensitive to changes in inclusion structure than either the CVN or DT tests. Therefore J_{Ic} tends to indicate

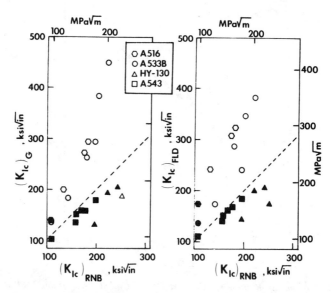

FIG. 13—*Comparison of K_{Ic} values obtained from J_{Ic} determinations and from CVN upper-shelf correlation. Dashed lines indicate equality. Solid points are valid J_{Ic} results.*

that there is more improvement by calcium treatment than would be indicated by these conventional tests.

2. The J_{Ic} determinations obtained by both the graphical method and first load drop technique were found to be the same for valid J_{Ic} tests.

3. It is suggested that the J_{Ic} specimen size requirement equation be modified to use a factor of 18 rather than 25 or use a new requirement which places more emphasis on the strength level of the steel, or use both.

4. The K_{Ic} determined by the Rolfe-Novak-Barsom CVN-K_{Ic} upper-shelf correlation was found to be conservative when compared with the K_{Ic} obtained from J_{Ic}, for invalid data. For valid data, the K_{Ic} from J_{Ic} gives a lower result than that predicted by the CVN correlation.

Acknowledgments

The author would like to acknowledge especially the contributions of J. Keith Donald, vice president, and David W. Schmidt, staff engineer, of the Del Research Division of the Philadelphia Suburban Corp. for performing the single-specimen J-integral tests. In addition, their general comments, suggestions, and assistance during this research program are greatly appreciated.

The guidance and comments provided by John A. Gulya during the experimental and writing aspects of this research are also appreciatively acknowledged.

Disclaimer

It is understood that the material in this paper is intended for general information only and should not be used in relation to any specific application without independent examination and verification of its applicability and suitability by professionally qualified personnel. Those making use thereof or relying thereon assume all risk and liability arising from such use or reliance.

References

[1] Wilson, A. D., "The Interaction of Advanced Steelmaking Techniques, Inclusions, Toughness and Ductility in A533B Steels," American Society for Metals, Technical Report System No. 76-02, 1976.
[2] Wilson, A. D., "The Effect of Advanced Steelmaking Techniques on the Inclusions and Mechanical Properties of Plate Steels," presented at American Institute of Mining Metallurgical and Petroleum Engineers Annual Meeting, Atlanta, Ga., March 1977, to be published.
[3] Wilson, A. D., "The Influence of Thickness and Rolling Ratio on the Inclusion Behavior in Plate Steels," presented at American Society for Metals Materials Conference, Chicago, Ill., Oct. 1977, to be published in *Metallography*, International Metallographic Society.

[4] Wilson, A. D., *Journal of Pressure Vessel Technology, Transactions*, American Society of Mechanical Engineers, Vol. 99, Series J, No. 3, Aug. 1977, pp. 459-469.
[5] Rice, J. R., *Journal of Applied Mechanics, Transactions*, American Society of Mechanical Engineers, Vol. 35, Series E, June 1968, pp. 379-386.
[6] Begley, J. A. and Landes, J. D. in *Fracture Toughness, ASTM STP 514*, American Society for Testing and Materials, 1972, pp. 1-23.
[7] Landes, J. D. and Begley, J. A. in *Fracture Toughness, ASTM STP 514*, American Society for Testing and Materials, 1972, pp. 24-39.
[8] Landes, J. D. and Begley, J. A. in *Fracture Analysis, ASTM STP 560*, American Society for Testing and Materials, 1974, pp. 170-186.
[9] Bucci, R. J., Paris, P. C., Landes, J. D., and Rice, J. D. in *Fracture Toughness, ASTM STP 514*, American Society for Testing and Materials, 1972, pp. 40-69.
[10] Rice, J. R., Paris, P. C., and Merkle, J. G. in *Progress in Flaw Growth and Fracture Toughness Testing, ASTM STP 536*, American Society for Testing and Materials, 1973, pp. 231-245.
[11] Logsdon, W. A. in *Mechanics of Crack Growth, ASTM STP 590*, American Society for Testing and Materials, 1976, pp. 43-60.
[12] Marandet, B. and Sanz, G., "Characterization of the Fracture Toughness of Steels by the Measurement with a Single Specimen of J_{Ic} and the Parameter K_{Bd}," presented at the Tenth National Symposium on Fracture Mechanics, American Society for Testing and Materials, 23-25 Aug. 1976.
[13] Logsdon, W. A. and Begley, J. A., *Engineering Fracture Mechanics*, Vol. 9, 1977, pp. 461-470.
[14] Clarke, G. A., "Recommended Procedure for J_{Ic} Determination," presented at the ASTM E24.01.09 Task Group Meeting, Norfolk, Va., American Society for Testing and Materials, March 1977.
[15] Clarke, G. A., Andrews, W. R., Paris, P. C., and Schmidt, D. W. in *Mechanics of Crack Growth, ASTM STP 590*, American Society for Testing and Materials, 1976, pp. 27-42.
[16] Merkle, J. G. and Corten, H. T., *Journal of Pressure Vessel Technology, Transactions*, American Society of Mechanical Engineers, Vol. 96, Series J, No. 4, Nov. 1974, pp. 286-292.
[17] Donald, J. K., "Rotational Effects on Compact Specimens," presented at the ASTM E24.01.09 Task Group Meeting, Norfolk, Va., American Society for Testing and Materials, March 1977; available from Del Research Division, Philadelphia Suburban Corp., 427 Main St., Hellertown, Pa. 18055.
[18] Barsom, J. M. and Rolfe, S. T. in *Impact Testing of Metals, ASTM STP 466*, American Society for Testing and Materials, 1970, pp. 281-302.
[19] Rolfe, S. T. and Novak, S. R. in *Review of Developments in Plain Strain Fracture-Toughness Testing, ASTM STP 463*, American Society for Testing and Materials, 1970, pp. 124-159.
[20] Begley, J. A. and Logsdon, W. A., "Correlation of Fracture Toughness and Charpy Properties for Rotor Steels," Westinghouse Research Laboratories Scientific Paper 71-1E7-MSLRF-P1, Pittsburgh, Pa., July 1971.
[21] Sailors, R. H. and Corten, H. T. in *Fracture Toughness, ASTM STP 514*, American Society for Testing and Materials, 1972, pp. 164-191.
[22] Paris, P. C., written discussion to Ref 6, pp. 20-21.

W. L. Server[1]

Static and Dynamic Fibrous Initiation Toughness Results for Nine Pressure Vessel Materials

REFERENCE: Server, W. L., "**Static and Dynamic Fibrous Initiation Toughness Results for Nine Pressure Vessel Materials,**" *Elastic-Plastic Fracture, ASTM STP 668*, J. D. Landes, J. A. Begley, and G. A. Clarke, Eds., American Society for Testing and Materials, 1979, pp. 493-514.

ABSTRACT: The upper-shelf toughness regime corresponds to the normal operating temperature for ferritic nuclear pressure vessels. However, actual fibrous initiation toughness data have not been available for a wide variety of heats of steels and weldments. In fact, much of the data available have assumed that fracture initiation occurred at maximum load (equivalent energy approach), which is *not* the case for most materials tested on the upper shelf. Multispecimen (25.4 mm thickness compact or bend) initiation tests were, therefore, performed in the upper shelf temperature region to determine J_{Ic} for five base metal (two A533B-1, one A508-2, and two A302B) and four weld metal (two manual arc and two submerged arc) heats of nuclear ferritic steel. All of the heats were investigated under quasi-static loading (loading time ~ 100 s) and under either closed-loop hydraulic loading (loading time ~ 100 ms) or controlled impact loading (loading time ~ 1 ms). For all materials investigated, crack initiation occurred prior to maximum load, and initiation toughness values (K_{Jc}) were up to 50 percent less than the equivalent energy (maximum load) toughness values. The initiation toughness results at 177°C also increased with decreasing loading time for all but one heat tested. HSST Plate 02 (A533B-1 steel) was tested at 71°C using all three loading rates; there were little differences observed for the Plate 02 tests due to loading rate, although the intermediate loading rate results were slightly lower than the rest. The data were analyzed using a simple statistical approach to obtain approximate confidence limits for the J_{Ic} values obtained. A comparison of J_{Ic} values obtained using three-point and nine-point fibrous crack length averages across the specimen thickness was made. The nine-point average generally gave higher values of J_{Ic}, but the variance about the regression line was not consistently lower than for the three-point average.

KEY WORDS: fractures (materials), mechanical properties, test, *J*-contour integral, pressure vessels, steels, statistics, initiation toughness, crack propagation

Fracture-safe design analyses are generally based upon a critical fracture parameter which is measurable in the laboratory and can be assumed

[1] Vice president, Fracture Control Corp. Goleta, Calif. 93017.

to be equivalent for both the structure and the laboratory specimen. The need for convenient, small specimens and relatively easy laboratory test procedures is obvious. In particular, linear elastic fracture mechanics (LEFM) and its critical fracture criterion, K_{Ic}, are widely accepted for fracture-safe design purposes. However, the laboratory method used to measure K_{Ic} per the ASTM Test Method for Plane-Strain Fracture Toughness of Metallic Materials (E 399-74) requires stringent limitations on minimum specimen size requirements. These size requirements may not be overly restrictive for high-strength materials, but for lower-strength steels the size requirements demand specimens which are often larger than the full section size of the structure. Therefore, the J-integral concept has been proposed in the United States [1][2] to provide an extension of LEFM for large-scale plastic behavior, both in the laboratory test and in the structure itself.

The J-integral is basically a two-dimensional, path-independent line integral applicable to both elastic and elastic-plastic response when coupled with a plastic deformation theory. The J-integral can also be interpreted in terms of a potential energy difference per unit thickness (∂U) between two identically loaded bodies having infinitesimally differing crack length (∂a), that is

$$J = -\left.\frac{\partial U}{\partial a}\right|_{\delta = \text{constant}} \quad (1)$$

Rice et al [3] have developed a simple, approximate expression for J when deeply cracked specimens are loaded in bending

$$J = \frac{2\int_0^{\delta_1} P d\delta}{Bb} \quad (2)$$

where the integral term is the area under the load-displacement ($P - \delta$) curve to some deflection δ_1, B is the specimen thickness, and b is the specimen ligament depth. The value of J at fracture initiation is defined as the critical fracture parameter, J_{Ic}. The original derivation of Eq 2 for three-point bend specimens referred only to the portion of the load-displacement curve due to a crack; therefore, the uncracked body energy should be subtracted out. However, recent analytical and empirical results indicate that Eq 2 is more accurate when the uncracked body energy is included [4,5]. Merkle and Corten [6] have also presented a correction for compact specimen testing which accounts for the tension component. Again, the inclusion of the compact specimen correction has not been shown to be empiri-

[2]The italic numbers in brackets refer to the list of references appended to this paper.

cally necessary. In the work that follows, the calculation of J is based on Eq 2 using the complete load-displacement curve; the effects of including the corrections are discussed at the end of the paper.

Experimental Procedure

Materials and Test Specimens

The nuclear pressure vessel materials used in this investigation are listed in Table 1 along with their relevant mechanical properties [7-12]. The materials were chosen based upon their relatively low level, within a material group, of equivalent energy (maximum load) fracture toughness as determined from instrumented precracked Charpy tests [13]. In particular, the heats EN (A302B steel) and BAS (submerged arc weld metal, Linde 80 flux) are very low upper-shelf materials with high copper levels (0.20 and 0.33 weight percent, respectively). The low initial shelves plus the high copper levels make these heats highly suspect after in-service neutron irradiation damage.

Compact and bend test specimens were machined at quarter thickness from plate and forging materials and not within 12.7 mm of the surfaces for weld metal materials. All specimens were notched and fatigue pre-cracked to a crack depth to specimen width ratio (a/w) of between 0.5 and 0.6 with the crack running in the rolling, forging, or welding direction (TL orientation; see ASTM E 399-74). The compact specimens were machined so that the clip gage could be mounted at the specimen load-line.

Testing and Analysis Procedure

The general procedure for the J initiation testing followed the early guidelines proposed by the ASTM Task Group on Elastic-Plastic Fracture Criteria [14]. The first eight heats of steel listed in Table 1 were tested under quasi-static loading (~ 100 s loading to J_{Ic}) using 25.4-mm-thickness compact specimens at 177°C. These same steels were then tested either at 177°C as dynamic 25.4-mm-thickness compact specimens (~ 100 ms loading to J_{Ic}; heats CJ, BBB, and BAS) or dynamic 25.4-mm-thickness bend specimens (~ 1 ms loading to J_{Ic}, heats EN, NA, EG, and EK). The HSST Plate 02 material was tested at room temperature under quasi-static loading and at 71°C under all three rates of loading using 25.4-mm-thickness specimens.

Compact specimens were loaded to different displacement values in a closed-loop MTS test machine (89-kN capacity) in displacement (ram) control for both static and dynamic tests. For the dynamic tests, no overshoot in load or deflection was allowed; therefore, a change in rate during the last few milliseconds of loading occurred. The real effect of this decrease in loading rate is not known, but there is definitely a difference in the test results as compared with static loading [15].

TABLE 1—Mechanical properties of the steels investigated (TL orientation).

Heat Designation	Material	Room Temperature Tensile				NDTT, °C	RT_{NDT}, °C	Charpy Upper Shelf Level, J	Comments
		Yield MPa	Ultimate MPa	Elongation, %					
EG [7,8]	A533B-1	486	645	24.8[a]		−12	−12	125	241-mm-thickness plate (double austenitized)
BBB [7,9]	A508-2	488[d]	626[d]	26.5[d]		−7	−7	152	254-mm-thickness forging (nozzle dropout)
EN [7,8]	A302B	465	625	25.2[a]		−18	...	60	152-mm-thickness plate (ASTM surveillance correlation heat)
NA [10]	A302B	477	616	...		−23	...	71	162-mm-thickness plate (special EPRI-NRL heat)
EK [7,8]	MMA weld metal (A533B-1 base metal)	556	632	26.2[b]		−46	−40	137	shielded metal arc weld, 127 mm thickness, E8018-NM filler/metal (6.4 mm electrode), 3.5 MJ/m heat input
BKM [7,9]	MMA weld metal (A508-2 base metal)	447[d]	538[d]	27.2[c,d]		−62	−62	193	manual metal arc weld, 127 mm thickness, E8015-C3 filler metal (4.8 mm electrode), 1.9 to 2.2 MJ/m heat input
CJ [7,11]	SA weld metal (A533B-1 base metal)	545[d]	630[d]	27.2[d]		−57	−57	174	submerged-arc weld, 300 mm thickness, Linde 0091 flux, 2.8 to 4.5 MJ/m heat input
BAS [7,9]	SA weld metal (A508-2 base metal)	490[d]	626[d]	24.0[c,d]		−34	−9	104	submerged-arc weld, 102 mm thickness, Linde 80 flux, 4.0 MJ/m heat input
HSST Plate 02 [12]	A533B-1	481	642	25.4[a]		−29	−15	137	305-mm-thickness plate

NOTE—NDTT = nil ductility transition temperature; RT_{NDT} = reference transition temperature.
[a]Based on a 50.8-mm gage length.
[b]Based on a 25.4-mm gage length.
[c]Based on a 31.8-mm gage length.
[d]Properties measured in the LT orientation.

The dynamic bend initiation technique developed for a drop tower impact test is shown in Fig. 1. Hardened-steel deflection stops were used to stop the falling tup at differing amounts of deflection. When the tup strikes the deflection blocks, a sudden increase in the load signal occurs, marking the stopping event. A typical load signal is shown in Fig. 2. It should be noted that a few millimeters of deflection can still occur due to the elastic brinelling in the stop blocks and the tup. However, the values of J were calculated from the load-time trace when the tup hit the deflection stop; these resulting J values are therefore slightly conservative (low). The drop tower mass for the 25.4-mm-thickness specimens was 961 kg, and the impact velocity was 1.41 m/s. This impact velocity meets the current requirements for reliable instrumented impact testing (see Ref *13* and *16-20* for review of these requirements).

After loading, each specimen was heat-tinted at 288°C for 15 to 30 min.

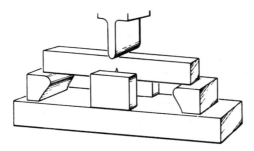

FIG. 1—*Schematic diagram of the drop tower stop-block arrangement.*

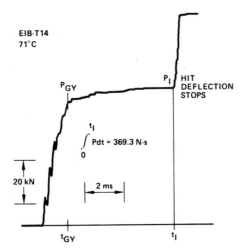

FIG. 2—*Impact test record for HSST Plate 02 specimen tested at 71°C.*

Specimens were then broken apart at $-70°C$ to reveal the amount of fibrous crack growth. The amount of crack growth was taken to include all extension from the end of the fatigue crack to the end of the heat-tinted marking (thus including the stretch zone). J values were then calculated using Eq 2. For the impact bend tests, no direct measure of displacement was possible, and the value of the load-displacement integral was determined from the load-time trace as follows:

$$W_1 = V_0 \int_0^{t_1} P\,dt \left[1 - \frac{V_0 \int_0^{t_1} P\,dt}{4E_0} \right] \quad (3)$$

where

W_1 = velocity-corrected energy value representing the total specimen plus machine energy consumed up to time t_1 (when the stop block is hit),
V_0 = initial impact velocity, and
E_0 = total energy available $(1/2\ MV_0^2)$.

The value of W_1 is then corrected for extraneous compliance contributions [21,22], giving a value of energy (E_1) to be used in Eq 2

$$E_1 = W_1 - \frac{P_1^2}{2} \left[\frac{V_0 t_{GY}}{P_{GY}} - \frac{V_0^2 t_{GY}^2}{8E_0} - \frac{C_{ND}}{EB} \right] \quad (4)$$

The time and load at general yield (t_{GY} and P_{GY}) are used to determine the total system compliance, and the known nondimensional specimen compliance (C_{ND}) from finite element and boundary collocation studies [23] is then used to correct for the elastic extraneous energy contribution.

Once the crack extensions (Δa) have been measured using either a three-point or a nine-point average for each specimen tested and the J values calculated, a plot of J versus Δa is constructed. The straight line representing crack blunting is assumed to be known with certainty and is drawn with a slope equal to $2\sigma_f$ (see Fig. 3), where σ_f is the flow stress indicative of the specimen testing temperature and loading rate (equal to the average of the yield and ultimate stresses) [14]. Values of σ_f indicative of impact loading were estimated from instrumented standard Charpy V-notch results [7-13], which were analyzed using extrapolated slipline field solutions which include the indentor [24]

$$\sigma_f = (0.0467\ \text{mm}^{-2}) \times \left(\frac{P_{GY} + P_M}{2} \right) \quad (5)$$

where P_{GY} and P_M are the general yield and maximum loads from the

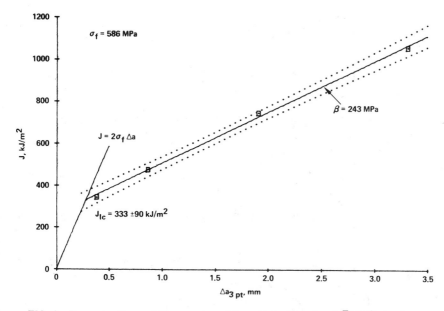

FIG. 3—*Regression line and 95 percent confidence limits (based on $J/\bar{J} \times S^2$) for dynamic bend results of heat EG at 177°C.*

instrumented Charpy trace. The values of σ_f for the dynamic compact tests were estimated by interpolating the static tensile results and the impact Charpy results on the basis of loading time [13]

$$\sigma_f = 0.6\,(\sigma_f,\text{ impact}) + 0.4\,(\sigma_f,\text{ static}) \tag{6}$$

The values of σ_f obtained at the appropriate test temperatures and loading rates are listed in Tables 2 and 3. The best line fit of the experimental J versus Δa values (not on the blunting line) which intersects with the $2\sigma_f$ blunting line describes the point of fibrous initiation (J_{Ic}). Typical plots for static and dynamic tests are shown in Figs. 3–5. The best line fit and confidence limits about the values of J_{Ic} were investigated by performing a statistical analysis of the data obtained. Also, the effects of using a three-point and a nine-point average for Δa were investigated. These analyses are presented in the next section.

Results and Discussion

Analysis of the Data

The J_{Ic} results obtained from three-point average Δa values were reported by Server et al [13] and are listed in Tables 2 and 3. The amount

TABLE 2—Results for eight heats tested at 177°C (B = 25.4 mm).

Heat	Material	Test Type	J_{Ic} from EPRI Report [13] kJ/m²	J_{Ic}, kJ/m²	σ_f, MPa	β, MPa	ϕ	S^2 (kJ/m²)²	$\dfrac{J_{Ic}}{J}$	$\dfrac{\Delta a_c}{\Delta a}$	Log Deviate J_{Ic} kJ/m²	Log Deviate Δa_c	J_{max} kJ/m²
EG	A533B-1	static CS[a]	131	115 →(125)[a]	521	140 (195)[a]	2	3230 (259)[a]	129 (35)[a]	62 (20)[a]	98 (37)[a]	101 (26)[a]	288
		impact[A] bend	298	→333 (389)	586	243 (329)	2	308 (1488)	156 (90)	54 (63)	37 (130)	26 (193)	928
		static[A] CS	233	→198 (226)	513	210 (284)	2	601 (1307)	58 (84)	34 (58)	81 (150)	96 (311)	471
BBB	A508-2	dynamic[A] CS	264	→293 (304)	581	212 (334)	2	1847 (1834)	119 (117)	72 (84)	122 (128)	86 (103)	565
		static CS	68	→64 (60)	503	25 (38)	2	94 (106)	26 (28)	6 (7)	15 (19)	1 (9)	82
EN	A302B	impact[A] bend	123	→113 (162)	548	77 (99)	2	1611 (989)	95 (82)	28 (28)	30 (33)	3 (4)	239
		static CS	63	→64 (63)	503	28 (42)	2	334 (495)	48 (59)	10 (16)	41 (36)	4 (3)	92
NA	A302B	impact[A] bend	105	→99 (142)	548	78 (101)	2	1200 (16992)	243 (329)	70 (112)	134 (241)	15 (46)	229

EK	MMA weld metal (A533B-1 base)	static CS	179	−168 (187)	541	150 (210)	2	974 (836)	82 (73)	42 (45)	83 (86)	36 (79)	307
		impact[A] bend	271	286 →(290)[b]	587	386 (693)	3 (2)[b]	39957 (5682)	343 (179)	228 (151)	487 (103)	177 (86)	833
		static[A] CS	289	→−300 (355)	460	230 (342)	1	3363 (5832)	523 (645)	347 (520)	790 (1616)	1181 (4405)	607
BKM	MMA weld metal (A508-2 base)	dynamic[A] CS	294	−278 (325)	529	255 (355)	2	1277 (1926)	99 (114)	64 (88)	86 (109)	53 (85)	596
	SA	static CS	238	−204 (231)	538	218 (295)	2	13 (231)	12 (45)	8 (34)	8 (47)	4 (37)	306
CJ	SA weld metal (A533B-1 base)	dynamic CS	224	172 →(229)	566	289 (415)	1	205 (1)	150 (10)	92 (8)	82 (6)	39 (4)	399
	SA	static CS	98	−96 (103)	498	125 (175)	3	103 (85)	15 (14)	7 (7)	14 (15)	7 (9)	220
BAS	SA weld metal (A508-2 base)	dynamic CS	156	−128 (140)	565	166 (228)	2	492 (673)	51 (59)	25 (35)	72 (89)	161 (356)	252

[a] Nine-point average value for Δa rather than three-point average above.
[b] One outlying data point was eliminated.
→ Indicates the value of J_{Ic} with the lowest average 95 percent confidence limits.
[A] Not all individual data points meet the $25J/\sigma_f$ specimen size criterion.
EPRI = Electric Power Research Institute.
CS = compact specimen.

TABLE 3—*Results for HSST Plate 02 (B = 25.4 mm).*

Test Temperature, °C	Test Type	J_{Ic} from EPRI Report [13], kJ/m²	J_{Ic}, kJ/m²	σ_f, MPa	β, MPa	ϕ	S^2, (kJ/m²)²	95% Confidence Limits at J_{Ic}				J_{max}, kJ/m²
								$\dfrac{J_{Ic}}{J}$	$\dfrac{\Delta a_c}{\Delta a}$	Log Deviate, J_{Ic}	Log Deviate, Δa_c	
										kJ/m²		
25	static CS^	236	–166 (207)a	577	246 (316)a	4	3846 (4748)a	88 (96)a	51 (65)a	96 (113)a	68 (100)a	400
	static CS^	313	273 –(288)	607	199 (283)	2	2934 (1704)	145 (107)	84 (73)	118 (97)	72 (101)	607
71	dynamic CS^	231	187 –(220)	746	284 (379)	3	2766 (517)	82 (35)	44 (22)	65 (37)	29 (17)	374
	dynamic bend^	315	–301b (316)b	839	231 (342)	3b	1559 (2238)	64 (77)	29 (42)	69 (66)	23 (24)	1100

aNine-point average value for Δa rather than three-point average above.
bOne outlying data point was eliminated.
—Indicates the value of J_{Ic} with the lowest average 95 percent confidence limits.
^Not all individual specimens meet the $25J/\sigma_f$ specimen size criterion.
EPRI = Electric Power Research Institute.
CS = Compact Specimen.

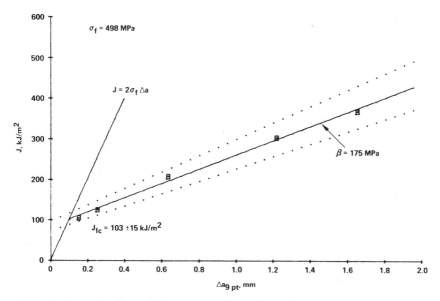

FIG. 4—*Regression line and 95 percent confidence limits (based on log deviate of J) for static compact results for heat BAS at 177°C.*

of crack extension was later remeasured to provide nine-point average values for comparison with the three-point average results. When the nine-point values were being averaged, it became obvious that some of the "equivalent" three-point values were different than those obtained earlier. Investigation of the measurement procedures revealed that the accuracy of location through the thickness of the earlier three-point values was ~0.15 mm. The later nine-point measurements were more accurate (<0.05 mm) with respect to thickness location. Since slightly different points of measurement were used for measuring the bowed crack fronts, some individual results differ markedly. However, the averages of the three-points in each case were usually within 0.10 mm.

The J_{Ic} values obtained using a least-squares linear fit to the "new" three-point and nine-point averages for Δa are listed in Tables 2 and 3. The first value is the three-point average value and the second (in parentheses) is the nine-point average. The slope of the least-squares fit (β), the number of degrees of freedom (ϕ) used for the fit (number of points minus 2), and the overall mean variance (S^2) were computed and are also listed in the tables. It is obvious that a best fit requires many data points and low variance about the line. The comparison of the two three-point average J_{Ic} values is further obscured by the manual curve fits (some of which had curvature) used to obtain the original values in Ref *13*. Due to these difficulties, no conclusions will be drawn with regard to the three-point average

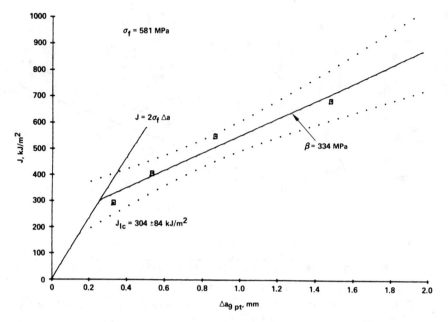

FIG. 5—*Regression line and 95 percent confidence limits (based on $\Delta a/\overline{\Delta a} \times S^2$) for dynamic compact results for heat BBB at 177°C.*

comparisons; the approach of a linear regression fit is a more consistent analysis than the manual curve fits.

A statistical analysis of the least-squares linear fit allows approximate confidence limits to be placed upon the intersection point of the two straight lines. If the variance of the data does not vary over the linear fit, the 95 percent confidence limits at J_{Ic} can be computed as [25]

$$J_{Ic} \pm tS \left\{ \frac{1}{n} + \frac{(\Delta a_c - \overline{\Delta a})^2}{\Sigma(\Delta a - \overline{\Delta a})^2} \right\}^{1/2} \tag{7}$$

where

n = number of points used to fit the linear line,
t = value of the t-distribution for $n - 2$ degrees of freedom,
$\overline{\Delta a}$ = average value of Δa from the data,
Δa_c = value of Δa at J_{Ic}, and
S = estimated value of the standard deviation taken as the square root of the variance about the line (S^2).

If the overall variance about the line was used, very large confidence limits developed for most cases. These large values are due to the few number of data points available for the fit and possibly due to a change in variance at higher levels of J and Δa.

The proper approach when the variance is not constant is to determine the variance distribution, to weight each individual data point by the inverse of the associated variance, and to then determine the regression parameters by minimizing the weighted sum of squares [25]. This approach would give different values for the regression parameters and would allow a weighted variance to be used in Eq 7. Due to the limited data, this approach was not possible. Instead, the overall variances were adjusted (weighted) without altering the regression parameters. Four schemes were used to weight the confidence limits to more realistic values:

1. Assume the variance is a linear function of J departing from the mean $\bar{J}[S^2 \times (J/\bar{J})]$.
2. Assume the variance is a linear function of Δa departing from the mean $\overline{\Delta a}[S^2 \times (\Delta a/\overline{\Delta a})]$.
3. Use a log deviate of J for computing the variance.
4. Use a log deviate of Δa for computing the variance.

The corrections of variance using linear functions related to $\bar{J}$ and $\overline{\Delta a}$ are straightforward and the results for two series over the entire data range are shown in Figs. 3 and 5, respectively. The log deviate approach for J weights the smaller J data by use of logarithms

$$\ln \bar{J} = \ln (\alpha + \beta \overline{\Delta a}) \qquad (8)$$

where α and β are the linear regression parameters. The variance is

$$V^* = \frac{\Sigma(\ln J - \ln \bar{J})^2}{(n - 1)} \qquad (9)$$

and the 95 percent confidence limits at J_{Ic} become

$$J_{Ic} \{\exp [t(V^*)^{1/2}]\} \qquad (10)$$

$$J_{Ic} \{\exp [t(V^*)^{1/2}]\}^{-1}$$

Results for the log deviate of J approach are shown in Fig. 4 over the entire data range. The log deviate approach for Δa requires that the original regression be computed as Δa on J; the variance and limits therefore reflect ranges in Δa rather than J. These Δa ranges are then converted to J limits by using the slope of the regression line. A similar form of limits as shown in Fig. 4 is obtained for the log deviate of Δa. Tables 2 and 3 list ± 95 percent confidence limits at J_{Ic} based on all four methods.

It is important to note that the confidence limits developed assume that the blunting line slope of $2\sigma_f$ is known with absolute certainty. If this line has some uncertainty (as it may have), the confidence limits would be inflated.

Results for Eight Heats Tested at 177°C

The results in Table 2 indicate that the nine-point J_{Ic} values are generally greater than the three-point values, even though the slopes of the lines (β) are much steeper. These higher values can be interpreted as indicating more plane-stress behavior, since the crack average includes surface measurements, while the three-point values may be indicative of more plane-strain behavior. The variances about the lines are not consistently lower for the nine-point average. It is evident that the two averaging techniques can result in different values of J_{Ic}. No one value for the 95 percent confidence limits at J_{Ic} gives consistently high or low values; the value to choose is rather arbitrary, but eliminating the highest and lowest values and taking an average of the two intermediate limits allows a qualitative evaluation of the ranges that J_{Ic} can be expected to have. This approach indicates that the deviation of J_{Ic} is generally lower for the three-point average values (see the arrows in Table 2 designating the J_{Ic} values with the lowest limits).

For all but one heat (BKM) the dynamic J_{Ic} values are higher than the static values, and the slope of the lines (β) increases with increasing loading rate. Also indicated in Table 2 are the values of J at maximum load (J_{max}). In many cases, only one test was carried beyond maximum load; the J_{max} result for materials which have more than one test beyond maximum load is computed as an average value. It is obvious that the maximum-load values are significantly higher than the initiation values.

Values of J_{Ic} can be converted to equivalent stress intensity values (K_{Jc}) by the equation

$$K_{Jc} = \left[\frac{EJ_{Ic}}{1 - \nu^2} \right]^{1/2} \quad (11)$$

where ν is Poisson's ratio. All J_{Ic} values in Tables 2 and 3 meet the current specimen size criterion for validity [14].

$$a, b, B \geq \frac{25J_{Ic}}{\sigma_f} \quad (12)$$

As indicated in Table 2, however, not all test points (usually the largest deflection results) meet the size criterion of Eq 12 for individual J results. It should be noted that K_{Jc} can be equivalent to large specimen linear elastic K_{Ic} values; however, the measurement point for J_{Ic} is not always the same point as for K_{Ic} [26], and statistical size effects can sometimes become important [13].

J_{Ic} values were chosen for either three-point or nine-point average crack advance based upon the lowest values of the average 95 percent confidence limits as discussed previously. These values were converted to equivalent K_{Jc} values using Eq 11. The K_{Jc} values are shown in Fig. 6 as a function of the Charpy V-notch upper-shelf energy level. Also shown superimposed on the graph are lines obtained for different yield stresses for the Rolfe-Novak upper-shelf correlation which was based on higher-yield-strength steels ($\sigma_y > 690$ MPa) [27]

$$\left(\frac{K_{Ic}}{\sigma_y}\right)^2 = 5\left[\frac{C_y}{\sigma_y} - 0.05\right] \qquad (13)$$

where K_{Ic} is in units of kips per square inch by square root inch, the yield stress (σ_y) is in units of kips per square inch, and the Charpy V-notch energy (C_v) is in units of foot pound force. It is interesting to note that there is good agreement between the data and Eq 13 for the static fracture results at a yield stress level of 400 MPa; the yield stress at 177°C for all the materials is near 400 MPa. The dynamic results agree favorably with the

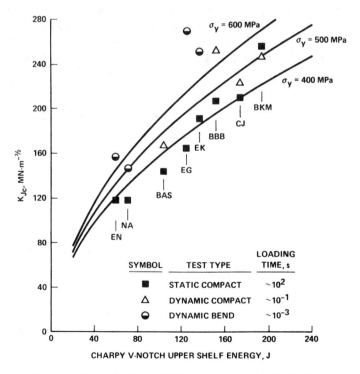

FIG. 6—*Initiation toughness results at 177°C compared with the Rolfe-Novak correlation.*

other yield stress lines drawn in Fig. 6; again, these yield stress levels are indicative of yield stresses at 177°C (using an approach similar to Eq 5 and 6 for yield only). Therefore, it appears that at 177°C the Rolfe-Novak correlation can be used to estimate levels of fracture toughness based upon Charpy V-notch energy and the yield stress. However, several of the data points fall below these lines, especially at the lower Charpy V-notch energy levels. Also, application of this approach to other temperatures on the upper shelf may be misleading. For example, Fig. 7 shows that static upper-shelf toughness results obtained from another program using heat CJ [28] and the results obtained here. Also, cleavage-initiated fracture results [7,11,13] are shown as a function of temperature. It appears that the fibrous fracture toughness shelf reaches a peak at the fracture mode transition and then decreases with increasing temperature, while the Charpy V-notch impact energy is relatively flat and fixed over this same temperature range (as is the yield stress). This trend in fracture toughness on the upper shelf has been observed elsewhere; for example, see Ref 26. Therefore, the good agreement of the data with the Rolfe-Novak correlation is perhaps fortuitous, although there is a basic trend indicative of the correlation.

The J_{max} values in Table 2 can also be converted to stress-intensity values using Eq 11. The values for K_{max} have been shown to be the same as equivalent energy toughness values obtained using energy to maximum load [29]. Since there was a large disparity between J_{Ic} and J_{max}, the stress-intensity factors will also vary according to the square root. Equivalent energy fracture toughness based upon maximum load produces a highly optimistic measure of upper-shelf fracture toughness.

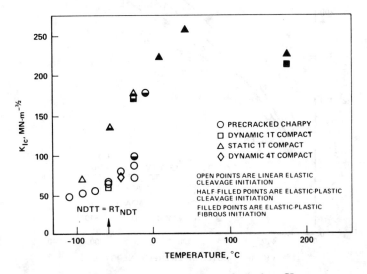

FIG. 7—*Initiation toughness results for heat CJ.*

HSST Plate 02 Results

The results for HSST Plate 02 were analyzed in the same manner as the eight heats tested at 177°C (see Table 3). The following observations were the same as for the other eight materials:
1. The nine-point average J_{Ic} values are larger than the three-point average values.
2. There is no consistent trend for the variance about the lines between the nine- and three-point crack averages.
3. Dynamic loading increases the slope of the regression line.
4. The J_{max} values are substantially higher than the J_{Ic} values.

Using the same approach as before for choosing J_{Ic} (lowest average confidence limits) and converting to K_{Jc} values (Eq 11) gives the results shown in Fig. 8. The 25°C static result is at the low end of the temperature range where fibrous initiation occurs. The 71°C static result increased by ~25 percent over the 25°C result. This increase is also consistent with the results shown in Fig. 7. It is likely that tests at 177°C would give a J_{Ic} value

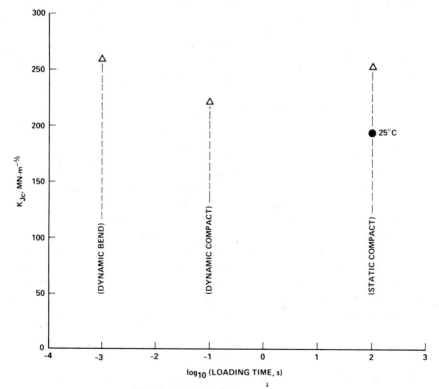

FIG. 8—*HSST Plate 02 results at 71°C.*

less than the 71°C result and probably close to the room-temperature value. The effect of loading rate at 71°C is somewhat puzzling. The impact bend toughness is only slightly larger than the static toughness, but the intermediate-rate dynamic compact toughness is lower than either the static or impact results.

Uncracked Body Energy and Compact Tension Component Corrections

The elastic uncracked body energy correction for the three-point bend test (including shear) can be calculated as

$$U_{\text{no crack}} = \frac{10 P_1^2}{EB} \tag{14}$$

The energy correction for the materials studied would be less than 10 kJ/m^2 (20 kJ/m^2 in terms of J_{Ic}). For heats EN and NA this correction would reduce the impact J_{Ic} values by ~20 percent, whereas for the other bend results the correction would be a decrease of less than 7 percent.

The tension component correction for the compact specimen [6] results in a revision of Eq 2

$$J = \gamma_1 \frac{2 \int_0^{\delta_1} P d\delta}{Bb} + \gamma_2 \frac{2 P_1 \delta_1}{Bb} \tag{15}$$

where γ_1 and γ_2 are variables dependent upon the a/w ratio. For the materials studied here (with $a/w \simeq 0.52$), the correction for J_{Ic} would be an approximate 20 percent increase.

The increase due to the compact specimen correction and the decrease due to the bend correction are shown by arrows in Fig. 9. Only in the cases of heats EN and NA is there a notable change in results—the results for the A302B steels show very little loading rate effect when the offsetting corrections are made. There is divided opinion among the technical community concerning these corrections (as stated earlier), and it is not evident from this work that the corrections are needed. Note, however, that the corrections were made for the already determined initiation values, not for the total original raw data; it is quite possible that the regression slopes, the variance about the lines, and possibly the J_{Ic} values would have changed had the corrections been made to the individual data points.

Summary

This paper has investigated the fibrous initiation toughness results for nine nuclear pressure vessel materials, in addition to the variation that

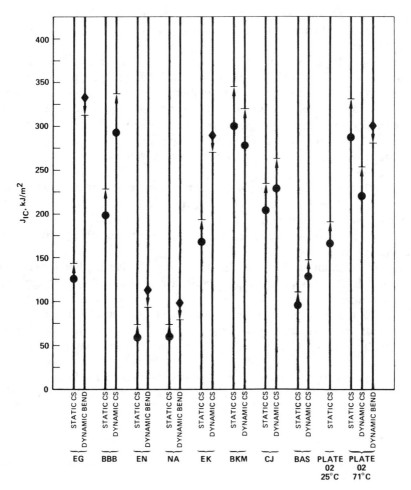

FIG. 9—*Effect of correcting the J_{Ic} values for elastic uncracked beam energy and tension component forces.*

these values may have. From the results presented, the following conclusions and observations can be made:

1. An experimental technique has been developed to measure J_{Ic} for fibrous initiation under impact three-point bend loading.

2. Dynamic loading increases the slope of the regression line (R-curve), and, in almost all cases, the J_{Ic} results for dynamic loading are higher than static values.

3. Maximum load values of J are significantly higher than the initiation J_{Ic} results.

4. There appears to be a functional relationship between initiation toughness and Charpy V-notch energy on the upper shelf. However, there also

appears to be a drop-off in toughness (once the maximum has been reached) while the Charpy level remains relatively unchanged over the same temperature range.

5. Three heats of materials had initiation stress intensity toughness results less than 150 MN-m$^{-3/2}$ under static loading. These heats were two A302B heats (EN and NA) and a submerged arc weld metal (heat BAS). Not only do these materials have low upper-shelf toughness, but two heats (EN and BAS) also have high copper levels, making them markedly susceptible to in-service neutron irradiation damage.

Summarizing, the areas of concern which should be resolved in the future are:

1. Whether three-point or nine-point average crack fronts (Δa) should be used for determining J_{Ic}. These two approaches appear to give different results, with the nine-point average giving higher results. However, the 95 percent confidence limits are generally lower for the three-point average regression line.

2. The certainty in knowing that the $2\sigma_f$ crack blunting line is generally correct. It seems best to use a σ_f indicative of the loading rate and test temperature.

3. The type of confidence limits to be used to determine the confidence of the straight-line curve fit with regard to magnitude (α) and slope (β). In particular, the distribution of the variance about the curve should be determined.

4. The necessity for correcting three-point bend data for elastic uncracked body energy and the compact data for tension component forces.

Acknowledgments

Support under Electric Power Research Institute Research Project RP 696-1 is gratefully acknowledged. The author is indebted to Dr. R. A. Wullaert and Dr. W. Oldfield for their discussions and reviews of the manuscript. Special thanks also goes to J. W. Sheckherd for his performance of the testing.

References

[1] Begley, J. A. and Landes, J. D. in *Fracture Toughness, ASTM STP 514,* American Society for Testing and Materials, 1972, pp. 1-20.
[2] Rice, J. R. in *Fracture,* H. Liebowitz, Ed., Vol. II, Academic Press, New York, 1968, pp. 191-311.
[3] Rice, J. R., Paris, P. C., and Merkle, J. G. in *Progress in Flaw Growth and Fracture Toughness Testing, ASTM STP 536,* American Society for Testing and Materials, 1973, pp. 231-245.
[4] Sumpter, J. D. G. and Turner, C. E. in *Cracks and Fracture, ASTM STP 601,* American Society for Testing and Materials, 1976, pp. 3-18.
[5] Robinson, J. N., "An Experimental Investigations of the Effect of Specimen Type on

the Crack Tip Opening Displacement and J-Integral Fracture Criteria," *International Journal of Fracture*, Vol. 12, No. 5, 1976, pp. 723-737.
[5] Merkle, J. G. and Corten, H. T., "A J Integral Analysis for the Compact Specimen, Considering Axial Forces as Well as Bending Effects," *Journal of Pressure Vessel Technology*, Nov. 1974, pp. 286-292.
[7] Wullaert, R. A., Oldfield, W., and Server, W. L. "Fracture Toughness Data for Ferritic Nuclear Pressure Vessel Materials; Task A," Final Report of Electric Power Research Institute on Research Project RP 232-1, EPRI NP-121, Electric Power Research Institute, April 1976.
[8] Server, W. L., Sheckherd, J. W., and Wullaert, R. A., "Fracture Toughness Data for Ferritic Nuclear Pressure Vessel Materials; Task B—Laboratory Testing, Final Report," EPRI NP-119, Electric Power Research Institute, April 1976.
[9] Van Der Sluys, W. A., Seeley, R. R., and Schwabe, J. E., "Determining Fracture Properties of Reactor Vessel and Forging Materials, Weldments, and Bolting Materials, Final Report," EPRI NP-122, Electric Power Research Institute, July 1976.
[10] Loss, F. J., Ed., "Structural Integrity of Water Reactor Pressure Boundary Components," NRL Report 8006, NRL NUREG 1, Naval Research Laboratory, Aug. 1976.
[11] Marston, T. U., Borden, M. P., Fox, J. H., and Reardon, L. D., "Fracture Toughness of Ferritic Materials in Light Water Nuclear Reactor Vessels, Final Report," EPRI 232-2, Electric Power Research Institute, Dec. 1975.
[12] Oldfield, W., Wullaert, R. A., Server, W. L., and Wilshaw, T. R., "Fracture Toughness Data for Ferritic Nuclear Pressure Vessel Materials; Task A—Program Office, Control Material Round Robin Program," Effects Technology, Inc. Report TR 75-34R, July 1975.
[13] Server, W. L., Oldfield, W., and Wullaert, R. A., "Experimental and Statistical Requirements for Developing a Well-Defined K_{IR} Curve," EPRI NP-372, Electric Power Research Institute, May 1977.
[14] ASTM Task Group E24.01.09 on Elastic-Plastic Fracture Criteria; Chairman, J. A. Begley and J. D. Landes.
[15] Logsdon, W. A. and Begley, J. A. in *Flaw Growth and Fracture, ASTM STP 631*, American Society for Testing and Materials, 1977, pp. 477-492.
[16] Ireland, D. R., Server, W. L., and Wullaert, R. A., "Procedures for Testing and Data Analysis," Effects Technology, Inc. TR 75-43, Oct. 1975.
[17] Server, W. L., Wullaert, R. A., and Sheckherd, J. W., "Verification of the EPRI Dynamic Fracture Toughness Testing Procedures," Effects Technology, Inc. TR 75-42, Oct. 1975.
[18] Server, W. L., Wullaert, R. A., and Sheckherd, J. W., in *Flaw Growth and Fracture, ASTM STP 631*, American Society for Testing and Materials, 1977, pp. 446-461.
[19] Server, W. L., "Impact Three-Point Bend Testing for Notched and Precracked Specimens," *Journal of Testing and Evaluation*, Vol. 6, No. 1, 1978, pp. 29-34.
[20] Oldfield, W., Server, W. L., Odette, G. R., and Wullaert, R. A., "Analysis of Radiation Embrittlement Reference Toughness Curves," Fracture Control Corp. FCC 77-1, Semi-Annual Progress Report No. 1 to the Electric Power Research Institute on Research Project RP 886-1, March 1977.
[21] Server, W. L., and Ireland, D. R. in *Instrumented Impact Testing, ASTM STP 563*, American Society for Testing and Materials, 1974, pp. 74-91.
[22] Server, W. L., Ireland, D. R., and Wullaert, R. A., "Strength and Toughness Evaluations from an Instrumented Impact Test," Effects Technology, Inc. TR 74-29R, Nov. 1974.
[23] Saxton, H. J., Jones, A. T., West, A. J. and Mamaros, T. C. in *Instrumental Impact Testing, ASTM STP 563*, American Society for Testing and Materials, 1974, pp. 30-49.
[24] Server, W. L., "General Yielding of Charpy V-Notch and Precracked Charpy Specimens," *Journal of Engineering Materials and Technology*, Vol. 100, 1978, pp. 183-188.
[25] Davies, O. L. and Goldsmith, P. L., Eds., *Statistical Methods in Research and Production*, Hafner, New York, 1972, p. 195.
[26] Landes, J. D. and Begley, J. A., "Recent Developments in J_{Ic} Testing," Westinghouse Scientific Paper 76-1E7-JINTF-P3, May 1976.
[27] Rolfe, S. T. and Novak, S. R. in *Review of Developments in Plane-Strain Fracture Toughness Testing, ASTM STP 463*, American Society for Testing and Materials, 1970, pp. 124-159.

[28] Borden, M. P. and Reardon, L. D., "Sub-Critical Crack Growth in Ferritic Materials for Light Water Nuclear Reactor Vessels," EPRI NP-304, Electric Power Research Institute, Aug. 1976.
[29] Merkle, J. G. in *Progress in Flaw Growth and Fracture Toughness Testing, ASTM STP 536*, American Society for Testing and Materials, 1973, pp. 264–280.

W. A. Logsdon[1]

Dynamic Fracture Toughness of ASME SA508 Class 2a Base and Heat-Affected-Zone Material

REFERENCE: Logsdon, W. A., "**Dynamic Fracture Toughness of ASME SA508 Class 2a Base and Heat-Affected-Zone Material,**" *Elastic-Plastic Fracture, ASTM STP 668,* J. D. Landes, J. A. Begley, and G. A. Clarke, Eds., American Society for Testing and Materials, 1979, pp. 515–536.

ABSTRACT: The American Society of Mechanical Engineers (ASME) Boiler and Pressure Vessel Code requires that dynamic fracture toughness data be developed for materials with specified minimum yield strengths greater than 345 MPa (50 ksi) to provide verification and utilization of the ASME specified minimum reference toughness K_{IR} curve. In order to qualify ASME SA508 Cl 2a pressure vessel steel [minimum yield strength equals 450 MPa (65 ksi)] per this requirement, dynamic fracture toughness tests were performed on three heats of base and heat-affected-zone (HAZ) material from both automatic and manual submerged-arc weldments. Linear elastic K_{Id} results were obtained at low temperatures while J-integral techniques were utilized to evaluate dynamic fracture toughness over the transition and upper shelf temperature ranges. Loading rates in terms of K were on the order of 2.2 to 4.4 × 10^4 MPa$\sqrt{m}$/s (2 to 4 × 10^4 ksi$\sqrt{in.}$/s). Tensile, Charpy impact, and drop weight nil ductility transition (NDT) tests were also performed. All dynamic fracture toughness values of SA508 Cl 2a base and HAZ material exceeded the ASME specified minimum reference toughness K_{IR} curve. Upper shelf temperature resistance curves obtained by the standard multiple-specimen J test technique, where each specimen was loaded dynamically to a specific displacement, and resistance curves obtained via specimens loaded dynamically to failure, yielded essentially identical J_{Id} values.

KEY WORDS: steel—A508, tensile, Charpy, dynamic, fracture, toughness, weldments, heat-affected-zone, crack propagation

The fail-safe performance of pressure-retaining vessels involved in nuclear applications (pressurized water reactors, etc.) can depend greatly on the ability of various structural materials to sustain high stress/strain in the presence of flaws. Pressure-retaining materials for vessels utilized in nuclear applications must comply with minimum dynamic fracture toughness stan-

---

[1] Senior engineer, Structural Behavior of Materials, Westinghouse R&D Center, Pittsburgh, Pa. 15235.

dards as set forth in Sections III and XI of the American Society of Mechanical Engineers (ASME) Boiler and Pressure Vessel Code [1].[2] In brief, for a particular selected material, the dynamic fracture toughness [which has been temperature corrected based on drop weight nil-ductility transition (NDT) tests and Charpy impact tests] [1,2] must lie above an ASME specified minimum reference toughness K_{IR} curve. This K_{IR} concept is based on lower-bound dynamic fracture toughness and crack arrest data generated on ASTM A533 Gr B Cl 1 and ASTM A508 Cl 2 pressure vessel steels and can be considered as a conservative representation of the dynamic fracture toughness of those pressure vessel materials with specified minimum yield strengths up to 345 MPa (50 ksi).

The present state of the art in nuclear pressure vessel technology calls for higher-strength materials such as ASME SA533 Gr A Cl 2 or ASME SA508 Cl 2a [minimum yield strengths equal 485 MPa (70 ksi) and 450 MPa (65 ksi), respectively]. The ASME Boiler and Pressure Vessel Code permits the use of higher-strength materials [greater than 345 MPa (50 ksi) minimum specified yield strength] for pressure vessels; however, Appendix G of the Code requires that dynamic fracture toughness data need be developed to enable verification and use of the ASME specified minimum reference toughness K_{IR} curve relative to these new materials.

To develop this data base relative to ASME SA508 Cl 2a pressure vessel steel, dynamic fracture toughness tests were performed on three heats of base and heat-affected-zone (HAZ) material. Linear elastic K_{Id} results were obtained at low temperatures while J-integral techniques were utilized to evaluate dynamic toughness over the transition and upper shelf temperature ranges. Support tests (tensile, Charpy impact, and drop weight NDT) were performed to permit a comparison of toughness results with the ASME specified minimum reference toughness K_{IR} curve.

Material, Mechanical Properties and Weld Parameters

ASME SA508 Cl 2a is a quenched-and-tempered vacuum-treated carbon and alloy steel typically utilized in forgings for nuclear pressure vessel applications such as vessel closures, shells, flanges, tube sheets, rings, heads, and similar components. Chemical compositions and heat treatments of SA508 Cl 2a base and HAZ material are outlined in Table 1. Parameters describing the automatic and manual submerged-arc weldments are presented in Table 2. Throughout this paper the weldments are identified as follows:

TO-material/TO-weld wire

Each weld was post-weld stress-relieved at 607°C (1125°F) for 3 to 3.5 h.

[2] The italic numbers in brackets refer to the list of references appended to this paper.

TABLE 1—Chemical compositions and heat treatments of SA508 Cl 2a pressure vessel steel.

TO Number	Base Plate or Weld Wire	Chemical Composition, weight %									
		C	Mn	P	S	Si	Ni	Cr	Mo	V	Cu	Al
4584	base plate	0.17	0.62	0.007	0.009	0.25	0.69	0.39	0.64	0.030	...	...
4585	base plate	0.17	0.63	0.010	0.008	0.25	0.70	0.40	0.64	0.031	...	...
5387	base plate	0.23	0.66	0.010	0.022	0.26	0.85	0.35	0.59	<0.03	...	...
5389	base plate	0.22	0.66	0.011	0.007	0.28	0.82	0.35	0.60	<0.03	...	...
3993	weld wire	0.05	0.94	0.020	0.010	0.36	1.62	0.01	0.26	0.01	...	...
4004	weld wire	0.06	1.22	0.020	0.010	0.29	1.63	0.01	0.24	0.01	...	...
4009	weld wire	0.04	1.10	0.015	0.010	0.25	1.65	0.01	0.23	0.01	0.02	...
4109	weld wire	0.16	2.04	0.015	0.007	0.05	0.63	0.06	0.50	<0.01	...	0.01

TO Number	Heat Treatment
4584 4585	heated to 893°C (1640°F), held 5 h, and water quenched; tempered 693°C (1280°F) and held 6 h, air cooled
5387 5389	heated to 860°C (1580°F), held 4 h, and water quenched; tempered 666°C (1230°F) and held 6 h, furnace cooled

all welds post-weld heat treated at 607°C (1125°F), held 3/3.5 h; plate thickness equaled 10.2 cm (4.0 in.) with top and bottom 1.3 cm (0.5 in.) not included in specimen layout

TABLE 2—*Parameters for submerged-arc weldments.*

Parameter	Automatic Welds 4585/4109 5387/4109 5389/4109	Manual Weld 4585 {3993, 4004, 4009}
Electrode size	TO-4109 0.40 cm (5/32 in.)	TO-3993 0.40 cm (5/32 in.) TO-4004 0.64 cm (1/4 in.) TO-4009 0.48 cm (3/16 in.)
Electrode type	MnMoNi	Type E-9018 M per ASME SFA 5.5
Flux type	Linde 0091	...
Current and polarity	TO-4109 600 DCRP	TO-3993 135 to 220 DCRP TO-4004 320 to 375 DCRP TO-4009 160 to 280 DCRP
Arc voltage	32	21 to 25
Travel speed	30 cm/min (12 in./min)	...
Welding position	downhand	...
Preheat temperature	121°C (250°F)	79 to 260°C (175 to 500°F)
Interpass temperature	260°C (500°F)	121 to 260°C (250 to 500°F)
Post-weld heat treatment	607°C (1125°F) hold 3/3.5 h	607°C (1125°F) hold 3/3.5 h
Inspection after fabrication	magnetic particle, radiography	magnetic particle, radiography

Tensile requirements for SA508 Cl 2a call for a minimum yield strength of 450 MPa (65 ksi), a range in ultimate strength of 620 to 795 MPa (90 to 115 ksi), and minimum total elongation and area reductions of 16 and 35 percent, respectively. Table 3 and Figs. 1 and 2 summarize the tensile properties of SA508 Cl 2a base plate and HAZ material. The HAZ material from weldment 5389/4109 displayed an ultimate strength slightly above the specified maximum. Compared with the base metal, the HAZ material typically demonstrated moderately superior strengths and elongations and inferior area reductions.

Dynamic fracture toughness data are typically plotted versus $T - RT_{NDT}$ for comparison with the ASME specified minimum reference toughness K_{IR} curve, where RT_{NDT} is defined as a reference temperature. The method for establishing a reference temperature is outlined in detail in Section III, Division I and Subsection NB-2331 of the ASME Boiler and Pressure Vessel Code [1]. Table 3 summarizes the drop weight NDT temperature and Charpy V-notch impact properties necessary to determine the reference temperatures relative to SA508 Cl 2a base and HAZ material. Charpy impact properties are also illustrated in Figs. 3 and 4. Drop weight NDT temperatures and Charpy V-notch impact properties of SA508 Cl 2a HAZ material were superior to those of the base material. In addition, SA508 Cl 2a HAZ material reference temperatures were defined by the drop weight NDT temperatures whereas base metal reference temperatures were defined by Charpy V-notch impact properties (in two cases by the energy absorbed and in one case by lateral expansion). These results are in direct contrast to those previously developed for SA533 Gr A Cl 2 base plate and weldments [3].

Experimental Procedures

All dynamic fracture toughness tests were performed on 2.5 cm-thick (1.0 in.) precracked compact toughness (CT) specimens with the exception of two base metal tests (TO-4584). The smaller CT specimens were tested on a servohydraulic MTS machine with load frame and load cell capacities of 22 680 kg (50 kips) and 9072 kg (20 kips), respectively. Dynamic capability was realized by employing a 341 litres/min (90 gpm) MTS Teem valve (two stage with feedback). Loading rates in terms of $\dot{K}$ were on the order of 2.2 to 4.4 × 10^4 MPa$\sqrt{m}$/s (2 to 4 × 10^4 ksi$\sqrt{in.}$/s). Load versus time, displacement versus time, and load versus displacement traces were recorded for each test. The larger CT specimens were tested in a facility previously described by Shabbits [4].

Some specimens tested at low temperatures were linear elastic and similar to those described by previous investigators [4-6]. The majority of test specimens, however, were in the elastic-plastic regime where J-integral test techniques applied [7-9]. Dynamic instrumented precracked Charpy tests have been previously employed to obtain dynamic fracture toughness values

TABLE 3—*Mechanical, drop weight, and impact*

TO-Number	Base or HAZ[a]	Nil Ductility Transition Temperature		50 ft·lb[b] Energy Temperature	
		°F	°C	°F	°C
4584	Base	0	−18	145	63
5387	Base	20	−7	115	46
5389	Base	50	10	130	54
4585/4109	HAZ (automatic)	−30	−34	15	−9
4585/3993, 4004 and 4009	HAZ (manual)	−10	−23	50	10
5387/4109	HAZ (automatic)	−20	−29	15	−9
5389/4109	HAZ (automatic)	−10	−23	−5	−21
ASTM requirements	Charpy impact at 21 °C (70 °F) = 48 J (35 ft·lb) (minimum average value of three specimens)				

NOTE— RT = reference temperature; NDT = nil ductility transition temperature.
[a] Heat-affected zone.
[b] 1 ft·lb = 1.356 J.

at upper shelf temperatures. Because crack growth initiation often occurs prior to the maximum load point and because the actual initiation point cannot be ascertained, dynamic instrumented precracked Charpy tests typically overestimate a material's dynamic fracture toughness at transition and upper shelf temperatures and as such were not included in this study [*3,10*].

The dynamic test techniques employed in this investigation can be divided into two categories: (1) load-to-failure and (2) dynamic resistance curve. These test techniques are described and illustrated in Ref *3* and will be briefly reviewed herein.

Load-to-Failure

All ASME SA508 Cl 2a specimens tested at temperatures below that where upper shelf fracture toughness behavior was first experienced were loaded dynamically to failure and sustained cleavage-controlled fractures. The onset of crack extension was abrupt and unambiguous. There was no stable growth. A sudden drop in the load deflection curve occurred at the fracture point. Inertial loading effects were negligible at the testing speed utilized. At low temperatures the load versus displacement records were linear and the fracture toughness was calculated directly from the failure load as outlined in the ASTM Test for Plane-Strain Fracture Toughness of Metallic Materials (E 399-74), although in some cases the specified size criterion was not met by the 2.5 cm-thick (1.0 in.) CT specimens.

properties of SA508 Cl 2a pressure vessel steel.

35-mil Lateral Expansion Temperature		RT$_{NDT}$		Mechanical Properties (room temperature)					
				σ_{ys}		σ_{ut}		Elonga-	Reduction
°F	°C	°F	°C	ksi	MPa	ksi	MPa	tion, %	in Area, %
160	71	100	38	88.3	608.8	105.5	727.4	21.7	62.4
90	32	55	13	83.2	573.6	99.3	684.6	20.5	57.0
105	41	70	21	89.1	614.3	105.9	730.2	20.4	59.2
20	−7	−30	−34	93.0	641.2	111.6	769.5	25.7	49.1
40	4	−10	−23	93.9	647.4	105.2	725.3	21.8	38.9
35	2	−20	−29	97.2	670.2	112.2	773.6	33.6	47.5
20	−7	−10	−23	101.9	702.6	117.8	812.2	35.0	55.0
				65	450	90 to 115	620 to 795	16	35

At transition temperatures, nonlinear load versus displacement records were observed although the specimen fractures were cleavage controlled. Fast fracture occurred at maximum load. For these tests, J was calculated from the estimation method outlined by Rice et al [11]. Corresponding K_{Id} values were calculated from the relationship between elastic-plastic and linear elastic fracture mechanics parameters [9]. The criterion for determining if a fracture was cleavage initiated consisted of evaluating as follows the amount of stretching (blunting) experienced by the specimen

$$\Delta \bar{a} \leq \frac{0.55 J_M}{\sigma_f}$$

where

$\Delta \bar{a}$ = average amount of stretching (blunting),
J_M = J calculated at the maximum load point, and
σ_f = flow stress midway between the material's yield and ultimate stresses.

For ferritic steels such as SA508 Cl 2a, compliance with the foregoing requirement indicates cleavage initiation; if $\Delta \bar{a}$ is larger, the mode of fracture initiation is fibrous. Thus in dynamic fracture toughness testing it is not uncommon to obtain nonlinear load versus displacement records

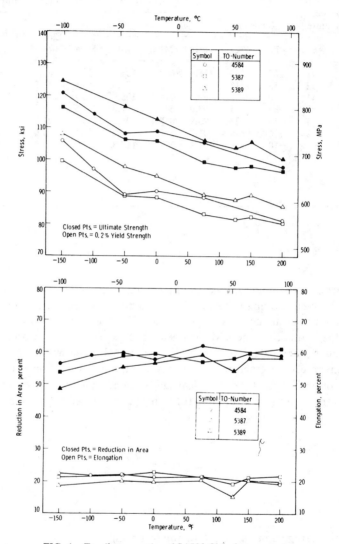

FIG. 1—*Tensile properties of SA508 Cl 2a base material.*

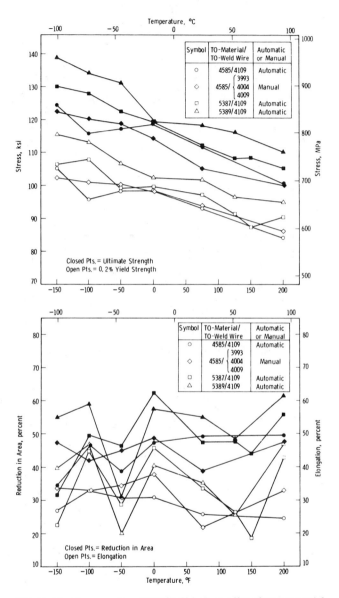

FIG. 2—*Tensile properties of SA508 Cl 2a heat-affected zone material.*

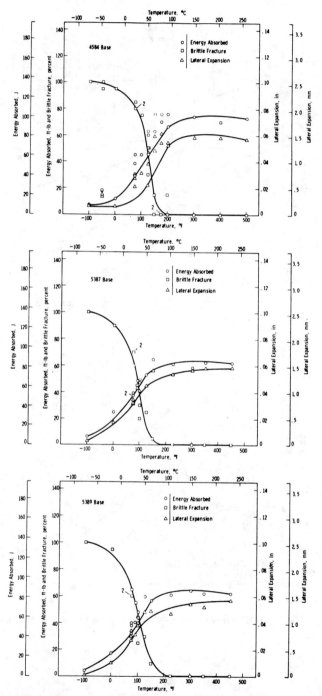

FIG. 3—*Charpy V-notch impact properties of SA508 Cl 2a base material.*

and cleavage-controlled fracture initiation without some form of stable fibrous crack growth (ductile tearing). Although rather infrequent, this fracture behavior can also occur under quasi-static loading rates—generally at lower temperatures and over a smaller temperature range for a given material.

At initial upper shelf temperatures, specimens loaded dynamically to failure experienced fractures which displayed a zone of ductile tearing followed by cleavage rupture. The point of fibrous crack initiation was not apparent from the load-displacement records, which often exhibited some load drop prior to fracture. Calculating a fracture toughness based on maximum load is clearly not related to the point of crack growth initiation. Crack growth may in fact occur prior to or after the maximum load. Therefore, it was not possible to obtain a dynamic fracture toughness value from a single specimen loaded-to-failure at upper shelf temperatures.

A schematic of this combined fracture behavior experienced by specimens loaded dynamically to failure at upper shelf temperatures is illustrated in Fig. 5. This schematic clearly illustrates the interaction of the two basic fracture processes. The only modification to this schematic as a result of dynamic loading is that the crosshatched zone of ductile tearing followed by cleavage rupture would span a larger temperature range. All of the tests loaded to failure at upper shelf temperatures for SA508 Cl 2a base or HAZ material [maximum temperature equaled 66°C (150°F)] displayed a region of ductile tearing followed by cleavage rupture. Increasing the maximum test temperature approximately 27°C (50°F) would have resulted in totally fibrous, ductile fractures. The purpose of applying the previously stated requirement for cleavage initiation would guarantee that a particular dynamic fracture toughness test result occurred prior to the crosshatched zone of ductile tearing followed by cleavage rupture.

Dynamic Resistance Curve

To obtain clearly defined dynamic fracture toughness values at upper shelf temperatures, it was necessary to employ a resistance curve test technique identical to that set forth by Landes and Begley in Ref *9* for quasi-static fracture toughness testing. This technique is applicable to the ductile tearing upper shelf fracture regime where the onset of crack growth cannot be ascertained from the appearance of the load-deflection record. Compact toughness specimens were dynamically loaded to a specific displacement (not to failure), unloaded, heat tinted, and broken open to reveal the amount of stable crack growth.

Results and Discussion

The dynamic fracture toughness values generated on ASME SA508 Cl 2a

526 ELASTIC-PLASTIC FRACTURE

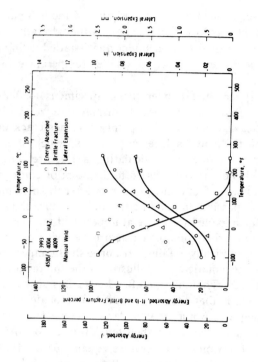

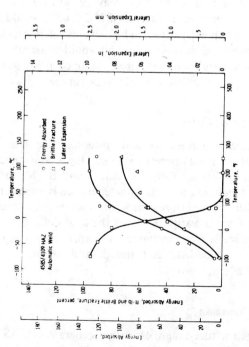

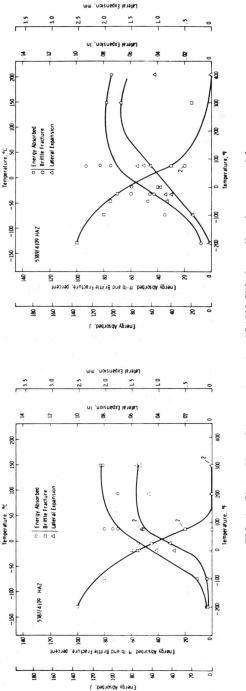

FIG. 4—*Charpy V-notch impact properties of SA508 Cl 2a heat-affected zone material.*

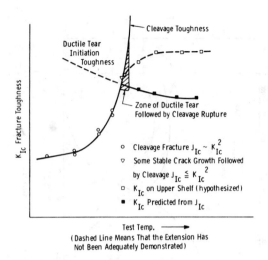

FIG. 5—*Schematic of K_{Ic} transition temperature curve.*

base and HAZ material are plotted versus $T - RT_{NDT}$ for comparison with the ASME specified minimum reference toughness K_{IR} curve in Figs. 6 and 7, respectively, and versus temperature in Fig. 8. Single upper shelf dynamic fracture toughness values generated via the dynamic resistance curve test technique on SA508 Cl 2a base metal are also included in Figs. 6 and 8. In all cases the dynamic fracture toughness of SA508 Cl 2a base and HAZ material exceeded the ASME specified minimum reference toughness K_{IR} curve. Gillespie and Pense previously developed quasi-static fracture toughness data on SA508 Cl 2a which also fell above the K_{IR} curve [12]. Therefore, this 450 MPa (65 ksi) minimum yield strength material is acceptable for nuclear pressure vessel structural applications from a dynamic fracture toughness standpoint.

The dynamic fracture toughness, drop weight NDT temperatures, and Charpy V-notch impact properties of SA508 Cl 2a HAZ material were superior to those of the base material. The fracture toughness behavior demonstrated by SA508 Cl 2a was quite unlike that previously reported for SA533 Gr A Cl 2, where at any given temperature the average base plate dynamic fracture toughness surpassed that of the weldments by approximately 30 percent [3]. Recall that the bases for defining reference temperatures relative to SA508 Cl 2a and SA533 Gr A Cl 2 pressure vessel steels (whether drop weight NDT temperatures or Charpy V-notch impact properties) were also in direct contrast. This toughness superiority displayed by SA508 Cl 2a HAZ material was not manifested as increased conservatism when the fracture toughness values were compared with the ASME specified minimum reference toughness K_{IR} curve. The superior drop weight NDT temperatures, Charpy impact properties, and resulting reference temper-

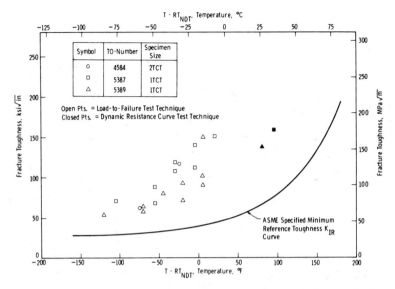

FIG. 6—*Fracture toughness versus* T − RT_{NDT} *for SA508 Cl 2a base material.*

atures displayed by the HAZ material actually penalized the HAZ dynamic fracture toughness values by shifting them such that the HAZ and base metal values both demonstrated the same degree of conservatism relative to the ASME specified minimum reference toughness K_{IR} curve.

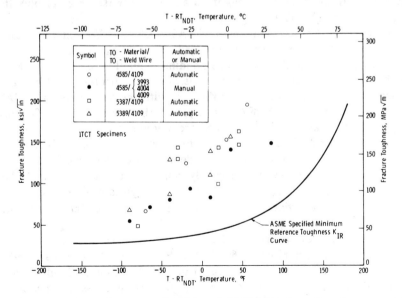

FIG. 7—*Fracture toughness versus* T − RT_{NDT} *for SA508 Cl 2a heat-affected zone material.*

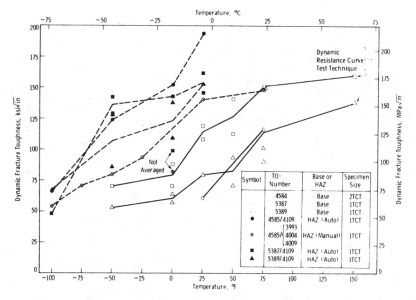

FIG. 8—*Dynamic fracture toughness of SA508 Cl 2a base and heat-affected zone material.*

Concerning SA508 Cl 2a base metal, superior dynamic fracture toughness was demonstrated by TO-5387, which also produced the lowest yield strength, ultimate strength, and reference temperature. TO-5389 displayed the lowest dynamic fracture toughness and ductility (see Fig. 1) plus the highest drop weight NDT temperature.

The dynamic fracture toughness of SA508 Cl 2a HAZ material manufactured utilizing automatic submerged-arc welding (4585/4109) substantially exceeded that of the corresponding manual weldment. The HAZ of this manual weldment also demonstrated the poorest ductility and Charpy impact properties. As was the case with the base metal, the HAZ material from weldment 5389/4109 displayed the lowest dynamic fracture toughness of the automatic submerged-arc weldments.

The SA508 Cl 2a HAZ dynamic fracture toughness data exhibited relatively large scatter. Some CT specimens demonstrated step-type crack fronts as the fracture plane, which normally remained in the HAZ material, searched out the path of least resistance. This typically produced higher dynamic fracture toughness values than when the fracture plane was identical with that of the fatigue precrack.

Resistance curves relative to the single upper shelf dynamic fracture toughness values generated at 66°C (150°F) on SA508 Cl 2a base metal (TO-5387 and TO-5389) are illustrated in Fig. 9. Based on Madison and Irwin's equation for estimating dynamic yield strength as a function of temperature and test speed [6,13], the dynamic yield strengths of SA508 Cl

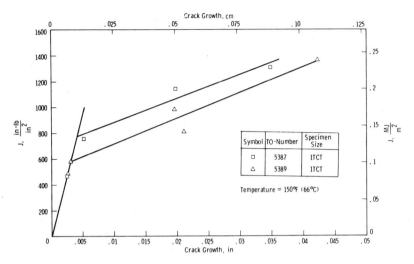

FIG. 9—*Standard J resistance curves for SA508 Cl 2a base material at a temperature of 66°C (150°F).*

2a base material (TO-5387 and TO-5389) increased by an average of only 4.9 percent over the static yield strength values. Therefore, the blunting lines in Figs. 9 and 10 were determined utilizing quasi-static yield and ultimate strength values. Note the slopes (dJ/da) of these two resistance curves (determined via least-squares linear regression) are nearly identical.

Further support relative to the dynamic resistance curve test technique is demonstrated through Fig. 10, which illustrates modified resistance curves developed on specimens loaded dynamically to failure over the upper shelf temperature range of 24 to 66°C (75 to 150°F). The true upper shelf dynamic fracture toughness of SA508 Cl 2a should be nearly constant over this small temperature range when determined via specimens which follow the ductile tear initiation toughness curve of Fig. 5 (that is, via the standard resistance curve test technique where specimens are loaded to specific displacements). Since ductile tearing occurred in each of the tests included in Fig. 10, the point on the load-deflection curves where fibrous crack growth first initiated was not obvious, and calculating individual J_{ID} values was impossible. As previously mentioned, each of the fracture surfaces displayed a region of fibrous, ductile tearing immediately adjacent to the precrack followed by an area of cleavage fracture. For comparison, Fig. 11 illustrates representative fracture surfaces from a series of test specimens loaded to specific displacements (heat tinted, TO-5387) and a series loaded dynamically to failure (TO-5389). Measuring the fibrous, ductile tearing type crack growth on the dynamically failed specimens (three-point average) and plotting it versus J (calculated based on the total area under the load-

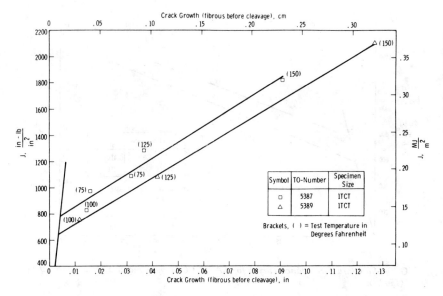

FIG. 10—*Modified J resistance curves for SA508 Cl 2a base material.*

deflection curve to abrupt failure) resulted in the modified resistance curves pictured in Fig. 10. A nine-point average measure of fibrous crack growth was impractical due to interaction of the shear lip formation with the crack extension adjacent to the specimen precrack. Note again that there is little heat-to-heat variation in resistance curve slope.

A direct comparison of the standard and modified resistance curves is illustrated in Fig. 12. When the identical average measure of fibrous crack extension is employed (three-point average), the standard and modified resistance curves are essentially identical in terms of both critical J (J_{Id}) and slope (dJ/da). Dynamic fracture toughness values derived from both the standard and modified resistance curves are compared in Table 4. Fracture toughness values obtained from these totally independent resistance curves are surprisingly similar. Obviously, no deceleration occurred for the SA508 Cl 2a tests loaded dynamically to failure, where the extent of ductile growth was fortunately marked by a change in fracture mode. The similarity of test results supports the contention that deceleration also did not unduly affect dynamic fracture toughness values in dynamically interrupted tests.

Conclusions

1. All dynamic fracture toughness values of ASME SA508 Cl 2a base and HAZ material exceeded the ASME specified minimum reference toughness

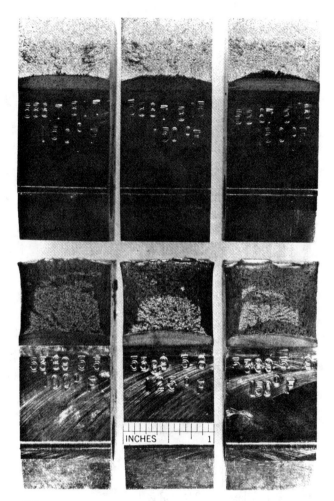

FIG. 11—*Fracture surfaces from a series of specimens loaded dynamically to specific displacements (TO-5387) and a series loaded dynamically to failure (TO-5389).*

K_{IR} curve. Therefore, this 450 MPa (65 ksi) minimum yield strength material is acceptable for nuclear pressure vessel structural applications from a dynamic fracture toughness standpoint.

2. The dynamic fracture toughness, ductility, and Charpy impact properties of SA508 Cl 2a HAZ material manufactured utilizing automatic submerged-arc welding substantially exceeded those of the corresponding manual weldment.

3. Upper shelf temperature resistance curves obtained by the standard multiple-specimen test technique (dynamically load each specimen to a

TABLE 4—*Dynamic fracture toughness values comparison.*

TO-Number	Standard Resistance Curve, 66°C (150°F)				Modified Resistance Curve, 24 to 66°C (75 to 150°F)			
	J_{Id}		K_{Id}		J_{Id}		K_{Id}	
	MJ/m²	in.-lb/in.²	MPa√m	ksi√in.	MJ/m²	in.-lb/in.²	MPa√m	ksi√in.
5387								
3-point avg	0.131	750	174.0	157.2	0.137	785	178.1	160.9
9-point avg	0.136	775	176.9	159.8				
5389								
3-point avg	0.102	580	153.1	138.3	0.112	640	160.8	145.3
9-point avg	0.102	580	153.1	138.3				

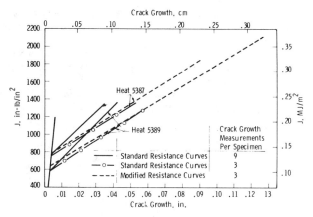

FIG. 12—*Standard and modified J resistance curves for SA508 Cl 2a base material.*

specific displacement and heat tint to mark the degree of stable crack growth) and resistance curves obtained via specimens loaded dynamically to failure (where a region of fibrous, ductile tearing adjacent to the precrack was observable due to a change in fracture mode) were essentially identical in terms of both critical J (J_{Id}) and slope (dJ/da). Therefore, deceleration did not unduly affect dynamic fracture toughness values in dynamically interrupted tests.

References

[1] *ASME Boiler and Pressure Vessel Code*, American Society of Mechanical Engineers, New York, 1974.
[2] *PVRC Recommendations on Toughness Requirements for Ferritic Materials*, Appendix 1, Derivation of K_{IR} Curve, WRC Bulletin 175, Welding Research Council, Aug. 1972.
[3] Logsdon, W. A. and Begley, J. A. in *Flaw Growth and Fracture, ASTM STP 631*, American Society for Testing and Materials, 1977, pp. 477-492.
[4] Shabbits, W. O., "Dynamic Fracture Toughness Properties of Heavy Section A533 Gr B Cl 1 Steel Plate," Technical Report No. 13, Heavy Section Steel Technology Program, Dec. 1970.
[5] Bush, A. J. in *Impact Testing of Metals, ASTM STP 466*, American Society for Testing and Materials, 1970, pp. 259-280.
[6] Paris, P. C., Bucci, R. J., and Loushin, L. L. in *Fracture Toughness and Slow-Stable Cracking, ASTM STP 559*, American Society for Testing and Materials, 1974, pp. 86-98.
[7] **Begley, J. A. and Landes, J. D.** in *Fracture Toughness, ASTM STP 514*, American Society for Testing and Materials, 1972, pp. 1-23.
[8] Landes, J. D. and Begley, J. A. in *Fracture Toughness, ASTM STP 514*, American Society for Testing and Materials, 1972, pp. 24-39.
[9] Landes, J. D. and Begley, J. A. in *Fracture Analysis, ASTM STP 560*, American Society for Testing and Materials, 1974, pp. 170-186.
[10] Stahlkopf, K. E., Smith, R. E., Server, W. L., and Wullaert, R. A. in *Cracks and Fracture, ASTM STP 601*, American Society for Testing and Materials, 1976, pp. 291-311.

[11] Rice, J. R., Paris, P. C., and Merkle, J. G. in *Progress in Flaw Growth and Fracture Toughness Testing, ASTM STP 536*, American Society for Testing and Materials, 1973, pp. 231-245.

[12] Gillespie, E. H. and Pense, A. W., "The Fracture Toughness of High Strength Nuclear Reactor Materials," Department of Metallurgy and Materials Science, Lehigh University, Bethlehem, Pa., March 26, 1976.

[13] Madison, R. B. and Irwin, G. R., *Journal of the Structural Division, Proceedings of the American Society of Civil Engineers*, Sept. 1971, pp. 2229-2242.

R. L. Tobler[1] and R. P. Reed[1]

Tensile and Fracture Behavior of a Nitrogen-Strengthened, Chromium-Nickel-Manganese Stainless Steel at Cryogenic Temperatures*

REFERENCE: Tobler, R. L. and Reed, R. P., **"Tensile and Fracture Behavior of a Nitrogen-Strengthened, Chromium-Nickel-Manganese Stainless Steel at Cryogenic Temperatures,"** *Elastic-Plastic Fracture, ASTM STP 668,* J. D. Landes, J. A. Begley, and G. A. Clarke, Eds., American Society for Testing and Materials, 1979, pp. 537-552.

ABSTRACT: J-integral fracture and conventional tensile properties are reported for an electroslag remelted Fe-21Cr-6Ni-9Mn austenitic stainless steel that contains 0.28 percent nitrogen as an interstitial strengthening element. Results at room (295 K), liquid-nitrogen (76 K), and liquid-helium (4 K) temperatures demonstrated that the yield strength and fracture toughness of this alloy are inversely related and strongly temperature dependent. Over the investigated temperature range, the yield strength tripled to 1.24 GPa (180 ksi) at 4 K. The fracture toughness, as measured using 3.8-cm-thick (1.5 in.) compact specimens, decreased considerably between 295 and 4 K. During plastic deformation at 295 K the alloy undergoes slight martensitic transformation, but at 76 and 4 K it transforms extensively to martensites. The amount of body-centered cubic (bcc) martensite formed during tension tests was measured as a function of elongation.

KEY WORDS: cryogenics, fracture, low-temperature tests, martensitic transformations, mechanical properties, stainless steel alloys, crack propagation

Recently, austenitic stainless steel strengths have been increased considerably by the substitution of nitrogen and manganese for nickel. In addition to providing interstitial and solid solution strengthening, these elements serve to increase austenite stability with respect to martensitic transformations. Compared with nickel, these elements are more abundant and less expensive. The alloy studied in this report, Fe-21Cr-6Ni-9Mn-0.3N (21-6-9),

---

*National Bureau of Standards contribution, not subject to copyright.
[1]Metallurgist and section chief, respectively, Cryogenics Division, Institute for Basic Standards, National Bureau of Standards, Boulder, Colo. 80302.

has a room temperature yield strength nearly twice that of AISI 304. Available tensile and impact data [1–4][2] suggest that the 21-6-9 alloy retains good toughness at low temperatures, leading to consideration of its use for applications benefiting from high strength and toughness.

Accordingly, 21-6-9 is currently being considered for such critical components as the coil form for the prototype controlled thermonuclear reaction superconducting magnets and the torque tube for rotating superconducting machinery. To insure satisfactory service life and to compare with other candidate materials, it is necessary to evaluate the fracture resistance of the alloy. This study presents the first fracture toughness data for this alloy.

Material

The electroslag remelted 21-6-9 austenitic stainless steel plate was processed and donated by Lawrence Livermore Laboratories, Livermore, Calif. The chemical composition (in weight percent) of this heat is 19.75Cr-7.16Ni-9.46Mn-0.019C-0.15Si-0.004P-0.003S-0.28N. This steel was soaked at 1366 K for 4 h, then cross-rolled from 30.5 by 30.5 by 10-cm (12.2 by 12.2 by 4-in.) slabs to 50 by 50 by 3.6-cm (20 by 20 by 1.44-in.) plate. Rolling was completed in 12 steps, using five 90-deg rotations. The final plate temperature after this hot rolling was 1089 K. Each plate was then annealed at 1283 K for 1½ h and air cooled, followed by an anneal at 1366 K for 1½ h and a water quench. The resultant hardness was Rockwell B92 and the average grain diameter was 0.16 mm (0.0064 in.).

Procedure

Tensile

Tension specimens were machined following the ASTM Standard Methods of Tension Testing of Metallic Materials (E 8-69). The reduced section diameter was 0.5 cm (0.1 in.) and gage length was 2.54 cm (1 in.). The tension axis was oriented transverse to the final rolling direction. Tests were performed at a crosshead rate of 8.3×10^{-4} cm/s, using a 44.5-kN (10 000 lb) screw-driven machine that was equipped with the cryostat assembly designed by Reed [5]. The tests at 295 K were conducted in laboratory air, whereas tests at 76 and 4 K used liquid nitrogen and liquid helium environments, respectively. Load was monitored with a commercial load cell while specimen strain was measured with a clip-on, double-beam, strain-gage extensometer. Yield strength was determined as the stress at 0.2 percent offset plastic strain.

---

[2]The italic numbers in brackets refer to the list of references appended to this paper.

Magnetic

To detect the amount of ferromagnetic, body-centered cubic (bcc) martensitic phase in the paramagnetic, face-centered cubic (fcc) austenitic matrix, a simple bar-magnet torsion balance was used [6]. Previous measurements on iron-chromium-nickel (Fe-Cr-Ni) austenitic steels established a correlation between the force required to detach the magnet from the specimen and the percent bcc martensite [6]. The same correlation was used for this study to estimate the amount of bcc martensite in the iron-chromium-nickel-manganese (Fe-Cr-Ni-Mn) alloy.

Fracture

The J-integral specimens were 3.78-cm-thick (1.488 in.) compact specimens of a geometry described in the ASTM Test for Plane-Strain Fracture Toughness of Metallic Materials (E 399-74). The specimen width, W, and width-to-thickness ratio, W/B, were 7.6 cm (3.0 in.) and 2.0, respectively. Other dimensions are shown in Fig. 1. The notch, machined parallel to the final rolling direction of the plate, was modified to enable clipgage attachment in the loadline.

The J-integral specimens were precracked at their test temperatures, using a 100-kN (22 480 lb) fatigue testing machine and cryostat [7]. All fatigue operations were conducted using load control and a sinusoidal load cycle at 20 Hz. Maximum fatigue precracking loads (P_f) were well below the maximum load of J tests (P_{max}), as indicated in Table 1. The maximum stress in-

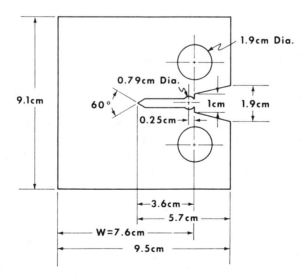

FIG. 1—*Compact specimen for fracture testing of Fe-21Cr-6Ni-9Mn alloy (1 cm = 0.4 in.).*

TABLE 1—*Precracking parameters for J-integral test specimens.*

Test Temperature, K	P_f/P_{max} × 100, %	K_f, MPa·m$^{1/2}$	Relative Crack Length, a/W	a_e/a
295	40 to 45	48 to 54	0.638	0.90 to 0.91
76	22 to 27	52 to 63	0.640	0.87 to 0.90
4	30 to 35	52 to 63	0.64 to 0.795	0.88 to 0.89

tensities during precracking (K_f), the final relative crack lengths (a/W), and the edge-crack-to-average-crack-length ratios (a_e/a) at each temperature are also listed in Table 1. After precracking, the specimens were transferred to a 267-kN (60 000 lb) hydraulic tension machine for fracture testing. Thus, the 76 and 4 K fracture specimens were warmed to room temperature between precracking and J testing at 76 and 4 K. This was necessary since the load limitations of the 100-kN (22 480 lb) fatigue machine precluded loading this alloy to fracture at low temperatures.

The J-integral tests followed a resistance curve technique similar to that described originally by Landes and Begley [8]. A series of nearly identical specimens was tested at each temperature. Each specimen was loaded to produce a given amount of crack extension. The specimens were then unloaded and heat tinted or fatigued a second time to mark the amount of crack extension associated with a particular value of J. The oxidized zone of crack extension (including blunting, plus material separation) could be identified and measured after fracturing the specimen into halves.

Using the approximation for deeply cracked compact specimens [9]

$$J = 2A/B(W - a) \qquad (1)$$

the value of J for each test was calculated from the total area, A, under the load-versus-deflection record. The values of J obtained at each temperature were plotted versus crack extension, Δa, which was measured at five locations equidistant across the specimen thickness, and averaged.

The critical value of the J integral, J_{Ic}, defined as the J value at the initiation of crack extension, was obtained by extrapolating a reasonable fit of the J-Δa curve to the point of actual material separation. An estimation of the plane-strain fracture toughness parameter, denoted $K_{Ic}(J)$, was made using [8]

$$K_{Ic}^2(J) = \frac{E}{1 - \nu^2}(J_{Ic}) \qquad (2)$$

where E is Young's modulus and ν is Poisson's ratio. At room temperature, $E \simeq 195$ GPa (28 306 ksi) and $\nu \simeq 0.287$; at 76 and 4 K, $E \simeq 203$ GPa

(29 467 ksi) and $\nu \simeq 0.278$, according to Ledbetter's measurements by an acoustic technique [10].

Results and Discussion

Tensile

The yield and tensile strengths, elongation, and reduction of area were obtained for the 21-6-9 alloy at 295, 76, and 4 K. These data are summarized in Table 2. The results from this study are combined in Figs. 2-4 with the unpublished results of Landon [1] for the same heat, also hot rolled and annealed, and with the results of Scardigno [2], Malin [3], and Masteller [4] for annealed bar stock. The spread of the Malin data represents results from both the longitudinal and transverse specimen orientations. Agreement is very good, except that the ultimate-strength data of Masteller are consistently higher than the average of the other data.

Typical stress-strain curves at each temperature are presented in Fig. 5. The pronounced discontinuous yield behavior at 4 K probably is associated with adiabatic specimen heating of the type described by Basinski [11]. Note that at 4 K the materials' specific heat is very low so that plastic deformation may cause significant heat evolution. Significant local heating is indicated, as the flow stress drops to stress levels less than sustained at 76 K. These load drops should not be attributed to martensitic phase transformations, for three reasons: (1) More extensive transformation was detected in this alloy at 76 K than at 4 K (see later discussion) and no discontinuities in the stress-strain mode at 76 K were observed; (2) load drops have been observed in both

TABLE 2—*Tensile properties of Fe-21Cr-6Ni-9Mn alloy.*

Temperature, K	Yield Strength, 0.2% Offset, GPa	Tensile Strength, GPa	Elongation, 2.5-cm Gage Length, %	Reduction of Area, %
295 K	0.350	0.696	61	79
	0.357	0.705	61	78
average	0.353 (51 ksi[a])	0.701 (102 ksi)	61	78
76 K	0.913	1.462	42	32
	0.886	1.485	43	41
average	0.899 (130 ksi)	1.474 (214 ksi)	43	37
4 K	1.258	1.633	16	40
	1.224	1.634	NA[b]	NA
average	1.241 (180 ksi)	1.634 (237 ksi)	16	40

[a]1 ksi = 6.894×10^{-3} GPa.
[b]NA = not available.

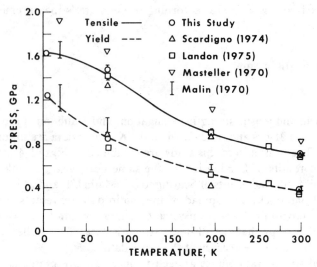

FIG. 2—*Summary of tensile and yield strength data as a function of temperature for the Fe-21Cr-6Ni-9Mn alloy (1 GPa = 145.16 ksi).*

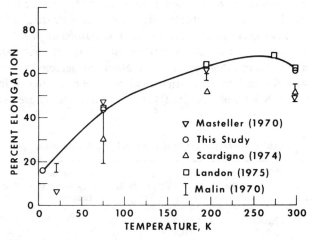

FIG. 3—*Summary of tensile elongation as a function of temperature for the Fe-21Cr-6Ni-9Mn alloy.*

metastable (for example, AISI 304) and stable (for example, AISI 310) austenitic stainless steels at 4 K and no distinction is apparent between the two alloy groups [*12*]; and (3) in austenitic steels the amplitude and frequency of the load drops at 4 K are a function of the strain rate [*12*] which would be expected if local heating were responsible.

Another indication of significant local heating is the rise of the reduction of area to values higher than obtained during 76 K tests. Specimens tested at

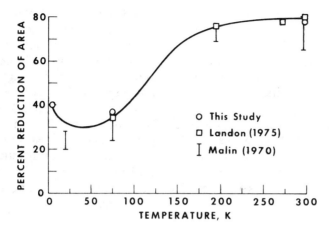

FIG. 4—*Summary of tensile reduction of area as a function of temperature for the Fe-21Cr-6Ni-9Mn alloy.*

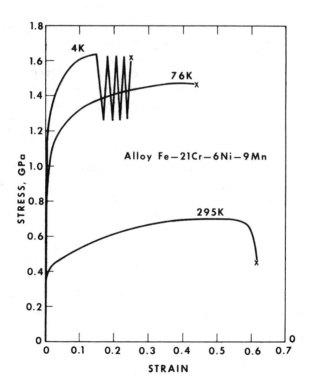

FIG. 5—*Stress-strain curves for the Fe-21Cr-6Ni-9Mn alloy at 295, 76, and 4 K (1 GPa = 145.16 ksi).*

4 K developed very local areas of increased plastic deformation, which resulted in sizable specimen necking prior to fracture. From Figs. 3 and 4, it is clear that the 21-6-9 alloy shows a significant decrease of ductility below 195 K, and tensile elongation decreases progressively between 195 and 4 K.

A primary advantage offered by this alloy is its high yield strength compared to other austenitic alloys. At room temperature the yield strength of the 21-6-9 alloy is about 0.38 GPa (55 ksi), compared with AISI 300 series (Fe-Cr-Ni) steel values of 0.21 to 0.25 GPa (30 to 35 ksi). The yield strength of the 21-6-9 steel approximately triples to a value of 1.24 GPa (180 ksi) as the temperature is decreased to 4 K. Similarly, the Fe-Cr-Ni austenitic alloys achieve values about double or triple their room temperature values of 0.42 to 0.76 GPa (60 to 110 ksi) at 4 K. Therefore, the strength advantage offered by the 21-6-9 alloy is greatest at low temperatures.

Fracture

The load-versus-load-line deflection curves for compact specimens at 295, 76, and 4 K are shown in Fig. 6. The curves at 295 K extended to larger deflections than indicated on the axis of the diagram. The fracture test data are tabulated in Table 3. At no temperature could valid K_{Ic} data be measured according to ASTM E 399-74. The 5 percent secant offset data are denoted K_Q because the thickness and crack front curvature criteria were not satisfied. Using $B \geq 2.5\,(K_Q/\sigma_y)^2$, a specimen thickness of 4.2 cm (1.7 in.) at 4 K is required, slightly larger than the 3.8-cm (1.5 in.) thickness tested. The crack front curvatures shown in Fig. 7 are also excessive. The surface crack lengths are 88 to 89 percent of the average of internal crack lengths, whereas 90 percent is specified in ASTM E 399-74 as the minimum deviation.

The J-versus-Δa results at room temperature are plotted in Fig. 8. Ductile tearing (slow, stable cracking) occurred at this temperature, and large apparent crack extensions were observed due to crack-tip deformation. Only in two specimens at the highest values of Δa was actual material separation noted. These two values fall on the same trend line as the specimen data that did not exhibit material separation. Furthermore, the recommended blunting line, $J = 2\Delta a \sigma_f$, does not match the experimental trend. Therefore the response of this extremely ductile material to J-integral tests at room temperature is inconclusive, with no well-defined J_{Ic} measurement point observable.

The room temperature behavior may result from failure to meet the J test specimen size criterion. According to the tentative criterion suggested by Landes and Begley [8], the specimen thickness for valid J_{Ic} measurements should satisfy the relationship

$$B \geq \alpha(J/\sigma_f) \qquad (3)$$

where α is 25 and σ_f is the average of the yield and tensile strengths. In the

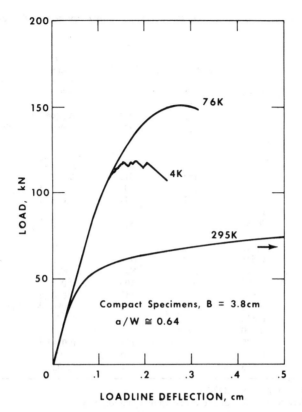

FIG. 6—*Typical load-deflection curves for compact specimens at 295, 76, and 4 K for annealed Fe-21Cr-6Ni-9Mn alloy (1 kN = 224.8 lb; 1 cm = 0.4 in.).*

tests at 295 K, we tentatively estimate the critical J values to be in the range 925 to 1350 kJ·m^{-2} (5285 to 7714 in.·lb·in.$^{-2}$). Using the flow stress value of 0.527 GPa (76.5 ksi), the J-integral results at room temperature are invalid for the specimen thickness tested here. A specimen thickness of 6.3 cm (2.5 in.) may be needed to insure valid data, according to Eq 3.

The J-resistance curve at 76 K is also shown in Fig. 8. The data fit a regular trend, with the exception of the point representing the largest observed crack extension (not shown). The curve drawn through the remaining data indicates that crack extension initiates at a J_{Ic} value of about 340 kJ·m^{-2} (1943 in.·lb·in.$^{-2}$). The corresponding value of $K_{Ic}(J)$, estimated using Eq 2, is 275 MPa·m$^{1/2}$ (250 ksi·in.$^{1/2}$).

At 4 K, the alloy approached linear-elastic behavior, but the results of the first three tests failed to satisfy the ASTM E 399-74 validity criteria for direct K_{Ic} measurements. Consequently, eight additional J tests were conducted and these results are included in Fig. 8. The J-Δa curve at 4 K is nearly horizontal, indicating a J_{Ic} value of about 150 kJ·m^{-2} (857 in.·lb·in.$^{-2}$); the

TABLE 3—*Fracture results for 3.8-cm-thick (1.5 in.) compact specimens of Fe-21Cr-6Ni-9Mn alloy.*

Temperature, K	a/W	$K_Q{}^a$ MPa·m$^{1/2}$	J, kJ·m^{-2}	Δa, cm	J_{Ic} kJ·m^{-2}
295	0.638	58	177	0.013[b]	
	0.636	61	744	0.051[b]	
	0.640	55	905	0.069[b]	between 905 and 1355
	0.635	63	1355	0.097[c]	
	0.642	50	1423	0.112[c]	
76	0.612	134	261	0.0	
	0.634	153	413	0.028	
	0.640	131	499	0.053	340
	0.637	137	674	0.079	
	0.645	130	788	0.091	
	0.643	130	698	0.198	
4	0.645	164	NA	NA	
	0.648	162	NA	NA	
	0.643	159	NA	NA	
	0.670	NA	100	0.0	
	0.655	167	147	0.020	
	0.670	158	149	0.080	150[d]
	0.656	160	162	0.076	
	0.750	NA	89	0.0	
	0.725	NA	191	0.0313	
	0.725	NA	274	0.105	
	0.755	NA	141	0.033	

[a] Calculated from ASTM E 399-74.
[b] Apparent crack extension due to crack-tip deformation only.
[c] Crack extension due to deformation and material separation.
[d] Incorrect J_{Ic} values were reported for this alloy in Ref *13*, due to transcribing errors.
NOTES: 1 in·lb·in.$^{-2}$ = 0.175 kJ·m^{-2}; 1 in. = 2.54 cm; 1 ksi·in.$^{1/2}$ = 1.099 MPa·m$^{1/2}$; NA = not available.

$K_{Ic}(J)$ estimate from Eq 2 is 182 MPa·m$^{1/2}$ (165 ksi·in.$^{1/2}$). Data comparison with other alloys is made in Fig. 9, which indicates that the 21-6-9 alloy offers relatively high toughness for its strength level. Therefore it is attractive for some applications at temperatures as low as 4 K.

Phase Transformations

After tension tests at 76 and 4 K, the deformed specimens were magnetic. Therefore, these specimens were measured, using bar-magnet torsion balance equipment [5], to correlate magnetic attraction with specimen reduction of area. The magnetic readings were converted to percent bcc martensite and the reduction of area converted to elongation, assuming constant volume. These data are plotted in Fig. 10. Typical microstructures are shown in Fig. 11.

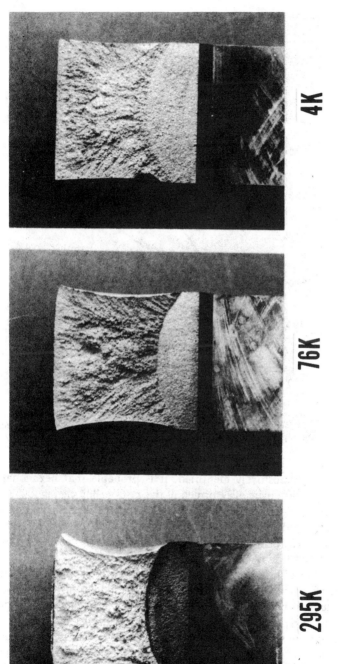

FIG. 7—Fracture characteristics of Fe-21Cr-6Ni-9Mn alloy at 295, 76, and 4 K.

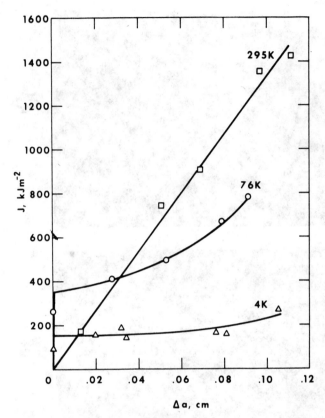

FIG. 8—*The J-integral as a function of crack extension at 295, 76, and 4 K for annealed alloy Fe-21Cr-6Ni-9Mn (1 in.·lb·in.$^{-2}$ = 0.175 kJ·m^{-2}; 1 cm = 0.4 in.).*

Although not positively identified, it is probable that hexagonal close-packed (hcp) martensite also formed in the 21-6-9 alloy during low-temperature deformation. The microphotographs after tensile deformation at 4 K show transformed regions which are parallel to the {111} slip band traces. These appear identical to the hcp areas identified in earlier research on AISI 304, an Fe-Cr-Ni alloy [6,12].

The amount of bcc martensite formed is large and only slightly less than that which is formed in AISI 304 at the same temperatures [6,12]. Permeability values of the order of 10 were measured in heavily deformed specimen portions at 76 K, but it is difficult to identify bcc martensite in the Fig. 11 photomicrographs. Normally, in austenitic stainless steels the bcc martensitic product has an acicular, plate-like morphology with the habit plane of the plate not {111}. Examination of specimen microstructures, typified by Fig. 11, indicate that only at {111} band intersections are the distinctive plate-like microstructures observed.

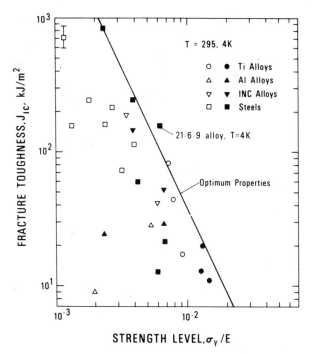

FIG. 9—*Mechanical properties of 21-6-9 alloy at 4 K, as compared with data for other alloys from Ref 13 (1 in. ·lb·in.$^{-2}$ = 0.175 kJ·m^{-2}).*

There is clear evidence that the amount of the transformation is suppressed, as a function of either stress or strain, as temperature is lowered from 76 to 4 K. This is similar to the Fe-Cr-Ni (AISI 304) alloy martensitic transformation behavior [6,12], where formation of the hcp martensitic phase was suppressed at temperatures between 20 and 4 K. Apparently, in the complicated energy balance affecting martensitic transformation for these alloy systems at low temperatures, the increase of flow stress and the decrease of dislocation mobility more than offset the gradually increasing free energy difference between the structures.

It is not clear that martensitic transformations are deleterious to material application. Normally, the stress levels used in service are less than the yield strength, and no martensitic transformations should occur. The complexities and concern usually are discussed when one considers welds and weld techniques. Chemical segregation and stress concentrations are then more likely, rendering particular sections less stable and, locally, stressed above the yield strength. In these situations martensitic products will form.

AISI 304 behaves in a similar manner; it is stable on cooling to low temperatures but transforms to hcp and bcc martensitic products during plastic deformation. But, unlike 21-6-9 alloy, the fracture toughness of annealed

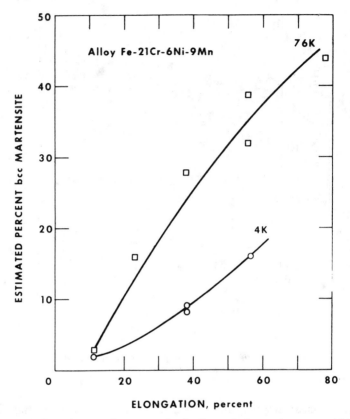

FIG. 10—*Estimated percent bcc martensite that forms during tension tests as a function of tensile elongation.*

AISI 304 remains extremely high at 4 K [*14*], implying that martensitic transformations are not necessarily detrimental to fracture toughness. This is less certain in the case of the Fe-Cr-Ni-Mn-N alloy, however, where the toughness rapidly decreases between 76 and 4 K. For appropriate safety of operation at 4 K, additional research is necessary to understand the effect of martensitic transformations on the fracture toughness of stainless steels.

Conclusions

1. The fracture toughness of the 21-6-9 austenitic stainless steel exhibits an adverse temperature dependence between 295 and 4 K, but retains a respectable J_{Ic} toughness of 150 kJ·m^{-2} (857 in.·lb·in.$^{-2}$) at 4 K. Linear-elastic behavior was approached at 4 K.

2. The yield strength of the 21-6-9 alloy is also strongly temperature

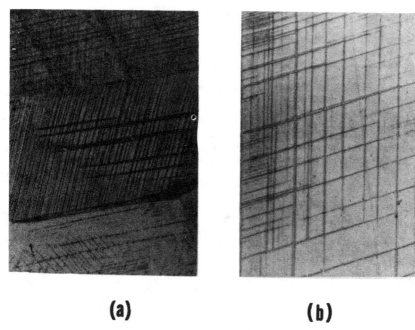

FIG. 11—*Microstructures of alloy 21-6-9 after deformation at 4 K. Bands lie on {111} austenitic planes and probably represent hcp and bcc martensite:* (a) × 440, (b) × 660.

dependent, tripling between room temperature and 4 K, and reaching a value of 1.24 GPa (180 ksi) at 4 K.

3. During plastic deformation at 76 and 4 K, bcc martensite was identified in increasing amounts as a function of strain. Suppression at 4 K, compared with 76 K, of the amount of bcc martensite was found.

Acknowledgments

The authors thank D. P. Landon, Lawrence Livermore Laboratories, for supplying the test material. Dr. R. P. Mikesell conducted the tension tests, R. L. Durcholz contributed technical assistance to tension, fracture, and metallographic preparation, and Dr. M. B. Kasen provided the photomicrographs.

References

[1] Landon, P. R., Unpublished data, Lawrence Livermore Laboratories, Livermore, Calif., 1975.
[2] Scardigno, P. F., M.Sc. degree thesis, Naval Postgraduate School, Monterey, Calif., AD/A-004555, 1974.
[3] Malin, C. O., NASA SP-5921(01), Technology Utilization Office, National Aeronautics and Space Administration, Washington, D.C., 1970.

[4] Masteller, R. D., NASA CR-72638(N70-27114), Martin Marietta Corp., Denver, Colo., 1970.
[5] Reed, R. P. in *Advances in Cryogenic Engineering*, Vol. 7, K. D. Timmerhaus, Ed., Plenum Press, New York, 1962, p. 448.
[6] Reed, R. P. and Guntner, C. J., *Transactions*, American Institute of Mining Engineers, Vol. 230, 1964, p. 1713.
[7] Fowlkes, C. W. and Tobler, R. L., *Engineering Fracture Mechanics*, Vol. 8, 1976, p. 487.
[8] Landes, J. D. and Begley, J. A. in *Fracture Analysis, ASTM STP 560*, American Society for Testing and Materials, 1974, pp. 170–186.
[9] Rice, J. R., Paris, P. C., and Merkle, J. G. in *Progress in Flaw Growth and Fracture Toughness Testing, ASTM STP 536*, 1973, pp. 231–245.
[10] Ledbetter, H. M., *Materials Science and Engineering*, Vol. 29, 1977, p. 255.
[11] Basinski, Z. S., *Proceedings of the Royal Society*, London, England, Vol. A240, 1957, p. 229.
[12] Guntner, C. J. and Reed, R. P., *Transactions*, American Society for Metals, Vol. 55, 1962, p. 399.
[13] Tobler, R. L. in *Fracture 1977*, D. M. R. Taplin et al, Eds., University of Waterloo Press, Waterloo, Ont., Canada, 1977, p. 839.
[14] Reed, R. P., Clark, A. F., and van Reuth, E. C., Eds., *Materials Research for Superconducting Machinery III*, AD-A012365/3WM, National Technical Information Service, Springfield, Va., 1975.

W. H. Bamford[1] and A. J. Bush[1]

Fracture Behavior of Stainless Steel

REFERENCE: Bamford, W. H. and Bush, A. J., "**Fracture Behavior of Stainless Steel**," *Elastic-Plastic Fracture, ASTM STP 668*, J. D. Landes, J. A. Begley, and G. A. Clarke, Eds., American Society for Testing and Materials, 1979, pp. 553–577.

ABSTRACT: An experimental program has been carried out to characterize the fracture properties of austenitic stainless steel piping and plate material. Characterization was in terms of the J-integral, and several specimen types were tested, including compact specimens, center-cracked panels, and three-point bend specimens.

Several methods of monitoring crack extension were used in the program, including unloading compliance, electrical potential, and acoustic emission in addition to the multiple-specimen heat tinting method used for baseline data. These methods are compared and evaluated in detail.

In addition to determining J_{Ic} values for the material, Paris's proposed tearing modulus is evaluated, and various proposed specimen size requirements are discussed.

KEY WORDS: fracture properties, stainless steel, J-integral, toughness, piping, tearing modulus, compliance, crack propagation.

The fracture behavior of reactor coolant piping is an important consideration in assessing the integrity of a nuclear reactor system. The piping of interest here is very large—73.66 to 83.82 cm (29 to 33 in.) in diameter and 5.08 to 7.62 cm (2 to 3 in.) in thickness. The piping carries an internal pressure of 15.52 MPa (2250 psi) and is subject to various thermal and bending loadings as well.

This reactor coolant piping is manufactured of stainless steel—either forged or centrifugally cast. Because of its extensive ductility, quantitative characterization of the fracture properties of stainless steel has not been possible until recently, with the development of the J-integral. A test program was carried out to characterize the J_{Ic} properties of two types of stainless steel piping material. Three different specimen types were tested, and data were prepared in accordance with recommended ASTM procedures [1].[2] In addition, several methods of monitoring crack extension were evaluated, and a brief discussion is provided on size requirements. Although

---

[1] Senior engineer, Westinghouse Nuclear Energy Systems, and senior engineer, Westinghouse R&D Laboratories, respectively, Pittsburgh, Pa.

[2] The italic numbers in brackets refer to the list of references appended to this paper.

the majority of the data were obtained for piping materials, additional tests were performed on 304 stainless steel plate material, which showed equivalent results for J_{Ic} and leads to the conclusion that the results of this program apply to 304 and 316 stainless steel in general.

Because of its extensive ductility, stainless steel is a particularly good material for evaluating the adequacy of the proposed J_{Ic} testing methods. Comments are made on test methods, data presentation, and validity criteria.

Experimental Program

Materials and Specimens

Three materials were tested in the program, two types of stainless steel reactor coolant piping and one heat of 304 stainless steel plate.[3] The majority of the specimens tested were machined from production heats of reactor coolant piping material, one of forged 304 and the other of centrifugally cast 316 stainless steel. The chemistry and heat treatment of the materials are given in Tables 1 and 2.

Three specimen types were tested, as shown in Fig. 1. These were compact specimens, three-point bend specimens, and center-cracked panels. Most of the specimens were 5.08 cm (2 in.) thick, with only a few of the compact specimens 2.54 cm (1 in.) in thickness. The tests were conducted at room temperature and 316°C (600°F). A summary of the combinations of materials, geometries, and test conditions is provided in Table 3.

Compact specimens were machined with cracks oriented in both the axial and circumferential directions for both the piping materials, and for the forged piping 2.54-cm (1.0 in.) compact specimens were oriented with the crack propagating in the through-thickness direction. No directional affects were observed for the cast piping material, so through-thickness tests were not done. Because of size limitations of the actual piping, the three-point bend specimens and center-cracked panels were all machined with the crack propagating circumferentially. The specimen orientations are shown in Fig. 1.

All the specimens were precracked in air at room temperature prior to testing, following the guidelines of ASTM Test for Plane-Strain Fracture Toughness of Metallic Materials (E399-74). The precracking was done with a sinusoidal tension-tension loading in all cases except one. To minimize the crack front curvature of the precrack in one of the bend bars, it was precracked in compression. The crack front produced was much straighter, but since the technique appeared to influence the results (as seen in Fig. 14), it was discontinued.

---

[3]This plate was supplied by L. A. James of Hanford Engineering Development Laboratory.

TABLE 1—Chemical composition of steels tested.

Alloy	C	Si	S	P	Mn	Cr	Ni	Mo	Co	Cu	N
Cast 316 (CF8M) ASTM A351 SW specimens	0.05	0.74	0.022	0.013	0.86	19.9	9.5	2.55	0.06	...	...
Forged 304 A376-304N[a] CI specimens	0.048	0.38	0.025	0.024	1.84	18.42	9.51	0.53	0.06	...	0.12
304 plate material Allegheny Ludlum heat 55697	0.060	0.49	0.010	0.019	0.86	18.26	9.43	0.18	...	0.21	0.032

[a] ASME Code Case 1423-1.

TABLE 2—*Heat treatment and tensile properties.*

Material	Condition	Temperature, °C[a]	0.2% Yield Strength, ksi[b]	Tensile Strength, ksi	Elongation, %	Reduction in Area, %
Cast 316 (CF8M) ASTM A351	annealed at 1121°C for 4 h water Quenched U.S. Pipe&Foundry Heat A3391234	RT[c] 343	46.0 22.9	75.4 64.5	... 41.0	... 55.5
Forged 304 A376-304N	annealed at 1052°C for 1 h water Quenched Cameron Iron Heat K2980	RT 316	42.6 23.6	88.6 71.8	... 44.3	... 66.0
AISI304 plate	annealed at 1093°C for 1 h water Quenched Allegheny Ludlum Heat 55697	RT 427	27.8 20.2	86.3 62.9	70.2 35.5	84.7 67.5

[a] °F = °C × 9/5 + 32.
[b] 1 ksi = 6.90 MPa.
[c] RT = room temperature.

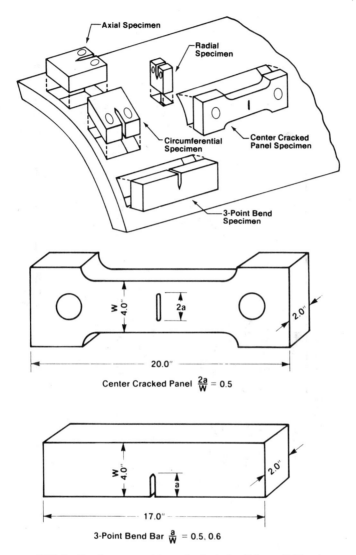

FIG. 1—*Specimen geometries and orientation (1 in. = 2.54 cm)*.

Test Apparatus and Procedure

Because of the high ductility of stainless steel in the temperature range of interest, the fracture properties are best discussed in terms of an elastic-plastic parameter, and for this study the J-integral was chosen. Efforts were made to determine J_{Ic} in a manner consistent with the recommended procedure under development presently by ASTM [1]. This procedure involves the determination of J_{Ic} for the material and condition of interest

558 ELASTIC-PLASTIC FRACTURE

TABLE 3—J_{Ic} fracture toughness results.

Material	Orientation	Specimen	Temperature[a]	J_{Ic}[b] (in.-lb/in.^{2})
304 forged stainless (CI specimens)	axial	2T-compact	RT	4449
	circumferential	3-point bend	RT	>4000
	circumferential	2T-compact	316°C	2569
	axial	3-point bend	316°C	2737
	circumferential	1T-compact	316°C	2308
316 cast stainless (SW specimens)	radial	2T-compact	RT	4293
	axial	3-point bend	RT	>4000
	circumferential	center-cracked panel	RT	4568
	circumferential	2T-compact	316°C	1933
	axial	3-point bend	316°C	428 (?) [~1200][c]
	circumferential			
304 plate (J specimens)	circumferential	1T-compact	316°C	1500

NOTE: All results here were obtained from multiple specimen tests.
[a] RT = room temperature; 3.6°C = 600°F.
[b] To convert from in.lb/in.^{2} to MJ/m^2, multiply by 0.0001751.
[c] This value of J_{Ic} was calculated by assuming that J_{Ic} occurs at the same amount of crack extension as that for compact specimens of the same material and thickness, tested at the same temperature.

by plotting of J versus subcritical crack extension. The value of J_{Ic} was determined from this plot at the point corresponding to zero apparent crack extension. Producing such a plot of J versus crack extension can be accomplished by testing multiple specimens to produce different amounts of stable crack growth, and heat tinting the specimens to mark the crack. It is also possible to produce such a plot from a single specimen, provided a reliable method can be found for determining crack length without breaking the specimen.

All specimens were tested in electrohydraulic test machines, with both load and load line displacement recorded during the test. In addition, the experimental apparatus included several methods for crack length determination, including acoustic emission, electrical potential, and elastic compliance. The electric potential and compliance methods were used with the three-point bend tests, and the acoustic emission and electric potential methods were used in conjunction with the center-cracked panel tests. Although these methods were investigated, the primary method used for obtaining data was the multiple-specimen technique, where a series of specimens was tested to different amounts of crack extension, and the specimens were then heat tinted and broken apart so the stable crack extension could be measured.

Heat Tinting—Since the heat tinting method has long been successful, it was used to obtain baseline data. In the heat tinting method, multiple specimens are used in which specimens are (1) loaded to some predetermined displacement to obtain an estimated crack length Δa, (2) unloaded, (3) heat tinted to mark the crack advance, and finally (4) cooled and broken apart by further loading in order to expose the heat tinted surface for the actual measurement of Δa. In the 24°C (75°F) series of tests, heat tinting was done by placing the specimens in a furnace at 316°C (600°F) for a minimum of 4 h, and for the 316°C (600°F) series the heat tinting was done at 427°C (800°F) for a minimum of 4 h.

Loads were measured with a combination load cell-loading tool for the bend tests, while the center-cracked panel and compact specimen tests employed the testing machine load cell outputs directly. Displacements were measured at the centerline of the loading tool to eliminate bending effects for the bend tests, and by a clip gage mounted across the center of the crack for the center cracked panels. The compact specimens employed a clip gage mounted over the specimen front to read load line displacement.

Electrical Potential—To explore other ways of determining crack initiation and advance in the stainless steel material, the electrical potential method was also tried. This method is shown schematically in Fig. 2 [1]. In this technique a constant current is applied to the specimen during loading. As the crack advances, the resistance of the specimen increases and the change is measured as an increase in the electrical potential. In the present tests both load and displacement versus electrical potential

560 ELASTIC-PLASTIC FRACTURE

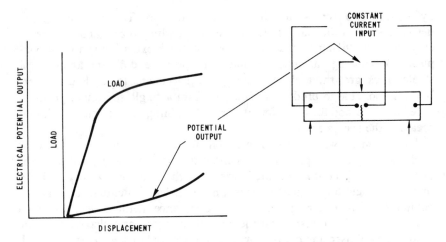

FIG. 2—*Electrical potential method for measuring crack advance.*

were recorded simultaneously and a record of the curves obtained is shown in Figs. 3b and 3c. The region for the start of crack advance as determined from the heat tinting tests already described is shown on the curves. As Figs. 3b and 3c show, there is no clearly defined correlation between the electrical potential curve slope changes and crack initiation for these materials. Use of the electric potential method for this test would have resulted in an implied J_{Ic} value much lower than the true value, as shown in Fig. 3c.

Similar results were obtained when the technique was applied to the center-crack panel test. In this case the implementation was somewhat more difficult, because the loading pins were electrically insulated from the specimen. A plot of electrical potential versus load for Specimen SW-35 is provided in Fig. 4. This specimen was tested to failure, and again the electrical potential output underpredicted the onset of crack extension by a considerable amount, as shown in the figure.

The shortcomings of the electrical potential method in predicting the onset of crack extension for these materials were not altogether unexpected, because the extensive plasticity developed in the specimens alters the resistivity of the material, and this probably led to the premature signal.

Acoustic Emission—The acoustic emission method was used to attempt to determine the onset of crack extension for one center-cracked panel, Specimen SW-35. The method relies on the fact that crack growth results in the release of energy, some of which is in the acoustic frequency range. High ultrasonic frequencies are generally measured, to minimize interference from rubbing and other mechanical sources of noise. In spite of this, mechanical interference is a significant problem.

The test setup is shown schematically in Fig. 5. An acoustic emission sensor was mounted on the face of the center-cracked panel above the

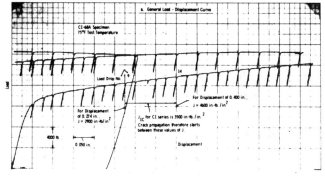

a General Load Displacement Curve

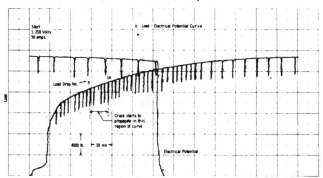

b Load-Electrical Potential Curve

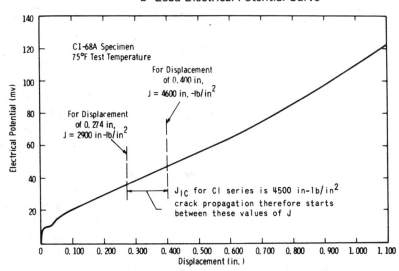

c Displacement-Electrical Potential Curve

FIG. 3—*Electrical potential test results for three-point bend specimen Cl-68 (1 in. = 2.54 cm; 1 lb = 0.4536 kg; 1 lb/in.^{2} = 6.895 kPa).*

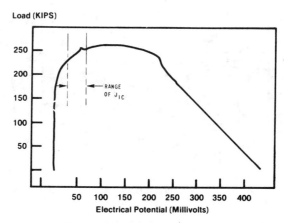

FIG. 4—*Electrical potential test results for center-cracked panel specimen SW-35 (1 kip = 4448 N)*

crack tip at one end, held in place with a spring loading and acoustically coupled to the specimen with conductive grease. Output from the sensor was amplified and then fed through a rate meter, which is actually an averaging device. The rate meter averages the pulses over discrete periods of time so that they can be mechanically recorded; the pulses actually occur over such short periods that the recording response is not fast enough to pick them up.

The results of the test are also presented in Fig. 5, where the acoustic count rate is displayed as a function of load. The figure shows a large increase in count rate at about 1.0 MN (225 000 lb). This value is somewhat below the true J_{Ic} for the material as measured with the multiple-specimen tests (shown in Fig. 9). While this result is somewhat disappointing, it was not altogether unexpected. Stainless steel is a particularly poor material for acoustic emission, because it is not only a low emitter but also a poor transmitter of acoustic noise. Another important factor which undoubtedly influenced the results is the extensive plasticity developed in the specimen, which also produces acoustic emission.

Elastic Compliance—The elastic compliance method is illustrated schematically in Fig. 6 [2]. During the loading cycle, load drops of approximately 10 percent are made at various intervals. The changes in the slope of the linear portion of the load-displacement curve during the load drop should be a measure of any changes in crack length. Because the slope change could not be measured with sufficient precision on the general or conventional load-displacement curve, as shown in Fig. 6a, a second curve shown in Fig. 6b was recorded simultaneously with greatly amplified scales. To facilitate the amplification in the present test series, most of the elastic

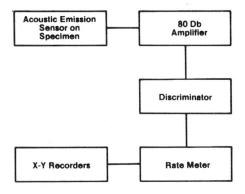

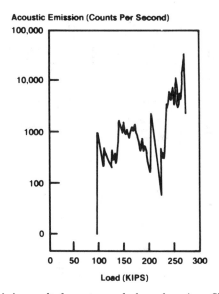

FIG. 5—*Acoustic emission results for center cracked panel specimen SW-35 (1 kip = 4448 N).*

contribution was electronically subtracted from the curve using a special instrumentation package developed for this purpose.

Because of the precision required in the amplified curve, hystersis in the output of the loading and displacement measuring system must be kept to a minimum. Therefore, to measure load, rather than use the testing machine load indicator, a combination load cell-loading tool was designed and used for the bend tests. To measure displacements, various methods were tried. A three-point beam system, developed for making bend bar compliance measurements, was first tried. In this system, three strain-

564 ELASTIC-PLASTIC FRACTURE

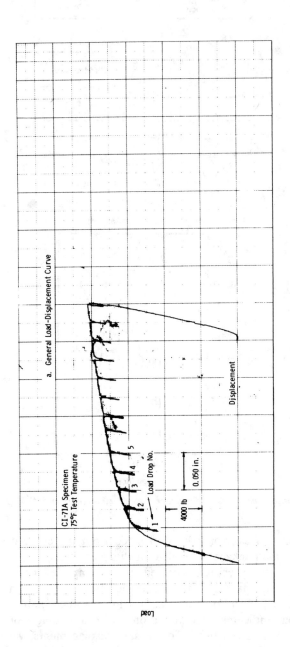

a. General Load-Displacement Curve

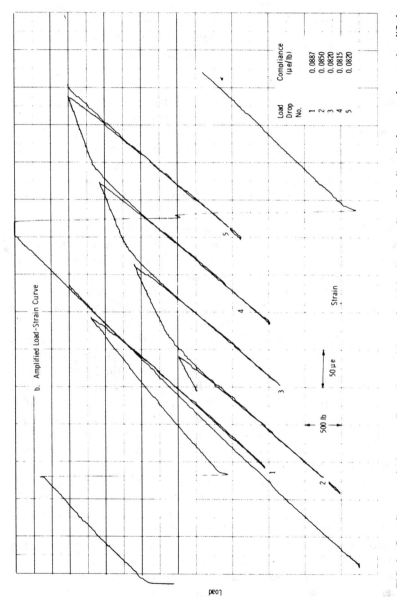

FIG. 6—*Compliance method curves. General curve (a) was obtained using centerline of loading displacement beams. Amplified curve (b) was obtained using strain gage on top surface of specimen 3.175 cm (1.25 in.) from loading tool (1 in. = 2.54 cm; 1 lb = 0.4536 kg).*

gaged cantilever beams are used—one beam contacts the specimen at the centerline of loading point and the other two are placed over the support points. The strain gages on the beams were wired in a Wheatstone bridge configuration so that only the vertical displacement of the beam relative to the support points was recorded. Other methods used to determine displacements were strain gages mounted both above the crack tip and near the center loading point on the compression surface of the bend specimens. Also, a clip gage was placed across the crack mouth opening.

A typical set of load displacement curves obtained using the mouth opening clip gage is shown in Fig. 6b. The curves shown are considerably reduced in scale for presentation, but are representative of the type of traces obtained for all the four measurement systems. The strain gage on the side of the specimen was the only one to show any increase in hysteresis. All of the systems showed an initial decrease in displacement or strain over the first few load drop cycles, indicating a pseudo-decrease in crack length, except for the case where a strain gage was placed near the top of the crack.

To investigate whether or not the decrease in strain or displacement shown in the compliance curve may have been caused by support conditions, a high-pressure lubricant was applied to the ASTM E 399-74 recommended rollers [3] used to support the specimen. Since electrical potential measurements were also scheduled, insulators at the supports would be required. Therefore, while investigating support conditions, Micarta plates along with the lubricant were used, during some of the tests. When using the lubricant, the rollers, instead of rolling, slid to the back of the support block. Rather than have the rollers slip during the early part of the test and affect the curves, the rollers were placed against the back support at the beginning of the test. To prevent brinelling of the specimen, hardened steel plates were placed between the rollers and the specimen. Regardless of the support conditions, the initial decrease in compliance occurred. For large displacements, the curves having the least hystersis and smoothest appearance were obtained using the lubricant; therefore, lubricant was used for all of the tests reported here.

Application of the compliance method to the bend tests was singularly unsuccessful, and an adequate explanation for this behavior was not obtained. It appears to be related to the extremely large displacements for the bend bars combined with extensive plasticity present in the tests. Further complications arose from the fact that the data points follow the blunting line very closely, with no sharp deviation at all, as is discussed later.

Recent tests with three-point bend specimens of a high-strength steel have verified the adequacy of the compliance method used in the present tests. Results showed excellent agreement with multiple-specimen tests, and tend to support the contention that the extensive plasticity and large displacements present in these tests led to the lack of success.

The compliance method was also applied to a 5.08-cm-thick (2.0 in.) compact specimen, with considerably more success. The same techniques were used as previously explained, and the results are shown in Fig. 7. In this case, no pseudo-decrease in crack length was obtained, and the results of the test agreed very well with the multiple-specimen test results, as summarized in Fig. 8.

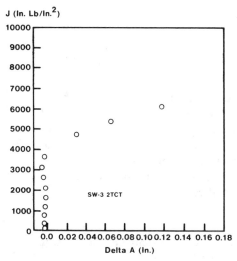

FIG. 7—*Compliance method results for compact specimen SW-3 (1 in. = 2.54 cm; 1 in. lb/in.^{2} = 0.0001751 MJ/m^2).*

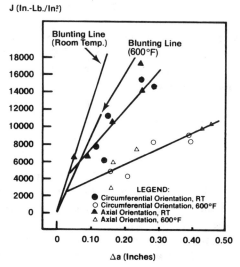

FIG. 8—*J_{Ic} determination, 316 cast stainless steel, compact specimens (1 in. = 2.54 cm; 1 lb/in.^{2} = 0.0001751 MJ/m^2).*

Results

The data were analyzed by plotting the J-integral values obtained as a function of crack extension for each material type and specimen type.

The value of J for the compact specimens was calculated from the expression proposed by Merkle and Corten [4] which is presently recommended by ASTM [1]

$$J = \alpha_1 \frac{2A}{Bb} + \alpha_2 \frac{2P\delta}{Bb} \tag{1}$$

where

α_1, α_2 = coefficients developed by Merkle and Corten [4] to account for the tension component in the compact specimen; the values of α_1 and α_2 are functions of the crack depth of the specimen,
A = area under the load versus load point displacement curve,
B = thickness of the specimen,
b = remaining ligament of the specimen,
P = final load value, and
δ = final load point displacement.

For the three-point bend specimens the J-integral was calculated from the expression originally developed by Bucci, et al [5]. Using the same symbols as Eq 1, the expression is given by

$$J = \frac{2A}{Bb} \tag{2}$$

The expression for J for the center-cracked panel specimens was based on the estimation method proposed by Rice et al [6]. The expression results from the summation of the linear elastic strain energy release rate G added to the plastic portion of the loading, and is

$$J = G + \frac{A^*}{bB} \tag{3}$$

where G is the linear elastic strain energy release rate, and the value of A^* is the area under the load displacement curve between that curve and a straight line drawn from the origin to the point of interest. The remaining symbols are the same as defined in Eq 1.

The cracks tended to lead somewhat in the center of the specimen, and so measurement of crack advance was accomplished by two methods, an area averaging method, where the area of advance was actually measured and divided by the specimen width, and a nine-point averaging method.

These two methods gave very consistent results, so the nine-point averaging was used. The J_{Ic} value was determined by a least-squares best fit of the data to a straight line and analytical determination of its intersection with the so-called "blunting" line, given by

$$J = 2 \sigma_0 \Delta a \qquad (4)$$

where

Δa = crack extension,
σ_0 = flow stress = $\frac{1}{2} (\sigma_y + \sigma_u)$,
σ_y = 0.2 percent offset yield stress, and
σ_u = ultimate strength.

Test results showed that the stainless steels investigated are extremely tough, and consistent in their properties. Results for compact specimen tests of the cast 316 stainless steel, Fig. 8, show that there is no effect of orientation on the results, although both the slope of the J versus Δa curve and the J_{Ic} value are somewhat temperature dependent. The experimentally determined J_{Ic} values for all the steels tested are summarized in Table 3. Center-cracked panel tests conducted at room temperature are summarized in Fig. 9 and show remarkable consistency with both the slope of the curve and J_{Ic} value obtained from the compact specimens, as shown in Tables 3 and 4.

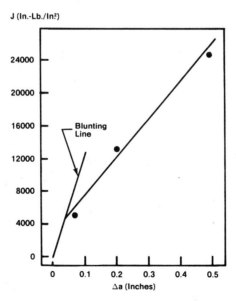

FIG. 9—J_{Ic} determination, 316 cast stainless steel, center-cracked panels (1 in. = 2.54 cm; 1 in. lb/in.^{2} = 0.0001751 MJ/m^2).

TABLE 4—Determination of several validity criteria, and other parameters.

Material	Specimen	Temperature[a] °C	B,[b] in.	$25 \dfrac{J_{Ic}}{\sigma_o}$	J_{Ic},[c] in.-lb/in.^{2}	$\dfrac{dJ}{da}$, in.-lb/in.^{3}	$\dfrac{E}{\sigma_o^2} \dfrac{dJ}{da}$	$\dfrac{b}{J_{Ic}} \dfrac{dJ}{da}$	$\dfrac{b}{J_{Ic}} \sigma_o$
304 forged stainless	compact	RT	2.0	1.70	4450	28 600	188.1	12.9	29.5
	3-point bend	RT	2.0	1.52	>4000	107 400	706.3	67.2	40.9
	compact	316	2.0	1.35	2569	20 700	231.1	16.1	37.1
	3-point bend	316	2.0	1.43	2737	49 000	547.0	39.4	38.3
	compact	316	1.0	1.21	2308	34 800	388.5	15.1	20.7
316 cast stainless	compact	RT	2.0	1.77	4293	47 400	364.1	22.1	28.3
	3-point bend	RT	1.75	1.65	>4000	81 700	627.5	50.7	37.6
	center crack panel	RT	2.0	1.88	4568	45 500	349.5	10.0	13.3
	compact	316	2.0	1.04	1933	19 000	225.1	19.7	47.9
	3-point bend	316	1.75	0.65	~1200	51 700	612.6	93.9	84.1
304 plate stainless	compact	316	1.0	0.87	1500	24 300	333.8	16.2	28.7

[a] RT = room temperature; 3.6 °C = 600 °F.
[b] To convert to cm, multiply by 2.54.
[c] To convert from in.lb/in.^{2} to MJ/m^2 multiply by 0.0001751.

Tests of the forged 304 stainless piping material again showed a temperature dependence of the data, although less pronounced than that of the cast piping, as seen in Fig. 10. The orientation of the specimens again appeared to have little effect on the results, although some scatter is evident. To determine further whether orientation was important, radially oriented 2.54-cm-thick (1.0 in.) compact specimens were machined, and test results are shown in Fig. 11. It can be seen that there is a very slight effect, in that J_{Ic} at 316°C (600°F) is somewhat lower for the radial direction while the slope of the J versus Δa curve is somewhat higher. However, a definite conclusion as to orientation effect cannot be reached because of the scatter in the data.

Compact specimens were also machined from 304 stainless plate material, and these 2.54-cm-thick (1.0 in.) specimens were tested at 316°C (600°F). Results are shown in Fig. 12, and indicate that the J_{Ic} value for this material is somewhat lower than for the piping steels tested at the same temperature, although the slope of the J versus Δa curve is slightly higher. Also, much less scatter is evident in these data.

Considerable difficulty was encountered in interpreting the data obtained from the three-point bend specimens, as shown in Fig. 13 and 14. The data display less scatter than the compact specimens, and have the same trends, in that the slope of the J versus Δa line decreases with temperature. Unlike the compact specimens, however, the slope of the J versus Δa line is nearly equal to that of the blunting line for both materials at

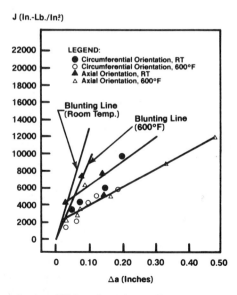

FIG. 10—J_{Ic} determination, 304 forged stainless steel, compact specimens (1 in. = 2.54 cm; 1 in.lb/in.^{2} = 0.0001751 MJ/m^2).

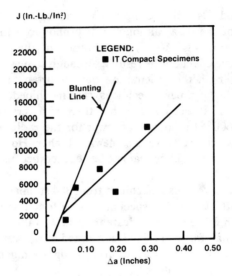

FIG. 11—J_{Ic} determination, 304 forged stainless steel, radial orientation, compact specimens (1 in. = 2.54 cm; 1 in. lb/in.^{2} = 0.0001751 MJ/m^2).

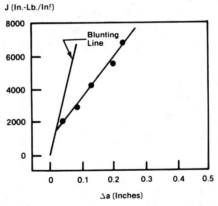

FIG. 12—J_{Ic} determination, 304 stainless steel plate, compact specimens (1 in. = 2.54 cm; 1 in. lb/in.^{2} = 0.0001751 MJ/m^2).

room temperature. This is a remarkable result, because it indicates that the crack has little or no influence on the failure of these specimens, that instead the specimen simply tears apart. This would be understandable if similar behavior were observed for the other specimen types, because it is well known that stainless steel is not particularly notch sensitive. But this implies that the fracture behavior of this material is somewhat geometry dependent, at least when portrayed in the manner presently recommended [1]. An alternative explanation is that the present blunting line concept

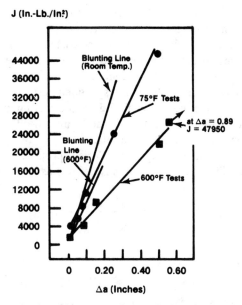

FIG. 13—J_{Ic} determination, 316 cast stainless steel, three-point bend specimens (1 in. = 2.54 cm; 1 in. lb/in.^{2} = 0.0001751 MJ/m^2).

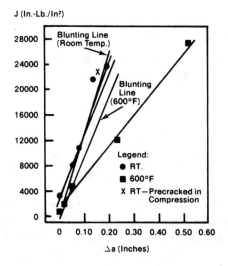

FIG. 14—J_{Ic} determination, 304 forged stainless steel, three-point bend specimens (1 in. = 2.54 cm; 1 in. lb/in.^{2} = 0.0001751 MJ/m^2).

needs to be modified to cover the full range of material fracture characterization, and to include specimen geometry effects.

Results of the bend specimen tests are consistent for the two materials and where portrayed in the recommended manner produce J_{Ic} values which are much different than results from compact specimens and center-cracked panels in all but one case. The bend specimen tests were very carefully done, and no anomalies could be found in the testing procedures, so it appears that the data presentation methods may need improvement for this type of test with very ductile materials.

As a sidelight to this investigation, Paris's proposed tearing modulus [7] was evaluated. The tearing modulus is defined as

$$t = \frac{E}{\sigma_0^2} \frac{dJ}{da}$$

where

$E =$ Young's modulus,
$\sigma_0 =$ flow stress, and
$dJ/da =$ slope of the J versus Δa curve.

Results of this calculation are given in Table 4, and show that the tearing modulus is quite large for this material. However, it is certainly not independent of specimen geometry, as has been proposed. This finding agrees with conclusions reached by several other investigators for other materials.

Comments on Size Requirements

The proposed size requirements of ASTM [1] apply to both the specimen thickness, B, and the remaining ligament, b, and are

$$B, b > 25J/\sigma_0$$

where $\sigma_0 =$ flow stress $= \frac{1}{2}(\sigma_y + \sigma_u)$.

All the specimens tested meet this requirement at J_{Ic} except the 2.54-cm (1.0 in.) radially oriented compact specimens, as shown in Table 4. This does not mean to imply, however, that the tests all meet the proposed size requirements, which apply to individual specimens. As may be seen in Table 4, the specimens are close to the limit of the requirements even at J_{Ic}, so many of the specimens do violate the criteria. As seen in Fig. 8 through 14, there is no apparent change in the fracture behavior once the proposed requirement is violated. The data remain on the same straight line, and may even display less scatter at longer crack lengths. This implies that the size requirement may be too restrictive for very ductile materials. Further, the very high strain hardening of these austenitic stainless steels implies an enhancement of the dominance of the crack tip singular field, and thus a lessening of the size requirement.

Several authors have recently proposed other criteria for applicability of the J-integral to characterization of elastic-plastic fracture. Even though the limits of these criteria are not yet well developed, it is of interest to calculate the parameters involved.

Hutchinson and Paris [8] proposed that one important requirement would be

$$w = \frac{b}{J}\frac{dJ}{da} \gg 1$$

where the symbols have been previously defined, to ensure that proportional loading takes place in the specimen. This parameter has been evaluated at $J = J_{Ic}$, and results are summarized in Table 4, showing that the parameter ranges from 13 to greater than 90 for the tests reported here. An interesting point is that the highest values of this parameter were obtained for the three-point bend specimens. All the data obtained for the stainless steels tested appear to meet this criterion.

McMeeking and Parks [9] have also proposed a criterion for J dominance of the crack tip field for a specimen, which will be called Q in this work

$$Q = \frac{b\sigma_0}{J}$$

where the symbols have been previously defined.

This parameter is also tabulated in Table 4. McMeeking and Parks claim that the parameter Q should be much greater than 200 for a center-cracked panel, but for bend type specimens the value of Q need not be nearly as high, although they make no quantitative recommendation. Table 4 shows clearly that the center-cracked panel does not meet their proposed value, but, since no recommendations were made for other specimens, the numbers are provided for information only.

Conclusions

1. Three specimen types were tested at two different temperatures, room temperature and 316°C (600°F). Of these, the most efficient specimen was found to be the compact specimen, although good agreement was obtained between compact and center-cracked panel specimens. The three-point bend specimens gave results which were inconsistent and difficult to inter-

pret, and thus should be avoided for characterizing very ductile materials according to the presently recommended practice. Note that these specimens have been found to be quite adequate for characterizing materials which do not harden extensively.

2. The only suitable methods for obtaining J versus crack extension information on very ductile materials were found to be the unloading compliance method and the multiple-specimen heat tinting technique.

3. The presently recommended procedures for data interpretation produce consistent results for compact specimens and center-cracked panels, but may need to be improved for three-point bend specimen results for ductile materials. The proposed validity criteria appear to be too restrictive for ductile materials, a conclusion which is supported by the consistent specimen behavior before and after violating the proposed requirement. Further evidence is provided by consideration of the validity criterion recently proposed by Hutchinson and Paris [8], which the specimens clearly meet.

4. Results of the tests show that the three materials were all very tough at both room temperature and 316°C (600°F), with J_{Ic} equal to about 0.79 MJ/m²(4500 in. lb/in.²) at room temperature, and ranging from 0.26 to 0.40 MJ/m² (1500 to 2500 in.-lb/in.²) at the higher temperatures. It is also important to note that J_{Ic} is a very conservative measure of the fracture resistance of this material, since considerable stable crack growth occurs prior to fracture. In one specimen, for example, a value of $J = 8.40$ MJ/m² (48 000 in.-lb/in.²) was sustained without failure.

Acknowledgment

The authors wish to express their appreciation for the helpful advice received from Jim Begley, John Landes, and Garth Clarke during the testing. Also thanks are due to Lou Ceschini, who performed the center-cracked panel tests, and to Andy Manhart, who assisted with the electrical potential and acoustic emission measurements.

References

[1] "Recommended Practice for the Determination of J_{Ic}" as detailed in correspondence from G. A. Clarke to ASTM Task Group E24.01.09 dated 10 March 1977.
[2] Clarke, G. A., Andrews, W. R., Paris, P. C., and Schmidt, D. W. in *Mechanics of Crack Growth, ASTM STP 590*, American Society for Testing and Materials 1976, pp. 27-42.
[3] *ASTM Book of Standards*, Part 10, American Society for Testing and Materials, 1976, pp. 471-490.
[4] Merkle, J. G. and Corten, H. T., *Transactions*, American Society of Mechanical Engineers, *Journal of Pressure Vessel Technology*, Series J, Vol. 96, No. 4, Nov. 1974, pp. 286-292.
[5] Bucci, R. J., Paris, P. C., Landes, J. D., and Rice, J. C. in *Fracture Toughness, ASTM STP 514*, American Society for Testing and Materials, 1972, pp. 40-69.

[6] Rice, J .R., Paris, P. C., and Merkle, J. C., in *Progress in Flaw Growth and Fracture Toughness Testing,* American Society for Testing and Materials, *ASTM STP 536,* 1973, pp. 231-245.
[7] Paris, P. C., Tada, H., Zahoor, A., and Ernst, H., this publication, pp. 5-36.
[8] Hutchinson, J. W. and Paris, P. C., this publication, pp. 37-64.
[9] McMeeking, R. M. and Parks, D. M., this publication, pp. 175-194.

Applications of Elastic-Plastic Methodology

G. G. Chell[1]

A Procedure for Incorporating Thermal and Residual Stresses into the Concept of a Failure Assessment Diagram

REFERENCE: Chell, G. G., "**A Procedure for Incorporating Thermal and Residual Stresses into the Concept of a Failure Assessment Diagram,**" *Elastic-Plastic Fracture, ASTM STP 668,* J. D. Landes, J. A. Begley, and G. A. Clarke, Eds., American Society for Testing and Materials, 1979, pp. 581–605.

ABSTRACT: The Failure Assessment Curve proposed by Harrison, Loosemore, and Milne has been rederived and interpreted in terms of an equivalent J-integral analysis. Comparison with computed values of *J* indicates that the curve is a good approximation to a lower-bound failure criterion for mechanical loading. The J-integral interpretation enables thermal, residual, and secondary stresses to be included within the concepts of the Failure Assessment Diagram. A procedure is introduced which transforms points on a failure diagram obtained, for instance, from a J-integral analysis, into approximately equivalent points on the Failure Assessment Diagram. This procedure is particularly useful for failure assessments involving thermal, residual, or secondary stresses where a plastic collapse parameter is not definable. The procedure assumes that a plastic stress intensity factor can be estimated. A method of assessing the severity of a mechanical load superposed on an initial constant load is also presented. Examples showing the applications and advantages of the technique are given.

KEY WORDS: failure criterion, assessment curve, J-integral analysis, thermal stresses, residual stresses, secondary stresses (fracture), assessment diagram, fracture (materials), elastic-plastic, post-yield, crack propagation

Nomenclature

L Generalized load
L_a Applied load
L_1 Plastic collapse load
L_f Load at fracture

[1] Fracture Mechanics Project leader, Materials Division, Central Electricity Research Laboratories, Kelvin Avenue, Leatherhead, Surrey, U. K.

L_K Fracture load determined using linear elastic fracture mechanics (LEFM)
L_i Initial loading
σ Generalized stress $= AL/Bw$
σ_a Applied stress
σ_1 Plastic collapse stress
σ_f Fracture stress
$\sigma_a{}^p$ Relaxed stress determined elastic-plastically
$\sigma_a{}^e$ Relaxed stress determined linear elastically
$\bar{\sigma}$ Flow stress
Y Geometric function
a Crack size
a' Effective crack length determined using Irwin's first-order plasticity correction
K_1 Stress intensity factor
K_p Plastic stress intensity factor $= \sqrt{E'J}$ or K_1 determined using effective crack length a'
K_c Fracture toughness
α Plastic constraint factor
J J-integral
J_1 J determined linear elastically, $= K_1{}^2/E'$
E Young's modulus
ν Poisson's ratio
K_r K_1/K_c, a failure assessment coordinate
$K_r{}^f$ Value of K_r at failure
$K_r{}^{f,Q}$ Value of $K_r{}^f$ lying on the Failure Assessment Curve immediately above the failure assessment point Q
$K_r{}^i$ Value of K_r due to initial loading
S_r σ_a/σ_1, a failure assessment coordinate
$S_r{}^f$ Value of S_r at failure
$S_r{}^{f,Q}$ Value of S_r corresponding to the point Q
$S_r{}^i$ Value of S_r due to initial loading
d Displacement at loading point
d_a Component of displacement due to uncracked body
B Thickness of component
W Width of component
A Geometric constant

Recently a procedure for assessing the integrity of cracked components in the linear elastic and post-yield fracture regimes has been proposed [1].[2] To simplify elastic-plastic fracture analyses a Failure Assessment Diagram was introduced. This enables the integrity of cracked plant to be ascertained

---

[2] The italic numbers in brackets refer to the list of references appended to this paper.

through two separate calculations based on the two extremes of fracture behavior, namely, linear elastic and fully plastic. These two calculations provide a point on the Failure Assessment Diagram, and the relative position of this point to a Failure Assessment Curve determines the integrity of the structure. If the point falls below the Failure Assessment Curve, the structure is safe; if on or above it, failure is predicted. The Failure Assessment Curve interpolates between the critical conditions necessary for fracture in the two extremes and is based on the fracture equation given in Ref 2 as modified and generalized in Ref 3.

The Failure Assessment Diagram provides a valuable contribution to a very complex problem. It not only reduces difficult concepts to an easily comprehensible pictorial representation, but also bypasses the need to perform detailed elastic-plastic calculations. Furthermore, progress has been made in validating the application of the Failure Assessment Diagram for real problems. Experimental data are available which demonstrate that the Failure Assessment Curve is a lower bound, and hence safe [1].

Although it is recommended that the Failure Assessment Diagram be used together with lower-bound data and conventional engineering safety factors, and therefore the detailed form of the Failure Assessment Curve is not important, it is still valuable to know how the Failure Assessment Curve compares with other possible curves based, for example, on a J-integral analysis. Furthermore, at the present time it is not at all clear how failure assessments which involve thermal or residual loadings are to be included on the Failure Assessment Diagram. These loadings do not contribute to the plastic collapse load as formally defined in plastic limit load theory, but may contribute strongly to fracture in the post-yield regime. It is the purpose of this paper to go some way toward answering these questions.

Analytical Basis of the Failure Assessment Diagram

The interpolation fracture formula used [3] is

$$L_f = L_1 \frac{2}{\pi} \cos^{-1} \left\{ \exp \left(- \frac{\pi^2 L_k^2}{8 L_1^2} \right) \right\} \quad (1)$$

where

L_f = generalized fracture load, for example, pressure,
L_1 = value of the generalized load corresponding to plastic collapse, and
L_k = fracture load determined from linear elastic fracture mechanics (LEFM).

Both L_1 and L_k depend on crack length. Equation 1 can be written in terms of a generalized stress, σ (the load L divided by a geometric constant of the dimensions of length squared) and to show the explicit dependence on the fracture toughness, K_c, as

$$\sigma_f = \sigma_1 \frac{2}{\pi} \cos^{-1}\left\{\exp\left(-\frac{\pi^2 K_c^2}{8aY^2\sigma_1^2}\right)\right\} \quad (2)$$

where a is the crack length. Y is a function dependent on flaw shape and size, structural geometry, and loading system, such that for an applied stress σ_a, the stress intensity factor, K_1, is given by

$$K_1 = \sigma_a \sqrt{a}\, Y \quad (3)$$

The suffixes f and 1 have the same meaning as for loads L. Equation 2 can be rearranged to read

$$K_c^2 = \frac{8}{\pi^2} aY^2 \sigma_1^2 \ln \sec\left(\frac{\pi \sigma_f}{2\sigma_1}\right) \quad (4)$$

In defining a plastic stress intensity factor, K_p, for an applied stress σ_a, as

$$K_p^2 = \frac{8}{\pi^2} aY^2 \sigma_1^2 \ln \sec\left(\frac{\pi \sigma_a}{2\sigma_1}\right) \quad (5)$$

it can be seen from Eq 4 that at fracture, when $\sigma_a = \sigma_f$

$$K_p = K_c$$

K_p incorporates the effects of plasticity. It can also be related to the J-integral [4] through the equation

$$K_p = \sqrt{E'J} \quad (6)$$

where $E' = E$, Young's modulus for plane stress, or $E' = E/(1 - \nu^2)$, where ν is Poisson's ratio, for plane strain. Hence, when fracture is governed by crack tip events, that is, characterized by K_c, the fracture Eq 1 can be interpreted as a post-yield fracture expression based upon an approximate functional form for the J-integral, given by Eqs 6 and 5 [5]. In this context Eqs 5 and 6 were proposed and successfully employed in obtaining valid fracture toughness values from invalid specimen test data [6].

Dividing both sides of Eq 4 by K_1^2, using Eq 3, and inverting the result gives

$$K_1/K_p = \left[\frac{8\sigma_1^2}{\pi^2\sigma_a^2} \ln \sec\left(\frac{\pi\sigma_a}{2\sigma_1}\right)\right]^{-1/2} \quad (7)$$

With $K_p = K_c$, $\sigma_a = \sigma_f$, and K_1 evaluated for σ_f, Eq 8 represents the Failure Assessment Curve [1], that is

$$K_1/K_c = \left[\frac{8\sigma_1^2}{\pi^2\sigma_f^2} \ln \sec\left(\frac{\pi\sigma_f}{2\sigma_1}\right)\right]^{-1/2} \quad (8)$$

It can be seen that Eq 7 has the same functional relationship between K_1/K_p and σ_a/σ_1 as that given by the Failure Assessment Curve for the dependence of K_1/K_c on σ_f/σ_1. Hence a failure curve may be obtained from any elastic-plastic theory which relates K_1/K_p or, equivalently, $(J_1/J)^{1/2}$ to σ_a/σ_1, where J_1 is the linear elastic value K_1^2/E'. This forms the basis of the present paper.

The Failure Assessment Diagram

This is constructed with respect to the two axes K and S where the coordinates of points are [1]

$$K_r = K_1/K_c$$

$$S_r = \sigma_a/\sigma_1$$

The distance of each point (K_r, S_r) from the origin is linearly proportional to the load characterizing parameter σ, as can be seen from the definition of K_r and S_r. Equation 8 is a curve on the diagram corresponding to the loci of points (K_r^f, S_r^f) which coincide with the onset of failure (Fig. 1). Therefore any loading or crack size which produces a point under the curve is safe; conversely, failure will occur if it is on or outside. If (K_r, S_r) is below the curve, then failure can occur either by increasing the load or the crack size. In the case of the former, the fracture load for a given crack length is easily determined using the property that the point (K_r, S_r) is linearly proportional in σ to its distance from the origin. In the case of the latter, the path traversed by the point (K_r, S_r) as it moves toward the Failure Assessment Curve will be called the path to failure. Such a path AB is shown in Fig. 1 as the crack length increases from a_1 to a_8. Once a path to failure has been calculated for a given loading, other paths to failure, or parts of them, for other loadings can be easily calculated using the same linear proportionality as before, and for each point on the path determining the corresponding point for the new loading. Such a path $A'B'$ corresponding to half the loading that generates the path AB is shown in Fig. 1.

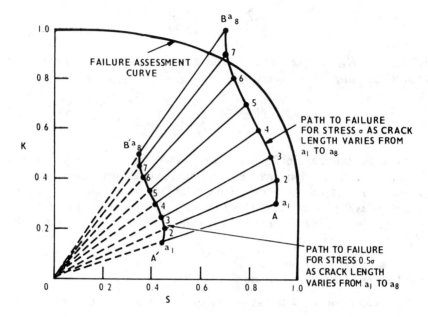

FIG. 1—*Failure Assessment Diagram.*

Examples of Failure Curves

Mechanical Loading

The post-yield solutions in Refs 2 and 3 are based on the Bilby, Cottrell, and Swinden [7]-Dugdale [8] (BCS-D) model of a yielded crack in an infinite body subject to uniform applied stress. Hence in this case, Eq 8, with σ_1 identified as the ultimate tensile strength, represents the Failure Assessment Curve. An analytical solution for the penny-shaped crack in tension is available [9] and can be expressed in terms of K_p [10] so that the equation of the failure curve can be written down directly as

$$K_1/K_p = (\sigma_a/\sigma_1) \{ 2[1 - (1 - \sigma_a^2/\sigma_1^2)^{1/2}] \}^{-1/2} \qquad (9)$$

This curve is shown in Fig. 2a.

There are several computed J against load curves in the literature. These have been used, together with the elastic solutions, to obtain failure curves for the following geometries: three-point bend with span-to-width ratios of 4 [11] and 8 [12], center-cracked plate in tension [13], and a crack emanating from a hole in a plate in tension [13]. All the computations were in plane strain and the results are shown in Fig. 2a. Plane-stress solutions for a

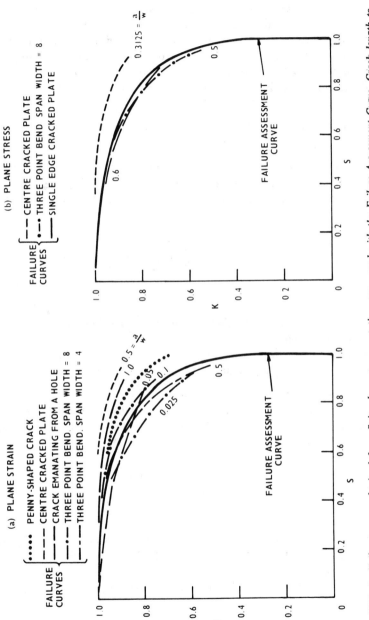

FIG. 2—*Failure curves obtained from finite-element computations compared with the Failure Assessment Curve. Crack length to plate width (a/w) or hole radius (a/r) ratios are indicated. The BCS-D model solution for a penny-shaped crack is also shown.*

single-edge cracked plate in tension,[3] three-point bend with span-to-width ratio of 8, and a center-cracked plate are shown in Fig. 2b.

From Fig. 2 it is clear that using a J analysis does not produce a universal failure curve. Associated with each structure is a set of failure curves depending on crack shape, size, and loading. However, the Failure Assessment Curve does provide a good approximation to a lower-bound curve. Furthermore, the differences between the curves are insignificant compared with the uncertainties inherent in an assessment of any real structure.

Fixed-Grip Loading

Using Eqs 6 and 5 to represent J, the effects of fixed-grip loading on K_p have been determined [5]. The result is given by Eq 5, where the symbols have the same meaning but now σ_a depends on the fixed displacement d and crack length. This dependence is obtained by solving the following equation for σ_a

$$d = d_a + \frac{B}{A} \frac{\partial}{\partial \sigma} \int_0^a J \, da \qquad (10)$$

where

$d_a =$ displacement due to the uncracked body subject to the effective, relaxed load L_a,
$B =$ thickness of the body, and
$A =$ a geometric constant such that $\sigma_a = L_a/A$ and $J = K_p^2/E'$.

The elastic solution is obtained by writing $J = K_1^2/E'$.

The dependence of σ_a on d and a obtained linear elastically differs from that obtained elastic-plastically. The failure curve is thus represented by the equation

$$K_1/K_p = \left[\frac{8}{\pi^2} \left(\frac{\sigma_1}{\sigma_a^e} \right)^2 \ln \sec \left(\frac{\pi \sigma_a^p}{2\sigma_1} \right) \right]^{-1/2} \qquad (11)$$

where superscripts e and p signify the values of σ_a which satisfy Eq 10 in the elastic and elastic-plastic cases, respectively. The results of evaluating Eq 11 subject to Eq 10 are shown in Fig. 3 for center-cracked plate and three-point bend geometries. For these, σ_1 is given by

$$\sigma_1 = \bar{\sigma} \, (1 - a/w)$$

---

[3] Neale, B. K., private communication.

and

$$\sigma_1 = 2\bar{\sigma}(1 - a/w)^2$$

respectively, where $\bar{\sigma}$ is a flow stress. Further, $S_r = \sigma_a^e/\sigma_1$ to maintain the linear relationship between loading (either stress σ_a^e or displacement d in this case) and the distance of the point (K_r, S_r) from the origin. The center-cracked plate results are given for two normalized gage lengths L^*, that is, separation of loading pins divided by half the plate width w.

From Fig. 3 it is clear that now the failure curves not only depend strongly on the specimen geometry and crack length, but also on the gage length over which the displacements are imposed. Two things are immediately striking about Fig. 3. The first is that S_r can exceed 1 without failure occurring. This is due to the definition of S_r in terms of the elastically determined relaxed stress σ_a^e. The presence of plasticity further relaxes the induced stress, so that although σ_a^e may exceed σ_1, σ_a^p cannot. The second is that K_1/K_p can exceed 1. This can happen when the displacement loading is such that K_1 decreases with increasing crack length. Since, in a sense, plasticity increases the effective crack length (for example,

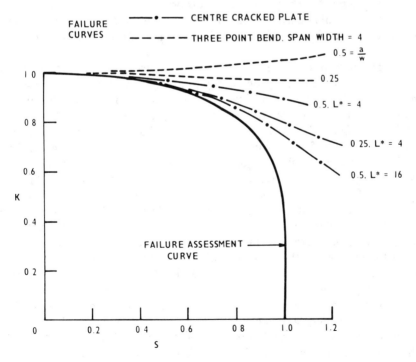

FIG. 3—*Failure curves for fixed-grip loading.*

Irwin's first-order plasticity correction procedure involves adding a term proportional to K_1^2 to the original crack length, and treating the new crack linear elastically), K_p will be less than K_1 if the latter decreases with crack length. This is the case for three-point bend [15] and explains the apparent anomalous behavior.

The foregoing examples demonstrate that cracked structures subject to secondary loading can be described in terms of the failure curve concept, but that in some cases the Failure Assessment Curve may be far too conservative as a lower bound.

Approximate Transformation of Failure Assessment Points

To avoid the risk of overconservatism in, for example, secondary loading, and to take into account thermal or residual loading, it could be argued that every structure should be looked at independently with reference to its own failure curve. This is unsatisfactory since it would involve the generation of a multiplicity of failure curves to cover every situation. This can be avoided, however, if failure assessments can be referred to a single curve. Thus it is useful to produce a means of transforming failure assessment points from a given failure diagram to corresponding points on the Failure Assessment Diagram. The transformation should not depend on the failure coordinate S_r; this parameter is not defined for self-equilibrated thermal and residual loadings since these do not affect the plastic collapse load.

Let all quantities that refer to the actual failure diagram be denoted by primes. A typical diagram containing a failure curve $A'B'$ and an assessment point P' with coordinates (K_r', S_r') is shown in Fig. 4a. If the applied loading is increased, failure will occur at the point P_f' with coordinates $(K_r^{f'}, S_r^{f'})$. A simple graphical procedure for locating an equivalent point P and failure point P_f on the Failure Assessment Diagram that avoids specifying a value for S_r' is easily constructed by putting $K_r = K_r'$ and $K_r^f = K_r^{f'}$, drawing in the line $0P_f$, and hence finding the coordinates S_r and S_r^f as shown in Fig. 4b, where superscript f signifies that K_r and S_r are evaluated at the failure load. This procedure is satisfactory provided the failure curve $A'B'$ is known, but this requires a comprehensive elastic-plastic fracture analysis for its determination. Furthermore, as just mentioned, if the applied loading is thermal or residual, the failure curve cannot be constructed at all because S'_r is not known. These two disadvantages limit the scope of this simple graphical technique. However, an approximate means of obtaining an equivalent failure assessment point to P' on the Failure Assessment Diagram, which does not involve either a detailed knowledge of the failure curve $A'B'$ or S_r', is possible, provided that a pessimistic evaluation of the elastic-plastic stress intensity factor K_p is performed at the load level corresponding to P'. This pessimism is

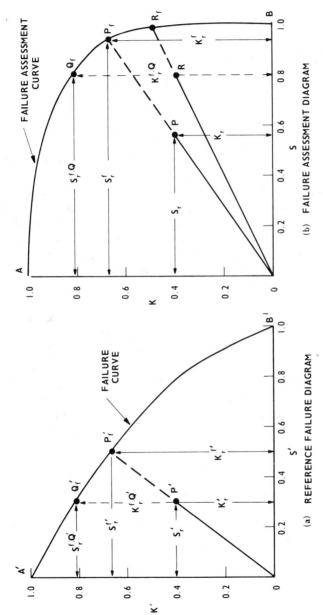

FIG. 4—*Schematic representation of procedure that transforms the failure assessment point P' to an approximately equivalent point R on the Failure Assessment Diagram.*

compatible with normal engineering safety practice, as well as the safety procedures recommended when using the Failure Assessment Diagram [1].

Since, as can be seen from Eq 7, a failure curve is obtained by plotting K_1/K_p as a function of applied load, it is clear that the point $Q_f{}'$ which lies on the failure curve $A'B'$ immediately above the point P' (Fig. 4a) has coordinates $(K_r^{f,Q'}, S_r^{f,Q'})$ where $S_r^{f,Q'} = S_r{}'$ and $K_r^{f,Q'} = K_1/K_p$. The equivalent point to $Q_f{}'$ on the Failure Assessment Diagram is Q_f with coordinates $(K_r^{f,Q'}, S_r^{f,Q'})$ (Fig. 4b), where $S_r^{f,Q}$ can be obtained graphically. Hence an approximately equivalent point to P' on the Failure Assessment Diagram is given by the point R with coordinates $(K_r{}', S_r^{f,Q})$ and this will be a pessimistic point provided that $OR_f/OR < OP_f{}'/OP$, where R_f is the point where the line OR cuts the Failure Assessment Curve.

This approximate procedure will, of course, be exact if the actual failure curve is coincident with the Failure Assessment Curve. The accuracy of the approximations and the degree of pessimism therefore depend on the assumption that the Failure Assessment Curve provides a realistic lower-bound failure criterion. The results just given indicate that for mechanical and secondary loading this is the case. The procedure will not work if $K_1/K_p > 1$, a situation which may occur if K_1 decreases with increasing crack length. In these cases a pessimistic assessment can be made using linear elastic fracture mechanics and putting $S_r^{f,Q} = 0$, so that failure is predicted when $K_r = 1$.

Even when thermal and residual stresses are self-equilibrated within the substructure subject to a failure analysis, they can still contribute substantially to the J-integral. Within the context of the BCS-D model of a crack with plastic yielding, thermal stresses can result in a value of K_p which greatly exceeds K_1 [5,18]. This conclusion has recently been verified by elastic-plastic finite-element calculations performed on a center-cracked sheet subjected to thermal loading [17]. Hence thermal and residual stresses can contribute to elastic-plastic failures by reducing the tolerance of the structure to mechanical loading. Since the failure parameter S_r is a measure of elastic-plastic effects (zero for purely elastic, one for purely plastic), its definition must be modified to incorporate thermal and residual stresses. This can be accomplished using the transformation technique developed in the foregoing in order to obtain an apparent value of S_r, and hence an equivalent failure assessment point, on the Failure Assessment Diagram.

The graphical procedure outlined for determining the point R is straightforward. However, it is computationally convenient to have an analytical method for obtaining the coordinates of the approximate assessment point R. The Failure Assessment Curve is described by the relationship

$$K_r^f = F(S_r^f) \tag{13}$$

where $F(S_r)$ is the function on the right-hand side of Eq 7 or 8. For a given value of $K_r{}^f$, $S_r{}^f$ can be obtained as the solution of Eq 13. A solution which is 96 percent accurate is given by

$$S_r{}^f = \left[1 - \exp\left\{ \frac{-24}{\pi^2}\left[\frac{1}{(K_r{}^f)^2} - 1\right] \right\} \right]^{1/2} \quad (14)$$

Thus, putting $K_r{}^f = K_r{}^{f,Q'}$ in Eq 14 gives the coordinate $S_r{}^{f,Q}$ and hence the coordinates of the point R as $(K_r{}', S_r{}^{f,Q})$.

Example of Transformation Procedure: Secondary Loading

Secondary stresses may arise, for example, when one part of a structure is heated or cooled with respect to another part and, as a consequence, impose boundary displacement loading on the other part. An example of the use of the transformation procedure for such a case is given in the following.

Consider a center-cracked plate subject to fixed displacement d. The relevant details concerning the values of the initial stress, fracture toughness, etc. are shown in Fig. 5. The plate contains a sharp defect of length-to-width ratio (a/w) of 0.2. Two things are required: first, the path to failure, so that the critical defect size can be determined under the given displacement loading, and second, the maximum displacement that can be imposed before the initial defect will result in failure.

In the first part of the calculation, K_1/K_p is evaluated for $a/w = 0.2$, 0.3, 0.4, and 0.5 using Eqs 10 and 11. The transformed points given by $K_r{}' (= K_1/K_c)$ and $S_r{}^{f,Q}$ (obtained from Eq 14 with $K_r{}^f = K_1/K_p$) are then plotted on the Failure Assessment Diagram (Fig. 5a). It is clear that failure occurs at the point P_1 corresponding to $a/w = 0.5$ on the path to failure. The second part is easily calculated. The distance of the transformed point, corresponding to $a/w = 0.2$, from the origin is measured on the Failure Assessment Diagram. The ratio of this distance divided by the length of the line to the Failure Assessment Curve which passes through the origin and the transformed point ($0P_2$) is determined. The original displacement of 0.3 mm multiplied by the inverse of this ratio gives the maximum allowable displacement to be 0.4 mm. The points on the diagram corresponding to displacements of 0.375 and 0.45 mm are also shown in Fig. 5a for comparison.

The path to failure and the effect of increasing displacement are shown in Fig. 5b with respect to the reference failure diagram. Also shown are the failure curves corresponding to $a/w = 0.5$ and 0.2. The failure points $P_1{}'$ and $P_2{}'$ are equivalent to the two points P_1 and P_2 in Fig. 5a. It can be seen that in this case the approximate transformation procedures predict a displacement at failure in good agreement with the value of about 0.4 mm obtained using the reference failure diagram.

594 ELASTIC-PLASTIC FRACTURE

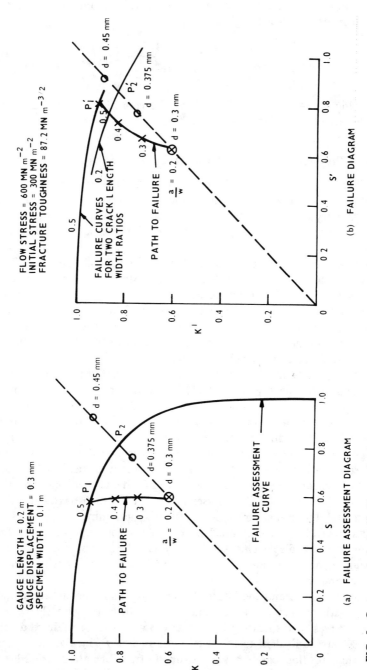

FIG. 5—*Center-cracked plate subject to fixed-grip displacement. Schematic representation of failure due to increase in crack length at constant displacement* (X) *and increase in displacement at constant crack length* (O). *Points* P_1 *and* P_2 *correspond to failure* (a) *and are the transformed equivalents to* P_1' *and* P_2' (b).

Initial Loads

A situation which may often occur in practice is when a load L is added to a structure which already experiences a constant initial load L_i which is independent of L. Here an assessment is required with respect to the load L. The initial load could arise, for example, from residual stresses in a weld that has not been stress relieved. If L_i were applied mechanically, then the plastic collapse load, L_l, would depend on L_i. If the initial loading were by self-equilibrated thermal or residual stresses, then the plastic collapse load would be independent of L_i. In both cases, however, L_i would produce a failure assessment point on the Failure Assessment Diagram. The coordinates of this point ($K_r{}^i$, $S_r{}^i$) for mechanical loads can be obtained following the procedures advocated in Ref 1. For thermal or residual stresses they can be determined using the transformation procedure after evaluation of the ratio K_1/K_p. Such an assessment point P is shown in Fig. 6a. Any additional loading will cause the assessment point P to move due to an increase in K_r plus a further shift in S_r. It is therefore proposed that relative to the added load L, the initial loading has the effect of moving the origin from 0 to the point P while still retaining the same Failure Assessment Curve centered on the origin at 0, the differences being that the K-axis records a nonzero value when $L = 0$ due to the loading L_i, and the new S-axis, which still goes from 0 to 1, is contracted into the reduced length $1 - S_r{}^i$ [PB'' in Fig. 6b, where a B'' corresponds to the point ($K_r{}^i$, 1)] relative to the axis $0B$. The new Failure Assessment Curve relative to the new coordinate system is $A'B'$, and the assessment point P' for the load L is linearly proportional to its distance PP' from the new origin at P. The logic of this construction can be seen when the load L_i is of the same type as the added load L. In this case it can be shown that the point P' with coordinates ($K_r{}'$, $S_r{}'$) determined with respect to the origin at P is, for assessment purposes, equivalent to the point P' with coordinates (K_r, S_r) determined with respect to the origin 0 on the Failure Assessment Diagram.

Examples of Initial Loading

Thermal and Mechanical Loading

Recently the effects of thermal loading on a center-cracked plate have been calculated using an elastic-plastic finite-element program [17]. The plate, of width $2w$, contained a crack of length $a/w = 0.26$ and was subjected to a uniform applied tensile stress σ_a superposed on the thermally induced stress $\sigma(x)$

$$\frac{\sigma(x)}{\bar{\sigma}} = 0.6 - 1.8(x/w)^2$$

596 ELASTIC-PLASTIC FRACTURE

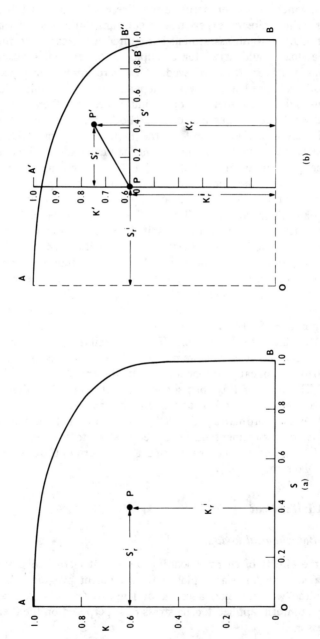

FIG. 6—*Assessment point P for initial constant load (a) and how it forms the origin for an assessment point based on an additional load, and part of the Failure Assessment Curve A'B'.*

where $\bar{\sigma}$ equals the yield stress and the center of the plate is the origin of coordinates. Since both J_1 and J were determined as a function of σ_a, a failure curve, shown in Fig. 7a as a dashed line, can be drawn where S_r is defined in terms of the mechanical load. Also shown, for reference, is the failure curve corresponding to the case of mechanical loading only (dotted line) as well as the Failure Assessment Curve (full line).

Following the procedures outlined in the previous section, all points on the failure curve AB can be transformed to points lying on a failure curve $A'B''$ on a new failure diagram with origin P and axes K' and S' (Fig. 7b). The coordinates of P with respect to the origin at 0 are (K_r^i, S_r^i) where the value of K_r^i will depend on the value of K_c, but the value of S_r^i will be given by the transformation procedure applied to the ratio K_1/K_p when $\sigma_a = 0$. In the present case the ratio is 0.98, so that the value of S_r^i obtained either graphically or using Eq 14, with $K_r^f = 0.98$, is approximately 0.3. Using this value, the transferred failure curve, $A'B''$, was constructed and, as can be seen in Fig. 7b, it is very similar to the part of the Failure Assessment Curve $A'B'$ which traverses the new coordinate system.

To illustrate the failure assessment procedure, consider the following example, which is to find the applied stress at failure for the given initial thermal loading and crack length. The material properties and plate dimensions are given in Fig. 7. When $\sigma_a = 0$, $K_1 = 40.3$ and hence the initial assessment point, P, forming the origin of the new axes, has the coordinates ($K_r^i = 0.403$, $S_r^i = 0.3$) since, from the foregoing, $K_r^f = K_1/K_p = 0.98$. When $\sigma_a = \sigma_1 = 440$ MNm^{-2}, $K_1 = 105.5$ MNm$^{-3/2}$ and the assessment point coordinates with respect to the new axes K' and S' are ($K_r' = 1.055$, $S_r' = 1$). This is shown as the point P' in Fig. 7b. The approximate procedure based upon the Failure Assessment Curve $A'B'$ gives the failure point as P_f, and hence the failure stress, σ_a^f, is (PP_f/PP') $\times$ 440 MNm^{-2} = 292 MNm^{-2}. The equivalent assessment points on the reference failure diagram (Fig. 7a) are also marked P, P_f', and P', and the failure stress is calculated to be 277 MNm^{-2}, in good agreement with the approximate answer. The value of 277 MNm^{-2} could have been obtained, of course, from Fig. 7b as ($PP_{f'}/PP'$) $\times$ 440 MNm^{-2}.

The predictions of LEFM are obtained by taking the failure curve to be the line $K_r = 1$. Thus from Fig. 7 the failure stress is (PP''/PP') $\times$ 440 MNm^{-2} = 402 MNm^{-2}, considerably in excess of the value obtained from the reference curve.

Residual and Mechanical Loading

The same procedure can be followed here as for the thermal loading. Consider the initial residual stress

$$\frac{\sigma(x)}{\bar{\sigma}} = 0.8 + 1.0667 \left|\frac{x}{w}\right| - 4\left(\frac{x}{w}\right)^2$$

598 ELASTIC-PLASTIC FRACTURE

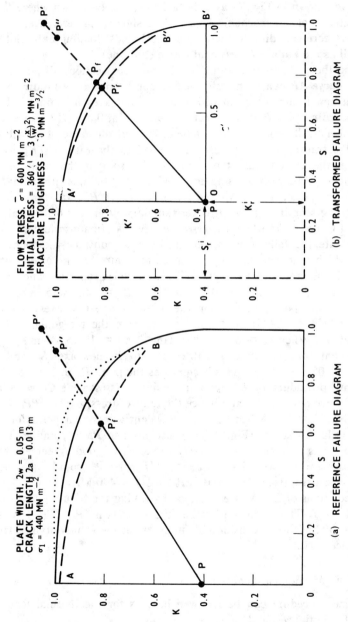

FIG. 7—*Example of initial loading illustrating the transformation procedure: a center-cracked plate subjected to thermal loading and an additional uniform tensile stress. The figures show the Failure Assessment Curve (full line) and the calculated curves for mechanical and thermal loading (dashed line) and mechanical loading (dotted line).*

which is symmetric about the center of a plate of width $2w$ and the additional mechanical tensile stress σ_a. An elastic-plastic solution to this problem can be obtained using the strip yielding model solutions for plates of finite width [*18*]. These model solutions will give, in general, pessimistic values of K_p as a function of σ_a for the center cracks of varying sizes, and hence enable pessimistic reference failure curves to be determined. These are shown in Fig. 8 for $a/w = 0.05$, 0.2, and 0.5, together with the material constants and plate dimensions. Using the procedures outlined in the foregoing, the three failure curves can be transformed into equivalent curves on the Failure Assessment Diagram.

Figure 8*b* shows these curves superposed on part of the Failure Assessment Diagram where the S axis has, for convenience, been magnified by two. Each of the curves has an associated origin corresponding to the coordinate $S_r{}^i$, which has the values 0.885, 0.81, and 0 for the crack sizes $a/w = 0.05$, 0.2, and 0.5, respectively. The $K_r{}^i$ coordinates are 0.260, 0.547, and 0.792, corresponding to K_1 values of 26.0 MNm$^{-3/2}$, 54.7 MNm$^{-3/2}$, and 79.2 MNm$^{-3/2}$ when $\sigma_a = 0$. Relative to the transformed axes (K' and S' for $a/w = 0.05$ and K'' and S'' for $a/w = 0.2$), the failure assessment points, P' and Q', evaluated for the plastic collapse stresses, 475 MNm^{-2}, 400 MNm^{-2}, are ($K_r{}' = 0.558$, $S_r{}' = 1$) and ($K_r{}'' = 1.06$, $S_r{}'' = 1$).

The equivalent points on the reference failure diagram are shown in Fig. 8*a*. The approximate transformation procedure then gives the following failure stresses: for $a/w = 0.05$, $\sigma_a{}^f = (PP_f/PP') \times 475$ MNm^{-2} = 405 MNm^{-2} compared with the reference value of $(PP_{f'}/PP') \times 475 = 266$ MNm^{-2}; for $a/w = 0.2$, $\sigma_a{}^f = (QQ_f/QQ') \times 400 = 159$ MNm^{-2} compared with the reference value of $(QQ_f/QQ') \times 400 = 150$ MNm^{-2}. In this case the failure stresses are higher than the reference failure stresses and to a large extent this is probably due to the pessimistic failure curves resulting from an analysis based upon strip yielding model solutions. The failure stresses obtained using linear elasticity are 1191 MNm^{-2} for $a/w = 0.05$ and 353 MNm^{-2} for $a/w = 0.2$.

In Fig. 9 is shown a series of paths to failures which were determined using the transformation procedure and the results from the residual stress example. The three paths (dashed lines) correspond to applied stresses, σ_a, of 0, 100, and 200 MNm^{-2}. From these paths to failure it can be seen that when $\sigma_a = 100$ MNm^{-2}, failure is predicted for a crack length $a/w = 0.34$, and when $\sigma_a = 200$ MNm^{-2}, the critical defect size is $a/w \simeq 0.16$, assuming as before that $K_c = 100$ MNm$^{-3/2}$. The results obtained from the strip yielding model are $a/w \simeq 0.33$ and 0.11, respectively. Again, as in the case of failure loads, the shape of the failure curve for small crack sizes means that the approximate transformation procedure is optimistic. The critical defect sizes obtained from a linear elastic analysis are approximately $a/w \simeq 0.45$ and 0.31 for the two stress levels, showing

600 ELASTIC-PLASTIC FRACTURE

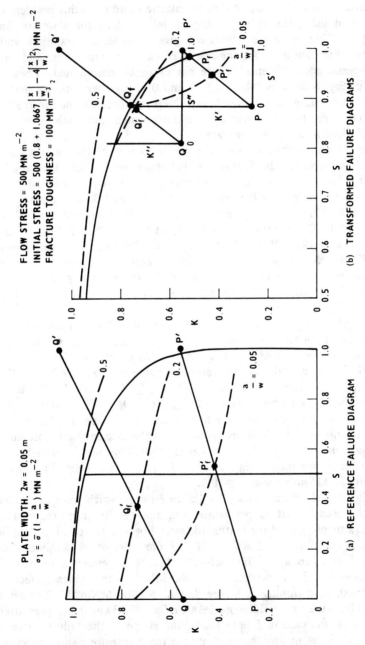

FIG. 8—*Example of initial loading: a center-cracked plate subjected to a uniform tensile stress superposed on a residual stress. The dashed lines are failure curves for different crack lengths, the full curve the Failure Assessment Curve.*

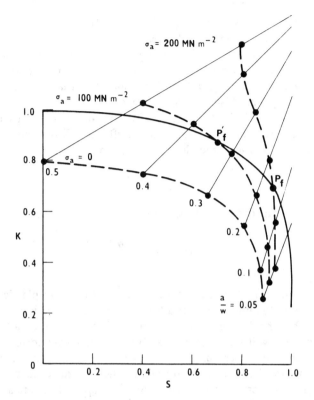

FIG. 9—*Construction of paths to failure (dashed lines) for three different stress levels superposed on an initial residual stress.*

that the transformation procedure is again much superior to this approximation.

It is interesting to note that because the origins, to which assessments are referred, are dependent on the defect size, the paths to failure are no longer similar as the applied load is increased. This is in contrast to the case when only a single mechanical load is applied (see Fig. 1).

Discussion

From the two examples it can be seen that the approximate transformation procedure based upon the Failure Assessment Curve can, in some circumstances, produce an optimistic value for the failure load. This is particularly true if the initial loading is so severe that the ratio K_1/K_p evaluated without additional mechanical loading is less than about 0.8 and if the actual failure assessment curve differs considerably in shape from the Failure Assessment Curve (see Fig. 2a; $a/w = 0.05$). However, the technique still produces answers which are considerably better than those given using linear elastic fracture mechanics LEFM.

One of the advantages and simplifying features of the Failure Assessment Diagram concept is that only the elastic and fully plastic solutions are necessary for a failure analysis. In order to maintain that principle, methods of specifying K_p based solely on elastic analyses are required, and the obvious method would be to adopt Irwin's first-order plasticity correction as a means of incorporating elastic-plastic effects. Within this approximation the value of K_p for a crack length a is given as

$$K_p = K_1(a')$$

where a' is an effective crack length given by

$$a' = a + \alpha \left(\frac{K_1(a)}{\bar{\sigma}}\right)^2 \tag{15}$$

and α is a constant which incorporates the constraint on plastic deformation. For initial loading by self-equilibrated thermal or residual stresses, it is unlikely that the degree of plastic deformation prior to the addition of a mechanical load will be so great that Irwin's first-order correction, along with a pessimistic choice for α, will not be adequate enough to give an upper bound to K_p. This estimate would then produce an upper-bound value for S_r^i. Suggested values for α are, perhaps, the plane-stress value of $\pi/16$ to be used for plane-strain problems, and double this value, $\pi/8$, to be used for plane stress or where loss of plastic constraint is possible. In the residual stress example, remembering that the strip yield model simulates plane stress, this choice would have produced a failure stress of approximately 325 MNm^{-2} compared with the reference value of 266 MNm^{-2}, and the previously calculated approximate value of 405 MNm^{-2}.

The foregoing problem is encountered only when the initial loading is thermal or residual. If it is mechanical, then the Failure Assessment Curve provides a realistic lower-bound failure criterion. Furthermore, the shift in the position of the origin for initial mechanical loading is determined using the Failure Assessment Curve procedures [1] and hence will already have built-in safety factors.

In tough materials, particularly those which fail by ductile mechanisms, stable crack growth may precede an instability leading to fast fracture. In these circumstances it is recommended that the toughness value at the initiation of crack growth be used in the evaluation of the assessment parameter K_r [1]. The problem of predicting the instability point is a very difficult one and depends not only on material properties but also on the loading system and the geometry of the component. In the case of ductile failures the problem has recently been discussed [*19*]. In terms of the Failure Assessment Diagram the effects of system loading conditions are reflected in the ratio K_I/K_p (compare the earlier section on fixed-grip loading) and hence the approximate transformation procedure adequately caters for these as regards an assessment based upon initiation.

Some of the advantages of basing failure assessments on the Failure Assessment Diagram rather than on an explicit elastic-plastic failure analysis have already been mentioned. Besides the simplicity of a diagrammatical representation, and the ease with which changes in the applied loading can be determined, the Failure Assessment Diagram also allows the effects of variations in fracture toughness and flow stress to be studied. For example, the effect of dividing the toughness (or flow stress) by two is to double the value of K_r (or S_r) and move the failure assessment point accordingly.

If initial loading is present, the same procedures apply for variations in toughness, but the effects of changes in flow stress on the apparent value of S_r due to the initial load are more complicated, but still relatively easily calculated. By applying the transformation procedure to initial loading, most of the advantages of the Failure Assessment Diagram can be recovered with respect to additional mechanical loading.

Summary of Failure Assessment Procedure

A Single Mechanical Load

The procedure in this case for obtaining a failure assessment point is outlined in Ref *1*. To allow for crack growth due to fatigue, for example, it is useful to draw a path to failure on the Failure Assessment Diagram for a particular load level. Other paths to failure are then easily constructed (see Fig. 1) and hence useful information extracted, such as the variation of critical defect size with applied load. For each crack length, only two calculations are necessary, a linear elastic determination of K_I to obtain K_r, and a limit analysis for σ_1 to obtain S_r. If the assessment point is above or on the Failure Assessment Curve, failure is predicted; if below, the structure is safe at the given load.

A Mechanical Load Superposed on a Constant Initial Mechanical Load

The same procedure as in the foregoing is followed to obtain a failure assessment point (K_r, S_r) for the constant initial load. This point then forms the origin of a new failure diagram where the K-axis goes from K_r to 1 and the S-axis goes from 0 to 1 and is contracted into the length $1 - S_r$. With respect to these new axes, the failure assessment procedure is then repeated for the additional loading to obtain an assessment point (K_r', S_r'). The magnitude of the additional load is directly proportional to the distance of the point (K_r', S_r') from the new origin.

A Mechanical Load Superposed on a Constant, Initial Thermal, or Residual Stress

The ratio $K_r^f = K_1/K_p$ due to the initial loading must first be determined. This can be accomplished using an elastic analysis and utilizing Irwin's first-order plasticity correction factor made pessimistic by a suitable choice of α in Eq 15. Then, either graphically or using Eq 14, the apparent value of S_r corresponding to K_r^f can be found. The point (K_r, S_r), where K_r is obtained from the elastic stress intensity factor due to the initial loading, now forms the origin of a new failure diagram based on the old K-axis and a reduced S-axis, as in the case of initial mechanical loads described in the foregoing. The assessment point due to any additional mechanical load with respect to the new axes is then found as before.

Conclusions

1. The Failure Assessment Curve provides a realistic lower-bound failure criterion for most mechanical loading situations.
2. The curve can be interpreted as being equivalent to an elastic-plastic analysis based upon an approximate functional form for the J-integral.
3. Using this interpretation, failure curves can be constructed from any elastic-plastic analysis which relates the ratio of elastic to plastic stress intensity factors to the applied stress divided by the collapse stress.
4. This interpretation also allows thermal, residual, and secondary stresses to be included within the framework of a failure diagram.
5. In the presence of initial loading, either by mechanical loads or by thermal and residual stresses, a failure assessment point can be found which forms the origin of coordinates with respect to a new failure diagram based on part of the Failure Assessment Curve. The assessment of any additional loading must be made with respect to a set of axes centered at this new origin of coordinates.
6. This procedure allows thermal and residual stresses to be incorporated within the concept of the Failure Assessment Diagram.

Acknowledgment

The author would like to thank Dr. I. Milne for helpful discussions and Dr. I. L. Mogford for his comments on the manuscript.

This work was performed at the Central Electricity Research Laboratories and is published by permission of the Central Electricity Generating Board.

References

[1] Harrison, R. P., Loosemore, K., and Milne, I., "Assessment of the Integrity of Structures Containing Defects," CEGB Report No. R/H/R6, Central Electricity Generating Board, U.K., 1976.
[2] Heald, P. T., Spink, G. M., and Worthington, P. J., *Materials Science and Engineering*, Vol. 10, 1972, p. 129.
[3] Dowling, A. R. and Townley, C. H. A., *International Journal of Pressure Vessels and Piping*, Vol. 3, 1975, p. 77.
[4] Rice, J. R. in *Mathematical analysis in the mechanics of fracture*, Vol. 2 (H. Liebowitz, Ed.), Academic Press, New York, 1968, p. 191.
[5] Chell, G. G. and Ewing, D. J. F., *International Journal of Fracture Mechanics*, Vol. 13, 1977, p. 467.
[6] Chell, G. G. and Milne, I., *Materials Science and Engineering*, Vol. 22, 1976, p. 249.
[7] Bilby, B. A., Cottrell, A. H., and Swinden, K. H. in *Proceedings*, Royal Society, Vol. A272, 1963, p. 304.
[8] Dugdale, D. S., *Journal of the Mechanics and Physics of Solids*, Vol. 8, 1960, p. 100.
[9] Keer, L. M. and Mura, J. in *Proceedings*, 1st International Conference on Fracture, T. Yokobori, T. Kawasaki, and J. L. Swedlow Eds., Published by Japanese Society for Strength and Fracture of Materials, Tokyo, Vol. 1, 1965, p. 99.
[10] Chell, G. G., *Engineering Fracture Mechanics*, Vol. 9, 1977, p. 55.
[11] Hayes, D. J. and Turner, C. E., *International Journal of Fracture*, Vol. 10, 1974 p. 17.
[12] Sumpter, J. D. G. and Turner, C. E., work reported by P. Chuahan in General Electric Co. Report No. W/QM/1974-14, 1974.
[13] Sumpter, J. D. G., "Elastic-Plastic Fracture Analysis and Design Using the Finite Element Method," Ph.D. thesis, Imperial College, London, U.K., 1973.
[14] Andersson H., *Journal of the Mechanics and Physics of Solids*, Vol. 20, 1972, p. 33.
[15] Chell, G. G. and Harrison, R. P., *Engineering Fracture Mechanics*, Vol. 7, 1975, p. 193.
[16] Chell, G. G., *International Journal of Pressure Vessels and Piping*, Vol. 5, 1977, p. 123.
[17] Ainsworth, R. A., Neale, B. K., and Price, R. H. in *Proceedings*, Conference on the Tolerance of Flaws in Pressurized Components, Institution of Mechanical Engineers, London, U.K., 16–18 May 1978.
[18] Chell, G. G., *International Journal of Fracture Mechanics*, Vol. 12, 1976, p. 135.
[19] Paris, P. C., Tada, H., Zahoor, A., and Ernst, H., "A Treatment of the Subject of Tearing Instability," USNRC Report NUREG-0311, National Research Council, Aug. 1977.

J. D. Harrison,[1] *M. G. Dawes,*[1] *G. L. Archer,*[1] *and M. S. Kamath*[1]

The COD Approach and Its Application to Welded Structures

REFERENCE: Harrison, J. D., Dawes, M. G., Archer, G. L., and Kamath, M. S., **"The COD Approach and Its Application to Welded Structures,"** *Elastic-Plastic Fracture, ASTM STP 668,* J. D. Landes, J. A. Begley, and G. A Clarke, Eds., American Society for Testing and Materials, 1979, pp. 606–631.

ABSTRACT: The crack opening displacement (COD) approach has, since its inception as a fracture initiation parameter in yielding fracture mechanics, gained increasing acceptance both as a viable research tool and an engineering design concept. In the United Kingdom, The Welding Institute has pioneered the application of COD in the structural fabrication industry largely through the development of the COD design curve. This paper is a representation of the philosophy underlying design curve applications and illustrates the practical significance of COD by drawing on case studies from various welded structures.

Following a brief appraisal of the origins of the design curve, the paper outlines procedures for the use of COD in design, that is, either as a basis for material selection or in setting up acceptance levels for weld defects. The reliability of a small-scale test prediction from the design curve has been investigated on a statistical basis from a survey of more than 70 wide-plate tension test results in which the material had also been categorized by COD. Specific practical examples are then discussed covering the various types of application, material selection defect assessment, and failure investigation.

Structures included in these examples are offshore rigs, oil and gas pipelines, pressure vessels, etc., with special emphasis on the manner in which small-scale COD test results are translated to the structural situation.

Finally, the paper includes information on the considerable range of structures to which the concept has been applied during the past five years.

KEY WORDS: mechanical properties, fracture test, crack initiation, toughness, crack opening displacement J-integral, elastic-plastic cracking (fracturing), fracture properties, structural steels, design, crack propagation

---

[1] Head of Engineering Research, principal research engineers, and research engineer, respectively, The Welding Institute, Abington Hall, Abington, Cambridge, England.

Nomenclature

- a Depth of surface crack or half height of buried crack
- a_{cr} Critical value of a for unstable fracture
- a_{max} Maximum allowable value of a
- $\bar{a}$ Half length of through-thickness rectilinear crack
- $\bar{a}_{cr}$ Critical value of $\bar{a}$ for unstable fracture
- $\bar{a}_{max}$ Maximum allowable value of $\bar{a}$
- B Section or specimen thickness
- c Half length of buried or surface crack
- E Young's modulus
- e Strain
- e_Y Yield strain $= \sigma_Y/E$
- G_I Mode I crack extension force
- J The J-contour integral
- K_I Mode I stress intensity factor
- K_{Ic} Critical plane-strain stress intensity factor
- L Plastic constraint factor
- m Plastic stress intensification factor
- M_o, M_π Correction factors for buried cracks at the edge of the crack nearest to a free surface due to that free surface and due to the remote free surface
- M Correction factor for buried cracks $= M_o \times M_\pi$
- M_t Finite thickness correction factor for surface cracks
- M_s Free-surface correction factor for surface cracks
- p Depth below surface of buried defects
- R Radius of holes
- T Wall thickness of cylindrical vessels
- s a_{crit}/a_{max}
- $\bar{s}$ $\bar{a}_{crit}/\bar{a}_{max}$
- W Half width of CNT and DENT specimens
- δ Crack tip opening displacement (COD)
- δ_c Critical value of δ
- δ_m δ at first attainment of maximum load plateau in bend test
- π Constant ≈ 3.142
- σ Applied stress
- σ_1 Effective stress $= (\sigma \times \text{SCF}) + \sigma_R$
- σ_Y Uniaxial yield stress
- Φ Nondimensional COD $= \delta E/2\pi\sigma_Y \bar{a}$
- Φ_2 Complete elliptic integral of second kind.
- SCF   Elastic stress concentration or, where localized uncontained yielding occurs, the strain concentration factor

The theoretical and experimental basis for crack opening displacement (COD) as a fracture characterizing parameter in yielding fracture mechanics is described by Dawes [1][2] elsewhere in this publication. In order to place the application of the COD approach in context it is convenient to think in terms of the brittle-to-ductile transitional behavior encountered with rising temperature in structural steels. If single-edge notched bend specimens of thickness B equal to that of the structure, of width $W = 2B$, and with notch depth $a = B$ (the preferred COD test specimen geometry), are tested over a range of temperatures, the following behavior may be expected. At low temperature, failure will occur under elastic conditions, the test will give a valid value of K_{Ic} to the ASTM Test for Plane-Strain Fracture Toughness of Metallic Materials (E 399-74), and the result may be applied in structural analysis using linear elastic fracture mechanics (LEFM). With increasing temperature, the toughness will rise until the K_{Ic} measurement capacity of the specimen (the greatest possible capacity for the given thickness) is exceeded and failure will occur only after significant yielding. COD will then be determined from the test record and the result may be applied using yielding fracture mechanics as explained later.

With further increases in temperature, the material will behave in a fully ductile manner so that the specimen does not fracture but fails by a simple plastic instability. In such cases, structural failure will also be by plastic instability and this will be assessed by limit load analysis.

Proposals for a weld defect assessment method based on a continuous approach covering these three regimes are currently in an advanced stage of drafting by a British Standards Committee. The approach may be summarized as follows:

Specimen Behavior	Structural Behavior	Analysis Method
K_{Ic}	elastic	LEFM
COD	contained yielding	YFM
fully plastic	plastic instability	limit load

The current paper deals only with the proposed method of application of yielding fracture mechanics. Because of the greatly increased complexity of rigorous elastic-plastic analyses compared to LEFM, the approach is simplified and employs a 'design curve' which is semi-empirical. This curve is considered to be conservative and makes possible swift assessments in practical situations, but more accurate analyses of specific problems are, of course, possible.

[2] The italic numbers in brackets refer to the list of references appended to this paper.

Derivation of the Design Curve

The evolution of the COD design curve has been described in detail by Dawes and Kamath [2]. The basis was established by Burdekin and Stone [3], who studied the extension of the Dugdale strip yield model into the general yielding regime.

The design curve takes the form of a relationship between the nondimensional COD, Φ, and the ratio of applied strain to yield strain, e/e_Y; Φ is defined as $\delta_c/2\pi e_Y \bar{a}_{max}$. The applied strain, e, is taken as the local strain which would exist in the vicinity of the crack, were the crack itself not present. The significance of $\bar{a}_{max}$ should be stressed. The design curve has always been intended, as the name implies, to be one which can be used directly in design. Its purpose is to give conservative predictions of the size of defect which can be allowed to remain in a structure without repair and it is not intended to predict criticality. Thus, $\bar{a}_{max}$ should be smaller than the critical defect size $\bar{a}_{cr}$.

The original curve of Burdekin and Stone was changed to take account of later experimental findings, first by Harrison et al [4], then by Burdekin and Dawes [5], and was finally set out in its current form by Dawes [6]. This is given by

$$\Phi = \left(\frac{e}{e_Y}\right)^2 \quad \text{for} \quad \frac{e}{e_Y} \leq 0.5 \tag{1a}$$

$$\Phi = \left(\frac{e}{e_Y}\right) - 0.25 \quad \text{for} \quad \frac{e}{e_Y} \geq 0.5 \tag{1b}$$

It can be shown that, for small-scale yielding, δ is related to G_I by

$$G_I = m\sigma_Y \delta$$

where the plastic stress intensification factor m is equal to 1 for plane stress.
Hence

$$\delta = \frac{G_I}{\sigma_Y} = \frac{K_I^2}{\sigma_Y E} = \frac{\sigma^2 \pi a}{\sigma_Y E}$$

or

$$\Phi = \frac{\delta}{2\pi e_Y \bar{a}} = \frac{1}{2}\left(\frac{\sigma}{\sigma_Y}\right)^2 = \frac{1}{2}\left(\frac{e}{e_Y}\right)^2 \tag{2}$$

Thus Eq 1a has a factor of safety of 2 on defect size based on the plane stress equivalence between δ, K, and G at low stresses.

Method of Application

For situations where the effective ratio of defect size to plate width $\bar{a}/W$ is less than about 0.1 and where the nominal design stress σ is less than the yield stress of the base material, Dawes [6] proposed that Eq 1 could be rewritten as follows in terms of stress

$$\bar{a}_{max} = \frac{\delta_c E \sigma_Y}{2\pi\sigma_1^2} \quad \text{for} \quad \frac{\sigma_1}{\sigma_Y} \leq 0.5 \tag{3a}$$

$$\bar{a}_{max} = \frac{\delta_c E}{2\pi(\sigma_1 - 0.25\sigma_Y)} \quad \text{for} \quad \frac{\sigma_1}{\sigma_Y} \geq 0.5 \tag{3b}$$

where σ_1 is the total pseudo-elastic stress in the vicinity of the defect. While σ_1 may be above yield, the structure itself may still behave in a predominantly elastic manner. This is because the yielding in the zone under consideration is contained by the surrounding elastic material. In welded structures, contained yield occurs as a result of residual stresses which may themselves be equal to the yield stress and may be additive to the applied stress. Contained yielding also occurs at stress concentrations where pseudo-elastic stresses may be well above yield.

For general applications, the values of σ_1 as given in Table 1 were suggested.

Part-Through Surface and Buried Defects

The design curve was originally formulated on the basis of through-thickness defects. Dawes [6] suggested that part-through cracks could be dealt with by assuming that, for contained yielding situations, the parameters governing flaw shape effects would be the same as those under

TABLE 1—*Total pseudo-elastic stress values.*

Crack Location	Weld Condition	σ_1
Remote from stress concentrations	stress relieved[a]	σ
	as welded	$\sigma + \sigma_Y$[b]
Adjacent to stress concentrations	stress relieved[a]	SCF[c] $\times \sigma$
	as welded	(SCF $\times \sigma$) + σ_Y[b]

[a] Here it is assumed that post weld heat treatment (PWHT) has eliminated all the residual stresses. Often this will not be so and it is prudent to make some allowance for the residual stress remaining after PWHT.

[b] It has been assumed that residual stresses of yield point magnitude will exist in as-welded structures. While this is true for stresses along the weld, transverse residual stresses can often be assumed to be lower than yield in specific cases.

[c] Strain concentration factor.

linear elastic conditions. It was realized that this approach could not be justified rigorously, but it seems unlikely that elastic-plastic solutions for the part-through crack will be available for some time to come. The following LEFM expression was used to describe a semi-elliptic surface crack

$$K_1 = \frac{M_t M_s \, \sigma\sqrt{\pi a}}{\Phi_2} \tag{4}$$

For the equivalent through thickness crack of length $2\bar{a}$

$$K_1 = \sigma\sqrt{\pi \bar{a}}$$

Thus

$$\frac{\bar{a}}{B} = \frac{a}{B}\left(\frac{M_t M_s}{\Phi_2}\right)^2 \tag{5}$$

Values of

$$\left(\frac{M_t M_s}{\Phi_2}\right) = f\left(\frac{a}{B}, \frac{a}{2c}\right)$$

were taken from a survey by Maddox [7] and a/B is plotted against $\bar{a}/B$ in Fig. 1.

With the exception of deep surface cracks, Fig. 1 agrees closely with formulas proposed more recently by Newman [8].

For buried elliptical cracks, the equivalent equation to (5) is

$$\frac{\bar{a}}{B} = \frac{a}{2(p+a)}\left(\frac{M}{\Phi_2}\right)^2 \tag{6}$$

M was calculated as the product of the magnification factors M_π and M_o applicable to the stress intensity factor at the end of the minor axis which approaches nearest to a free surface. M_o is the magnification factor at that point due to the presence of the near surface and M_π is that due to the presence of the more remote free surface. M_π and M_o were taken from the work of Shah and Kobayashi [9]. For $a/c = 0$, M was derived from Feddersen's relationship [10]

$$M = \left(\sec\frac{\pi a}{B}\right)^{1/2}$$

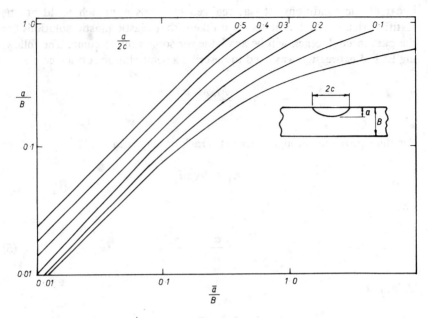

FIG. 1—*Relationships between surface crack dimensions and equivalent through-thickness crack dimension $\bar{a}$.*

Figure 2 is a plot of $a/(p + a)$ against $\bar{a}/[2(p + a)]$ or $\bar{a}/B$ for buried defects.

Recent Experimental Justification for the Design Curve

The implementation of the COD approach through the simple design curve, which takes into consideration the effects of residual stresses and geometric stress concentrations, has found wide application to welded structures. However, because of its semi-empirical origins and inherent simplicity, the design curve, as already stated, predicts maximum allowable crack sizes and not critical crack sizes, with a margin of saftey only *vaguely estimated* as being approximately 2.0. It was decided, therefore, to carry out an assessment of the COD design curve by making a comparison of the allowable crack sizes predicted by the small-scale COD tests and the critical crack sizes at fracture in wide-plate tests [11]. From a survey of the published literature and work carried out at The Welding Institute, a total of 73 sets of small- and large-scale tests was compiled. The results were then analyzed on a statistical basis. The main steps and observations from these analyses are summarized in the following.

Initially, the test data were processed, as shown in Fig. 3, to give safety

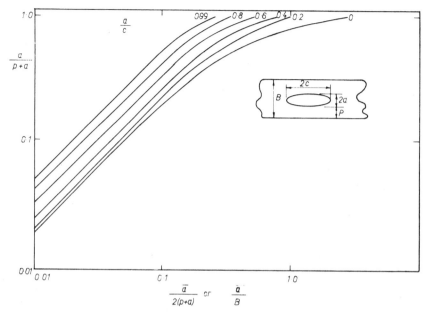

FIG. 2—*Relationships between buried crack dimensions and equivalent through-thickness dimension ā.*

factors $\bar{s}$ and s for through-thickness and surface cracks, respectively. A comparison between the predicted allowable and the critical crack sizes obtained is shown in Fig. 4 with some of the important probability levels indicated. This shows that, on average, the design curve has a built-in factor of safety of approximately 2.5, and the maximum allowable size derived from the curve implies a 95.4 percent probability of survival with respect to the wide-plate test. However, when the scatter in results is taken into consideration, there appears to be little scope for modifying the shape of the design curve.

An examination of the variables in the wide-plate tests drew attention to the influence of residual stresses on brittle fracture. For situations where it was reasonable to assume no residual stresses, for example, plain plate and some stress-relieved weldments only, Fig. 5a shows that critical values of Φ were generally well below the design curve. However, when the results for as-welded plates are added to the plot and if residual stresses are still assumed to be zero (Fig. 5b), it can be seen that a significant proportion of the as-welded plate specimen results fall to the left of the design curve. In other words, the failure stress assumed was lower than that to be expected from the design curve for the known value of Φ. The as-welded wide-plate results fall within the same general scatter band as those for plates assumed to be free from residual stress and were thus safely predicted by the design

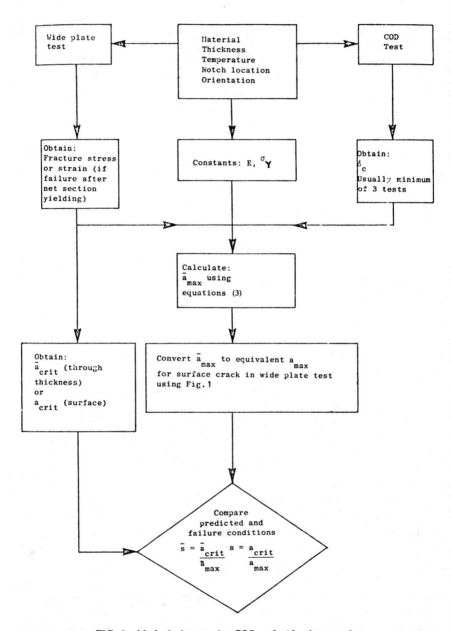

FIG. 3—*Method of processing COD and wide-plate test data.*

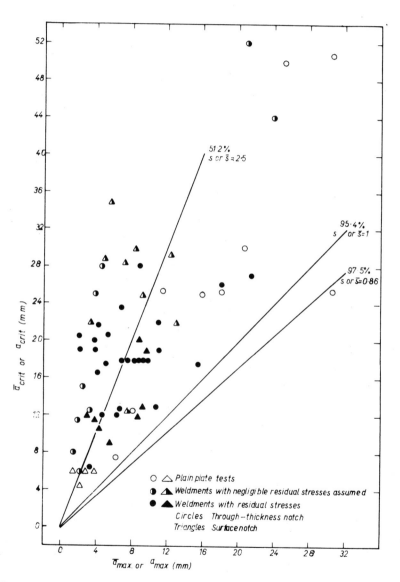

FIG. 4—*Comparison of critical and maximum allowable crack sizes showing probability levels and safety ratios.*

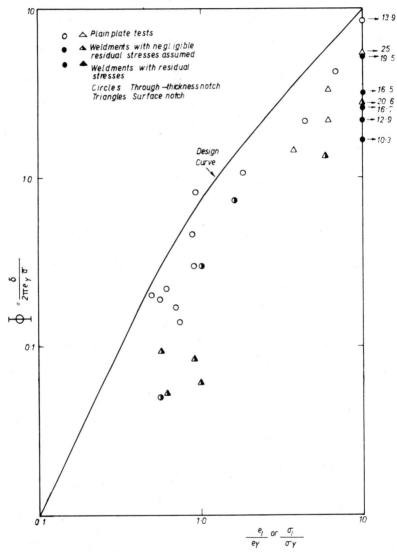

(a) plain plate and weldments with negligible residual stress effects (e.g.: stress relieved welds.)

FIG. 5—*Design curve relationships between nondimensional COD and applied strain and stress (normalized), with experimental COD/wide-plate test results.*

HARRISON ET AL ON COD APPROACH 617

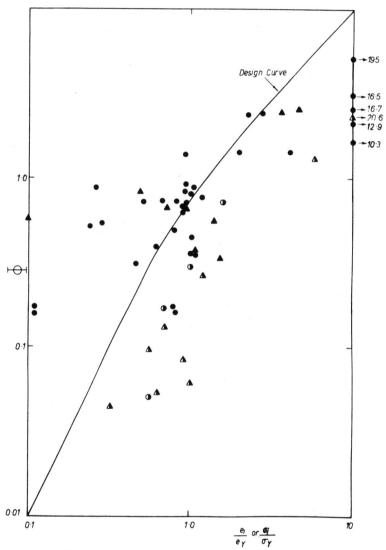

(b) All weldments, but residual stress neglected in design curve calculation. Includes weldments from figure 5(a)

FIG. 5—*Continued.*

618 ELASTIC-PLASTIC FRACTURE

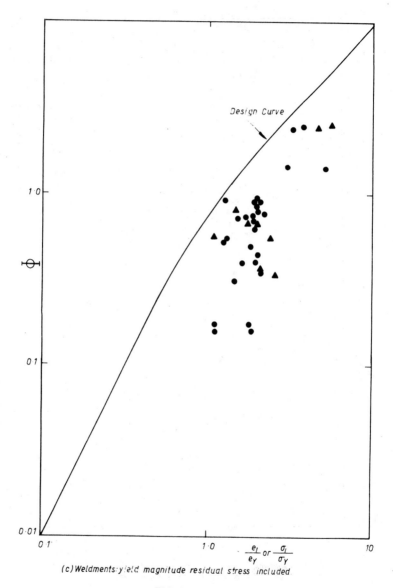

(c) Weldments: yield magnitude residual stress included.

FIG. 5—*Continued.*

curve only when full-yield residual stresses were assumed to be active (Fig. 5c).

The assessment also showed that when residual stresses were present there were no significant differences in the average factors of safety for through-thickness and surface cracks. In the absence of residual stresses, however, the results suggested that the factors of safety were slightly higher in the case of surface cracks.

Numerical Assessments of the Design Curve

Sumpter [12] and Sumpter and Turner [13] report the results of elastic-plastic finite-element analyses. Some of these were for an elastic perfectly plastic material, but some assumed a work-hardening law approximately equal to that for the ASTM Specifications for Pressure Vessel Plates, Alloy Steel, Quenched and Tempered, Manganese-Molybdenum and Manganese-Molybdenum-Nickel (A 533B-74). The latter were compared by Sumpter with the COD design curve. The following geometries were studied.

1. Edge cracked plate $a/W = 0.1$,
2. Crack at the edge of a hole of radius R, $a/R = 0.05, 0.1$, and 1.0, and
3. Radial crack at the bore of a pressurized cylinder of thickness T, $a/T = 0.03$.

The results are plotted in Fig. 6. For the edge-cracked plate there was close agreement between the finite-element analysis results and the design curve.

For cracks at the edge of a hole, the design curve was shown to be conservative, but not excessively so for ratios of crack length to hole radius, a/R, of 0.05 and 0.10. As stated in the section dealing with the application of the design curve, the recommended procedure is to calculate the value of σ_1 as SCF $\times$ σ. Sumpter shows that this procedure becomes very pessimistic for the unusual case of very long cracks at the edge of a hole with $a/R = 1.0$. However, as Burdekin and Dawes originally suggested, it is more realistic for $a/R > 0.2$ to assume that the crack is one of total length $a + 2R$ in a stress field equal to the membrane stress. If this procedure is adopted for the results for $a/R = 1.0$, the plot of Φ against e/e_Y comes closer to that for $a/R = 0.1$, but remains very conservative. The mean factor of safety on Φ between the results for $a/R = 0.05$ and 0.1 and the design curve for a given value of e/e_Y is 2.0.

For the radially cracked cylinder, a comparison was made with the design curve for one ratio of crack depth to cylinder wall thickness, $a/T = 0.03$. The design curve was again found to be conservative with factors of safety on Φ of 4.5 at $e/e_Y = 0.6$, 2.5 at $e/e_Y = 1.0$, and 1.2 at $e/e_Y = 1.6$.

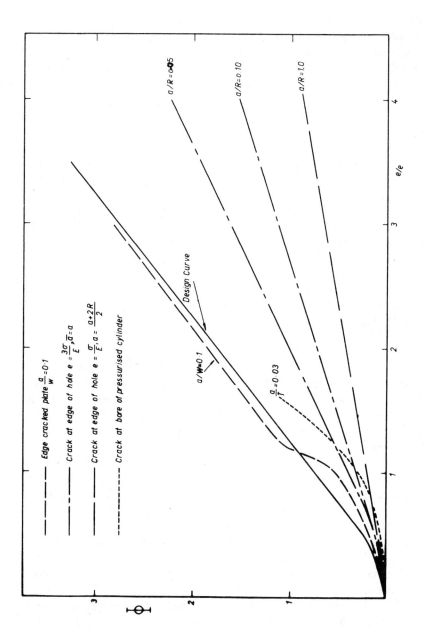

FIG. 6—*Comparison between Sumpter's finite-element results [12] and the COD design curve.*

The *J* Design Curve

Begley et al [14] have proposed a *J* design curve which is in essence very similar to the COD design curve. The similarity between the two approaches has been discussed by Merkle [15].

The design curve of Ref *14* takes the form

$$\frac{J}{E\pi \bar{a} e_Y^2} = \left(\frac{e}{e_Y}\right)^2 \quad \text{for} \quad \frac{e}{e_Y} \leq 1.0 \tag{7a}$$

and

$$\frac{J}{E\pi \bar{a} e_Y^2} = \frac{2e}{e_Y} - 1 \quad \text{for} \quad \frac{e}{e_Y} \geq 1.0 \tag{7b}$$

It will be seen that these are similar in character to Eq 1. It is generally stated that $J = m\, \sigma_Y\, \delta$, where *m* is a plastic stress intensification factor which ranges from about 1.0 to 2.0. Substituting for *J* and assuming $m = 1.0$, Eqs 7a and 7b reduce to

$$\Phi = \frac{1}{2}\left(\frac{e}{e_Y}\right)^2 \quad \text{for} \quad \frac{e}{e_Y} \leq 1.0 \tag{8a}$$

and

$$\Phi = \frac{e}{e_Y} - 0.5 \quad \text{for} \quad \frac{e}{e_Y} \geq 1.0 \tag{8b}$$

For $m = 2.0$ they reduce to

$$\Phi = \frac{1}{4}\left(\frac{e}{e_Y}\right)^2 \quad \text{for} \quad \frac{e}{e_Y} \leq 1.0 \tag{9a}$$

and

$$\Phi = \frac{1}{2}\left(\frac{e}{e_Y} - 0.25\right) \quad \text{for} \quad \frac{e}{e_Y} \geq 1.0 \tag{9b}$$

These are plotted for comparison with the design curve in Fig. 7.

It should be borne in mind that, while the COD design curve is intended to be conservative and is empirically based on the results of wide-plate tests on specimens where it may be assumed that *m* varied, the curve of Begley et al is intended to predict critical conditions. Viewed in this light, it is felt that the two approaches are in reasonable agreement.

622 ELASTIC-PLASTIC FRACTURE

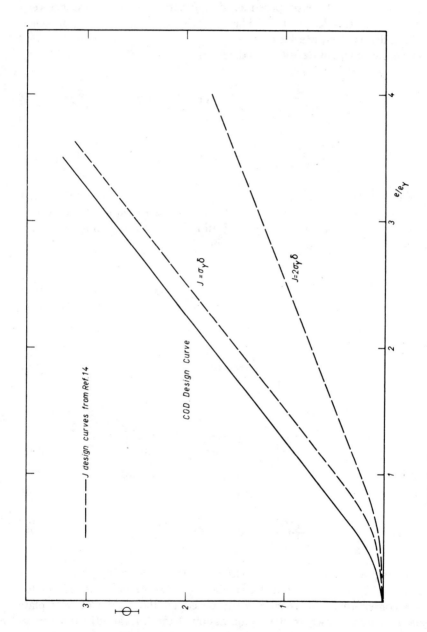

FIG. 7—*Comparison between the COD design curve and curves based on Ref 14.*

Experience in the Practical Application of the COD Design Curve

The design curve approach has been applied in recent years to a great range of structures in a variety of contexts.

The applications may be broadly classified into four major groups:
1. Material selection (design stress and defect levels predetermined).
2. Acceptance levels for weld defects (decisions regarding known defects or fixing defect acceptance standards where material and design stress are predetermined).
3. Fixing allowable stress (material and inspection level predetermined).
4. Failure analysis.

Lists of some of the applications with which the authors have been concerned over the past five years are given in Tables 2-6. These are by no means exhaustive. Some specific examples are discussed in greater detail in the following.

Material Selection

If the design stress is known and a size of defect which might escape detection by nondestructure testing (NDT) is assumed, the design curve may be used to determine the required level of toughness. Table 2 lists welding consumable manufacturers whose products have been tested by The Welding Institute in order to establish whether toughness levels fixed in this way have been achieved. Table 3 lists a range of structures where this approach has been used. Two specific examples are given in the following.

Offshore Production Platforms—The production platforms for British Petroleum's (BP) Forties Field in the North Sea involved a quantum jump in size over the great majority of similar structures. Because of the greater depths in the North Sea and the severity of the environment, structural steels of higher strength (320 N/mm^2 yield, 500 N/mm^2 tensile strength) and increased thickness (60 to 100 mm) were employed. Would such structures be safe if the welds in stress concentration regions at the massive intersections (nodes) remained in the as-welded conditions, or should they be post-weld heat treated (PWHT) as would be mandatory for pressure vessels built of similar materials? It was assumed that surface and buried defects up to 12.5 and 25 mm deep, respectively, might escape detection

TABLE 2—*Welding consumable manufacturers for whom COD tests have been carried out with the objective of meeting specific requirements.*

Arcos	BOC Murex	Esab
Big 3 Lincoln UK	GKN Lincoln	Kobe Steel
Lincoln Electric	West Falische Union	Varios Fabrieken
Metrode	Oerlikon	Phillips

TABLE 3—*Cases where the COD design curve has been used as a basis for material selection.*

Class of Structure	Company	Project
Pipelines	Aramco	LNG[a] pipeline
	Aramco	spiral weld gas pipeline
	Bechtel International	offshore gas pipeline
	BP	crude-oil pipeline
	BP	automatic MIG[b] girth welds for Ninian and Forties field lines
	Brown and Root	offshore flare line
	Brown and Root	weld overlay on offshore riser pipe
	Canadian Artic Gas	natural gas pipeline
	Conoco	offshore gas pipeline
	Hoesch	pipeline steel
	Gasunie	thick-walled pipeline
	Shell	oil pipeline girth welds
Offshore structures	BP	selection of weld procedures and decision regarding post-weld heat treatment
	Conoco	crane pedestals
	Highland Fabricators	production platform
	Laing offshore	production platform
	McDermott	production platform
	Phillips Petroleum	general specification for offshore structures
	Redpath Dorman Long	production platform
	Shell	general specification for production platforms
	Shell	jack-up rig
Pressure vessels and boilers	Air Products	cryogenic vessels (9%Ni steels)
	Clarke Chapman	boiler drum (decision regarding post-weld heat treatment)
	Internation Combustion	boiler drum
Nuclear	Central Electricity Generating Board	stainless steel weld for AGR[c] boiler
	Nuclear Power Company	steam drum SGHWR[d]
	The Nuclear Power Group	circulator outlet gas duct/liner weld in AGR
Miscellaneous	Allis Chalmers	general purposes
	Aramco	plates for low-temperature service
	Ove Arup	cast steel weldments for building frame
	Capper Neil	LPG[e] storage tanks
	Capper Neil	oil storage tanks
	Central Electricity Generating Board	penstocks for pumped storage scheme
	Chicago Bridge & Iron	LNG storage tanks
	High Duty Alloys	crash barriers
	High Duty Alloys	repair procedures for 12 000-t extrusion press
	Johns & Waygood	high-rise building
	Kockums	high heat input welding for ships
	Lindsey Oil Refineries	oil storage tanks
	Uddeholms	9%Ni steel for LNG tanks
	Whessoe	5-m low-speed wind tunnel
	Whessoe	oil storage tanks

[a] Liquefied natural gas.
[b] Metal inert gas.
[c] Advance gas-cooled reactor.
[d] Steam-generated hot-water reactor.
[e] Liquefied petroleum gas.

TABLE 4—*Cases where the COD design curve has been used to fix acceptance levels for weld defects and inspection sensitivity.*

Company	Project
Clarke Chapman	electroslag welds in boiler drum
Central Electricity Generating Board	oil storage tanks
Elliott	compressor rotor
Gasunie	gas pipelines
Shell	oil pipelines
Shell	offshore pressure vessels

TABLE 5—*Assessment of weld defects.*

Class	Company	Project
Pipelines	BP	Aleyaska pipeline
	BP	refinery pipes
	BP	ethelyne pipeline
	Danish Weld Inst	spiral-welded oil products line
	Gasunie	gas pipelines
	Metallurgical Consultants	spiral-welded pipe
	Unit Inspection	spiral-welded pipe
Offshore structures etc.	Conoco	offshore crane pedestals
	Occidental	deck modules
	Shell	flash welds in anchor chain links
Pressure vessels and boilers etc.	British Gas	gas bullets
	Clarke Chapman	steam and mud drums
	ICI	hydro desulphurizer
	Richard Ross	nozzle welds
	Whessoe	pressure vessels
Miscellaneous	Colocotronis	supertanker
	Dorman Long (South Africa)	pumped storage scheme penstocks
	Cleveland Potash	mine shaft lining

TABLE 6—*Failure investigations.*

Company	Structure Type
Bechtel International	pile for offshore platform
BP	oil tankers
Conoco	offshore gas pipeline
Conoco	rolled beams for deck module
Noble Denton	leg for offshore structure
Shell	pile for offshore platform
South of Scotland Electricity Board	steam riser pipe
South of Scotland Electricity Board	feedwater header forging

by NDT in these complex structures. Photoelastic analysis of a typical node using the frozen stress method indicated a maximum SCF of 8, and the nominal design stress was 75 N/mm² or approximately ¼ σ_Y. The required COD value determined from the design curve is given by Eq 3b as

$$\delta_c = \frac{2\pi \bar{a}_{max}}{E} (\sigma_1 - 0.25\,\sigma_Y)$$

Hence, if the nodes are as-welded

$$\sigma_1 = (\text{SCF} \times \sigma) + \sigma_R = (8 \times \tfrac{1}{4}\,\sigma_Y) + \sigma_Y = 3\sigma_Y$$

and substituting for σ_Y and $\bar{a}_{max}$ gives

$$\delta_c = 0.37 \text{ mm}$$

COD tests were carried out at the design temperature of $-10°C$ on the parent steel, BS 4360 Grade 50D, which gave a minimum $\delta_m = 0.49$ mm, but the best weld metal, out of the total of 17 tested, gave only $\delta_c = 0.12$ mm.

It was clear that it would be necessary to heat treat the nodes in order to obtain the required defect tolerance. It after PWHT, σ_R is assumed to be zero, σ_1 becomes $8 \times \tfrac{1}{4}\,\sigma_Y = 2\sigma_Y$. This gives a required COD of

$$\delta_c = 0.24 \text{ mm}$$

Not only did PWHT lower the required COD, but it significantly increased the toughness of the weld metal. Five weld metals were found giving satisfactory toughness and one had a minimum value of $\delta_m = 0.49$ mm.

This study (which has been described more fully elsewhere [16]) showed quite clearly that it was necessary to heat treat the node regions to ensure safety of the complete structures. This decision had a marked effect on the design philosophy adopted.

Specification of Toughness of Girths Welds in Large-Diameter Pipeline for Service in Arctic Regions—Normally girth welds are not highly stressed in the longitudinal direction and hence fracture risks are very low. When the line goes through areas liable to subsidence or earthquakes or both, however, high longitudinal bending stresses can develop. In this particular case, strains up to 0.5 percent due to the aforementioned causes were envisaged, which meant that weldments with good fracture resistance were needed, particularly in view of the associated low ambient temperatures (down to $-40°C$). Semiautomatic and manual welding processes were considered in conjunction with pipe to API-5LX70 and 19-mm wall thickness. The fracture resistance of the various regions of the heat-affected zones (HAZs) and weld metals was thoroughly examined by the COD test at the minimum service and other selected temperatures on specimens taken from welds made under field conditions. These tests demonstrated clearly that the fracture toughness of the HAZs was adequate at all temperatures, but that there could be difficulties in the weld metals, both manual and semiautomatic.

In order to judge the validity of the maximum allowable flaw sizes predicted from the COD results, a series of full-size bend tests was carried out in a specially made rig. The girth welds in the pipe lengths contained crack-like defects of suitable size in the center of the weld deposits, which were made under typical field conditions. The weld area was placed at the position of maximum bending moment in the rig and cooled to the required temperature. Load was slowly applied up to failure and the maximum applied strain measured by electric-resistance strain gages. After failure, the depth of the actual defect a_{cr} was measured and compared with the predicted maximum allowable depth a_{max}. In the calculations it was assumed that the peak tensile residual stress level transverse to the girth welds would be between 0.5 and 0.75 e_Y. Using these values of residual stress, the minimum COD values from the small-scale tests, the measured applied strain values, and Eqs 3b and 5, ratios of a_{cr}/a_{max} between 2 and 3 were obtained, which is consistent with the normal experience from wide-plate tests as indicated in Fig. 4.

Acceptance Levels for Weld Defects

Table 4 lists a number of cases where the design curve has been used to fix NDT requirements at an early stage in design or construction. Table 5 lists cases where the concept has been used to assess the significance of known defects where repair was felt to be undesirable for reasons of cost, delivery, or because of the possibility of introducing more harmful defects in the course of the repair.

Defects in the Alyeska Pipeline—As a result of disclosures to the press of falsification of radiographs of the girth welds in the Alyeska crude oil pipeline, all the X-ray films for the part of the line completed at that stage were reexamined. This audit indicated that there were defects larger than those permitted by the construction code, API-1104, in some 2955 of the 30 000 welds audited. The defect acceptance levels in API-1104 are set in order to maintain a certain level of workmanship and bear no relationship to the performance of pipelines in service. Nevertheless, the code had been adopted by the Department of Transportation (DOT) as a basis for licensing the pipeline. The pipeline company through BP asked The Welding Institute to help to develop a submission to DOT for waivers to the code. This case was based on the design curve and, in terms of the numbers of specimens tested (some 450), represented the largest single case of its application.

The effects of weld procedure (three different procedures), position around pipe circumference, and notch orientation were studied. Nine notch positions and orientations with respect to the weld metal and HAZ were investigated. Tests were carried out at 0, −12, −18, −29, and −40°C. In fact −12°C was chosen as the design basis to allow for the possibility of cold pressurization during start-up. Because of the considerable variety of microstructures sampled, there was wide scatter in the COD values, but a lower bound of 0.1 mm was used. At −40°C, one specimen gave a result as low as 0.025 mm. Although these COD values are relatively low, no specimen gave a result which could possibly be interpreted as a valid K_{Ic} value for the specimen thickness of 12.7 mm (0.5 in.). The stresses considered in the analysis included pressure, thermal, bending due to expansion and self-weight, earthquake, and residual. It was found as a result of the analysis that none of the defects required repair.

The National Bureau of Standards carried out an independent assessment [*17*]. In Ref *17* an analysis is reported by Begley, McHenry, and Read based on a different interpretation of the COD test. This suggested that the Alyeska proposals were conservative by factor of about 1.2 to 2.0. However, it is believed that the Begley, McHenry, and Read approach is aimed at predicting critical conditions while the design curve already incorporates a factor of safety of 2.5 on the best estimate for criticality. Thus the two approaches seem to be in good agreement.

The DOT accepted in principle the case for waivers, but asked for a further safety margin of 2 to allow for problems of X-ray interpretation. On this basis a small number of repairs were required.

In fact, many of the defects were repaired because the negotiations of the case were time-consuming and because the pipeline company could not afford to waste this amount of time when the proposal might have been rejected. The important point, however, is that a major U.S. Government authority, that is, DOT, accepted the principle of using a yielding fracture mechanics analysis to derive defect acceptance levels in a large pipeline project. Furthermore, some defects in a section of pipe crossing the Koyukuk River, which would have cost about $5 million to repair, were accepted.

Fixing Allowable Stresses

The use of fracture mechanics to fix design stresses is less common, but one case involving liquefied natural gas (LNG) storage tanks can be cited.

A program of COD and wide-plate tests was carried out on weldments in 9 percent nickel steel, 18 mm thick, mostly at −164°C, which was the minimum design temperature. The main objective of the work was to see if the design stress level could be safely raised from the API-620Q value of 196.5 N/mm^2 to 290 N/mm^2. Two plate materials, A553 and A353, were examined and two suitable weld metals. The COD tests showed that all the materials were fully ductile at $-164\,°C$. Values of $\bar{a}_{max}$ between 7 and 19 mm were obtained for a design of 290 N/mm^2 using COD values at maximum load in conjunction with equations 3b and 5 and making conservative assumptions about effects of distortion and residual stresses. Through-thickness defects 20 to 25 mm long were incorporated in various weld regions in wide-plate specimens and tested at $-164\,°C$. The fact that all the plates failed at stresses above the yield stress of the weld metal indicated that the approach was very conservative. The work was carried out about five years ago and, in view of the complete ductility of the materials involved, it is now thought that a limit load approach, as suggested in the introduction, would be more appropriate and less conservative. Nevertheless, it was clearly demonstrated that an increase in design stress level to 290 N/mm^2 was reasonable in terms of tolerance to severe fabrication defects. This resulted in a significant reduction in the cost of the LNG tanks.

Failure Investigation

A number of instances where structural failure conditions have been compared with predictions from the design curve are listed in Table 6. These represent some of the most interesting applications of the design curve, but space permits only one example to be discussed in more detail. The example chosen is the Cockenzie boiler drum. This has already been discussed sometime ago by Burdekin and Dawes [5] and by Ham [19], but it may now be reassessed in the light of the revised design curve and of the defect shape corrections in Fig. 1.

The failure [19] during hydrostatic test at nominal stress level of 0.55 σ_Y occurred from a large semi-elliptical surface defect 81 mm deep by 325 mm long at the edge of an attachment. The material, which was a low-alloy structural steel, had a thickness of 141 mm, a yield stress of 376 N/mm^2, and minimum COD value of 0.25 mm in the stress-relieved condition.

Substitution of the foregoing values into the design curve gives $\bar{a}_{max} = 74$ mm that is, $\bar{a}_{max}/B = 0.52$. From Fig. 1, $a_{max}/B = 0.49$, giving a maximum depth of 69 mm for a surface defect with aspect ratio 0.25.

The factor of safety in this case is 1.17. This is smaller than usual, but it could be influenced by two factors which were ignored in the analysis, both of which would tend to increase it. First, residual stress was assumed to be zero. It is probable, however, that some residual stress will have remained since the material was thick and the geometry was complex.

Second, the defect was close to a nozzle and the local stress may have been elevated because of this. The details available are not sufficient for either of these possible effects to be assessed; however, use of the design curve still gives a reasonable explanation of the failure.

Conclusions

The COD test is a useful method of studying the fracture toughness of materials in the transition region between linear elastic behavior, where K_{Ic} should be used, and fully ductile behavior, where a limit load approach is appropriate. It was concluded from the statistical analysis of 73 sets of tests, where predictions from the design curve were compared with the results of large-scale tests, that the average inherent safety factor is approximately 2.5. The analysis also revealed a 95 percent probability of the predicted allowable crack size being smaller than the critical crack size. It was shown that the approach described is comparable with design curves derived from finite-element analyses and J-analyses. It was concluded from the several practical examples described that the design curve can be successfully used in at least three different ways: (1) selection of materials during initial design stage, (2) specification of maximum allowable flaw sizes at design or after fabrication to establish the necessity for repairs, and (3) failure analysis.

References

[1] Dawes, M. G., this publication, pp. 306-333.
[2] Dawes, M. G. and Kamath, M. S., "The Crack Opening Displacement (COD) Design Curve Approach to Crack Tolerance," Conference on the Significance of Flaws in Pressurised Components, Institution of Mechanical Engineers, London, England, May 1978.
[3] Burdekin, F. M. and Stone, D. E. W., *Journal of Strain Analysis,* Vol. 1, No. 2, 1966, p. 194.
[4] Harrison, J. D., Burdekin, F. M., and Young J. G., "A Proposed Acceptance Standard for Weld Defects Based Upon Suitability for Service," 2nd Conference on the Significance of Defects in Welded Structures, The Welding Institute, London, England, 1968.
[5] Burdekin, F. M. and Dawes, M. G., "Practical Use of Linear Elastic and Yielding Fracture Mechanics with Particular Reference to Pressure Vessels," Conference on Application of Fracture Mechanics to Pressure Vessel Technology, Institution of Mechanical Engineers, London, England, May 1971.
[6] Dawes, M. G., *Welding Journal Research Supplement,* Vol. 53, 1974, p. 369s.
[7] Maddox, S. J., *International Journal of Fracture Mechanics,* Vol. 11, No. 2, April 1975, pp. 221-243.
[8] Newman, J. C. in *Part-Through Cracks Life Prediction,* American Society for Testing and Materials, 1979.
[9] Shah, R. C. and Kobayashi, A. S., *International Journal of Fracture Mechanics,* Vol. 9, No. 2, 1973, p. 133.
[10] Feddersen, E. E. in *Discussion to Plane Strain Fracture Toughness Testing of High-Strength Metallic Materials, ASTM STP 410,* 1967, p. 77.
[11] Kamath, M. S., "The COD Design Curve: An Assessment of Validity using Wide Plate Tests," The Welding Institute Members Report 71/E, 1978, to be published.
[12] Sumpter, J. D. G., "Elastic-Plastic Fracture Analysis and Design Using the Finite Element Method," Ph.D. thesis, University of London, Dec. 1973.
[13] Sumpter, J. D. G. and Turner, C. E., "Fracture Analysis in Areas of High Nominal Strain," 2nd International Conference on Pressure Vessel Technology, San Antonio, Tex., Oct. 1973.
[14] Begley, J. A., Landes, J. D. and Wilson, W. K. in *Fracture Analysis, ASTM STP 560,* American Society for Testing and Materials, 1974, pp. 155-169.

[15] Merkle, J. G., *International Journal of Pressure Vessels and Piping*, Vol. 4, No. 3, July 1976, pp. 197–206.
[16] Harrison, J. D. in *Performance of Offshore Structures*, Series 3, No. 7, Publication of the Institution of Metallurgists, London, England, 1977.
[17] Berger, H. and Smith, J. H., Eds., "Consideration of Fracture Mechanics Analysis and Defect Dimension Measurement Assessment for the Trans-Alaska Oil Pipeline Girth Welds," National Technical Information Service Report PB-260-400, Oct. 1976.
[18] Harrison, J. D. and Carter, W. P., "The Use of 9%Ni Steel for LNG Application," *Proceedings*, Conference on Welding Low Temperature Containment Plant, The Welding Institute, London, England, Nov. 1973.
[19] Ham, W. M., Discussion to Conference on Practical Application of Fracture Mechanics to Pressure Vessel Technology, Institution of Mechanical Engineers, London, England, May 1971.

H. I. McHenry,[1] D. T. Read,[1] and J. A. Begley[2]

Fracture Mechanics Analysis of Pipeline Girthwelds*

REFERENCE: McHenry, H. I., Read, D. T., and Begley, J. A., **Fracture Mechanics Analysis of Pipeline Girthwelds,"** *Elastic-Plastic Fracture, ASTM STP 668,* J. D. Landes, J. A. Begley, and G. A. Clarke, Eds., American Society for Testing and Materials, 1979, pp. 632–642.

ABSTRACT: Size limits for surface flaws in pipeline girthwelds are calculated on the basis of fracture mechanics analysis. Parameters for the analysis were selected from data on a 1.22-m-diameter (48 in.), 12-mm-thick (0.46 in.) pipe welded by the shielded metal-arc process. The minimum fracture toughness of the welds as determined by the crack opening displacement (COD) method was 0.1 and 0.18 mm (0.004 and 0.007 in.), depending on the flaw location. The yield strength of the welds was 413 MPa (60 ksi). Because the toughness to yield strength ratio was high, elastic-plastic fracture mechanics analysis methods were required to determine critical flaw sizes. Four approaches were employed: (1) a critical COD method based on the ligament-closure-force model of Irwin; (2) the COD procedure of the Draft British Standard Rules for Derivation of Acceptance Levels for Defects in Fusion Welded Joints; (3) a plastic instability method based on a critical net ligament strain developed by Irwin; and (4) a semi-empirical method that uses plastic instability as the fracture criterion developed by Kiefner on the basis of full-scale pipe rupture tests. Allowable flaw sizes determined by the Draft British Standard method are compared with the critical flaw sizes calculated using critical-COD and plastic instability as the respective fracture criteria. The results for both axial- and circumferential-aligned flaws vary significantly depending on the analysis model chosen. Thus, experimental work is needed to verify which model most accurately predicts girthweld behavior.

KEY WORDS: carbon-manganese steel, fracture mechanics, fracture toughness, pipeline, radiographic inspection, weld flaws, weldments, crack propogation

During the summer of 1976, the National Bureau of Standards (NBS) assisted the Office of Pipeline Safety Operations (OPSO) of the Department of Transportation (DOT) in the evaluation of fracture mechanics as a method of calculating allowable flaw sizes in pipeline girthwelds. The

---

*Contribution of the National Bureau of Standards; not subject to copyright.
[1]Metallurgist and physicist, respectively, National Bureau of Standards, Boulder, Colo. 80302.
[2]Associate professor, Ohio State University, Columbus, Ohio.

approach taken by NBS was to calculate allowable flaw sizes in a specific pipeline in accordance with guidelines established by OPSO [1].[3] The material property data, the pipeline stresses, and the analysis approaches used are described in this paper. The overall NBS program has been described in a report to the DOT [2].

Pipeline Weldment Properties

The pipeline evaluated was built from 1.22-m-diameter (48 in.), API 5LX-65 steel pipe having wall thicknesses of either 12 or 14 mm (0.46 or 0.56 in.). It was welded by the manual shielded metal-arc process using AWS E7010G and E8010G electrodes. The mechanical properties of several pipeline segments containing girthwelds were evaluated by NBS [2], the Welding Institute [3], and Cranfield Institute of Technology [4]. The test results were used as the basis for selecting conservative values for use in the analysis as summarized in the following.

The tensile properties of interest in the analysis were the yield strength (σ_y), the flow strength ($\bar{\sigma}$), and Young's modulus (E). The yield strength values used were 448 MPa (65 ksi) for the base metal and 413 MPa (60 ksi) for the weld metal. The flow strength value was taken as the yield strength plus 68.9 MPa (10 ksi) as recommended by Kiefner et al [5] on the basis of extensive pipeline studies. Young's modulus was 208 GPa (30.2 × 10^3 ksi) for both the base metal and the weld metal.

The fracture toughness of the pipeline weldments was determined for notch locations in the weld metal, the heat-affected zone (HAZ), and the base metal at temperatures ranging from -40 to $0\,°C$ (-40 to $32\,°F$). The tests were conducted using the crack opening displacement (COD) method by the British Welding Institute [3] and by Cranfield Institute of Technology [4]. The COD fracture-toughness values used to establish allowable-flaw-size curves were 0.1 mm (0.004 in.) for nonplanar weld defects, which tend to be randomly located in the weld, and 0.18 mm (0.007) for planar defects, which tend to be on the inside surface. These values were the minimum values obtained for the applicable notch orientations at $-12\,°C$ ($+10\,°F$), 10 deg C (18 deg F) below the minimum exposure temperature possible during service.

Pipeline Operating Stresses

The maximum credible stresses in the appropriate orientations were used in the critical flaw size calculations. Girthweld flaws are typically oriented circumferentially, and consequently the axial stresses are used in the analysis. Arc burns are typically spots or axially aligned drags, and flaw growth would be caused by the hoop stresses.

[3]The italic numbers in brackets refer to the list of references appended to this paper.

Axial stresses in the pipeline during service are caused by the internal pressure, thermal expansion, and earthquake loadings. These stresses are superimposed on a pipe bending stress due to soil settlement. The maximum credible stress in the axial direction is 398 MPa (57.7 ksi) caused by the hypothetical condition of extended winter shutdown, followed by full pressurization. The maximum credible stress includes a bending stress caused by 15 cm (6 in.) of soil settlement in 30 m (100 ft) of pipe length. The maximum axial stress of 398 MPa (57.7 ksi) was used for critical flaw size calculations for weld defects.

Hoop stresses in the pipeline are caused exclusively by internal pressure. Maximum hoop stresses are 72 percent of σ_y, 322 MPa (46.8 ksi), during normal operation; 80 percent of σ_y, 358 MPa (52.0 ksi), during surges; and 95 percent of σ_y, 425 MPa (61.8 ksi), during hydrotest. The maximum credible stress during pipeline operation, that is, the stress of 358 MPa (52.0 ksi) due to a pressure surge, was used for critical crack size calculations for arc burns.

Analysis Methods

Critical flaw sizes were calculated using four distinct fracture-mechanics analysis methods and the appropriate maximum-credible-stress and material-property information. The fracture-mechanics models were (1) the critical-COD method, (2) the Draft British Standard method, (3) the plastic-instability method, and (4) the semi-empirical method. Each of these methods is described in the following. In each method, weld flaws are assumed to be equivalent to surface cracks equal in size to the weld defect.

Critical-COD Method

This model is based on the critical (COD) concept. Crack extension that could cause leakage occurs when the COD value at the crack tip (designated δ) exceeds a critical value: the COD fracture toughness. δ is calculated using a ligament-closure-force model developed by Irwin [6] and based on plasticity-corrected linear-elastic theory. In this approach, the surface crack is modeled as a through-thickness crack in a wide plate coupled with closure forces due to the ligament. The opening of a through crack of length, l, in a plate under a remote tensile stress, σ, is given by.

$$\delta = 2l\sigma/E \qquad (1)$$

For a surface crack, the opening of Eq 1 is reduced by the remaining ligament. The effect of ligament depth can be estimated by a closing force distributed over the face of the crack. Assuming the ligament is yielded, the total closing force, F_c, is

$$F_c = l(t - a)\bar{\sigma} \qquad (2)$$

where

a = crack depth,
t = pipe thickness, and
$\bar{\sigma}$ = flow strength.

Distributing this closing force over the crack-face area, lt, gives a closing stress, σ_c, on the equivalent through crack of

$$\sigma_c = (1 - a/t)\bar{\sigma} \qquad (3)$$

This closing stress opposes the remote stress, σ. The resultant opening of the surface crack is then

$$\delta = 2l(\sigma - \sigma_c)/E \qquad (4)$$

To account for the additional crack opening due to crack tip plasticity, the effective crack length, which includes Irwin's [7] plasticity correction, r_y, is used in place of l. The resulting expression when Eq 3 is substituted into Eq 4 becomes

$$\delta = \frac{2(l + 2r_y)}{E}\left[\sigma - \left(1 - \frac{a}{t}\right)\bar{\sigma}\right] \qquad (5)$$

where $r_y = (1/2\pi)(K/\sigma_y)^2$ and $K = \sigma\sqrt{\pi l/2}$ = stress intensity factor. The residual stresses can be accounted for by assuming they are of the self-equilibrating type resulting from weld shrinkage. In this case, one can assume that yield point stresses act over a distance comparable to the weld size or pipe thickness. If so, a displacement, Δ, of $\Delta = \sigma_y t/E$ will relieve the residual stress. An approximation of the contribution of such a residual stress to δ is simply to add Δ to the applied δ, or equivalently to subtract Δ from δ_c. For σ_y = 414 MPa (60 ksi), E = 208 GPa (30.2 × 10³ ksi), and t = 13 mm (0.51 in.), Δ = 0.025 mm (0.001 in.). Thus, when using Eq 5, failure occurs when the sum of δ plus Δ exceeds the fracture toughness, δ_c. The same value of Δ was used to account for the residual stresses in 12- and 14-mm-thick (0.46 and 0.56 in.) pipe.

Draft British Standard Method

This procedure is described in the Draft British Standard Rules for

Derivation of Acceptance Levels for Defects in Fusion Welded Joints [8]. The model is based on an "allowable COD" concept; crack extension will not occur if the flaw size is limited by the relationship

$$\delta_c = \frac{2\pi \bar{a}\, \sigma_y}{E} \left(\frac{\sigma + \sigma_r}{\sigma_y} - 0.25 \right) \qquad (14)$$

where $\bar{a}$ is an allowable-flaw-size parameter based on a conservative interpretation of wide-plate test results. The residual stress, σ_r, is assumed to be equal to the yield stress, σ_y. The relationship between $\bar{a}$ and the flaw dimensions, a and l, and the pipe thickness, t, is given in Fig. 1. This figure relates the surface flaw dimensions, a and l, to the half-length, $\bar{a}$, of an equivalent through-thickness flaw for the case of flat plates. In this study, curvature effects were neglected and Fig. 1 was used directly for all circumferential flaws and for axial flaws up to 8 cm (3.2 in.) in length.

Plastic-Instability Method

This model applies to circumferential flaws and was developed by Irwin [9] on the basis of investigations of net ligament fractures from part-through

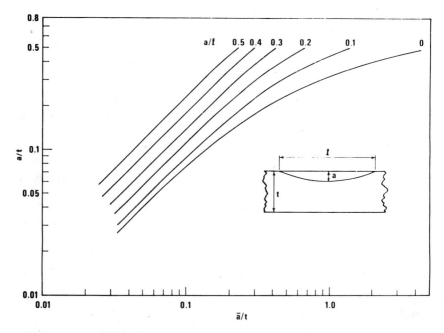

FIG. 1—*Draft British Standard relationship between actual flaw dimensions and the parameter $\bar{a}$ for surface flaws.*

cracks in flat plates of X-65 line pipe and estimated corrections for bulging effects in pressurized cylinders. Plastic instability leading to rupture occurs when the net ligament strain, ϵ_n, reaches a critical value, ϵ_c

$$\epsilon_n = \frac{\delta_0 - \left(a - \frac{t}{2}\right)\theta}{t - a} \qquad (15)$$

where δ_0 = COD at the mid-thickness and $\theta/2$ is the rotation of the crack surface due to bulging. The failure condition selected on the basis of the flat-plate tests [10] was $\epsilon_c = 0.18$. Details regarding the evaluation of δ_0 and θ on the basis of shell theory for the specific geometry, yield strength, and applied stresses applicable to the pipeline are given by Irwin [9].

Semi-Empirical Method

This model applies to axial flaws and was developed by Kiefner et al [5] on the basis of full-scale pipe rupture tests. Plastic instability leading to rupture occurs when the applied stress reaches a critical value related to the flaw size, material flow strength, $\bar{\sigma}$, and pipe dimensions.

$$\sigma = \bar{\sigma}\frac{1 - a/t}{1 - a/tM} \qquad (16)$$

$$M = \left(1 + \frac{0.628l^2}{Dt} - \frac{0.0034l^4}{D^2t^2}\right)^{1/2} \qquad (17)$$

where D is the pipe diameter.

Results and Discussion

Critical and allowable (per the Draft British Standard) flaw sizes were calculated using the applicable fracture-mechanics models, material-property data, and pipeline operating stresses for circumferential and axial flaws. The results are plotted in figures as critical-flaw-size curves with flaw depth as the y-axis and defect length as the x-axis.

Since all flaws are considered surface cracks, the principal differences in the three types are orientation and location. Flaw orientation determines whether the applicable stresses are axial or hoop. Flaw location is used to establish the applicable minimum fracture-toughness, 0.1 mm (0.004 in.) for randomly located flaws and 0.18 mm (0.007 in.) for surface flaws.

Calculated sizes of circumferential flaws are plotted in Fig. 2 for each of the applicable analysis methods. Allowable flaw sizes determined by the

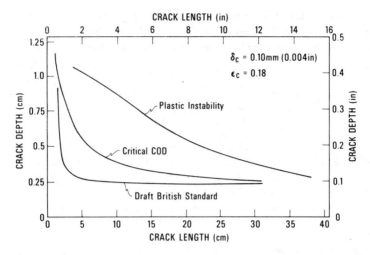

FIG. 2—*Comparison of allowable circumferential flaw sizes determined by the Draft British Standard method with critical circumferential flaw sizes determined using critical-COD and plastic instability as the respective fracture criteria.*

Draft British Standard method are compared with two critical flaw size curves determined using critical COD and plastic instability as the respective fracture criteria. For the critical-COD and the Draft British Standard methods, a toughness value of 0.1 mm (0.004 in.) was used. This is the minimum toughness measured [3,4] for through-thickness notches in the weldment and is applicable to randomly located flaws such as porosity and slag inclusions. For the plastic instability method, a critical ligament strain of 0.18 mm (0.007 in.) was used as the failure criterion and all flaws were located on the exterior surface, the worst-case location when bulging is considered.

Results of flaw size calculations using higher weld toughness are shown in Fig. 3, where a critical-COD value of 0.18 mm (0.007 in.) was used. This is the minimum toughness measured [3,4] for surface notches in the weldment and is considered applicable to surface defects such as lack of penetration and lack of root fusion.

In Fig. 4, calculated sizes of axial flaws are plotted for each of the applicable analysis methods. Here, as in Fig. 2, allowable flaw sizes determined by the Draft British Standard method are compared with two critical-flaw-size curves using critical COD and plastic instability (the semi-empirical curve) as the respective fracture criteria. For the critical-COD and Draft British Standard methods, a toughness value of 0.18 mm (0.007 in.) was used. The surface notch toughness [0.18 mm (0.007 in.)] was used because the only axial-aligned flaws considered were arc burns on the surface. For analysis purposes, arc burns were considered surface cracks of length equal to the arc burn length and depth equal to the depth estimated from a

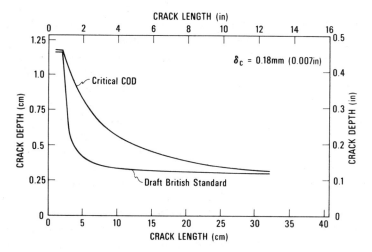

FIG. 3—*Comparison of allowable circumferential flaw sizes determined by the Draft British Standard method with critical circumferential flaw sizes determined by the critical-COD method for the case $\delta_c = 0.18$ mm (0.007 in.).*

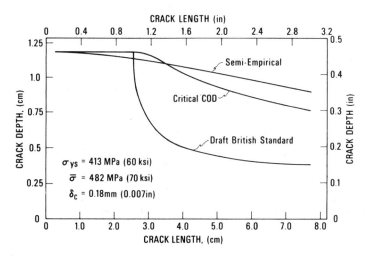

FIG. 4—*Comparison of allowable axial flaw sizes determined by the Draft British Standard method with critical axial flaw sizes determined using critical-COD and plastic instability (the semi-empirical curve) as the respective failure criteria.*

metallographic correlation discussed elsewhere [2]. The semi-empirical results were obtained using a flow stress value of 482 MPa (70 ksi) to calculate plastic collapse. Notice that the crack length axis in Fig. 4 extends to 8 cm (3.2 in.) instead of 40 cm (16 in.) as used in Fig. 2 and 3. The shorter axis was used because the short arc-burn lengths were of principal interest.

The smallest flaw sizes were calculated using the Draft British Standard method. This method is intended to calculate allowable flaw sizes (flaw sizes regarded as safe) and contains conservative assumptions and safety factors absent in the other models. The principal conservative elements in this method include the relationship between toughness, stress level, and crack size; the treatment of crack aspect (depth to length) ratio; and the assumed residual-stress level. The relationship between toughness, stress level, and crack size is based on a lower-bound interpretation of wide-plate test results; failure does not occur under the conditions used (in this investigation) to calculate "critical" flaw size. Linear-elastic theory is used to calculate flaw-shape effects; beneficial effects of stress redistribution due to plastic strain in the ligament are neglected. Under linear-elastic conditions, the stress intensity at the leading edge of a surface crack rapidly increases in severity between aspect ratios of $a/l = 0.5$ and 0.1. Thus, the most severe conditions are attributed to relatively short cracks, and the Draft British Standard curves characteristically become asymptotic at crack lengths less than 50 mm (2 in.) Residual stresses are assumed to equal the yield strength of the weld and this stress system is added to the applied stress. This conservatism is partly offset by the empirical nature of the stress/flaw-size/toughness relationship.

The largest flaw sizes were calculated by Irwin's plastic-instability model for weld flaws and by Kiefner's semi-empirical model for arc burns. The failure criterion for both these models is plastic collapse of the ligament. The credibility of both models is enhanced by their relationship to large-scale test results on pipeline steels. Irwin's results apply to flaws at the exterior surface, the most severe location when bulging is considered. Buried or internal flaws, which are more common, can be approximately 30 percent longer before reaching critical size. Further calculations by Irwin and Albrecht [9] show that using a critical strain of 0.12 instead of 0.18 reduces the allowable length by less than 25 percent and would not significantly affect the relative position of the curve.

Flaw sizes calculated using the critical-COD model fall between those of the Draft British Standard and the plastic-instability models. The results are higher than the Draft British Standard results because failure occurs at a critical flaw size instead of at an allowable flaw size. The results are lower than those obtained by the plastic-instability models because the critical-COD level usually is reached at stable values of ligament strain. Care should be taken in using the critical-COD model because the difference between the flow strength of the material and the applied stress strongly influences the position of the curves as shown in Fig. 5.

Conclusions and Recommendations

The critical flaw sizes calculated using fracture mechanics vary significantly depending on the fracture criterion chosen, that is, critical crack size, allowable crack size, or plastic instability. Thus, experimental work is needed to verify which fracture criterion most accurately models girthweld behavior. In addition, further analytical development is needed to improve the models evaluated.

The critical-COD and the Draft British Standard methods yield similar results for short deep flaws and for long shallow flaws. Thus, tests are

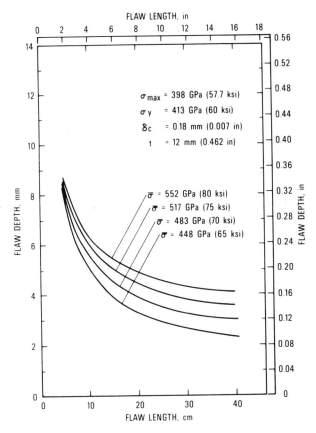

FIG. 5—*Effect of changes in the difference between the flow stress and the applied stress on allowable flaw sizes calculated by the critical-COD method.*

needed to determine flaw shape effects under conditions of ligament yielding for the a/l range of 0.5 to 0.01.

The plastic-instability analyses indicate that relatively large flaws are required for plastic collapse of the ligament. For welds with COD toughness levels less than 0.2 mm (0.008 in.) crack extension is generally predicted to occur prior to plastic collapse. Experiments are needed to establish the governing conditions for each failure mode.

Acknowledgment

This work was sponsored by the U.S. Department of Transportation, Office of Pipeline Safety. The authors wish to express appreciation to Lance Heverly of OPSO, the project monitor; to Drs. Richard P. Reed and Maurice B. Kasen of NBS, the task leaders; to Harold Berger of NBS, the program manager, and to G. M. Wilkowski of Battelle, who critically reviewed the manuscript.

References

[1] Office of Pipeline Safety Operations notice in the Federal Register, Aug. 13, 1976.
[2] *Consideration of Fracture Mechanics Analysis and Defect Dimension Measurement Assessment for the Trans-Alaska Oil Pipeline Girth Welds,*" H. Berger and J. H. Smith, Eds., NBSIR 76-1154, National Bureau of Standards, Gaithersburg, Md., Oct. 1976.
[3] Harrison, J. D., "COD and Charpy V Notch Impact Tests on Three Pipeline Butt Welds Made in 1975," Welding Institute Report LD 22062/5, July 1976.
[4] Spurrier, J. and Hancock, P., "Crack Opening Displacement and Charpy Impact Testing at Cranfield Institute of Technology for British Petroleum Trading Co., Ltd., "Cranfield Institute of Technology, Cranfield, U.K., July 1976.
[5] Kiefner, J. F., Maxey, W. A., Eiber, R. J., and Duffy, A. F. in *Fracture Toughness, ASTM STP* 536, American Society for Testing Materials, 1973, pp. 461-481.
[6] Irwin, G. R., "Fracture Mechanics Notes," Lehigh University, Bethlehem, Pa., 1969.
[7] "Fracture Testing of High Strength Sheet Materials," First Report of Special ASTM Committee, ASTM Bulletin, American Society for Testing and Materials, Jan. 1960.
[8] Draft British Standard Rules for the Derivation of Acceptance Levels for Defects in Fusion Welded Joints, British Standards Institution, London, U.K., Feb. 1976.
[9] Irwin, G. R. and Albrecht, P. in *Consideration of Fracture Mechanics Analysis and Defect Dimension Measurement Assessment for the Trans-Alaska Oil Pipeline Girth Welds,* H. Berger and J. H. Smith, Eds., NBSIR 76-1154, National Bureau of Standards, Gaithersburg, Md., Vol. 2, Appendix D, Oct. 1976.
[10] Irwin, G. R., Krishna, G., and Yen, B. T., Fritz Engineering Laboratory Report 373.1, Lehigh University, Bethlehem, Pa., March 1972.

L. A. Simpson[1] *and C. F. Clarke*[1]

An Elastic-Plastic R-Curve Description of Fracture in Zr-2.5Nb Pressure Tube Alloy

REFERENCE: Simpson, L. A. and Clarke, C. F., **"An Elastic-Plastic R-Curve Description of Fracture in Zr-2.5Nb Pressure Tube Alloy,"** *Elastic-Plastic Fracture, ASTM STP 668,* J. D. Landes, J. A. Begley, and G. A. Clarke, Eds., American Society for Testing and Materials, 1979, pp. 643–662.

ABSTRACT: An R-curve approach was investigated with the aim of establishing a means of predicting critical crack lengths in Zr-2.5Nb pressure tubes using small fracture-mechanics specimens. Because of the elastic-plastic nature of the fracture process and limitations on the maximum specimen size, conventional R-curve methods were not applicable. The crack growth resistance was therefore expressed in terms of the crack opening displacement (COD) and R-curves were plotted for several sizes of specimens and crack lengths at 20°C and at 300°C. The effect of hydrogen on R-curve behavior at these two temperatures was investigated as well.
Conventional clip-gage methods were not suitable for this work. Crack length was determined from electrical resistance, and COD, at the actual crack front, was determined from photographs of the specimens taken during testing. Crack length and specimen size had little, if any, effect on the R-curve shape. A method for expressing crack growth resistance in terms of the J-integral was also investigated and appears to be consistent with the COD approach. The effects of hydrogen and temperature on R-curve shape are consistent with their known effects on the mechanical behavior of Zr-2.5Nb. Finally, predictions of critical crack length in pressure tubes obtained by matching R-curves to crack driving force curves are consistent with published burst-testing data.

KEY WORDS: crack propagation, fracture, R-curves, metals, zirconium, pressure tubes, potential drop

The CANDU[3] nuclear reactor system uses cold-worked Zr-2.5Nb pressure tubes as the primary coolant containment. At present, critical crack length data are obtained by burst testing full-size tube sections [1].[2] The work described here is part of a program to develop a framework for predicting

---
[1] Research officer and research technician, respectively, Materials Science Branch, Atomic Energy of Canada Ltd., Whiteshell Nuclear Research Establishment, Pinawa, Man., Canada.
[2] The italic numbers in brackets refer to the list of references appended to this paper.
[3] *CAN*ada *D*euterium *U*ranium.

critical crack lengths in these tubes from small fracture-mechanics specimens. The advantages of using small specimens are material conservation, experimental convenience, and a more suitable specimen geometry for studying the micromechanisms of the fracture process.

The feasibility of using small specimens depends on the development of a geometry-independent fracture criterion. The typical pressure tube, as used in the Pickering reactor design, has a mean diameter of 10.7 cm and a wall thickness of 4.1 mm. For the most serious type of defect, a through-thickness crack lying in the axial direction, propagation will occur under near-plane stress conditions so that a plane-strain linear elastic fracture-mechanics (LEFM) approach is not feasible. Early attempts at establishing a fracture criterion for zirconium alloys dealt with δ_{max}, the crack opening displacement (COD) at instability or maximum load. Some limited success in predicting tube behavior was achieved with this criterion [2,3]; however, there were also discrepancies which suggested that δ_{max} was geometry dependent [4]. More recently, the first author [5] confirmed this geometry dependence but also suggested that δ_i, the COD at crack initiation, may be geometry independent. Subsequent to initiation, however, Zr-2.5Nb tolerates considerable slow, stable crack growth under rising load and δ_i is therefore unnecessarily conservative as a fracture criterion. For the same reason, the use of the J-integral at crack initiation is unsuitable.

R-Curve Methods

An R-curve, briefly, is a plot of the resistance to further crack extension in a specimen undergoing slow, stable crack growth, against the extent of this stable crack extension. It has been suggested [6] that the R-curve for a material of fixed thickness is geometry independent. If this is so, the failure condition for any geometry can be determined from the point of tangency of the R-curve with the plot of crack driving force against crack length for that geometry. These techniques and concepts are well documented in Ref 6 and many other papers in the literature dealing with R-curves and will not be repeated here.

The geometry independence of the R-curve is still a debatable concept and should be established for a particular material. For example, work by Adams [7] on two high-strength aluminum alloys suggests that the R-curve depends on specimen configuration. Thus, one aim of this work is to assess the geometry dependence of R-curves for Zr-2.5Nb.

Traditionally, the crack growth resistance, K_R, has been calculated using LEFM equations for the stress-intensity factor and the effective crack length (corrected for plastic zone contribution) for a particular type of specimen. The stress-intensity factor has significance only if the in-plane specimen dimensions of crack length and ligament size are large compared with the plastic zone size. The ASTM Recommended Practice for R-curve Determination (E 561-75T) states that the uncracked ligament size should exceed $(4/\pi)$ $(K_{max}/\sigma_y)^2$ where K_{max} is the maximum K level in the test and σ_y is the yield stress. Thus very large specimens are required for calculations of K_R by LEFM equations from measurements on tough materials. Table 1 gives the yield strengths of Zr-2.5Nb at 20 and 300°C (maximum reactor operating temperature) and the minimum in-plane specimen dimensions calculated from the foregoing criterion, assuming (conservatively) $K_{max} = 100$ MPa/m$^{1/2}$.

Test specimens must be cut from flattened pressure-tube material to obtain the relevant microstructure and mechanical properties. The diameter of the tubes limits the practical specimen size to about 60 mm although, because of the large nonuniform deformations experienced in flattening large specimens, sizes of the order 35 mm are preferred. Thus the LEFM approach was not suitable for determining R-curves in this work.

Recently a number of investigations have considered the use of elastic-plastic fracture parameters such as COD [8,9] and the J-integral [8-11] to describe crack-growth resistance. For steel, McCabe [8] converted COD, δ, to an effective K_R via

$$K_R = m(E \times \sigma_y \times \delta)^{1/2} \tag{1}$$

where

$E =$ Young's modulus, and
$m =$ constant $= 1.0$.

The validity of Eq 1 should be verified for a particular material as various derivations of Eq 1 give m values between 1 and 2.

While the J-integral is not well defined for situations in which the crack-tip region is unloaded, attempts have been made to calculate it subsequent to stable crack growth [8-11]. The usual assumption is made that the J value of a specimen following some crack extension from a to Δa is the same as in a nonlinear elastic specimen of initial crack length $a + \Delta a$ loaded to the same value of load or displacement or both with no crack extension. A valid J can be calculated for the latter specimen, so the problem reduces to calculating

TABLE 1—*Minimum in-plane specimen dimensions for LEFM calculation of K_R[a].*

Temperature, C°	σ_y, MPa	$\frac{4}{\pi}\left(\frac{K_{max}}{\sigma_y}\right)^2$ mm
20	800	20
300	533	45

[a]$K_{max} = 100$ MPa/m$^{1/2}$.

the J values for the equivalent specimens for various crack lengths. Garwood et al [10,11] have developed a convenient method for calculating J values following stable crack extension in deeply cracked compact tension or bend specimens from a single load (P)-load point deflection (δ_p) curve. For small increments of crack growth, they derive

$$J_n = J_{n-1} \frac{W - a_n}{W - a_{n-1}} + 2 \frac{U_n - U_{n-1}}{B(W - a_{n-1})} \qquad (2)$$

$J_n = J$ at nth point on P-δ_p curve

where

$W =$ specimen width,
$B =$ specimen thickness,
$U_n =$ area under the P-δ_p curve up to a point, n, on the curve, and
$a_n =$ crack length at point, n, on the P-δ_p curve.

$$J_o = \frac{2U_o}{B(W - a_o)} \qquad (3)$$

where

$U_o =$ area under the P-δ_p curve up to crack initiation (or any arbitrary point prior to initiation), and
$J_o =$ corresponding initial value of J.

With this equation, J_n can be calculated from a single load-load point deflection curve provided crack extension is simultaneously monitored.

In this work the techniques for applying these R-curve methods to pressure-tube material are developed and the effect of temperature and hydrogen content on R-curve behavior are examined. An initial assessment is also made of the ability of R-curves to predict pressure-tube failure.

Experimental

Specimen Preparation

Factors affecting the choice of specimen size were:
1. The need to test at 300°C in a furnace.
2. The need to cut specimens directly from pressure tubes to obtain the relevant material condition.
3. The need to minimize deformation imparted to the specimens when flattened.

The compact tension specimen (CTS), Fig. 1, was chosen for this study in

FIG. 1—*Compact tension specimens used in this study.*

three sizes specified by their width (W) dimension of 17, 34 and 68 mm. The other dimensions are in the proportions recommended in the ASTM Test for Plane-Strain Fracture Toughness of Metallic Materials (E 399-72), except for thickness, which, after machining, was 3.75 to 3.80 mm for all specimens. Most of the testing was done on 34-mm specimens with some 17- and 68-mm specimens tested at 20°C to study geometry (size) effects. Two crack length ranges were studied as well with $a/W \approx 0.3$ and 0.6. Fatigue precracks were initiated in all specimens using maximum stress-intensity factors less than 20 MPa/m$^{1/2}$.

Some specimens were gaseously hydrided at 400°C to levels of 200 µg/g to study the effect of hydrogen on R-curve behavior. (The hydriding conditions were chosen to have a minimal effect on structure. Hydrogen exists as zirconium hydride when present in excess of its solubility limit of ~1 µg/g at 20°C and ~65 µg/g at 300°C [5,12] and under certain conditions is a factor in causing embrittlement.)

Measurement of Crack Length

Most R-curve determinations in the past have used compliance measurements to follow crack extension. The need to test at elevated temperatures prevented the use of the clip gages necessary for this approach, and our experience in using the d-c potential drop method to follow hydrogen-induced subcritical crack growth [13] suggested that this technique would be highly suitable, particularly as it gives a continuous reading of crack extension. A constant current (~10 A for 34-mm specimens) was passed through the specimen during the test via the screw-in copper leads shown in Fig. 2. The potential drop across the crack was monitored using zirconium leads and a chart recorder with 100-µV full-scale sensitivity.

The change in potential drop with crack extension is reported to be linear in a/W for the CTS geometry [14] for $0.3 < a/W < 0.7$. This was confirmed for our 34-mm specimens by following a fatigue crack in which the fracture surface was periodically marked (~ every 1 mm) by overloading. The results from two specimens are shown in Fig. 3. By using the slope of this curve, the amount of crack extension in any specimen was calculated from the change in potential drop from the start of the test. This technique is capable of detecting crack extension of less than 10 µm [13], which is more than adequate for the present task.

COD Measurement

In most R-curve studies to date, where COD measurements were required, they were determined from clip-gage readings at the crack mouth. These calculations usually assumed that the specimen rotated about a fixed center in the ligament [8,15]. Preliminary testing on Zr-2.5Nb specimens indicated

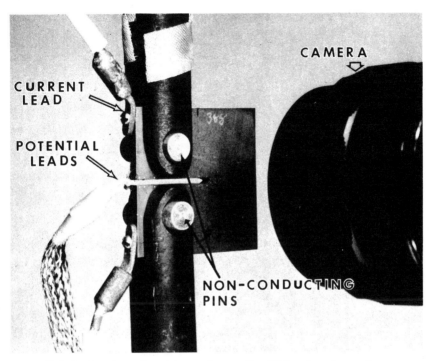

FIG. 2—*Experimental arrangement for recording COD and crack extension.*

that no such fixed center existed [5], even when the reduction in ligament size with crack extension was accounted for. Therefore, COD was measured directly on each specimen by photographing pairs of microhardness indentations on opposite sides of the crack mouth as the specimen was loaded [15]. For each load level, crack mouth displacements were measured, plotted against distance from the original crack tip as in Fig. 4, and a line was drawn through to the apparent center of rotation on the abscissa. The intercept at the original crack front was the COD at that point; however, our interest was in the COD at the actual crack tip. This was found by marking the position of the crack front, as determined by the potential drop data, on each line (load level) in Fig. 4. Joining these points yielded a locus of the actual COD during the test.

R-Curve Determination

Using the COD measurements just described and Eq 1, R-curve determinations were carried out at 20 and 300°C for specimens containing as-received hydrogen (~10 μg/g) and 200 μg/g hydrogen. The 300°C tests were done in a furnace containing a window to allow photographic recording of

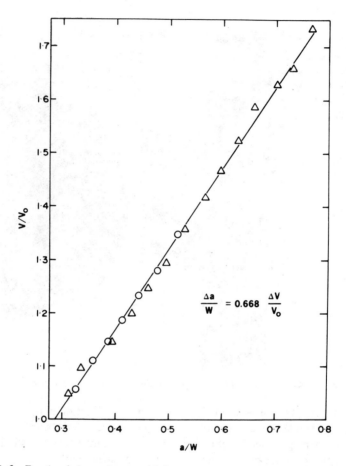

FIG. 3—*Fractional change in potential drop across compact tension specimen versus crack extension.*

COD. Specimens were loaded well past maximum load in all cases except for hydrided material at 20°C, where instability occurred shortly after maximum load. The specimens were heat-tinted at 300°C (if they had not already been tested at that temperature) prior to final fracture to identify the region of slow stable crack growth on the fracture surface. The amount of stable crack growth was measured and used in conjunction with the total change in potential drop to check the calibration of Fig. 3.

J-Integral Determination

J-integral values were determined from plots of load versus load-point displacement (the latter can easily be determined from plots similar to Fig. 4)

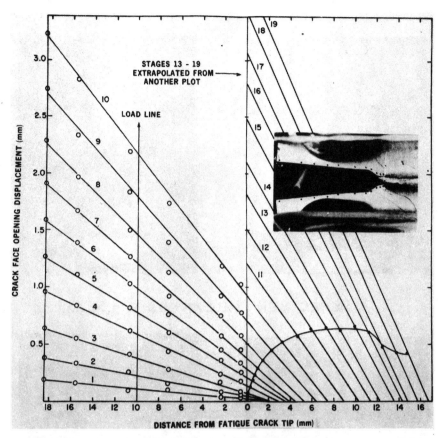

FIG. 4—*Method for determining COD from displacement measurements on crack face. Each numbered straight line represents a set of displacements for a given load level. The curved line on the right indicates the magnitude and position of the COD at the actual crack tip. Inset shows microhardness indentations used for crack face displacement measurements.*

at each loading stage of the test. The areas under the $P\text{-}\delta_p$ curves were measured with a planimeter. These determinations were confined to the deeply cracked specimens since Eq 2 assumes predominately bending conditions.

Results and Discussion

Stable Crack Development

Fracture surfaces of typical specimens after varying amounts of stable crack growth are shown in Fig. 5. While these specimens all contained as-received hydrogen levels (~ 10 μg/g) and were tested at 20°C, they are also

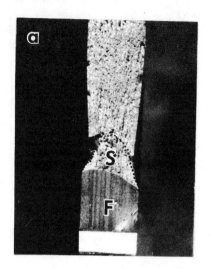 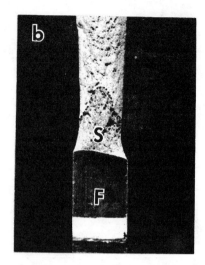

SPECIMEN THICKNESS = 3.75 mm

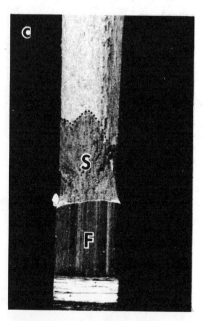 

FIG. 5—*Typical morphology of slow stable cracking in Zr-2.5Nb at various stages of development.* (F = fatigue crack surface; S = slow stable crack surface; end of stable crack marked for clarity.)

typical of the tests at 300°C on both nominal and hydrided material. The first notable feature is that the fatigue precracks were not always straight but were often restrained near one surface. This was observed in about 50 percent of the specimens and is attributable to residual stresses created by the tube-flattening treatment. Annealing the specimens prior to fatigue precracking was ruled out because of the possibility of altering the cold-worked structure of the material and the apparent insensitivity of R-curve shape to the initial fatigue-crack configuration.

As shown in Fig. 5, stable crack development commenced with initiation near the specimen midsection. The crack assumed a triangular shape as it tunneled forward and, after propagating a distance roughly equal to the specimen thickness, the crack front tended to straighten. This was accompanied by a transition from flat fracture to fracture on planes inclined about 45 deg to the original crack plane (slant fracture). When the ligament size was reduced to about 5 mm, the fracture surface became flat again.

Stable crack growth continued well past maximum load in all tests, except for the hydrided specimens at 20°C, which failed abruptly soon after maximum load. In these latter specimens, tunneling was more pronounced and development of slant fracture did not occur, although small shear lips formed at the specimen surfaces. While it was tempting to associate the flat fracture surface with plane-strain conditions, examination of the fracture surface in Fig. 6 revealed that splitting occurred along the hydride platelets, which were mostly oriented at right angles to the crack front and crack plane (that is, platelet normals were in the specimen thickness direction [5]). This

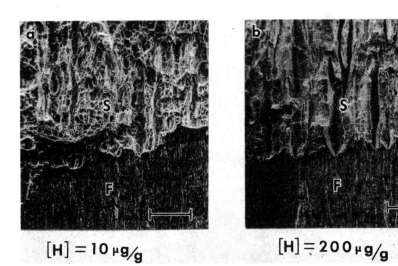

FIG. 6—*Effect of hydrogen on fracture morphology at 20°C.* (F = *fatigue crack surface;* S = *slow stable crack surface.*) *Bar indicates 100 µm.*

splitting is a common effect in hydrided Zr-2.5Nb near room temperature and causes the specimen to delaminate into a number of parallel thin specimens. Microscopic examination reveals that fracture of the individual lamina is by ductile tearing [5].

R-Curves Geometry Independence

It was mentioned earlier that the choice of $m = 1$ in Eq 1 was somewhat arbitrary. A test of validity would be to compare K_R calculated from Eq 1 with K_{LEFM}, the stress intensity factor calculated from the LEFM analysis for the compact tension specimen, in the early stages of loading, where LEFM analysis has validity. This is done for several specimens in Fig. 7 where K_{LEFM} for a particular loading stage is plotted against K_R calculated from Eq 1. To extend the valid range of K_{LEFM}, it is calculated using the plastic zone-corrected crack length (by adding the plastic zone size to the actual crack length). While there is some scatter, the data in Fig. 7 are distributed uniformly about the line representing $m = 1$. This suggests that $m = 1$ is an appropriate choice to make K_R, as calculated from Eq 1, compatible with an LEFM calculation of the crack growth resistance.

The *R*-curves calculated from Eq 1 are plotted in Figs. 8a to 8e. The 20°C data for nominal hydrogen are divided between Figs. 8a and 8b, representing shallow and deeply cracked 34-mm specimens, respectively. Figure 8a also includes results from two 68-mm specimens, and Fig. 8b includes a typical

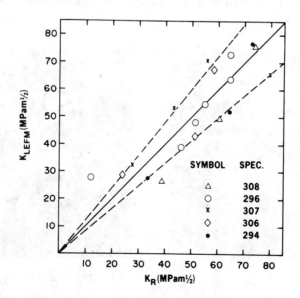

FIG. 7—*Comparison of* K_R *as calculated by LEFM methods with values calculated from COD. Solid line:* m = 1.0; *dashed lines:* m = 1.0 ± 20 *percent.*

result from a 17-mm specimen. Figures 8a and 8b suggest a crack length dependence of R-curve shape in the latter stages of its development, with the shallow-cracked specimens attaining higher plateau levels (for 34-mm specimens). When the shallow cracks were extended to depths typical of those in the deeply cracked specimens, the K_R values fell off toward the plateau values of the latter. This, combined with the tendency of deep cracks to revert to flat fracture, suggests that the crack may interact with the stress-free back surface of the specimen when the ligament size approaches approximately 10 mm. The two 68-mm specimens in Fig. 8a show the opposite crack length effect (higher plateau for deep crack).

The 34-mm specimens showing high plateau levels in Fig. 8a developed single, slant-fracture surfaces turned more than 45 deg from the original crack plane. This presents some concern over the interpretation of the K_R values since the fracture contains a Mode III component and COD is measured in the loading direction, *not* normal to the crack plane.

No clear effect of crack length was observed for either set of tests at 300°C (Figs. 8d and 8e); in fact, in Fig. 8d, the shallow-cracked specimen yields a slightly lower R-curve than the others, that is, an opposite effect of crack length to that suggested at 20°C (for 34-mm specimens) by Fig. 8a and 8b. The R-curves for hydrided material at 20°C were also insensitive to crack length (Fig. 8c). Because of this, and because the differences reported for as-received material at 20°C are significant only at the later stages of R-curve development (where some question exists about the meaning of COD), we conclude that R-curves in Zr-2.5Nb may well be geometry (size) independent. Certainly, further investigation is justified to resolve this question convincingly, possibly using a completely different specimen geometry.

R-Curves—Effect of Temperature and Hydrogen

The R-curves at 300°C for as-received and hydrided material are indistinguishable. In the as-received material, all the hydrogen would have been in solution, whereas hydrides would have been present in the hydrided material. Thus hydride has no effect at this temperature. At 20°C, hydride has a definite embrittling effect, demonstrated by the R-curves in Fig. 8c, which are much flatter and have a lower plateau value than the others. This behavior is consistent with earlier work on the fracture properties of zirconium alloys [1,16], where hydrogen up to 400 µg/g has a detrimental effect only below about 150°C.

An interesting comparison can be made here with McCabe's results [8] for a carbon-manganese steel, which suggest that the R-curve is temperature independent above the Charpy transition temperature. Hydrogen causes an upward shift in the transition temperature in Zr-2.5Nb [17] from just below room temperature to about 150°C for a concentration of 200 µg/g. Thus all tests except those at 20°C with 200-µg/g hydrogen were carried out in the

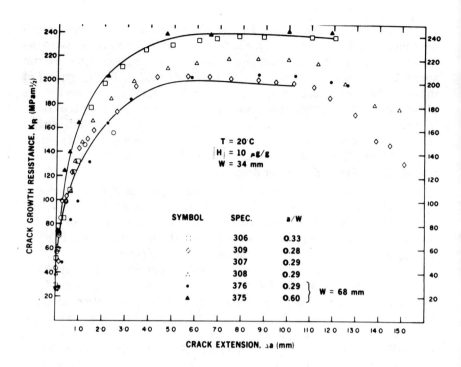
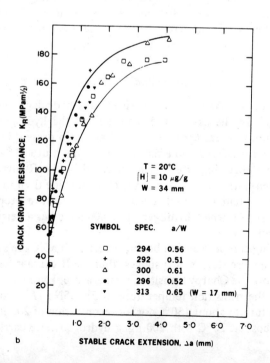

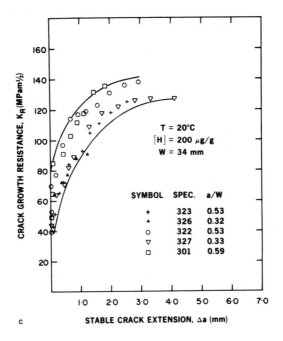

c

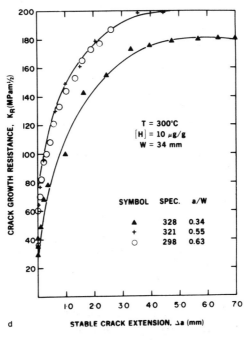

d

FIG. 8—*R-curves for Zr-2.5Nb specimens determined from COD.*

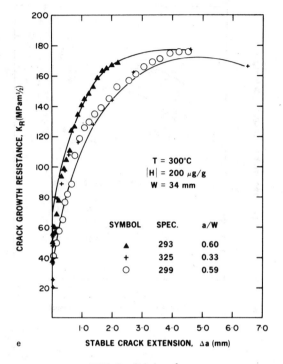

FIG. 8—Continued.

upper-shelf region and, as for the steel results, the R-curves were temperature independent.

J-Integral Measurements

The J-integral was calculated as a function of crack extension, using Eq 2, for all the deeply cracked 34-mm specimens. The exact physical significance of J as measured in this manner is not completely clear. The critical assumption by Garwood et al [10] is that the difference between the energy under the actual load-displacement curve and that for the hypothetical specimen loaded to the same load and deflection is the energy taken up by crack growth. They admit that no proof of this assumption exists. Also, the accuracy of the J calculation will be dependent on minimizing the segments of crack growth between calculation points to some as yet undefined optimum value. Because of these uncertainties, the credibility of the J-integral results will simply be discussed in terms of their self-consistency with the K_R data obtained by COD measurements.

The initial value of J, J_o, was not the initiation value as chosen by Garwood and Turner [11]. Their analysis is equally valid if J_o is chosen anywhere in the linear portion of the load-displacement curve (that is, prior to initiation) and this procedure was followed here. Plots of J_n versus Δa yield the same qualitative shapes as the K_R plots in Figs. 8a to 8e. A general comparison between the two methods is obtained by plotting J_n/σ_y against δ in Fig. 9. After some initial curvature, there is a generally good straight-line correlation between these two parameters which can be described by the equation

$$J_n = 1.1\sigma_y(\delta - \delta_0) \qquad (4)$$

where $\delta_0 \approx 0.032$ mm. Thus, after some initial loading, J_n is linearly dependent on δ. Calculations of J based on simple yield models [18] suggest that

$$J = \sigma_y \delta \qquad (5)$$

In the early stages of loading, the primary contribution to J will be elastic [19] and

$$J \propto \delta_p^2 \propto \delta^2 \qquad (6)$$

the latter proportionality resulting from the method of COD measurement used here (Fig. 4). This is consistent with the initial curvature in Fig. 9.

Thus the calculation of J beyond crack initiation, using Eq 2, is in good agreement with the foregoing relationships between J and COD. The factor of 1.1 could easily be accounted for by work hardening, since Eq 5 is based on perfectly plastic behavior. Further work is underway to substantiate the use of J_R as a fracture criterion, including the calculation of J in a cracked

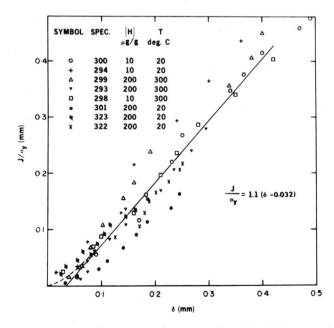

FIG. 9—*Relationship between J measurements and COD.*

tube body as a function of crack length and pressure, using finite-element analysis. J_R curves will then be used for critical crack length prediction via a tangency condition similar to that used for K_R curves.

Critical Crack Length Prediction

The ultimate test of K_R as a fracture criterion is in its ability to predict the failure condition of a structure, in this case the Zr-2.5Nb pressure tubes. To determine the critical crack length at operating pressure, it is necessary to place the R-curve such that it is just tangent to the crack driving force curve as in Fig. 10. The crack driving force for an axially cracked tube is given by [20]

$$K = \left[\frac{8a\,\sigma_y^2}{\pi} \times \ln\left(\sec\frac{\pi M\sigma}{2\sigma_y}\right) \right]^{1/2} \qquad (7)$$

where

a = crack half-length,
σ = hoop stress in pressure tubes, and
M = magnification factor due to tube curvature.

A "flow stress" is often used in place of the yield stress, σ_y, in Eq 7, which takes into account work-hardening. It is not used here because its selection is somewhat arbitrary, and, because the material is cold-worked, it has a negligible effect on the crack driving force curve over the range of crack lengths of interest. M is given by

$$M = 1 + 1.255\,\frac{a^2}{Rt} - 0.0135\,\frac{a^4}{(Rt)^2} \qquad (8)$$

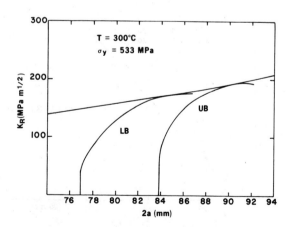

FIG. 10—*Critical crack length prediction for pressure tubes at 300°C obtained by matching upper (UB) and lower (LB) bounds of R-curves with crack driving force curve.*

where

R = tube radius, and
t = tube wall thickness.

In Fig. 10, the data for hydrided and as-received material at 300°C are combined and represented by R-curves corresponding to the upper and lower bounds of the scatter. An operating hoop stress, in the tube, of 125 MPa is assumed and the crack driving force (Eq 7) is plotted against axial crack length, $2a$, for a yield stress of 533 MPa. The lower-bound R-curve indicates that a 75-mm crack will grow stably as the operating stress is applied until it just reaches criticality at operating pressure and a length of 84 mm. Similarly, the upper-bound curve predicts that an 84-mm crack would become unstable if loaded to operating stress. Thus the data predict a range of critical crack lengths between 75 and 84 mm. Ideally this should be compared with burst-testing data on identical material. Because the pressure tubes are extruded hot, and undergo some cooling during the process, the strength can vary 20 percent over the length of the tube [21]. Yield strengths were not provided with the burst-testing data [1,17], which were obtained several years ago, and the choice of 533 MPa may be inappropriate. Most of the burst data pertain to irradiated material which will also cause significant changes in yield stress. In spite of these difficulties, the lower bound of the burst-testing data [17] indicates a critical crack length of 70 to 75 mm at 125 MPa, in excellent agreement with the R-curve prediction. Certainly further work is justified. The next logical step is to make direct comparisons by cutting compact tension specimens from previous burst sections, eliminating the effect of variations in material properties.

Summary and Conclusions

Because of the considerable amount of stable crack growth which Zr-2.5Nb will tolerate, an R-curve approach seems to be the most appropriate to use in the development of a fracture criterion. The geometry independence of the R-curve has not been unequivocally established in this work. However, the small crack length dependence at 20°C for as-received material is reversed at 300°C (Fig. 8*d*) and both dependencies may therefore reflect material variations rather than geometry effects. The calculations of K_R using COD are fully consistent in the LEFM limit. The expression of crack-growth resistance in terms of J_R is also compatible with the COD measurements, which suggests that the J-integral approach may be an equally appropriate way of deriving an R-curve. The potential-drop method has proved to be an accurate and convenient means of following stable crack extensions, especially at elevated temperatures where displacement-gage methods are unsuitable.

Finally, the effect of temperature and hydrogen on the R-curves themselves is fully consistent with the established effects of these parameters on material properties. The application of an R-curve to the prediction of critical crack length in a tube, while preliminary, shows promise and warrants further development.

References

[1] Langford, W. J. and Mooder, L. E. J., *Journal of Nuclear Materials*, Vol. 39, 1971, pp. 292-302.
[2] Fearnehough, G. D. and Watkins, B., *International Journal of Fracture Mechanics*, Vol. 4, 1968, pp. 233-243.
[3] Henry, B., "La Prevision des Conditions Critiques de Rupture de Tubes de Pression en Zr-2.5% Nb par le Critère de l'élargissement critique de Fissure," Euratom Report EUR 5017f, Ispra, 1973.
[4] Pickles, B. W., *Canadian Metallurgical Quarterly*, Vol. 11, 1972, pp. 139-146.
[5] Simpson, L. A., "Initiation COD as a Fracture Criterion for Zr-2.5% Nb Pressure Tube Alloy" in *Fracture 1977*, Vol. 3, D. M. R. Taplin, Ed., University of Waterloo Press, Waterloo, Ont., Canada, 1977.
[6] McCabe, D. E. and Heyer, R. H. in *Fracture Toughness Evaluation by R-Curve Methods, ASTM STP 527*, American Society for Testing and Materials, 1973, pp. 17-35.
[7] Adams, N. J. in *Cracks and Fracture, ASTM STP 601*, American Society for Testing and Materials, 1976, pp. 330-345.
[8] McCabe, D. E. in *Flaw Growth and Fracture, ASTM STP 631*, American Society for Testing and Materials, 1977, pp. 245-266.
[9] Tanaka, K. and Harrison, J. D. "An R-Curve Approach to COD and J for an Austenitic Steel," British Welding Institute Report No. 7/1976/E, July 1976.
[10] Garwood, S. J., Robinson, J. N., and Turner, C. E., *International Journal of Fracture*, Vol. 11, 1975, pp. 528-530.
[11] Garwood, S. J. and Turner, C. E., "The Use of the J-Integral to Measure the Resistance of Mild Steel to Slow Stable Crack Growth" in *Fracture 1977*, Vol. 3, D. M. R. Taplin, Ed., University of Waterloo Press, Waterloo, Ont., Canada, 1977.
[12] Kearns, J. J., *Journal of Nuclear Materials*, Vol. 27, 1968, pp. 64-72.
[13] Simpson, L. A. and Clarke, C. F. "The Application of the Potential Drop Technique to Measurements of Sub Critical Crack Growth in Zr 2.5% Nb," Atomic Energy of Canada Ltd., Report No. AECL 5815, 1977.
[14] McIntyre, P. and Priest, A. H., "Measurement of Sub Critical Flaw Growth in Stress Corrosion, Cyclic Loading and High Temperature Creep by the DC Electrical Resistance Technique," Bisra Open Report MG/54/71, British Steel Corp., London, 1971.
[15] Ingham, T., Egan, G. R., Elliott, D., and Harrison, T. C. in *Practical Applications of Fracture Mechanics to Pressure Vessel Technology*, R. W. Nichols, Ed., Institution of Mechanical Engineers, London, 1971, pp. 200-208.
[16] Watkins, B., Cowan, A., Parry, G. W., and Pickles, B. W. in *Applications-Related Phenomena in Zirconium and Its Alloys, ASTM STP 458*, American Society for Testing and Materials, 1969, pp. 141-159.
[17] Ells, C. E. in *Zirconium in Nuclear Applications, ASTM STP 551*, American Society for Testing and Materials, 1974, pp. 311-327.
[18] Rice, J. R. in *Fracture*, H. Liebowtiz, Ed., Academic Press, New York, 1968, Chapter 3, pp. 191-311.
[19] Knott, J. F., *Fundamentals of Fracture Mechanics*, Wiley, New York, 1973, pp. 170-171.
[20] Kiefner, J. F., Maxey, W. A., Eiber, R. J., and Duffy, A. R. in *Progress in Flow Growths and Fracture Toughness Testing, ASTM STP 536*, American Society for Testing and Materials, 1973, pp. 461-481.
[21] Evans, W., Ross-Ross, P. A., LeSurf, J. E., and Thexton, H. E., "Metallurgical Properties of Zirconium-Alloy Pressure Tubes and Their End Fittings for CANDU Reactors," Atomic Energy of Canada Ltd., Report No. AECL-3982, Sept. 1971.

B. D. Macdonald[1]

Correlation of Structural Steel Fractures Involving Massive Plasticity

REFERENCE: Macdonald, B. D., "**Correlation of Structural Steel Fractures Involving Massive Plasticity,**" *Elastic-Plastic Fracture, ASTM STP 668,* J. D. Landes, J. A. Begley, and G. A. Clarke, Eds., American Society for Testing and Materials, 1979, pp. 663–673.

ABSTRACT: A three-dimensional, elastic-plastic fracture strength correlation for A36 and HSLA structural steel connections containing discontinuities was determined. The fracture specimens comprised beam-column connections in which one column flange contained a mid-thickness plane of discontinuity. Beam loading or direct tension applied normal to the column face imposed tensile load transfer around the boundaries of the discontinuity. Fracture extension was mixed mode (crack opening and edge sliding), and inclined toward the free surface on the web side of the column flange containing the discontinuity. Successful correlation for these specimens was accomplished with a plastic stress singularity strength model, if the discontinuity was sufficiently large. The average singularity strengths at ultimate load were 64.6 MNm$^{-3/2}$ (58.7 ksi in.$^{1/2}$) for HSLA steels, and 53.7 MNm$^{-3/2}$ (48.8 ksi in.$^{1/2}$) for A36 steel. The percent coefficient of variation was 6.4 percent for HSLA steels and 8.4 percent for A36 steel.

KEY WORDS: fracture (materials), failure, cracking (fracturing), elastic theory, plastic theory, tensile properties, stress-strain diagrams, bend tests, analyzing steels, structural steels, crack propagation.

The problem that initially required this research was the need to determine the residual strength of moment connections in which one column flange contains a mid-thickness plane of discontinuity, or lamination, shown cross-hatched in Fig. 1a. The ultimate aim of this research is to develop a practical methodology for evaluating the residual strength of cracked structural steel elements. Presently available evaluations are often grossly conservative because they do not adequately account for one or more of the following modes of behavior observed in cracked structural steel components:

[1]Research engineer, Research Department, Bethlehem Steel Corp., Bethlehem, Pa. 18016.

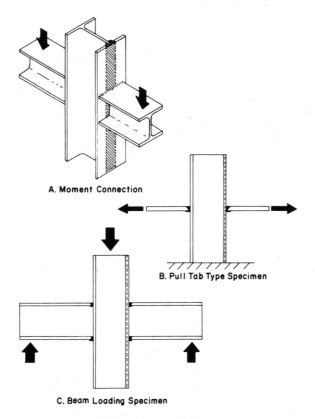

FIG. 1—*Fracture specimens*.

1. Structural steel elements do not generally operate in the plane-strain regime, which is the cracked-structure response quantified by many currently accepted residual strength calculations.

2. Massive plasticity may accompany final fracture, and such plasticity is not properly represented in existing residual strength models.

3. Strong three-dimensional effects cannot be adequately estimated with the two-dimensional analysis common to current cracked-structure models.

By combining three existing concepts, a correlation was developed between residual strength tests on structural steel subassemblages which did not behave in plane strain, exhibited massive plasticity, and were strongly three-dimensional. The concepts used are:

1. Three-dimensional elastic-plastic finite-element stress analysis.
2. Strength of the plastic stress singularity in the neighborhood of a material discontinuity.
3. Maximum tensile stress theory of fracture.

Specimens and Tests

The fracture specimens comprised beam-column connections in which one column flange contained a naturally occurring mid-thickness plane of discontinuity, shown cross-hatched in the moment connection sketch of Fig. 1a.

Twenty-five welded beam-column specimens were tested [1].[2] Nine specimens were three-plate welded column sections, 14 were rolled column sections with one flange removed and replaced by a plate containing a discontinuity, and two were as-rolled column sections without discontinuities. Initial discontinuity widths were 0 to 120 mm (0 to 4.74 in.) in 203 to 406 mm (8 to 16 in.) flange widths. Flanges were A36 and the HSLA steels, A572 Gr50 and A588, 24 to 51 mm ($15/16$ to 2 in.) in thickness. Column flange through-thickness loading was applied by beam tension flanges or tension pull tabs 203 to 356 mm (8 to 14 in.) in width and 14 to 32 mm ($9/16$ to $1 1/4$ in.) in thickness.

Twenty-two pull-tab-type specimens and three beam-loading specimens, Fig. 1b and 1c, respectively, were tested. Several specimens were subjected to axial column load in addition to pull-tab loading, and several others contained tension stiffeners recommended by the American Institute of Steel Construction (AISC) [2]. Residual strength correlation was obtained for specimens in which the initial discontinuity width was large enough to precipitate failure by unstable extension of the discontinuity.

Crack-Tip Considerations

Hilton and Hutchinson [3] have presented the concept of using plastic stress singularity strength, K, to predict plastic fracture instability in a cracked Ramberg-Osgood material. They contend that, as in linear elastic fracture mechanics, K would attain a critical value at the onset of fracture instability. They also contend that if the dominant crack-tip singularity were known, then K could be determined with the aid of finite-element (FE) stress analysis.

Figure 2 identifies a crack-tip polar coordinate system originating at the normal to the edge of the discontinuity. It lies in the mid-plane of the pull tab or the beam tension flange, and the Z-direction is parallel to the edge of the discontinuity. Hutchinson [4] derived the $r^{-1/2}$ stress singularity for a bilinear hardening material, Fig. 3, assuming all the material surrounding the crack tip to yield. The $r^{-1/2}$ stress singularity was assumed to be valid for the multilinear hardening (MLH) material, Fig. 3, used in the present study.

---

[2] The italic numbers in brackets refer to the list of references appended to this paper.

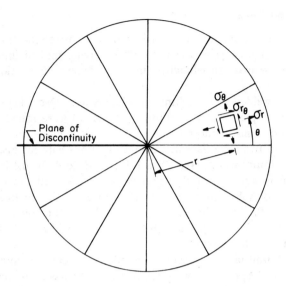

FIG. 2—*Crack-tip coordinates.*

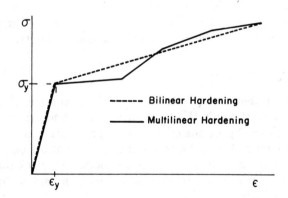

FIG. 3—*Stress-strain behavior.*

Finite-Element Analysis

Three-Dimensional Finite-Element Model

Complicated specimen geometry and plasticity observed in the tests indicated the need for elastic-plastic FE modeling. A two-dimensional elastic-plastic FE model was tried but was not representative of observed test behavior. That study used plane-strain boundary conditions on the column and plane-stress conditions on the pull tab.

A three-dimensional elastic-plastic FE model was used to analyze the laminated column flange, the column web, and the beam tension flange or the pull tab, Fig. 4. The column flange material outside the width of the pull tab was found to be unstressed and was deleted from the model. Three mutually perpendicular planes of symmetry divided the specimen through the web center, web mid-thickness, and pull tab mid-thickness. Symmetry boundary conditions were established wherever these planes touch the model. The FE models (shaded area in Fig. 4) were analyzed using ANSYS, a general-purpose large-scale computer program for the solution of structural and mechanical engineering problems. The model contained 341 elements and 670 nodes. The 293 nonsingular elements were 8-node isoparametric bricks and the 48 singular elements are described in the following. No lamination extension was allowed during loading of the FE model since no slow stable crack growth was found in the post-failure sectioned structures.

The FE model side elevation, Fig. 5, shows the four levels of elements established along the length of the column; t is the specimen pull tab or beam tension flange thickness. The cluster of lines in the flange shows the elements at the tip of the planar discontinuity. Twenty-nine elements comprised the pull tab. The FE models were loaded at the pull tab extremity to the nominal stress at failure determined from the fracture tests. The elastic-plastic iterative solution was based on the initial stress method. The largest plastic strain increment in the last iteration was 5

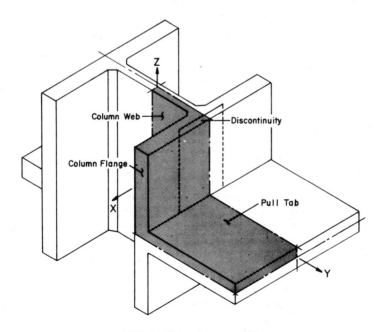

FIG. 4—*Finite-element model.*

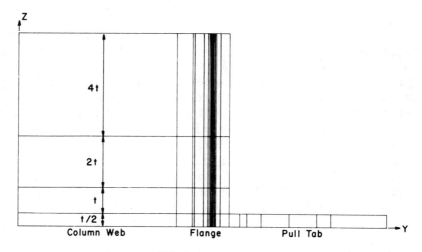

FIG. 5—*Side elevation.*

percent of the elastic strain. The von Mises equivalent stress was used as a measure of yielding.

The plan view of the FE model of the column is shown in Fig. 6. Each level of elements along the length of the column contained 78 elastic-plastic elements. Twelve singularity elements in each level, shown shaded in Fig. 6, surrounded the crack front. The radial extent of these elements was in constant proportion, 0.0133, to the initial discontinuity width.

These wedge-shaped singularity elements have been adapted from Tracey's [5] three-dimensional elastic element to include a five-point MLH approximation, Fig. 3, to the engineering stress-strain autographic record of the tensile test of each material used. The singularity elements were used to determine the plastic stress singularity strength along the crack front. The normal and shear stresses in the $r - \theta$ plane varied as $r^{-1/2}$. The Z stresses were nonsingular, and the Z shear stresses were insignificantly small in the singularity elements.

Results

The FE solution for the plastic yield zone at the mid-plane of the pull tab is shown in Fig. 7 for one test. The yield zone intersected the free surface of the column flange on the side toward the column web. This was corroborated by the whitewash spalling observed during the same fracture test, as shown in Fig. 8.

Figure 7 also shows that the region ahead of the discontinuity and slightly toward the web side of the column flange behaved elastically. Observed fracture instability was directed toward this elastic region, as was anticipated by Sih [6]. Note that mixed-mode fracture (crack opening and edge sliding) occurred despite the symmetry of the problem.

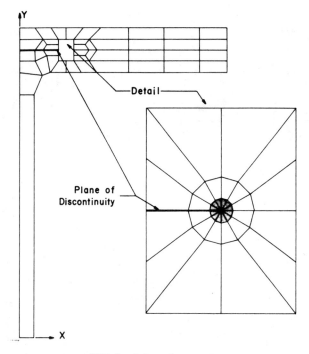

FIG. 6—*Column cross section.*

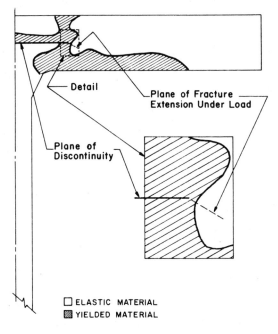

FIG. 7—*Yield zone.*

670 ELASTIC-PLASTIC FRACTURE

FIG. 8—*Fracture test showing whitewash spalling.*

Mixed-mode fracture response was taken into account by using the maximum stress theory of fracture [7]. Therefore, it was assumed that the plane of fracture extension was indicated by the vanishing of $\sigma_{r\theta}$. Let

$$\sigma_\theta = K(\theta)/(2\pi r)^{1/2}$$

where $K(\theta)$ is dependent on the boundary conditions and orientation with respect to the plane of discontinuity. It is assumed that at ultimate load

$$K_f = K(\theta)\,|_{\sigma_{r\theta}=0} = \sigma_\theta\,|_{\sigma_{r\theta}=0}\,(2\pi r)^{1/2}$$

Thus, the plastic stress singularity strength is assumed to take on a critical value, K_f, with incipient fracture. Stresses in the singularity elements, shown shaded in Fig. 6, were evaluated at the element centroid. Hence, r equals the radial distance from the crack front to that element centroid for purposes of the FE crack-front calculations.

The variation of $\sigma_{r\theta}$ for the FE model of one fracture test is shown in Fig. 9. The plane of expected fracture extension is indicated by $\sigma_{r\theta} = 0$ at about $\theta = -30$ deg. The actual failure angle was about $\theta = -45$ deg for

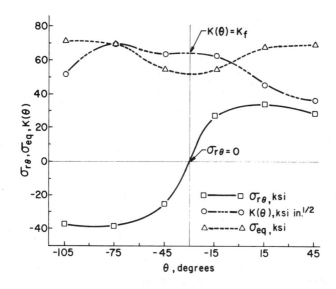

(1 ksi = 6.9 MNm^{-2}, 1 ksi in$^{1/2}$ = 1.1 MNm$^{-3/2}$)

FIG. 9—*Crack-tip behavior.*

this test, as shown in Fig. 10. The variation of $K(\theta)$ for this test is also shown in Fig. 9. Along the plane $\sigma_{r\theta} = 0$ the plastic singularity strength, $K(\theta)$, was assumed to take on its critical value, K_f, in this case about 68.8 MNm$^{-3/2}$ (62.5 ksi-in.$^{1/2}$). The tendency for fracture extension rather than material flow along this plane was also indicated by the von Mises equivalent stress, σ_{eq}, exhibiting a relative minimum, where $\sigma_{r\theta} = 0$. Recall that the predicted direction of fracture extension pointed toward the elastic region nearest to the crack border, as was shown in Fig. 7. This proximity of the elastic region was consistent with the local minimum in σ_{eq}.

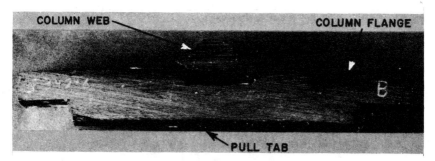

FIG. 10—*Fracture test cross section.*

Fracture Model Results

Plastic singularity strength, K_f, at ultimate load plotted versus initial flaw width, Fig. 11, shows the fracture correlation for the two types of steel tested. The A36 data represent four different heats of steel, and the HSLA data represent six different heats. The average plastic singularity strengths at ultimate load were 64.4 $MNm^{-3/2}$ (58.7 $ksi\text{-}in.^{1/2}$) for HSLA steels, and 53.2 $MNm^{-3/2}$ (48.4 $ksi\text{-}in.^{1/2}$) for the A36 steel. The present coefficient of variation was 6.4 percent for the HSLA steels, and 8.4 percent for the A36 steel.

Conclusions and Suggestions for Future Research

The combination of (1) three-dimensional elastic-plastic FE stress analysis, (2) the plastic stress singularity strength for a crack in an MLH material, and (3) the maximum tensile stress theory of fracture yields a good model of the fracture tests performed. This analysis is currently being applied to beam and axially loaded fracture toughness specimens in order to find a consistent interpretation of large-scale plasticity and through-

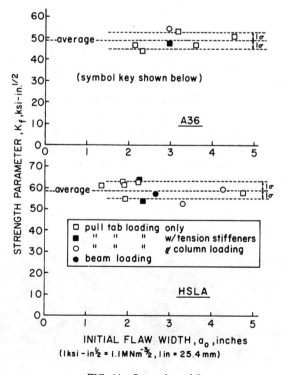

FIG. 11—*Strength model.*

thickness effects. Future work should include the strength analysis of notch toughness specimens of other geometries by this method. Low-cycle fatigue crack growth problems may also be amenable to this analysis when large-scale plasticity is a primary influence. The ultimate objective of such research is to develop a practical methodology for evaluating the strength of structural steel elements when a crack is present.

Acknowledgment

This research was sponsored by the American Iron and Steel Institute and monitored by the Task Force, Project 164. E. L. Meitzler and R. L. Kieffer were instrumental in the execution of this work.

References

[1] Macdonald, B. D., "Effect of Laminations on Moment Connections," submitted for publication in the American Society of Civil Engineers, *Journal of the Structural Division*.
[2] *Manual of Steel Construction*, 7th ed., American Institute of Steel Construction, New York, 1970, pp. 5-40.
[3] Hilton, P. D. and Hutchinson, J. W., *Engineering Fracture Mechanics*, Vol. 3, 1971, pp. 435-451.
[4] Hutchinson, J. W., *Journal of the Mechanics and Physics of Solids*, Vol. 16, 1968, pp. 13-31.
[5] Tracey, D. C., *Nuclear Engineering and Design*, Vol. 26, 1973, pp. 1-9.
[6] Sih, G. C. and Macdonald, B. D., *Engineering Fracture Mechanics*, Vol. 6, 1974, pp. 361-386.
[7] Sih, G. C. and Liebowitz, H., Eds., *Fracture*, Vol. 2, 1968, p. 94.

J. G. Merkle[1]

An Approximate Method of Elastic-Plastic Fracture Analysis for Nozzle Corner Cracks*

REFERENCE: Merkle, J. G., "**An Approximate Method of Elastic-Plastic Fracture Analysis for Nozzle Corner Cracks,**" *Elastic-Plastic Fracture, ASTM STP 668*, J. D. Landes, J. A. Begley, and G. A. Clarke, Eds., American Society for Testing and Materials, 1979, pp. 674–702.

ABSTRACT: Two intermediate test vessels with inside nozzle corner cracks have been pressurized to failure at Oak Ridge National Laboratory (ORNL) by the Heavy Section Steel Technology (HSST) Program. Vessel V-5 leaked without fracturing at 88°C (190°F), and Vessel V-9 failed by fast fracture at 24°C (75°F) as expected. The nozzle corner failure strains were 6.5 and 8.4 percent, both considerably greater than pretest plane-strain estimates. The inside nozzle corner tangential strains were negative, implying transverse contraction along the crack front. Therefore, both vessels were reanalyzed, considering the effects of partial transverse restraint by means of the Irwin β_{Ic} formula. In addition, it was found possible to accurately estimate the nozzle corner pressure-strain curve by either of two semi-empirical equations, both of which agree with the elastic and fully plastic behavior of the vessels. Calculations of failure strain and fracture toughness corresponding to the measured final strain and flaw size are made for both vessels, and the results agree well with the measured values.

KEY WORDS: fracture mechanics, fracture toughness, elastic-plastic analysis, fracture strength, pressure vessels, nozzles, cracks, stress concentrations, crack propagation

Nomenclature

A_1, A_2 Terms from which the real root of Eq 17, a cubic equation, is calculated, dimensionless

a Crack depth, cm (in.)

---

*Work done at Oak Ridge National Laboratory, operated by Union Carbide Corp. for the Department of Energy; this work funded by U.S. Nuclear Regulatory Commission under Interagency Agreements 40-551-75 and 40-552-75. By acceptance of this article, the publisher or recipient acknowledges the U.S. Government's right to retain a nonexclusive, royalty-free license in and to any copyright covering the article.

[1] Senior development specialist, Oak Ridge National Laboratory, Oak Ridge, Tenn. 37830.

B Plate thickness, cm (in.)
C Linear elastic fracture mechanics (LEFM) shape factor based on local stress, dimensionless
C_n LEFM shape factor based on nominal stress, dimensionless
E Modulus of elasticity, MPa (ksi)
E_s Strain-hardening tangent modulus, MPa (ksi)
K_I Mode I elastic crack-tip stress-intensity factor, MN·m$^{-3/2}$ (ksi $\sqrt{\text{in.}}$)
K_{Ic} Plane-strain fracture toughness, MN·m$^{-3/2}$ (ksi $\sqrt{\text{in.}}$)
K_{Icd} Fracture toughness measured with a specimen of thickness d and calculated from the test data by the equivalent energy procedure, MN·m$^{-3/2}$ (ksi $\sqrt{\text{in.}}$)
K_c Non-plane-strain fracture toughness, MN·m$^{-3/2}$, (ksi $\sqrt{\text{in.}}$)
K_t Elastic stress concentration factor, dimensionless
K_ϵ Inelastic strain concentration factor, dimensionless
K_σ Inelastic stress concentration factor, dimensionless
M Initial slope of the pressure-strain curve, MPa (ksi)
p Pressure, MPa (ksi)
p^*_f Elastically calculated failure pressure, MPa (ksi)
p_{GY} Gross yield pressure, MPa (ksi)
r_c Nozzle corner radius of curvature, cm (in.)
r_i Inside radius of vessel cylinder, cm (in.)
r_m Mid-thickness radius of vessel cylinder, cm (in.)
r_{ni} Inside radius of nozzle, cm (in.)
r_o Outside radius of vessel cylinder, cm (in.)
r_z Effective nozzle radius, cm (in.)
t Thickness of vessel cylinder, cm (in.)
β Shell analysis parameter, dimensionless
β_{Ic} Plane-strain plastic zone size parameter, dimensionless
β_c Non-plane-strain plastic zone size parameter, dimensionless
δ_c Calculated crack opening displacement, mm (in.)
ϵ Notch root strain, dimensionless
λ Applied strain, dimensionless
λ_d Stress-strain parameter, dimensionless
λ_f Failure strain, dimensionless
λ_{fo} Calculated failure strain for plane-strain conditions, dimensionless
λ_s Strain at the onset of strain hardening, dimensionless
λ_Y Yield strain, dimensionless
ν Poisson's ratio, dimensionless
ρ Notch root radius, cm (in.)
σ_h Nominal hoop stress in vessel cylinder, MPa (ksi)
σ_Y Yield stress, MPa (ksi)

The development of fracture mechanics methods of analysis has made it possible to quantitatively examine a given structural design and material selection to determine if there are sufficient margins between the specified flaw sizes, material properties and loading conditions, and those that could cause failure. In the case of a welded steel pressure vessel, two types of situations involving flaws need to be considered in a fracture safety analysis. The first is a flaw attempting to propagate out of an embrittled region, wherever one might exist, and the second is the attempted unstable extension of a flaw growing by fatigue in sound material. Precautions against the first type of failure (the nonarrest of a propagating crack) are based on defining the size and shape of a boundary surrounding the embrittled region in sound material and treating this boundary as the size of a crack that must arrest. This is the concept underlying the use of the reference flaw size and the reference (crack arrest) fracture toughness in nuclear pressure vessel design [1].[2] Precautions against the second type of failure (static initiation of a crack formed and growing by fatigue in sound material) can be based on fracture mechanics analysis methods that use the static initiation fracture toughness. Methods for considering, by analysis, the possible stable growth of cracks under monotonically increasing loads are now being developed [2-4], but the analysis to be discussed here does not include this phenomenon explicitly. Instead, stable crack growth will be treated approximately by using a maximum load fracture toughness determined from a test specimen in which some stable crack growth may have occurred before failure. Depending on the method of analysis, the amount of stable crack growth that may occur in the structure before failure may also be estimated, based on test data, and added to the original crack size. The reasonableness of this approach will be evaluated by comparing calculations with experimental data obtained from two Heavy Section Steel Technology (HSST) Program intermediate pressure vessel tests.

Statement of the Problem

The particular fracture prevention problem being considered here is that of preventing the unstable extension of a crack formed and growing by fatigue at the inside corner of a nozzle in a pressure vessel under vessel internal pressure loading. It will be shown that a relatively simple method of analysis can provide useful approximate results for this type of problem, provided that two important features of the problem, both of which have been observed experimentally, are considered. The first important feature is the dependence of the inside nozzle corner pressure-strain curve on the elastic stress concentration factor of the nozzle corner and on the gross yield pressure of the vessel cylinder. The second important feature is the apparent beneficial effect of transverse contraction at the inside nozzle

---

[2] The italic numbers in brackets refer to the list of references appended to this paper.

corner, under vessel internal pressure loading, on the toughness governing the extension of an inside nozzle corner crack.

The type of crack being considered is assumed to lie in the plane containing the axis of both the nozzle and the vessel (the longitudinal plane), because the inside nozzle corner stress concentration factor for pressure loading, here defined as the ratio of the peak nozzle corner stress to the average cylinder hoop stress, is known to be a maximum in this plane. In addition, cyclic pressure experiments have shown that fatigue cracks form first at this location [5]. The problem is relevant to the fracture safety analysis of nuclear pressure vessels because cracks formed by thermal fatigue have occurred around the inside corners of Boiling Water Reactor (BWR) feedwater nozzles [6]. Previous example calculations have also shown that inside nozzle corner cracks of sufficient initial size can grow appreciably by fatigue [7], thus increasing the importance of developing fracture analysis methods for such flaws. Since local yielding is permitted at nozzle corners by the ASME Code design rules, provided that rules regarding low-cycle fatigue prevention can also be satisfied [8], it is clear that satisfactory estimates of strength for vessels containing nozzle corner flaws cannot be made with only linear elastic fracture mechanics (LEFM) methods of analysis. Therefore, there is a need for elastic-plastic fracture analysis methods, simple enough for code application, by which such estimates of strength, in terms of load, can be made. The objective of this paper is to demonstrate a means of calculating the conditions for stable and unstable crack extension at the inside corner of a nozzle in a pressure vessel, under internal pressure loading, using some experimentally based approximations that appear to be both physically rational and reasonably accurate.

Experimental Results and Implications

Two intermediate test vessels containing A508 Class 2 forged nozzles with fatigue-sharpened inside nozzle corner cracks, designated Vessels V-5 and V-9, have been tested to failure by the HSST Program, which is managed for the U.S. Nuclear Regulatory Commission by the Oak Ridge National Laboratory (ORNL). The design of these vessels is shown in Fig. 1, and a general view of two intermediate test vessels as delivered, one of which contains a nozzle, is shown in Fig. 2. The data pertinent to the analysis of Vessels V-5 and V-9, except for the fracture toughness properties of the nozzle materials, are listed in Table 1. The tests were performed by ORNL, and a detailed report on the testing procedures, analyses, and experimental results is available [9].

Each vessel contained one fatigue-sharpened surface crack, approximately 3.05 cm (1.2 in.) deep, in the inside nozzle corner nearest to the vessel head, as indicated in Fig. 1. Each flaw was prepared by first sawing a

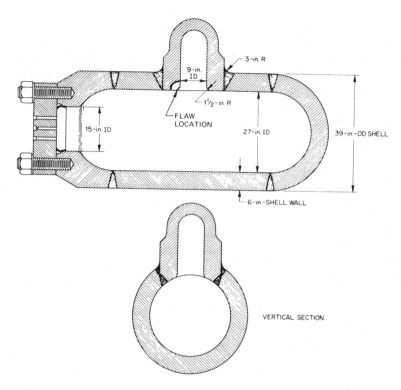

FIG. 1—*Design dimensions for intermediate test vessel with 22.86-cm-ID (9-in.) test nozzle (1 in. = 2.54 cm).*

FIG. 2—*General view of two HSST Program intermediate test vessels, showing bolted-on closure head used for all vessels and welded-in nozzle used for Vessels V-5 and V-9.*

TABLE 1—Reference data for analysis of HSST Program intermediate test vessels V-5 and V-9.

	Vessel V-5	Vessel V-9
Nozzle material	A508, Class 2 forging steel, base metal	A508, Class 2 forging steel, base metal
Nozzle NDT[a] temperature	−12°C (+10°F) (assumed, based on V-1 data)	−12°C (+10°F) (assumed, based on V-1 data)
Vessel test temperature	88°C (+190°F)	24°C (+75°F)
Expected fatigue-sharpened flaw depth at inside nozzle corner	3.05 cm (1.2 in.)	3.05 cm (1.2 in.)
Tensile properties of nozzle material at 24°C (75°F):		
Yield stress	425 MPa (61.6 ksi)	474 MPa (68.8 ksi)
Ultimate stress	553 MPa (80.2 ksi)	609 MPa (88.3 ksi)
Strain at maximum load	8.9%	9.0%
Total elongation	16.8%	18.1%
Gage length	3.175 cm (1.250 in.)	3.175 cm (1.250 in.)
Reduction in area	68.3%	70.3%
Original specimen diameter	0.4509 cm (0.1775 in.)	0.4509 cm (0.1775 in.)
Room temperature tensile and drop-weight NDT properties of cylinder material:		
material	A508, Class 2	A533, Grade B, Class 1
yield stress	500 MPa (72.5 ksi)	475 MPa (68.9 ksi)
ultimate stress	654 MPa (94.8 ksi)	574 MPa (83.3 ksi)
total elongation	not reported	28%
gage length	not reported	5.08 cm (2.0 in.)
NDT temperature	not reported	−51°C (−60°F)
Nozzle dimensions:		
inside radius	11.43 cm (4.5 in.)	11.43 cm (4.5 in.)
thickness	15.24 cm (6.0 in.)	15.24 cm (6.0 in.)
Cylinder dimensions:		
inside diameter	68.58 cm (27.0 in.)	68.58 cm (27.0 in.)
thickness	15.24 cm (6.0 in.)	15.24 cm (6.0 in.)
Charpy V-notch impact energy of nozzle material at 24°C (75°F)	no data	90.8 J (67 ft-lb)

[a] NDT = nil ductility transition.

20-mm-deep (0.80-in.) slot across the nozzle corner; then welding a steel boss over the opening of the slot; next applying cyclic hydraulic pressure to the notch cavity through a hole drilled in the boss until ultrasonic measurements made from the outside nozzle corner, in the notch plane, indicated sufficient fatigue flaw growth; and finally removing the weld boss by flame cutting and grinding. This difficult procedure required cutting, welding, and grinding to be done by a worker inside the vessel, a process requiring special equipment and safety precautions as described in more detail in Ref 9. The pretest ultrasonic estimates of crack front depth and shape for Vessels V-5 and V-9 were quite similar [9]. The pretest ultrasonically estimated crack front configuration for Vessel V-9 is shown in Fig. 3. The inflections in the crack front shape are believed to be due to the effects of the weld boss. Their effects on the test results, which are believed to be minor, will be discussed later.

Static fracture toughness data for the nozzle material of Vessel V-5 were obtained before the test using precracked Charpy V-notch (PCCV) [9] and a combination of 0.85T and 2.0T compact specimens [10]. Fracture toughness values at maximum load were calculated for each specimen, from its load-displacement diagram, by the equivalent-energy procedure [11]. This calculation procedure was justified by the known substantial agreement between J-integral and equivalent-energy toughness calculations

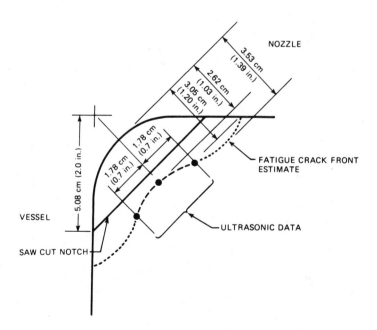

FIG. 3—*Pretest estimate of fatigue crack front position in the inside nozzle corner of intermediate test vessel V-9, based on ultrasonic data.*

for the same points on the load-displacement diagrams of notched beams and compact specimens [12]. Nevertheless, the omission of stable crack growth measurements from these data did lead to disadvantages in their application. At 93°C (200°F), the maximum load toughness values obtained from the PCCV and 0.85T compact specimens ranged from 159 to 236 MN·m$^{-3/2}$ (145 to 215 ksi $\sqrt{\text{in.}}$). At the same temperature, the two 2.0T specimens tested [9,10] gave toughness values of 245 and 265 MN·m$^{-3/2}$ (223 and 241 ksi $\sqrt{\text{in.}}$). Considering the range of data for each specimen size and the increase in both stable crack growth and static upper-shelf toughness values at maximum load with increasing specimen size generally observed in resistance curve testing, the latter value of 265 MN·m$^{-3/2}$ (241 ksi $\sqrt{\text{in.}}$) was selected as the toughness value to be used for analyzing the flawed 15.2-cm-thick (6-in.) Vessel V-5 nozzle forging.

Static and dynamic fracture toughness data for the nozzle material of Vessel V-9 were obtained before the test, using precracked Charpy V-notch and a combination of 0.85T, 1.5T, and 2.0T compact specimens [9,10]. These data are plotted versus temperature in Fig. 4. A vessel test temperature of 24°C (75°F), which is below the dynamic upper-shelf temperature, was selected in order to produce a fast-running fracture as a test result. The toughness data shown in Fig. 4 indicated that there was no consistent effect of specimen size on the static fracture toughness of the Vessel V-9 nozzle material at 24°C (75°F). This is because (1) the 1.5T specimens gave values near the middle of the static toughness range, (2) both greater and lesser values were obtained from smaller specimens, and (3) the minimum and maximum values were obtained from the 2T specimens. Consequently, it was decided to make static initiation calculations for three toughness values covering the full range of the values measured at 24°C (75°F): 159, 220, and 298 MN·m$^{-3/2}$ (145, 200, and 271 ksi $\sqrt{\text{in.}}$). Because the steepest part of the dynamic fracture toughness transition curve occurs at 24°C (75°F) and the range of dynamic values extends from below to above the range of static values, both stable crack growth and "popins" were considered possible [9].

The result of the test of Vessel V-5 at 88°C (190°F) was a leak without a fracture, which occurred at a pressure of 183 MPa (26 600 lb/in.^{2}). The position of the crack front was measured continuously during the test by an ultrasonic sensor located on the outside surface of the nozzle directly opposite the fatigue-sharpened crack front. Stable crack growth was first detected at a pressure of 124 MPa (18 000 lb/in.^{2}), and above that pressure the crack front continued to advance stably until it penetrated the outer surface near the ultrasonic crystal [9]. The point of leakage was barely visible and there was no visible distortion of the vessel. A closeup view of the point of leakage in Vessel V-5 is shown in Fig. 5. The result of the test of Vessel V-9 at 24°C (75°F) was a fast fracture as expected because of the test temperature selected for that purpose. Ultrasonic data did

682 ELASTIC-PLASTIC FRACTURE

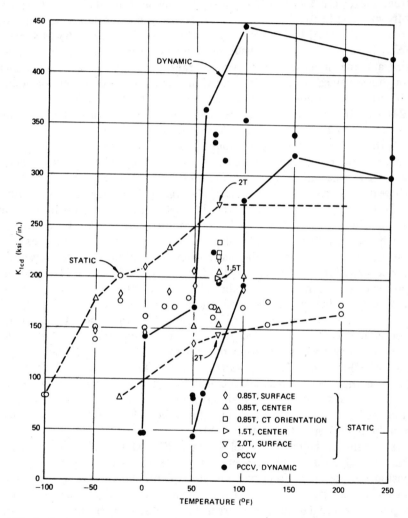

FIG. 4—*Static and dynamic* K_{Icd} *values for Vessel V-9 nozzle material* [1 in. = 2.54 cm; 1 ksi $\sqrt{in.}$ = 1.0988 MN·$m^{-3/2}$; °C = 5/9 (°F − 32)].

FIG. 5—*Closeup view of leak point adjacent to ultrasonic base block on nozzle of Vessel V-5 (arrow shows flaw penetration to surface).*

FIG. 6—*Closeup view of fractured nozzle in Vessel V-9; test temperature was 24°C (75°F).*

indicate that some stable crack growth occurred before failure, commencing at 145 MPa (21 000 lb/in.^{2}) and totaling about 1.27 cm (0.5 in.) just before failure at 185 MPa (26 900 lb/in.^{2}). A closeup view of the fractured nozzle in Vessel V-9 is shown in Fig. 6.

The circumferential strain values measured on the outside surfaces of the cylinders of Vessels V-5 and V-9, which are shown plotted in Fig. 7, indicate that the cylinders of both vessels were fully yielded at failure. The strains measured at the inside nozzle corners opposite the flaws, for Vessels V-5 and V-9, are shown plotted in Fig. 8. The nozzle corner strains at failure for Vessels V-5 and V-9 were 6.5 and 8.4 percent, respectively. Both of these strains are remarkably large compared with the maximum previously measured strain tolerance of the same material for a 4.75-cm-deep (1.87-in.) flaw in the cylindrical region of an intermediate test vessel [*13*], which was 2 percent.

The flaw region of Vessel V-5 has not yet been sectioned for post-test

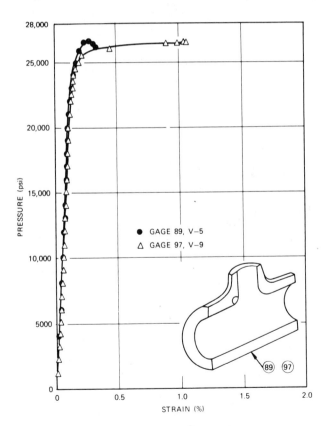

FIG. 7—*Pressure versus outside circumferential strain in vessel cylinder for intermediate test vessels V-5 and V-9 (1 lb/in.^{2} = 6895 Pa)*.

686 ELASTIC-PLASTIC FRACTURE

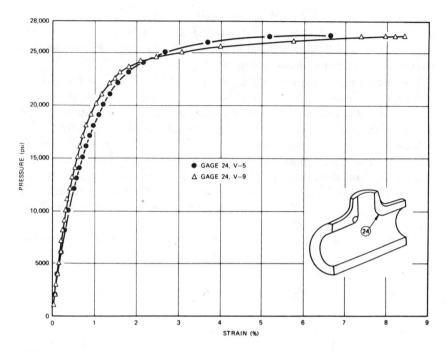

FIG. 8—*Pressure versus inside uncracked nozzle corner circumferential strain for intermediate test vessels V-5 and V-9 ($1\ lb/in.^2 = 6895\ Pa$).*

examination, but the fracture surfaces containing the original nozzle corner flaw in Vessel V-9 have been separated, with the results shown in Fig. 9. The original fatigue-sharpened crack in Vessel V-9 was very close to the size and shape estimated by ultrasonics before the test [9] (see Fig. 3). Furthermore, stable crack growth, the extent of which can be seen in Fig. 9, did increase the average crack depth by about 1.27 cm (0.5 in.), and it also eliminated the inflections in the crack front shape before failure.

Pre- and post-test estimates of the circumferential strains at failure at the unflawed nozzle corners opposite the flaws in Vessels V-5 and V-9, and the corresponding pressures, were made by ORNL and by others, using several different methods of elastic-plastic fracture analysis, all of a semi-empirical nature [9]. All of the direct estimates of the nozzle corner failure strains were low, most by a wide margin. Of these estimates, those that did not assume plane-strain toughness conditions were more accurate than those that did [9]. Thus it was apparent that some aspect of nozzle corner geometry was causing the strain tolerances for nozzle corner cracks to be substantially greater than would be expected for the same size surface cracks in the cylinder of a pressure vessel, where it is known that they would be subjected to plane-strain conditions. In fact, the measured nozzle

FIG. 9—*Closeup view of flaw in fractured nozzle of intermediate test Vessel V-9.*

corner failure strains were closer to the previously measured failure strains for surface-flawed uniaxial tension bars near and in the upper-shelf temperature range [14]. Thus it was clear that the tendency of a plane-strain analysis to underpredict nozzle corner failure strains by a wide margin, for high toughness conditions, must be due to either an error in the LEFM portion of the calculation or to the assumption of full transverse restraint around the crack front. The possibility of large errors in the LEFM portion of the pretest estimates was subsequently dismissed because (1) calculations based on several different methods for estimating the LEFM shape factor for nozzle corner cracks had given similar results [9]; (2) the method used by ORNL, based on Derby's epoxy model test data [15], was confirmed by later photoelastic experiments [16]; and (3) the difference between shape factor values estimated from Derby's data [15] and those based on the solution for an edge crack extending from a hole in a plate [1] was explained by Embly [9,17] as being due to the effects of pressure in the crack, effects that are experimentally included in the former solution but analytically neglected in the latter. For this reason, the experimentally measured principal strains at the unflawed nozzle corners opposite the flaws in both vessels were examined closely (see Tables 2 and

3). Both sets of strain readings indicated the occurrence of considerable transverse contraction in the plane of the crack at the nozzle corner, thus implying that full transverse restraint does not exist for nozzle corner cracks at that location, under vessel internal pressure loading. This phenomenon will be discussed further in the section on analysis.

The pretest estimates of failure pressure for Vessel V-5 were based on an elastic-plastic nozzle corner pressure-strain curve calculated by the finite-element method [9]. However, this curve proved to be inaccurate with respect to the experimental data obtained for Vessel V-5, because it underestimated the elastic stress concentration factor and overestimated the pressures for given strains in the elastic-plastic range. Therefore, the pretest calculations for Vessel V-9 were based on the experimentally measured pressure-strain curve for Vessel V-5 shown in Fig. 8, and it was recognized that improved methods for estimating elastic-plastic nozzle corner pressure-strain curves would be required as part of any practical method of fracture analysis for nozzle corner cracks.

Analysis

The objectives of the post-test analysis developments to be discussed in this section were principally to develop an improved method for estimating

TABLE 2—*Principal stress and elastic stress-concentration factor values at the inside unflawed nozzle corner of intermediate test vessel V-9, calculated from experimental strain data.*

Pressure, MPa (ksi)	Stress, MPa			Remarks	K_t
	σ_1	σ_2	σ_3		
6.9 (1.0)	63	0.4	− 6.9	elastic	4.05
13.8 (2.0)	134	3.5	−13.8	elastic	4.33
34.5 (5.0)	302	6.2	−34.5	elastic	3.89
55.2 (8.0)	419	− 9.9	−55.2	yield	...
68.9 (10.0)	405	−47.8	−68.9	yield	...
75.8 (11.0)	399	−75.8	−75.8	yield (corner)	...

TABLE 3—*Stress-concentration factor estimates for identical nozzles in an intermediate test vessel and a reference calculational model of typical PWR[a] vessel design.*

Term	Intermediate Test Vessel with Nozzle	Reference Calculational Model of PWR Vessel
Nozzle mean radius, r	19.05 cm (7.5 in.)	19.05 cm (7.5 in.)
Cylinder mean radius, r_m	41.91 cm (16.5 in.)	229.24 cm (90.25 in.)
Cylinder thickness, t	15.24 cm (6.0 in.)	21.59 cm (8.5 in.)
β	0.484	0.174
K_t	4.16	2.71

[a] PWR = Pressurized Water Reactor

elastic-plastic nozzle corner pressure-strain curves, and to find one or more reasonable methods for considering the combined effects of nominal yielding and partial transverse restraint conditions on the criteria governing the extension of nozzle corner cracks. Two analysis procedures, differing principally only in the relationship used between flaw size, toughness, and failure strain in the elastic-plastic range, were developed. Both procedures use the same estimates of the nozzle corner pressure-strain curve, and the same LEFM relationship between vessel internal pressure and the crack-tip stress intensity factor, for elastic conditions. Both procedures also use the same equation for estimating the increase in fracture toughness due to less-than-full transverse restraint. For elastic-plastic conditions, one procedure uses LEFM to calculate the relation between failure strain, flaw size, and toughness, and the other procedure uses the tangent modulus method equations for bending. The latter method is an incremental application of Neuber's equation for estimating inelastic stress and strain concentration factors, and is described in detail in Ref *13*. In this paper, the former procedure (using LEFM based on strain) is used to estimate failure strains for the given toughness values, using the initial flaw sizes, and the latter procedure (using the tangent modulus method) is used to estimate toughness values for the given failure strains, using the flaw sizes measured at or near failure. The development of the analysis procedures, and the calculated results for intermediate test vessels V-5 and V-9, are discussed in the following. The calculated results are summarized in Table 4, which also lists the numbers of the equations and figures used to obtain the calculated results.

Pressure-Strain Curve Estimates

In principle, the nozzle corner pressure-strain curve should be bounded by two tangents, the first representing the initial elastic behavior of the nozzle at low pressures, and the second being the line of constant pressure that defines the gross yield pressure of the nozzle region. By comparing the measured nozzle corner pressure-strain curves for Vessels V-5 and V-9 shown in Fig. 8 with the measured pressure-strain curves for the vessel cylinders remote from the nozzles shown in Fig. 7, it can be seen that the gross yield pressures indicated by both figures are essentially the same. This is consistent with the assumption that nozzle design by the area replacement method specified by the ASME Code [*8*] serves to prevent the gross yield pressure of a nozzle region from becoming less than that of the cylinder into which the nozzle is inserted. Therefore, for estimating purposes, the gross yield pressure of a nozzle region designed by the area replacement method will be assumed to be identical to that of the cylinder into which the nozzle is inserted.

Previous comparisons between theory and experiment have shown that

TABLE 4—*Calculated failure strains and fracture toughness values for HSST Program intermediate test vessels V-5 and V-9 with nozzle corner cracks.*

	Equation or Figure No. or Both	Vessel V-5	Vessel V-9
Test conditions, material properties, and test results:			
test temperature	...	88°C (190°F)	24°C (75°F)
nozzle yield stress	...	425 MPa (61.6 ksi)	474 MPa (68.8 ksi)
initial crack depth	(measured)	3.05 cm (1.2 in.)	3.05 cm (1.2 in.)
measured nozzle corner failure strain	...	6.5%	8.4%
measured fracture toughness	...	265 MN·m$^{-3/2}$ (241 ksi $\sqrt{\text{in.}}$)	159 MN·m$^{-3/2}$ (145 ksi $\sqrt{\text{in.}}$)
			220 MN·m$^{-3/2}$ (200 ksi $\sqrt{\text{in.}}$)
			298 MN·m$^{-3/2}$ (271 ksi $\sqrt{\text{in.}}$)
Calculated failure strains and pressures by LEFM based on strain:			
flaw size, a	(measured)	3.05 cm (1.2 in.)	3.05 cm (1.2 in.)
flaw depth ratio, a/r_z	Eq 12	0.243	0.243
shape factor, C_n	Fig. 11	2.5	2.5
toughness ratio, K_c/K_{Ic}	Eqs 14 and 16	7.61	2.41, 4.28, 7.71
calculated failure strains:			
λ_{f0} (plane strain)	Eqs 3, 2, 4, 24, and 25	0.73%	0.44%, 0.61%, 0.82%
λ_f (non-plane strain)	Eq 26	5.6%	1.1%, 2.6%, 6.4%
calculated failure pressure	Eqs 5 or 8	180 MPa (26.1 ksi)	142 MPa (20.6 ksi)
			172 MPa (25.0 ksi)
			181 MPa (26.2 ksi)
Fracture toughness calculations by the tangent modulus method:			
pressure, p	(measured)	183 MPa (26.5 ksi)	185 MPa (26.9 ksi)
nozzle corner strain, λ	(measured)	5.2%	8.4%
flaw depth, a	(measured)	8.4 cm (3.3 in.)	4.50 cm (1.77 in.)
flaw depth ratio, a/r_z	Eq 12	0.668	0.358
shape factors:			
C_n	Fig. 11	1.79	2.18
C	Eqs 3, 2, and 13	0.430	0.524
toughness ratios:			
K_c/σ_Y	Eqs 28–32	20.51 $\sqrt{\text{cm}}$ (12.87 $\sqrt{\text{in.}}$)	22.82 $\sqrt{\text{cm}}$ (14.32 $\sqrt{\text{in.}}$)
K_c/K_{Ic}	Eqs 18–23	3.15	4.13
K_{Ic}/σ_Y	(using results obtained above)	6.50 $\sqrt{\text{cm}}$ (4.08 $\sqrt{\text{in.}}$)	5.51 $\sqrt{\text{cm}}$ (3.46 $\sqrt{\text{in.}}$)
fracture toughness, K_{Ic}	(multiply K_{Ic}/σ_Y by σ_Y)	277 MN·m$^{-3/2}$ (252 ksi $\sqrt{\text{in.}}$)	262 MN·m$^{-3/2}$ (238 ksi $\sqrt{\text{in.}}$)
crack opening displacement, δ_c	Eq 33	8.6 mm (0.34 in.)	11.9 mm (0.47 in.)

the gross yield pressure of an intermediate test vessel cylinder can be closely estimated by the equation

$$P_{GY} = 1.04 \, \sigma_Y \ln (r_o/r_i) \tag{1}$$

where r_o and r_i are the outer and the inner vessel cylinder radii, respectively. In Eq 1, the factor 1.04 is an empirical factor based on both intermediate test vessel and small-scale steel model test data, and the remainder of the equation is based on the Tresca (maximum shear stress) yield criterion. From Table 1, the room temperature yield stresses of Vessel V-5 and Vessel V-9 cylinder materials were 500 and 475 MPa (72.5 and 68.9 ksi), respectively. Therefore, assuming test temperature yield stresses of $\sigma_Y = 476$ MPa (69 ksi) for both vessel cylinders, and using $r_o/r_i = 1.44$, Eq 1 gives $P_{GY} = 182$ MPa (26.4 ksi).

Although pretest estimates of the elastic stress concentration factor of the nozzle corners in Vessels V-5 and V-9, based on both elastic finite-element analysis [18] and epoxy model strain-gage data [15], were approximately 2.9, the experimental strain data obtained from both vessels indicated a value close to 4. Apparently the finite-element mesh size used analytically and the strain gages used experimentally on the epoxy models were not small enough relative to the other nozzle dimensions to determine the true peak nozzle corner strain. The principal stresses calculated from the measured principal strains at low pressures on the unflawed inside nozzle corner of Vessel V-9 are listed in Table 2. These stresses were calculated from Hooke's law before yielding, and with the aid of the Tresca yield criterion after yielding [9]. Not only is the initial elastic stress concentration factor close to 4, but the intermediate principal stress is initially small and tends to become compressive, eventually equaling the vessel internal pressure after local yielding occurs. In addition, the measured values of the nozzle corner stress concentration factor for Vessels V-5 and V-9 were found to be consistent with an analysis derived by Van Dyke [19] for calculating the stresses around a circular hole in a cylindrical shell. The value of the elastic stress concentration factor of the hole, at the longitudinal plane, is given by Van Dyke's analysis as

$$K_t = 2.5 + \frac{9\pi}{4} \beta^2 \tag{2}$$

where

$$\beta^2 = \frac{r^2 \sqrt{12(1 - \nu^2)}}{8 r_m t} \tag{3}$$

and where

r = hole radius,
r_m = cylinder midthickness radius, and
t = cylinder thickness.

Applying Eqs 2 and 3 to the nozzle design shown in Fig. 1, both for the case of an intermediate test vessel cylinder and for a cylinder of typical reactor vessel dimensions, gives the results shown in Table 3. The value of K_t for the nozzle in an intermediate test vessel is 4.16, but the value of K_t for the same nozzle inserted into a typical reactor vessel is only 2.71, because of the influences of the cylinder mean radius and thickness, both of which occur as factors in the denominator of Eq 3.

Having resolved both the estimates of the gross yield pressure and the elastic stress concentration factor, two semi-empirical equations were developed for estimating the elastic-plastic nozzle corner pressure-strain curves of Vessels V-5 and V-9. The initial elastic slopes of these curves were both determined by using the calculated elastic stress concentration factor, and by assuming that the intermediate principal stress at the inside nozzle corner was compressive and equal to the vessel internal pressure. Thus the initial slope, M, of the nozzle corner pressure-strain curves was calculated from

$$M = \frac{E}{K_t \left(\frac{r_i}{t}\right) + 2\nu} \qquad (4)$$

For E = 2068 MPa·percent^{-1} (300 ksi·percent^{-1}), K_t = 4.16, and ν = 0.3, Eq 4 gives M = 208 MPa·percent^{-1} (30.12 ksi·percent^{-1}).

The first semi-empirical equation was based on the assumption that the slope of the pressure-strain curve decreases linearly with increasing pressure, and reaches zero at the gross yield pressure. The resulting equation is

$$p = p_{GY} \left(1 - e^{-(M\lambda/p_{GY})}\right) \qquad (5)$$

where λ is the nozzle corner strain. For the intermediate test vessel nozzle corners, substituting the values of p_{GY} and M determined from Eqs 1 and 4 gives

$$p = 26.4 \left(1 - e^{-1.141\lambda}\right) \qquad (6)$$

where p is in ksi and λ is in percent. Equation 6 is shown plotted in Fig. 10, which demonstrates that it fits the data from Vessel V-5 with considerable accuracy.

The second semi-empirical equation was based on plotting the measured

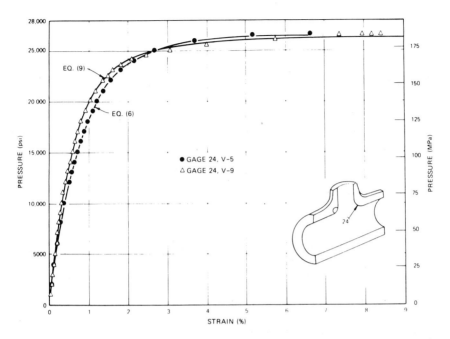

FIG. 10—*Comparison of calculated and measured nozzle corner pressure-strain curves for intermediate test vessels V-5 and V-9.*

pressure divided by the measured strain versus the measured pressure, for Vessel V-9, from which it was deduced that the two quantities plotted could be approximately related by the equation of an ellipse, namely

$$\left(\frac{p}{M\lambda}\right)^2 + \left(\frac{p}{p_{GY}}\right)^2 = 1 \tag{7}$$

Rearranging Eq 7 gives

$$p = \frac{p_{GY}}{\sqrt{1 + \left(\frac{p_{GY}}{M\lambda}\right)^2}} \tag{8}$$

Again, for the intermediate test vessel nozzle corners, substituting the values of p_{GY} and M obtained from Eqs 1 and 4 gives

$$p = \frac{26.4}{\sqrt{1 + \left(\frac{0.8765}{\lambda}\right)^2}} \tag{9}$$

where p is in ksi and λ is in percent. Equation 9 is shown plotted in Fig. 10, which demonstrates that it fits the data from Vessel V-9 with equal accuracy. Thus it appears that either or both of the simple semi-empirical expressions discussed in the foregoing can be used to obtain good estimates of elastic-plastic nozzle corner pressure-strain curves for use in elastic-plastic fracture strength calculations.

Fracture Analyses

Both methods of analysis developed make direct use of the LEFM solution for the problem being analyzed. Thus, for the intermediate test vessels with nozzle corner cracks, the experimental curve obtained by Derby [15] for a series of small, thick-walled epoxy model vessels, which were approximately geometrically similar to the intermediate test vessels, was used. This curve, shown in Fig. 11, gives the nondimensional LEFM

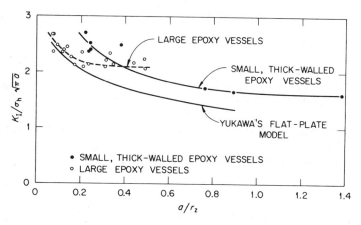

FIG. 11—*Summary of experimental results obtained from ORNL nozzle corner crack epoxy model fracture tests [15] and comparison with hole in flat plate approximation [1].*

flaw shape factor based on the nominal cylinder hoop stress, which is defined by

$$C_n = \frac{K_I}{\sigma_h \sqrt{\pi a}} \quad (10)$$

In Eq 10, σ_h is the nominal cylinder hoop stress, defined by

$$\sigma_h = p \left(\frac{r_i}{t}\right) \quad (11)$$

In Fig. 11, C_n is given as a function of a/r_z, where r_z is the effective nozzle radius defined by

$$r_z = r_{ni} + r_c (1 - 1/\sqrt{2}) \qquad (12)$$

where r_{ni} and r_c are the inside nozzle radius and the inside nozzle corner radius of curvature, respectively. For the intermediate test vessel nozzles, from Fig. 1, $r_{ni} = 11.43$ cm (4.5 in.) and $r_c = 3.81$ cm (1.5 in.), so that Eq 12 gives $r_z = 12.55$ cm (4.94 in.). For both methods of analysis, the LEFM shape factor based on the peak nozzle corner stress is calculated from

$$C = \frac{C_n}{K_t} \qquad (13)$$

where, for Vessels V-5 and V-9, $K_t = 4.16$ as determined previously.

The representation of the effects of partial transverse restraint on fracture toughness is the same in both methods of analysis. The concept underlying this part of the calculations is that the nominal strain in the direction tangent to the crack front, in the plane of the crack, is the primary agent of transverse restraint [13]. When this strain is zero, plane-strain toughness conditions prevail, but, when this strain is a contraction, the toughness is elevated above the plane-strain toughness. If the transverse contraction strain is approximately equal to or greater than that corresponding to uniaxial tension, the toughness elevation can be estimated from Irwin's empirical formula [20]

$$\frac{K_c}{K_{Ic}} = \sqrt{1 + 1.4\,\beta_{Ic}^2} \qquad (14)$$

For a through crack, β_{Ic} is defined by [20]

$$\beta_{Ic} = \frac{\left(\dfrac{K_{Ic}}{\sigma_Y}\right)^2}{B} \qquad (15)$$

where B is specimen thickness. However, for a part-through surface crack, an alternate definition

$$\beta_{Ic} = \frac{\left(\dfrac{K_{Ic}}{\sigma_Y}\right)^2}{2a} \qquad (16)$$

is used here, in order for the denominator in the expression for β_{Ic} to retain its identity as twice the distance from the point of greatest transverse restraint on the crack front to the nearest free surface, not including the crack surface [13].

Whereas Eq 14 is convenient for estimating the toughness elevation due to less than full transverse restraint when the plane-strain toughness is known, a rearrangement of Eq 14 is necessary for determining the plane-strain toughness when the known value of toughness is a non-plane-strain value. This rearranged equation is

$$\beta^3{}_{Ic} + (5/7)\beta_{Ic} - (5/7)\beta_c = 0 \tag{17}$$

where

$$\beta_c = \frac{\left(\frac{K_c}{\sigma_Y}\right)^2}{B} \quad \text{or} \quad \frac{\left(\frac{K_c}{\sigma_Y}\right)^2}{2a} \tag{18}$$

as appropriate. The solution to this equation is

$$\beta_{Ic} = A_1{}^{1/3} - A_2{}^{1/3} \tag{19}$$

where

$$A_1 = \sqrt{m^2 + 0.0135} + m \tag{20}$$
$$A_2 = \sqrt{m^2 + 0.0135} - m \tag{21}$$

and

$$m = (5/14)\beta_c \tag{22}$$

The toughness elevation is then determined from

$$\frac{K_c}{K_{Ic}} = \sqrt{\frac{\beta_c}{\beta_{Ic}}} \tag{23}$$

The estimate of nozzle corner failure strains by the method of LEFM based on strain begins with the combination of Eqs 10 and 11, rearranged and symbolically changed to read

$$p^*_f = \frac{K_{Ic}}{C_n \left(\frac{r_i}{t}\right)\sqrt{\pi a}} \tag{24}$$

In Eq 24, p^*_f is the elastically calculated failure pressure and K_{Ic} is the plane-strain fracture toughness. The failure strain for plane-strain conditions is calculated from

$$\lambda_{fo} = \frac{p^*_f}{M} \qquad (25)$$

The failure strain for non-plane-strain conditions is then calculated from

$$\lambda_f = \left(\frac{K_c}{K_{Ic}}\right) \lambda_{fo} \qquad (26)$$

where the ratio (K_c/K_{Ic}) is obtained from Eq 14. The estimated failure pressure is then calculated from Eq 9. Calculated results for Vessels V-5 and V-9 are shown in Table 4. The three values of failure strain and pressure listed for Vessel V-9 are those corresponding to the initial flaw size and the three measured fracture toughness values listed in the upper part of the table. For Vessel V-5, the calculated failure strain and failure pressure, based on the initial flaw size, are only slightly conservative, and the same is true of the strain and pressure corresponding to the maximum fracture toughness value measured for Vessel V-9. Noting the large differences between the plane-strain and the non-plane-strain estimates of failure strain for both vessels, it is clear that considering the effects of transverse restraint is essential to the accuracy of the analysis.

The calculations of the plane-strain fracture toughnesses corresponding to the measured values of nozzle corner strain and flaw size by the tangent modulus method were based on the directly measured flaw size at failure for Vessel V-9 (see Fig. 9), and the last ultrasonically measured flaw size in Vessel V-5 before the pressure began to decrease [9]. Note that the flaw in Vessel V-5 was 8.4 cm (3.3 in.) deep at a pressure of 183 MPa (26.5 ksi), and therefore underwent approximately 12.7 cm (5 in.) of stable crack growth during the last 0.7-MPa (100 lb/in.²) rise in pressure.

Because of the steep strain gradient in the nozzle corner region, the tangent modulus equations for the case of bending [13] were used for these toughness calculations. The derivation of these equations is given in Appendix H of Ref 13. Briefly, this method of analysis is based on the Neuber equation for inelastic stress and strain concentration factors

$$K_\sigma K_\epsilon = K_t^2 \qquad (27)$$

written in incremental form and then rearranged so that the increment in the notch ductility factor $d\epsilon\sqrt{\rho}$, where ϵ is notch root strain and ρ the notch root radius, appears on the left-hand side of the equation and only measurable quantities appear on the right-hand side. For a trilinearized stress-strain curve and the case of bending, with the applied strain in the strain-hardening range, the notch ductility factor increments were calculated from the equations given in the following [13]. For the elastic range

$$\Delta\epsilon\sqrt{\rho} = 2C\sqrt{a}\sqrt{E/E_s}\,\lambda_Y \tag{28}$$

For the transition range

$$\Delta\epsilon\sqrt{\rho} = 4C\sqrt{a}\sqrt{E/E_s}\,(\sqrt{\lambda_s \lambda_Y} - \lambda_Y) \tag{29}$$

For the strain-hardening range

$$\Delta\epsilon\sqrt{\rho} = 2C\sqrt{a}\left\{\sqrt{\lambda_f(\lambda_f + \lambda_d)} - \sqrt{\lambda_s(\lambda_s + \lambda_d)} \right. \\ \left. + \lambda_d \ln\left[\frac{\sqrt{\lambda_f} + \sqrt{\lambda_f + \lambda_d}}{\sqrt{\lambda_s} + \sqrt{\lambda_s + \lambda_d}}\right]\right\} \tag{30}$$

where

$$\lambda_d = \frac{\sigma_Y}{E_s} - \lambda_s \tag{31}$$

and where

λ_Y = yield strain,
λ_s = strain at the onset of strain hardening,
λ_f = applied or failure strain,
E = elastic modulus, and
E_s = strain-hardening tangent modulus.

For both vessels, the value of E_s was taken as 20.7 MPa·percent^{-1} (3.0 ksi·percent^{-1}), and λ_s was taken as 1.2 percent. The total values of $\epsilon_f\sqrt{\rho}$ were calculated by adding the values obtained from Eqs 28, 29, and 30, and the values of K_c/σ_Y were then obtained from [13]

$$\frac{K_c}{\sigma_Y} = \left(\frac{\sqrt{\pi}}{20\lambda_Y}\right)\epsilon_f\sqrt{\rho} \tag{32}$$

The values of K_{Ic}/σ_Y were then obtained by dividing the results of Eq 32 by the values of K_c/K_{Ic} obtained from Eq 23. The resulting toughness values for both vessels are listed in the lower part of Table 4. Both plane-strain toughness values compare well with the measured values for the two nozzle materials. The non-plane-strain toughness ratios, K_c/σ_Y, may look high, but the calculated crack-tip opening displacements listed at the bottom of Table 2, as calculated from

$$\delta_c = \frac{\sigma_Y}{E}\left(\frac{K_c}{\sigma_Y}\right)^2 \qquad (33)$$

are both very close to the crack mouth opening displacements measured at the pressures used for the calculations [9]. Thus the necessity for considering partial transverse restraint effects for nozzle corner cracks under vessel internal pressure loading is again indicated.

Discussion

The experimental data obtained from intermediate test vessels V-5 and V-9 revealed the need for improved accuracy in the representation of several factors involved in the fracture analysis of nozzle corner cracks. Although the LEFM relationship between vessel internal pressure and the crack-tip stress intensity factor was considered to be satisfactory, the finite-element method estimate of the nozzle corner pressure-strain curve made before the test of Vessel V-5 was not considered satisfactory, in either the elastic or the elastic-plastic ranges. Furthermore, the reasonableness of a method for extending LEFM into the elastic-plastic range for nozzle corner cracks required demonstration, and it was found that such a demonstration would require the consideration of transverse restraint effects on toughness as well as the effects of nominal yielding on crack-tip behavior per se.

The latter requirement was made evident by the tendency of pretest plane-strain analyses to underpredict nozzle corner flaw strain tolerances, for pressure loading, and the contraction strains measured on the unflawed inside nozzle corners of Vessels V-5 and V-9. Consequently, additional approximate non-plane-strain analyses were performed for both vessels with considerably improved results. These relatively simple calculations were performed by two partially different methods, namely, LEFM based on strain, and the tangent modulus method. In both methods of analysis, $C\sqrt{a}$ is a factor in the expression for the toughness corresponding to a certain strain and flaw size, and the other factor is a function of strain, uncracked geometry, and material properties. Two accurate analytical approximations for the pressure-strain curve were developed, and these approximations are useable in both methods of fracture analysis.

One difference between the two methods of analysis, as applied here, was that stable crack growth was neglected in one of the analyses, but was considered in the other. In estimating failure strains by the method of LEFM based on strain, the original crack sizes were used. Nevertheless, the results were slightly conservative. In calculating the toughnesses corresponding to given nozzle corner strain levels by the tangent modulus method, stable crack growth was considered, and the results compared well with the measured toughness values. It follows that stable crack growth should be considered when estimating failure strains by the latter method, in order to avoid unconservative results.

In developing approximations of the type presented in this paper, it is appropriate to recognize their limitations, and to anticipate possible improved approaches to the problem. One of the main limitations inherent in the analysis results presented in this paper is their dependence on equivalent-energy maximum load toughness values not accompanied by the corresponding stable crack growth values. It is possible that the large values of toughness developed by the nozzle corner cracks in intermediate test vessels V-5 and V-9 are the combined result of stable crack growth and partial transverse restraint. However, without toughness data including stable crack growth values, these effects cannot be separated. With such data, the analysis methods developed in this paper could still be used to calculate strength as a function of toughness and current flaw size.

Because the toughness elevation due to less-than-full transverse restraint appears to increase with increasing toughness, it also follows that near plane-strain conditions may exist for low toughness values. Thus the benefits of decreased transverse restraint cannot be taken for granted, and they should be better defined experimentally for low toughness conditions. For example, such experiments could be conducted by testing vessels containing nozzle corner flaws at or below their transition temperatures. By combining the data thus obtained with existing data, it could then be determined under what conditions and to what extent partial transverse restraint and stable crack growth can be relied upon to increase toughness values above those sufficient for plane-strain crack initiation, for nozzle corner flaws.

Conclusions

An approximate method of elastic-plastic fracture analysis has been developed for calculating the conditions governing the stable or unstable extension of an inside nozzle corner crack in a pressure vessel, under internal pressure loading. The approximations used in the analysis include (1) an estimate of the inside nozzle corner elastic-plastic pressure-strain curve, based on the elastic stress concentration factor of the nozzle corner and the fully plastic pressure of the vessel cylinder; (2) an estimate of the toughness elevation due to less-than-full transverse restraint, based on the Irwin β_{Ic} formula; and (3) one of two approximate elastic-plastic strain versus toughness relations, the first being LEFM based on strain, and the second being the tangent modulus method. The method of analysis is developed with the aid of experimental data from two HSST Program intermediate test vessels with inside nozzle corner cracks, both of which developed high fracture toughness values, and example calculations are made for both vessels. It is noted that the method of analysis could be improved by using resistance curve toughness data instead of equivalent-energy maximum load data, because the latter data do not permit an estimate of stable crack growth as a function of applied load. It is also noted that the effects of transverse restraint on toughness are expected to decrease as toughness decreases, and therefore that additional experiments on steel vessels under low toughness conditions, which have not yet been conducted, are desirable for examining the accuracy of the method under these conditions.

References

[1] PVRC Ad Hoc Group on Toughness Requirements, "PVRC Recommendations on Toughness Requirements for Ferritic Materials," WRC Bulletin 175, Welding Research Council, Aug. 1972.
[2] ASTM Task Group E24.01.09, "Recommended Procedure for J_{Ic} Determination," draft document dated 1 March 1977.
[3] Shih, C. F. et al, "Methodology for Plastic Fracture," Fourth Quarterly Progress Report to Electric Power Research Institute, SRD-77-092, General Electric Company, Schenectady, N. Y., 6 June 1977.
[4] Paris, P. C., Tada, H., Zahoor, A., and Ernst, H., this publication, pp. 5–36.
[5] Pickett, A. G. and Grigory, S. C., *Transactions*, American Society of Mechanical Engineers, *Journal of Basic Engineering*, Vol. 89(C), Dec. 1967, pp. 858–870.
[6] Stahlkopf, K. E., Smith, R. E., and Marston, T. U., *Nuclear Engineering and Design*, Vol. 46, No. 1, March 1978, pp. 65–79.
[7] Mager, T. R. et al, "The Effect of Low Frequencies on the Fatigue Crack Growth Characteristics of A533, Grade B, Class 1 Plate in an Environment of High-Temperature Primary Grade Nuclear Reactor Water," WCAP-8256, Westinghouse Electric Corp., Pittsburgh, Pa., Dec. 1973.
[8] ASME Boiler and Pressure Vessel Code, Section III, Division I, Nuclear Power Plant Components, 1974 edition.
[9] Merkle, J. G., Robinson, G. C., Holz, P. P., and Smith, J. E., "Test of 6-In.-Thick Pressure Vessels. Series 4: Intermediate Test Vessels V-5 and V-9 With Inside Nozzle Corner Cracks," ORNL/NUREG-7, Oak Ridge National Laboratory, Oak Ridge, Tenn., Aug. 1977.
[10] Mager, T. R., Yanichko, S. E., and Singer, L. R., "Fracture Toughness Characterization of HSST Intermediate Pressure Vessel Material," WCAP-8456, Westinghouse Electric Corp., Pittsburgh, Pa., Dec. 1974.
[11] Witt, F. J. and Mager, T. R., "A Procedure for Determining Bounding Values on Fracture Toughness K_{Ic} at Any Temperature, ORNL-TM-3894, Oak Ridge National Laboratory, Oak Ridge, Tenn., Oct. 1972.
[12] Merkle, J. G. and Corten, H. T., *Transactions*, American Society of Mechanical Engineers, *Journal of Pressure Vessel Technology*, Vol. 96, Series J, No. 4, Nov. 1974, pp. 286–292.
[13] Bryan, R. H. et al, "Test of 6-Inch-Thick Pressure Vessels. Series 2: Intermediate Test Vessels V-3, V-4 and V-6," ORNL-5059, Oak Ridge National Laboratory, Oak Ridge, Tenn., Nov. 1975.
[14] Grigory, S. C., *Nuclear Engineering and Design*, Vol. 17, No. 1, 1971, pp. 161–169.
[15] Derby, R. W., *Experimental Mechanics*, Vol. 12, No. 12, 1972, pp. 580–584.
[16] Smith, C. W., Jolles, M., and Peters, W. H., "Stress Intensities for Nozzle Cracks in

Reactor Vessels," VPI-E-76-25, Virginia Polytechnic Institute and State University, Blacksburg, Va., Nov. 1976.

[17] Embly, G. T., "Stress Intensity Factors for Nozzle Corner Flaws," Knolls Atomic Power Laboratory, Schenectady, N. Y., draft dated July 1974 (to be published).

[18] Krishnamurthy, N., in *Proceedings*, First International Conference on Structural Mechanics in Reactor Technology, Paper G 2/7, Vol. 4, Berlin, Germany, 1971.

[19] Van Dyke, P., *American Institute of Aeronautics and Astronautics Journal*, Vol. 3, No. 9, 1965, pp. 1733–1742.

[20] Irwin, G. R., Krafft, J. M., Paris, P. C., and Wells, A. A., *Basic Aspects of Crack Growth and Fracture*, NRL Report 6598, U.S. Naval Research Laboratory, Washington, D.C., 21 Nov. 1967.

M. M. Hammouda[1] and K. J. Miller[1]

Elastic-Plastic Fracture Mechanics Analyses of Notches

REFERENCE: Hammouda, M. M. and Miller, K. J., "**Elastic-Plastic Fracture Mechanics Analyses of Notches,**" *Elastic-Plastic Fracture, ASTM STP 668*, J. D. Landes, J. A. Begley, and G. A. Clarke, Eds., American Society for Testing and Materials, 1979, pp. 703–719.

ABSTRACT: Fatigue failures invariably start at a surface notch. Crack initiation is due to plasticity while crack propagation can continue in an elastically stressed material due to the crack generating its own crack tip plasticity.
From elastic-plastic analyses, it is possible to predict the effect of notch plasticity on the behavior of short propagating cracks. Elastic-plastic finite-element analysis solutions to cracks in notch fields indicate that a crack will initially propagate at a decreasing rate until the crack can generate a crack tip plasticity that is greater than the elastic threshold stress intensity condition.

KEY WORDS: notch root radius, notch depth, crack initiation, crack propagation, nonpropagating crack, threshold stress intensity factor, plain fatigue limit, notch fatigue limit, stress concentration factors, strength reduction factors, notch stress-strain field, crack tip stress-strain field, bulk stress-strain field, elastic-plastic finite-element analysis

Nomenclature

- D Notch depth
- e Notch contribution to fatigue crack length
- K Stress intensity factor
- ΔK Stress intensity factor range
- ΔK_{Th} Threshold stress intensity factor range
- K_T Theoretical elastic stress concentration factor
- l Fatigue crack length
- N Number of cycles
- N_f Number of cycles to failure
- dl/dN Fatigue crack growth rate
- Δ Length of plastic shear ear at crack tip
- σ Stress

[1] Research fellow and professor, respectively, Faculty of Engineering, University of Sheffield, Sheffield, U.K.

σ_e Fatigue limit for plain specimens
σ_0 Fatigue limit for a particular notch profile
σ_y Yield stress
ρ Notch root radius

Fatigue studies may be divided into two categories. The first is termed the high-strain, low-endurance regime. Here bulk plasticity occurs and the cycles required for crack initiation are negligible, all lifetime being concerned with propagating a crack to critical dimensions [1].[2] The second regime is that at low cyclic stresses and here plasticity is extremely localized. In this case, crack initiation will dominate the lifetime of the specimen or component [2].

In both regimes it is clear that the fatigue process is due to the material suffering irreversible, that is, plastic, deformation. In the case of smooth and flat surfaces, should the stress range level in the bulk material be very low, but just above the fatigue limit, the plasticity will be restricted to favorably oriented slip bands within one or two surface crystals and may be defined as microplasticity. In these circumstances, once a crack is initiated its growth will continue due to self-generated crack tip plasticity. This latter form of plasticity can be characterized by invoking linear elastic fracture mechanics (LEFM) analyses [3] that describe the elastic stress intensification at the crack tip.

In most components a crack is initiated at a stress concentration such as a notch, which generates bulk but localized plasticity. Such macroplasticity is grain size independent. The notch may be large or small, that is, a keyway or machining scratch. The notch geometry and the bulk stress field control the limits between the initiation and the propagation phases of the crack. Thus in order to understand the fatigue behavior of components it is necessary to study the interactive role of micro- and macroplasticity of defects in notches during the phases of crack initiation and propagation.

The present paper examines the behavior of short cracks in notches that are subjected to various levels of stress which induce differing degrees of plasticity. Elastic-plastic finite-element analyses are used to predict a safe bulk stress level below which notch plasticity will not cause fatigue failure although cracks will be initiated.

Previous Work

The role of plasticity in fatigue crack growth is best illustrated by reference to cracks in biaxially stressed plates. Consider a plate in the xy-plane, containing a through central crack whose normal is in the y-direction. Let the plate be subjected to a positive σ_y stress and also a σ_x stress that has

---

[2]The italic numbers in brackets refer to the list of references appended to this paper.

magnitudes $+\sigma_y$, 0, and $-\sigma_y$ to equate with equibiaxial, uniaxial, and shear loading conditions. It has been shown that for constant values of σ_y the crack tip elastic stress intensity factor K_1 is unaffected by the magnitude and sense of the stress σ_x applied on the plane perpendicular to the crack front [4]. Crack growth rates are affected [5], however, because the size, shape, and orientation of the crack tip plastic zone are dependent on the state of biaxial stress [4]. Knowing the size of the plastic shear ears at the crack tip, it is possible to predict crack growth rates for various degrees of biaxiality [6].

It follows that elastic solutions to notch problems need to be re-evaluated and modified to account for the various roles of notch and crack tip plasticity and also the state of stress biaxiality. For these reasons alone, theoretical stress concentration factors and their derivatives, for example, Neuber/Stowell equivalent strain concentration factors, are unsatisfactory since they do not admit to the presence of a crack and hence do not differentiate between initiation Stage I and Stage II propagation phases of the fatigue process [7]. Neither do they define the extent and strength of the notch stress-strain field and cannot predict size effects [8] or the phenomenon of nonpropagating cracks [9].

To overcome these limitations, some recent analyses [10,11] based on the physical processes of fatigue have been concerned with the propagation of cracks within notch fields to determine the boundary conditions for initation and nonpropagating cracks. In summary form, the conclusions reached on a basis that a crack in a notch field can be equated to a crack in a plain specimen when both cracks have the same velocity, that is the same crack tip condition (not necessarily the same ΔK field), were as follows:

1. The equivalent length of a fatigue crack of length l within the notch field can be stated as

$$L = l + e \tag{1}$$

where e increases from zero to the depth of the notch, D, as l increases from zero to the boundary of the notch field. Beyond the edge of the notch field, the equivalent crack length is simply defined as

$$L = l + D \tag{2}$$

It follows that the notch field is that which extends from the notch root to a point in the bulk of the material at which the effective crack length is given by Eq 2.

2. The depth of the notch field in uniaxially stressed plates is approximately given by

$$0.13 \sqrt{D\rho} \tag{3}$$

for a very wide range of notches of engineering importance. Here ρ is the notch root radius. For very sharp notches, $D \gg 0.13 \sqrt{D\rho}$ and the effective crack length can be approximated to D.

3. The equivalent length of a crack within the notch field is given by

$$\left[1 + 7.69 \sqrt{\frac{D}{\rho}}\right] l \tag{4}$$

and the stress intensity factor of a crack within the notch field can be defined as

$$K = \left[1 + 7.69 \sqrt{\frac{D}{\rho}}\right]^{0.5} \sigma \sqrt{\pi l} \tag{5}$$

It follows that the term prior to $\sigma \sqrt{\pi l}$ can be considered as a fatigue crack concentration factor which may be equated to the strength reduction factor K_f or theoretical stress concentration factor K_T, although neither of these terms allows for the presence of a crack. Outside the notch field

$$K = \sigma \sqrt{\pi(l + D)} \tag{6}$$

4. For sharply notched plates, an initiated crack will not propagate if the bulk stress level

$$\sigma < \frac{M \Delta K_{Th}}{\sqrt{D}} \tag{7}$$

where M is associated with the geometric factor of the stress intensification of cracks in bodies of different shape. For an edge nonpropagating crack where $l \ll W$, the term M is equal to 0.5, that is, $(1.12 \sqrt{\pi})^{-1}$.

These important conclusions are illustrated in Figs. 1 and 2. Equation (3) successfully combines both the size and shape effect of notches long known to affect the fatigue behavior of components and was derived for a very wide range of notch profiles by considering the interaction between the crack tip and notch elastic-stress fields. This present paper now considers the role of notch plasticity in the development of very short cracks.

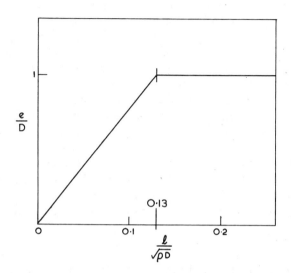

FIG. 1—*Notch contribution to fatigue crack length from linear elastic fracture mechanics analyses* [10].

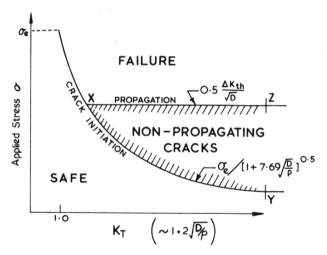

FIG. 2—*Fatigue regimes for notches with different elastic stress concentration factors* [11].

Present Work

The problem of understanding the behavior of the very short fatigue crack concerns the dominant role of plasticity in the very early crack growth regime. A very short fatigue crack is almost impossible to monitor in experimental growth rate studies, and so a theoretical crack growth analysis is required in order to assess the lifetime of the crack in this phase. Such a lifetime can be infinite in the case of an initiated but eventually nonpropagating crack. During this phase, notch plastic zones are bigger than the extent of crack tip plasticity and hence elasticity cannot describe crack tip conditions. Consider Fig. 3 and a very short fatigue crack. The bulk stress field controls the extent of the plastic zone, although it has minimal effect on the extent of the notch field. In this analysis it will be assumed that, immediately the crack is initiated, the plastic shear ears at the crack tip [12] will extend to the initial elastic-plastic boundary for the maximum tensile load applied. The length of the shear ears, Δ, during propagation can be determined from an elastic-plastic finite-element analysis for a crack

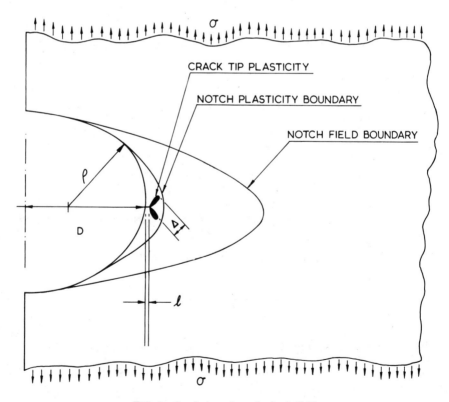

FIG. 3—*Crack tip and notch plastic fields*.

of any length. Figure 4 shows the type of finite-element idealization used in this study.

The mathematical procedure used in the elastic-plastic finite-element analysis was identical to that reported by Yamada et al [13] which invokes the Prandtl-Reuss equations for plastic flow in a von Mises material. The length of the shear ears is equated to the maximum extension of the plastic zone radiating from the crack tip.

Stress-strain analyses were carried out for plane stress conditions at two notch profiles, one shallow and one sharp. The first had a depth of 16 mm and a root radius of 16 mm. Analyses were for cracks of length 0, 0.035, 0.070, 0.105, 0.210, 0.430, 0.570, and 1.160 mm. The second notch had a depth of 16 mm and a root radius of 2.25 mm with crack lengths of 0, 0.035, 0.070, and 0.35 mm. The material stress-strain behavior was as follows: Young's modulus, 206 GN/m^2; cyclic yield stress, 176 MN/m^2 (that is, 0.6 of the monotonic yield stress, Ref 2); and work-hardening modulus (slope of the stress-strain curve beyond yield) zero for the first notch and 10330 MN/m^2 for the second notch. Such behavior is typical of yield and work-hardening behavior of the mild steel used both in the present experimental program and by Obianyor and Miller [14] to study the effect of stress overloads on threshold stress intensity values. The material has the following composition:

$$0.14C\text{-}0.58Mn\text{-}0.16Si\text{-}0.008N \text{ remainder Fe}$$

Figure 5 provides a schematic of the results of the finite-element analyses. At the LEFM threshold limit, which is here determined experimentally and is similar to that quoted in Ref 14, there is a known stress intensity field. This field characterizes a small plastic zone at the crack tip, the extent of which can be calculated to determine Δ by assuming a von Mises nonhardening material, plane stress conditions, and a yield condition equal to twice the cyclic yield stress. The line designated "no crack" represents the elastic-plastic boundary to which Δ will extend for just-initiated cracks. It follows that short nonpropagating cracks of length $l < l_3$ can develop in the shaded area due to notch plasticity and that the length which a nonpropagating crack can attain is stress-level dependent. For a crack to eventually stop, ΔK_{Th} must not be exceeded. Thus, according to LEFM, the critical nonpropagating crack size can be expected to decrease as stress range increases. However, Frost [15] has shown experimentally that nonpropagating crack length increases with increasing stress level. This conflict is due to the fact that early crack growth is due to plasticity and cannot be described by LEFM parameters. As stress level increases, the extent of the plastic zone increases and nonpropagating cracks in these zones can therefore be longer. Thus Point A represents the limiting stress level below which initiation will not take place, while Point B is an upper-bound

710 ELASTIC-PLASTIC FRACTURE

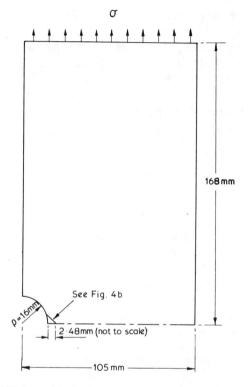

FIG. 4a—*Half plate, with circular edge notch*: $\rho = 16$ mm, $D = 16$ mm.

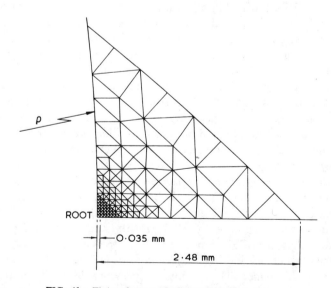

FIG. 4b—*Finite-element idealization of crack tip zone.*

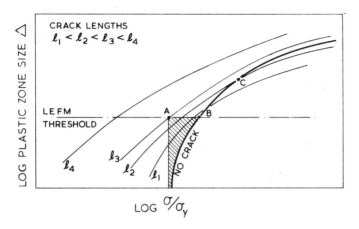

FIG. 5—*Schematic representation of the effect of increasing stress and crack length on crack tip plasticity for a given notch profile.*

stress-level solution for propagation to failure. At stress levels below Point A an existing flaw of length greater than l_3 is necessary for it to propagate to failure according to LEFM. At stress levels above Point B, failure will occur whether or not there preexists a flaw. Finally, Point C indicates that the same size of plastic shear ear exists when $l = 0$ and $l = l_2$ and that when $0 < l < l_2$ the crack is decreasing in growth rate because Δ is decreasing.

Figures 6 and 7 present finite-element analyses for the shallow and sharp notch profiles, respectively, and these can be compared with the schematic of Fig. 5.

To summarize, it is best to refer to Fig. 8. If a crack has a length sufficient to generate crack tip plasticity greater than that characterized by ΔK_{Th}, then the crack growth can be interpreted by a conventional LEFM analysis. To attain this critical crack length, it is necessary for crack growth to be a consequence of notch-generated plasticity. Should the plasticity not be of sufficient extent to develop this critical crack length, then a nonpropagating crack results. Figures 6 and 7 therefore define for a given notch geometry the stress boundaries between initiation and propagation of a crack based on a knowledge of the extent of crack tip plasticity Δ. The critical crack length is thus seen to be a function of applied stress level and notch profile.

It has been previously stated that it is difficult to experimentally monitor short crack growth behavior. To establish the validity of the present theory, however, it is possible to determine safe conditions for cracked specimens. Thus Fig. 9 presents the results of tests on prior edge-cracked mild steel specimens cycled in zero-tension. From these tests ΔK_{Th} was determined

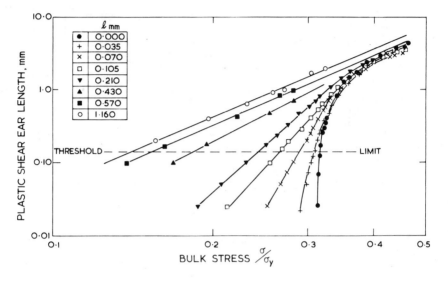

FIG. 6—*Elastic-plastic finite-element analysis results for cracks in a shallow-edge circular notch: $\rho = 16$ mm, D = 16 mm.*

as 5.35 MN/m$^{-3/2}$, which agrees with published work on the same material [14]. A second series of tests on edge notched plates, similar to that shown in Fig. 4a, determined the endurance limit, that is, that limit below which cracks were initiated but not propagated through the notch field and on to failure; see Fig. 10. From a knowledge of ΔK_{Th} and σ_0 it is possible to draw the limiting conditions for safety à la Kitagawa [16]; for example, see Fig. 11. In this latter figure are the data points of the present theoretical elastic-plastic finite-element analyses which are in close agreement with the experimentally determined boundary conditions, the former values being derived from Fig. 6 for plastic zone sizes equal to that corresponding to ΔK_{Th} in a simple cracked member.

Thus the safe stress levels for cracks of different length in various notched configurations can be determined from elastic-plastic finite-element analyses such as those shown in Figs. 6 and 7. Note that for very short crack lengths, LEFM analyses should not be employed. This is because LEFM can characterize the extent of plasticity only when (1) the plastic field is small in comparison with the elastic stress intensification field and (2) the extent of crack tip plasticity is physically meaningful. When cracks are less than 0.25 mm, the crack tip plastic zone size is measured in units of angstroms and hence LEFM characterization of fracture processes is no longer applicable.

It now remains to modify the Smith analysis [10] to account for plasticity effects on short crack growth rates. An assumption in the present approach

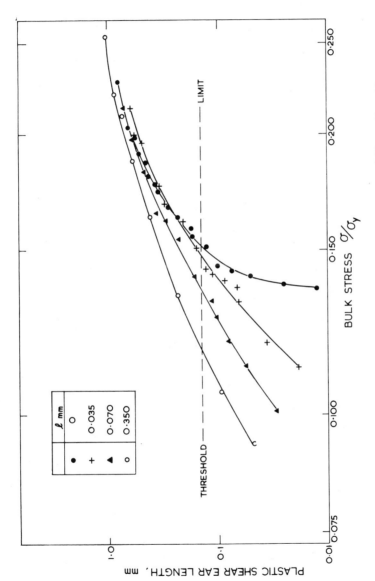

FIG. 7—*Elastic-plastic finite-element analysis results for a sharp-edge notch: $\rho = 2.25$ mm, $D = 16$ mm.*

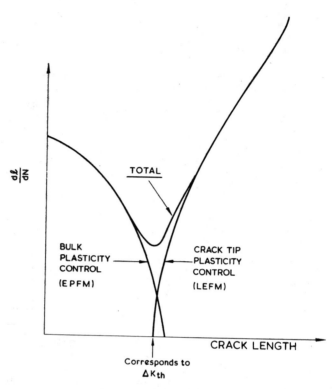

FIG. 8—*Plastic and elastic fracture mechanics characterization of fatigue crack growth.*

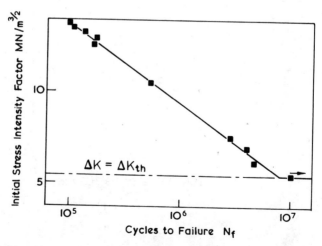

FIG. 9—*Experimental data for determining the threshold stress intensity factor.*

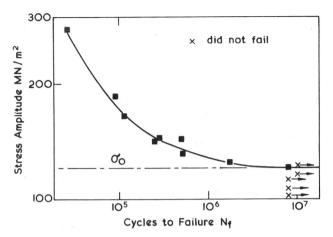

FIG. 10—*Experimental data for determining the shallow-notch fatigue limit:* $\rho = 16$ mm, $D = 16$ mm.

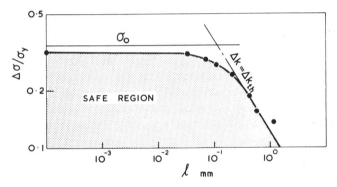

FIG. 11—*Comparison of experimental and theoretical (elastic-plastic finite element) results for the shallow notch.*

was that Δ for short cracks will extend to the elastic-plastic boundary at the root of the notch, and so, as the very short crack grows, the value of Δ will initially decrease, producing a decrease in crack growth rate until the crack is long enough to grow under LEFM control; see Fig. 8. Since e is determined from the equivalence of crack velocity, which is a function of crack tip plasticity conditions, this means that the notch contribution term e of Eq 1 will initially decrease while the crack is in the notch plastic zone. This effect can be determined from the crossover behavior of the curves of Figs. 6 and 7. While this effect is small, Fig. 12 also shows that e is a function of stress level since this controls the extent of plasticity and early growth rate.

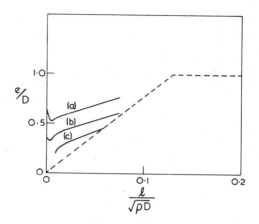

FIG. 12—*Notch contribution to fatigue crack length from elastic-plastic fracture mechanics analyses for different stress levels σ/σ_y:* (a) 0.45, (b) 0.35, (c) 0.31.

Discussion

Although the agreement between theory and experiment depicted in Fig. 11 is very good, continuum mechanics analyses, be they elastic or elastic-plastic, can be in error when crack lengths and growth rates are of the order of microstructural features such as grain size. The minimum finite-element size used in this study is 0.035 mm, that is, comparable to the grain size of mild steel. The ΔK_{Th} plastic zone size is 0.13 mm. It follows that while the present method cannot model the plastic behavior of a single grain, the very small mesh size can give some assessment of continuum plastic behavior around threshold conditions. It should be noted that the program was used only to determine the elastic-plastic boundary.

The form of the base curve ($l = 0$) in Fig. 5 is important. For a shallow notch, Fig. 6, the stress level is critical to the extent that, should a crack be initiated, then it will propagate to failure since threshold conditions are immediately exceeded as indicated by the steepness of the base curve. This condition is equivalent to Point X in Fig. 2. For a sharp notch, however (see Fig. 7), a crack may be initiated at a low stress value but it will not grow since the threshold is not exceeded. This is equivalent to Point Y in Fig. 2. Should the stress level be increased slightly, then an initiated crack may grow but still not propagate to failure unless the stress is increased to a level indicated by Point Z in Fig. 2, which approximates to Point B in Fig. 5. Thus, the fatigue failure of components from notches is seen to be a function of the applied stress level σ, the notch profile parameters ρ and D, and the material property, ΔK_{Th}.

Another aspect of Figs. 6 and 7 that is important concerns the crossover behavior of the "no crack" curve ($l = 0$). As an example, consider the sharp notch, Fig. 7, and a stress level σ equal to 0.2 σ_y. As soon as a crack is initiated, the plastic shear ear length decreases until the crack length is approximately 0.04 mm. However, it attains its original length when l equals 0.07 mm. It follows that the crack growth rate decreases and then increases as depicted in Fig. 8. Because the crack growth rate decreases, then the notch contribution factor e, based on an equivalence of crack growth rates, should decrease faster than the increase in the fatigue crack length l. The present studies show this to be the case, Fig. 12, for values of l up to about 0.1 mm for the shallow notch subjected to stress levels $\sigma > 0.3$ σ_y approximately. These studies also indicate as shown in Fig. 12 that even without a fatigue crack the notch has an equivalent crack length e which is stress-level dependent. In the present case, initial values of e/D are 0.36 and 0.64 for stress levels 0.35 σ_y and 0.45 σ_y, respectively, for the shallow notch. Thus, early life fatigue can have exceedingly high crack growth rates although the cracks are extremely small. In this regime, linear elastic fracture mechanics does not apply. It is recommended that in sharp notch situations designers use equivalent crack lengths derived from Eq 2; that is, they should assume that e/D is unity.

It is interesting to compare present results with other recent work. Jerram [17] concludes that the elastic stress intensity factor is not a suitable criterion for predicting crack initiation and that the fatigue crack length used in his paper to define initiation, namely, 10 μm (0.0004 in.) was too large. We would add that no elastic parameter is suitable; see Figs. 8 and 12. Ohji et al [18] studied, by finite-element methods, cracks emanating from notches, but these long cracks were 0.25 mm in length and therefore well beyond the notch stress field. They argued that nonpropagating cracks are due to crack closure and can be assessed by effective values of stress intensity factors or local strain range values or both. Thus they introduced a criterion which stated that a crack will continue to propagate if the magnitude of the strain range, at a certain characteristic distance ahead of the crack tip, exceeds a critical value. Thus the work of Obianyor and Miller [14] is relevant since they show that as a crack grows into a rapidly decreased stress field, due to the application of a prior overload, the threshold stress intensity factor increases, thus increasing the possibility of a nonpropagating crack. Now Kotani et al [19], like Jerram, examined stress-strain behavior ahead of the crack tip by invoking the Neuber relationship, but once again crack "initiation" lengths were such as to be propagating cracks and outside the notch stress field. Their work showed that specimens with stress concentrations had "initiation" lives greater than plain specimens on a basis of local stress range values. This was undoubtedly due to initially decreasing crack propagation rates in the notched specimens with cracks growing into much lower stress fields.

It appears that two classes of notch crack lengths have to be considered. The first concerns the very short crack whose initiation and early growth are not amenable to LEFM analyses, while the second type concerns the longer but still small crack that is amenable to LEFM if the stress level is high enough. Nevertheless, cyclic plasticity controls the birth and the early growth of both types of crack. The present work indicates that for the sharp notch a crack length of the order of 0.1 mm can develop at the notch root, due to plasticity, and failure will still not occur. This length is of the same order as the length of a crack necessary for the application of LEFM analyses; see Fig. 11. On the other hand, shallow notches require higher stress levels to develop the notch root plasticity that initiates cracks. Such plasticity is well contained within the notch field; compare the 0.03-mm-deep with the 2.08-mm-deep notch field. These short cracks are not amenable to LEFM analyses and the notch contribution factor e is stress level dependent since this controls the extent of plasticity. These initiated cracks will not cease propagation, however, because the threshold limit is easily exceeded (see Fig. 6) due to the higher stress levels and the strain concentration feature of the notch.

Finally, it should be noted that both classes of cracks will slow down as they come close to the elastic-plastic strain boundary.

All the foregoing work has confined itself to a two-dimensional appreciation of fatigue crack initiation and growth. Work is now continuing on a three-dimensional appreciation of crack growth of very short cracks at notch roots.

Conclusions

1. Notch root plasticity controls the early stage propagation of fatigue cracks in notches and in this regime LEFM analyses do not apply.

2. Elastic-plastic fracture mechanics can account for fatigue crack growth below the elastic threshold stress intensity condition by considering the interaction between crack tip and notch field plasticity.

3. Elastic-plastic fracture mechanics can account for decreasing crack growth rates and the production of nonpropagating fatigue cracks.

Acknowledgments

The authors would like to thank British Gas for providing a research scholarship to support M. M. Hammouda.

References

[1] Ham, R. K., in *Proceedings*, International Conference on Thermal and High Strain Fatigue, Institution of Metallurgists, London, England, 1967, pp. 55–79.
[2] Miller, K. J. and Zachariah, K. P., *Journal of Strain Analysis*, Vol. 12, No. 4, 1977, pp. 262–270.

[3] Irwin, G. R., *Transactions ASME, Journal of Applied Mechanics*, Vol. 24, 1957, p. 361.
[4] Miller, K. J. and Kfouri, A. P., *International Journal of Fracture*, Vol. 10, No. 3, 1974, pp. 393-404.
[5] Hopper, C. D. and Miller, K. J., *Journal of Strain Analysis*, Vol. 12, No. 1, 1977, p. 23.
[6] Miller, K. J., in *Proceedings*, "Fatigue 1977" Conference, Cambridge, England, *Metal Science Journal*, Vol. 11, Nos. 8 and 9, 1977, p. 432.
[7] Forsyth, P. J. E., in *Proceedings*, Symposium on Crack Propagation, Cranfield, England, 1961, p. 76.
[8] Coyle, M. B. and Watson, S. J., in *Proceedings*, Institution of Mechanical Engineers, Vol. 178, 1963, p. 147.
[9] Frost, N. E., Marsh, K. J., and Pook, L. P., *Metal Fatigue*, Oxford University Press, Oxford, England, 1974, p. 173.
[10] Smith, R. A. and Miller, K. J., *International Journal of Mechanical Sciences*, Vol. 19, 1977, pp. 11-22.
[11] Smith, R. A. and Miller, K. J., *International Journal of Mechanical Sciences*, Vol. 20, 1978, pp. 201-206.
[12] Tomkins, B., *Philosophical Magazine*, Vol. 18, No. 155, 1968, p. 1041.
[13] Yamada, Y., Yoshimura, N., and Sakuri, T., *International Journal of Mechanical Sciences*, Vol. 10, 1968, p. 343.
[14] Obianyor, D. F. and Miller, K. J., *Journal of Strain Analysis*, Vol. 13, No. 1, 1978, pp. 52-58.
[15] Frost, N. E., *The Engineer*, Vol. 200, 1955, pp. 464 and 501.
[16] Kitagawa, H. and Takahashi, S., in *Proceedings*, Second International Conference on Mechanical Behavior of Materials, Boston, Mass., 1976, p. 627.
[17] Jerram, K., "Fatigue Crack Initiation in Notched Mild Steel Specimens," Report 1972, Central Electricity Generating Board, RD/B/N1994.
[18] Ohji, K., Ogura, K., and Ohkubo, Y., *Engineering Fracture Mechanics*, Vol. 7, 1975, p. 457.
[19] Kotani, S., Koibuchi, K., and Kasai, K., "The Effect of Notches on Cyclic Stress-Strain Behaviour and Fatigue Crack Initiation," Report of the Mechanical Engineering Research Laboratory, Hitachi Laboratory, Tsuchiura, Japan.

W. R. Brose[1] and N. E. Dowling[1]

Size Effects on the Fatigue Crack Growth Rate of Type 304 Stainless Steel

REFERENCE: Brose, W. R. and Dowling, N. E., "**Size Effects on the Fatigue Crack Growth Rate of Type 304 Stainless Steel,**" *Elastic-Plastic Fracture, ASTM STP 668,* J. D. Landes, J. A. Begley, and G. A. Clarke, Eds., American Society for Testing and Materials, 1979, pp. 720-735.

ABSTRACT: Planar size effects on the fatigue crack growth rate of AISI Type 304 stainless steel characterized by linear-elastic fracture mechanics were experimentally investigated. Constant-load amplitude tests were conducted on precracked compact specimens ranging in width from 2.54 to 40.64 cm (1 to 16 in.). The da/dN versus ΔK data are compared on the basis of several size criteria which are intended to limit plasticity and thus enable linear-elastic analysis of the data. Also, the cyclic J-integral method of testing and analysis was employed in the fatigue tests of several specimens undergoing gross plasticity. The cyclic J crack growth rate data agree well with that from the linear-elastic tests. It is argued that an appropriate size criterion for linear-elastic tests must limit the size of the monotonic plastic zone and thus be based on K_{max}, the maximum stress intensity. While the size criteria considered vary widely in the amount of plasticity they allow, they provide comparable correlations of crack growth rate. Thus the use of the most liberal criterion is justified.

KEY WORDS: 304 stainless steel, fatigue crack growth rate, size effects, size criteria, plasticity, cyclic J-integral, crack propagation

In characterizing static fracture, the techniques of linear-elastic fracture mechanics can be applied if the cracked body is predominantly elastic at fracture. In a fracture toughness test, certain minimum specimen size requirements assure this condition by restricting the size of the crack-tip plastic zone with respect to that of the specimen. An analogous size requirement is probably necessary for fatigue crack growth data characterized by linear-elastic fracture mechanics.

Size effects in the areas of fracture and fatigue are actually of two kinds. The first concerns specimen or component thickness as it affects the

[1]Engineer and senior engineer, respectively, Structural Behavior of Materials Department, Westinghouse R&D Center, Pittsburgh, Pa. 15235.

amount of transverse constraint and thus the state of stress at the crack tip. The importance of thickness in the area of fracture is well documented. In fatigue, thickness effects reported in the literature are inconsistent. Increasing thickness has been observed to increase, decrease, and have no effect on growth rate [1,2].[2] Also, thickness can be considered to be a controlled variable in determining crack growth rate.

The other size effect involves what is known as planar size, generally identified as specimen width, the dimension perpendicular to the thickness direction and parallel to the crack plane. It is planar size which, for a given material, determines the degree of plasticity at a given K level in a cracked body. Size requirements for fatigue which limit plasticity and thus enable linear elastic analysis of the data must be related to planar size.

This paper is concerned with planar size effects on the fatigue crack growth rate of annealed AISI Type 304 stainless steel, a material widely used in the nuclear industry. Planar size effects are of interest in this material because of its relatively low monotonic yield strength. Relatively large specimens may be needed to obtain fatigue crack growth data under predominantly elastic conditions.

Test results are presented for specimens ranging a factor of 16 in size. Several size criteria are examined in terms of the degree of plasticity which they permit. Also examined is the effect of the observed plasticity on fatigue crack growth rate and thus the ability of these size criteria to produce size-independent correlations of growth rate. In addition to the linear-elastic tests, an elastic-plastic experimental and analytical technique is employed to generate fatigue crack growth data under conditions of gross plasticity. The data obtained by this technique on small specimens are compared with that from larger specimens under elastic conditions.

Plasticity and Size Criteria

Since planar size effects in linear-elastic fatigue crack growth are related to the development of plasticity, it is instructive to examine the stress-strain behavior near the tip of a growing fatigue crack in a ductile metal. Figure 1 schematically illustrates this behavior for an idealized elastic-perfectly plastic material. Two separate plastic zones are identified. Within the monotonic plastic zone, increments of yielding occur at the maximum point of each loading cycle while the cyclic behavior is still elastic. In the cyclic plastic zone, incremental yielding continues to occur on each loading cycle, and there is inelastic cyclic action as well. The size of these plastic zones can be estimated by the Irwin plastic zone size equation as shown in Fig. 1 and as explained by Paris [3]. For zero-to-tension loading, $R = 0$,

---

[2]The italic numbers in brackets refer to the list of references appended to this paper.

722 ELASTIC-PLASTIC FRACTURE

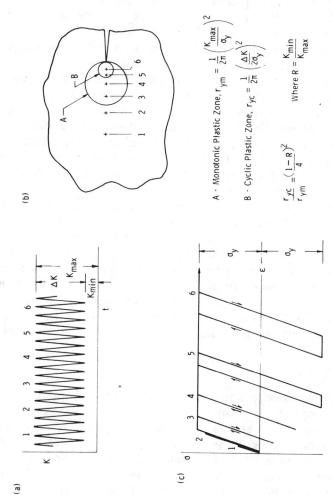

FIG. 1—*Stress-strain behavior at a point being approached by the tip of a growing fatigue crack. In (c), the reversed stresses and strains are shown for selected cycles only.*

the cyclic plastic zone is only one-fourth the size of the monotonic plastic zone. This difference in size increases as R increases.

Due to bending in the compact specimen, Fig. 2, the fully plastic condition occurs when the region of material yielded in monotonic tension spans about half the width of the uncracked ligament and meets with a similar zone of material yielded in compression and extending from the back of the specimen. The crack-tip cyclic plastic zone is thus only about one-eighth the size of the remaining specimen ligament at fully plastic yielding when $R = 0$.

Thus, as has been previously proposed [4], it appears reasonable to establish a size criterion for linear-elastic fatigue crack growth testing which would limit the amount of monotonic plasticity developed. One approach is to limit plasticity as reflected in the specimen load-deflection behavior which is directly measurable. A somewhat arbitrarily chosen form of such a criterion requires

$$V_{\text{plastic}} \leq V_{\text{max}}^e \tag{1}$$

where

$V_{\text{plastic}} = V_{\text{max}} - V_{\text{max}}^e$,
$V_{\text{max}} = $ maximum measured specimen deflection, and

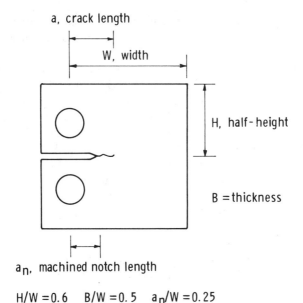

$H/W = 0.6 \quad B/W = 0.5 \quad a_n/W = 0.25$

FIG. 2—*Compact specimen geometry employed in linear-elastic tests.*

V_{max}^e = corresponding deflection calculated on the basis of elastic behavior [1,4].

Equation 1 can be rewritten as

$$V_{max} \leq 2 V_{max}^e \qquad (2)$$

The fatigue crack growth data presented later will be examined in light of this criterion.

It is not always convenient to measure specimen deflection response, and also a great deal of fatigue crack growth data exist without accompanying deflection measurements. A size criterion based on calculated quantities is thus desirable. Such a size criterion should be based on K_{max}, the maximum stress intensity in a loading cycle, rather than ΔK, the range of stress intensity. One such criterion is being considered for adoption in a forthcoming ASTM standard [4]. For compact specimens, it is required that

$$(W - a) \geq \frac{4}{\pi} \left(\frac{K_{max}}{\sigma_y} \right)^2 \qquad (3)$$

where $W - a$ is the uncracked ligament length (specimen width minus crack length) and σ_y is the 0.2 percent offset monotonic yield strength. Data on a 1310-MPa (190 ksi) yield strength 10Ni steel appear to conform well to this size criterion [4].

It can be argued that monotonic yield strength is not an adequate single-parameter index of the degree of plasticity developed in a cracked body, since it does not take into account the large difference in strain-hardening capacity different metals can exhibit. Considering two metals with the same static yield strength, the one with higher strain-hardening characteristics will indeed produce less extensive plasticity in a cracked body at the same stress intensity. A first-order correction for the effect of strain-hardening would be to substitute a value of flow stress, σ_{flow}, equal to the average of the yield and ultimate strengths, for σ_y in Eq 3. Both the σ_y and σ_{flow} size criteria will be applied to the data presented later on 304 stainless steel, a material which exhibits extensive strain hardening.

While the three size criteria presented in the foregoing are of primary interest in this paper, two other conditions of plasticity are here described. They represent possible upper and lower bounds in terms of plasticity within which any reasonable size criterion would fall. So-called nominal yielding occurs when the nominal stress or "$P/A + Mc/I$" stress reaches the yield, $\sigma_n = \sigma_y$. The other bound is the calculated condition of fully plastic limit load, $P = P_{LL}$. Rice's solution for this condition for the compact specimen, presented in Ref 5, is employed in this paper. Yield strength rather than flow stress was used in the equation for the results presented herein.

Experimental Procedure

The specimens tested in this study were machined from a 6.35-cm-thick (2.5 in.) rolled plate of solution-annealed 304 stainless steel. Additional test material information is given in Table 1. Specimens were taken from the plate in the longitudinal-transverse (L-T) orientation, defined in the ASTM Test for Plane-Strain Fracture Toughness of Metallic Materials (E 399-74). In this orientation, the specimen crack plane is perpendicular to the rolling direction of the plate.

The linear-elastic fatigue crack growth tests are now described. Compact specimens of the geometry shown in Fig. 2 were employed. In this paper, specimen size is identified by width. All other dimensions are as in Fig. 1 except that, as noted, the thickness sometimes differs. The largest specimen tested is shown in Fig. 3.

All testing was performed with servo-controlled, electrohydraulic test machines. Constant-amplitude tension-tension load control was employed with a ratio of minimum to maximum load, R, equal to 0.05, and frequencies in the range of 10 Hz. At the very end of tests, the frequency was decreased to facilitate crack length measurement. Deflections were measured 0.483 cm (0.190 in.) away from the specimen front face. In Fig. 2, the specimen front face is to the left of the load line. Crack growth was monitored visually using a calibrated traveling microscope in conjunction with scribe lines placed on the specimen surface. Crack length versus cycles data were reduced to crack growth rate versus stress intensity range using a seven-point incremental polynomial fitting technique [6] and a stress-intensity solution available in the literature [7].

The cyclic J tests were performed in accordance with the experimental and analytical techniques described by Dowling and Begley [8]. Test specimen geometry was that shown in Fig. 2 except that the notch was modified to permit deflection measurement at the load line, and the machined notch length was given by $a_n/W = 0.465$. Specimen width was 5.08 cm (2 in.).

Cyclic J tests are conducted under deflection control to a sloping line.

TABLE 1—*Test material information.*

Description: AISI 304 stainless steel; Jessop Steel Co. heat 24348
Condition: hot-rolled, annealed and pickled
Geometry: plate, 6.35 by 60 by 60 cm (2.5 by 24 by 24 in.)
Chemistry: 0.058C-1.48Mn-0.035P-0.012S-0.38Si-8.90Ni-18.15Cr-0.44Mo-0.17Co-0.57Cu-Fe remainder (values in weight %)
Tensile properties:[a]

Offset yield strength, MN/m² (ksi)	269 (39)
Ultimate tensile strength, MN/m² (ksi)	579 (84)
True fracture strength, MN/m² (ksi)	1920 (279)
Reduction in area, %	82
Charpy impact energy,[a] m-N (ft-lb)	320 (236)

[a]Longitudinal orientation with respect to rolling direction of plate.

726 ELASTIC-PLASTIC FRACTURE

FIG. 3—*Large compact specimen and fatigue test apparatus.*

This control condition is illustrated in Fig. 4 along with some of the load-deflection loops obtained in one test. A special analog control circuit was used to impose this condition in which neither load nor deflection amplitude is constant. Note that as the test progresses and crack length increases, the maximum load drops while the maximum deflection increases. The minimum deflection is always zero. The amount of cyclic plasticity

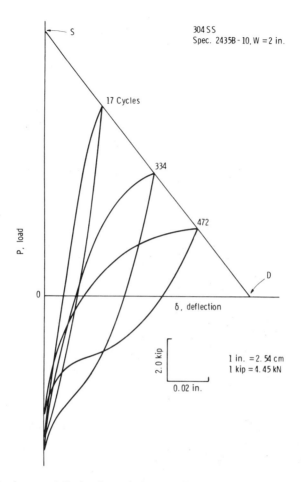

FIG. 4—*Load versus deflection loops during a cyclic J test under deflection control to a sloping line.*

gradually increases with crack length, producing gradually increasing values of ΔJ and crack growth rate. The cyclic J test technique is valuable in that it provides extensive cyclic plasticity without the large mean deflections, or specimen ratchetting, that can occur at the end of a load control test.

The method of calculation of ΔJ for a given loading cycle is shown in Fig. 5. It is based on the Rice et al [9] approximation for J, and employs the area under the load-deflection curve on the loading half-cycle. Only the area above the macroscopic crack closure load, which was considered to be at the point of inflection on the unloading half-cycle curve, was used. Further details on J in fatigue are given in Refs *8*, *10*, and *11*.

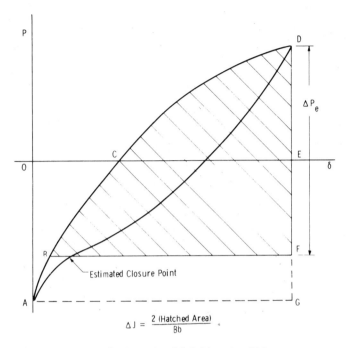

FIG. 5—*Operational definition of cyclic* J.

Results and Discussion

The measured specimen plasticity during the fatigue crack growth tests is now examined. Figure 6 compares the measured maximum specimen deflection with that predicted by elastic behavior for the test on one specimen. There is a plastic component of deflection through most of the specimen life, and the plastic deflection increases rapidly near the end of the test. While not shown in the figure, the measured deflection range, $\Delta V = V_{max} - V_{min}$, agreed closely with that predicted elastically for the duration of the test. This indicates that while the size of the region yielded monotonically becomes very large at the end of the test, the cyclic plastic zone remains small. These measurements concur with behavior expected from analytical considerations presented earlier.

The degree of plasticity permitted by each of the size criteria discussed earlier is also shown in Fig. 6. The stress-intensity range at which each criterion is violated is listed for each specimen in Table 2. Note that the order in which the size criteria are violated is unchanged for different specimen sizes. By definition, the plasticity permitted by the deflection criterion corresponds to V_{max} exceeding V_{max}^e by a factor of two. At the criterion based on yield strength, this factor is about 1.5, while for the

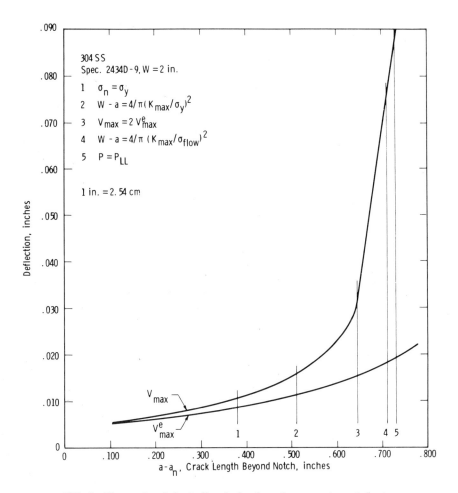

FIG. 6—*Measured and elastically calculated maximum specimen deflection.*

criterion based on flow stress, the factor is about four. The values of these plastic deflection factors at the three size criteria were roughly independent of specimen size.

The degree to which the plasticity allowed by the various size criteria affects fatigue crack growth rate is now examined. Figure 7 contains the da/dN versus ΔK data obtained in the linear-elastic tests. The shaded points differ from the unshaded points only in that the shaded points violate the size criteria based on yield strength, Eq 3. Figures 8 and 9 contain the same data except that the shaded points violate the size criteria based on deflections and on flow stress, respectively.

In each of these plots the data obtained from the cyclic J tests are also

TABLE 2—*Maximum ΔK for validity by five size criteria.*

Specimen Dimensions[a]		Criterion[a]				
W	B	(1)	(2)	(3)	(4)	(5)
2.54 (1)	0.5 W	22 (20)	26 (24)	35 (32)	39 (35)	42 (39)
5.08 (2)	0.5 W	30 (27)	36 (33)	46 (42)	52 (47)	54 (49)
10.16 (4)	1.27 (0.5)	38 (35)	44 (40)	52 (47)	66 (60)	69 (63)
10.16 (4)	0.5 W	38 (35)	44 (40)	58 (53)	66 (60)	69 (63)
40.64 (16)	5.08 (2)	67 (61)	81 (74)	92 (84)	114 (104)	127 (116)

(1) nominal yield: $\sigma_n \leq \sigma_y$.
(2) yield strength: $W - a \geq 4/\pi \, (K_{max}/\sigma_y)^2$.
(3) deflection: $V_{max} \leq 2 V_{max}^e$.
(4) flow stress: $W - a \geq 4/\pi \, (K_{max}/\sigma_{flow})^2$.
(5) fully plastic limit load: $P \leq P_{LL}$.
Units in centimetres (inches) or Mn·m$^{-3/2}$ (ksi$\sqrt{in.}$).
All tests, $R = 0.05$.
[a]ΔK values approximate.

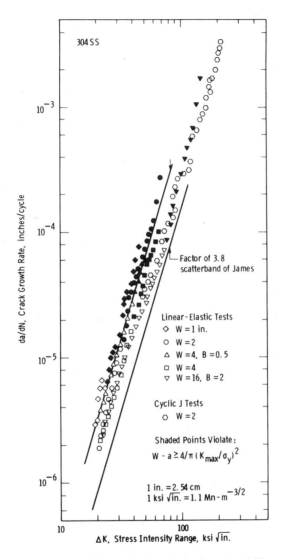

FIG. 7—*Comparison of data with size criterion based on yield strength.*

shown. Values of ΔJ were converted by $\Delta K = \sqrt{\Delta J \cdot E}$. The elastic-plastic data shown are actually a compilation of data from tests on four separate specimens covering different but overlapping ranges of ΔJ and crack growth rate.

Also shown on these plots are upper and lower scatterbands for a large body of fatigue crack growth data on annealed 304 stainless steel compiled by James [12]. These data represent 10 different specimen geometries

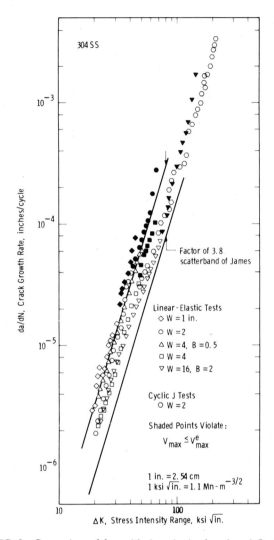

FIG. 8—*Comparison of data with size criterion based on deflections.*

and several heats of material. However, the specimens cover a narrow range of size and are mostly relatively small.

Consider all the linear-elastic test data except perhaps the last three points for each specimen at the high-growth-rate end. The spread in fatigue crack growth increases only slightly with ΔK, from a factor of 2.5 at the low end to a factor of about three at the high end. Also, the data agree quite well with the scatterband of James, which is based primarily on relatively small specimens. As ΔK increases, the difference in the degree of plasticity existing in specimens of different size increases. On this basis,

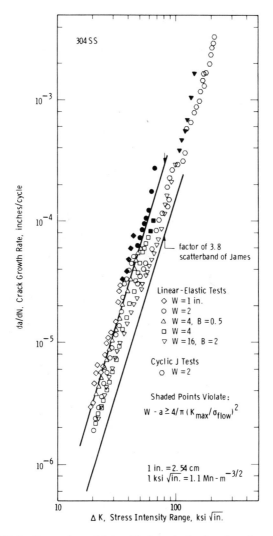

FIG. 9—*Comparison of data with size criterion based on flow stress.*

the experimental data indicate that the plasticity experienced in these tests has little effect on crack growth rate. It is not surprising, then, that the size criteria employed in Figs. 7, 8, and 9 provide comparable correlations of crack growth rate.

The data do show, however, a consistent layering with specimen size. At a given ΔK, crack growth rate increases as specimen size decreases. So it is likely that plasticity does have some effect here on growth rate. Perhaps the relative size of the cyclic plastic zone, which was always small in the subject tests, has first-order control of the validity of the linear-elastic analysis. A large degree of monotonic plasticity could have a secondary effect, though, and be a factor in the layering observed.

Since the linear-elastic test specimens represent a factor of four in thickness, a possible contribution to the spread in growth rate from this source should be considered. However, two previous investigators found no significant effect of thickness on growth rate in annealed 304 stainless steel. James [13] compared growth rates from foil specimens 0.254 mm (0.010 in.) thick with those from specimens of conventional thickness which correspond to the scatterbands already described. Shahiniam [2] tested specimens ranging in width from 0.762 to 2.54 cm (0.3 to 1 in.).

Finally, it is noted that the data obtained in the cyclic J tests on the 5.08-cm-wide (2 in.) specimens compare very well with those from the linear-elastic tests. This agreement is a positive reflection on the ability of the cyclic J method to characterize crack growth rate data under conditions of gross plasticity. Also, it is a further indication of the lack of a strong size effect due to plasticity in the linear-elastic tests. This is because the plasticity in the elastic-plastic tests is directly accounted for by the use of J, while it is not directly accounted for in the linear-elastic tests.

Regarding the linear-elastic data, duplicate tests, as well as data at other R's and starting stress intensities and on other geometries, would of course strengthen the conclusions reached here regarding 304 stainless steel. Elevated-temperature data on specimens of different sizes should also be obtained since this material is used widely in that situation. Of more importance, though, is that data on other high strain hardening materials be generated before conclusions regarding size effects on fatigue crack growth of 304 stainless steel are considered to have general applicability. An investigation of these variables is currently underway at Westinghouse.

Summary and Conclusions

In order to investigate planar size effects on the fatigue crack growth rate of annealed 304 stainless steel, compact specimens ranging a factor of 16 in width were tested using linear-elastic fracture mechanics techniques, as were compact specimens employing the cyclic J elastic-plastic technique. The following conclusions are offered.

1. Size criteria intended to limit plasticity during fatigue crack growth should restrict the size of the monotonic plastic zone with respect to specimen size, and should thus be based on K_{max} rather than on ΔK.

2. Monotonic plasticity did not appear to strongly affect crack growth rate in the subject material. Thus the three different size criteria considered, while allowing widely different amounts of plasticity, provided comparable correlations of crack growth rate.

3. Fatigue crack growth rates obtained under gross plasticity and characterized by J agreed well with those from linear-elastic tests.

4. Data at other load ratios, on other geometries, and especially on other strain-hardening materials, are needed before the observations on the utility of the various size criteria can be considered to have general applicability.

Acknowledgments

The experimental work was conducted in the Mechanics of Materials Laboratory under the direction of R. B. Hewlett. This laboratory is operated by the Structural Behavior of Materials Department managed by E. T. Wessel. A number of technicians participated in the experimental work, and the care exercised is greatly appreciated. Mr. P. J. Barsotti is especially thanked for preparing the oscillograph system for deflection measurement. This work was sponsored by the Westinghouse Advanced Reactor Division, Waltz Mill, Pa.

References

[1] Dowling, N. E. in *Flaw Growth and Fracture, ASTM STP 631*, American Society of Testing and Materials, 1977, pp. 131-158.
[2] Shahinian, P., *Nuclear Technology*, Vol. 30, Sept. 1976.
[3] Paris, P. C., *Fatigue—An Interdisciplinary Approach*, Syracuse University Press, Syracuse, N.Y., 1964, pp. 107-127.
[4] Hudak, S. J., Jr., Saxena, A., Bucci, R. J., and Malcolm, R. C., "Development of Standard Methods of Testing and Analyzing Fatigue Crack Growth Rate Data—Third Semi-Annual Report," AFML Contract F33615-75-5064, Westinghouse Research and Development Center, Pittsburgh, Pa., March 10, 1977.
[5] Hudak, S. J., Jr. and Bucci, R. J., "Development of Standard Methods of Testing and Analyzing Fatigue Crack Growth Rate Data—First Semi-Annual Report," AFML Contract F33615-75-5064, Westinghouse Research and Development Center, Pittsburgh, Pa., Dec. 16, 1975.
[6] Clark, W. G., Jr., and Hudak, S. J., Jr., "The Analysis of Fatigue Crack Growth Rate Data," Westinghouse Scientific Paper 75-9E7-AFCGR-P1, Aug. 26, 1975, to be published in *Proceedings*, 22nd Sagamore Army Materials Research Conference on Application of Fracture Mechanics to Design, Sept. 1975.
[7] Saxena, A. and Hudak, S. J., Jr., "Review and Extension of Compliance Information for Common Crack Growth Specimens," Westinghouse Scientific Paper 77-9E7-AFCGR-P1, May 3, 1977.
[8] Dowling, N. E. and Begley, J. A. in *Mechanics of Crack Growth, ASTM STP 590*, American Society of Testing and Materials, 1976, pp. 82-103.
[9] Rice, J. R., Paris, P. C. and Merkle, J. G. in *Progress in Flaw Growth and Fracture Toughness Testing, ASTM STP 536*, American Society of Testing and Materials, 1973, pp. 231-245.
[10] Dowling, N. E. in *Cracks and Fracture, ASTM STP 601*, American Society of Testing and Materials, 1976, pp. 19-32.
[11] Dowling, N. E. in *Cyclic Stress-Strain and Plastic Deformation Aspects of Fatigue Crack Growth, ASTM STP 637*, American Society for Testing and Materials, 1977, pp. 97-121.
[12] James, L. A. in *Atomic Energy Review*, Vol. 14, No. 1, 1976, pp. 37-86.
[13] James, L. A., "HEDL Magnetic Fusion Energy Programs Progress Report—January-March 1977," Westinghouse Hanford Co., Energy Research and Development Administration Contract EY-76-C-14-2170.

D. F. Mowbray[1]

Use of a Compact-Type Strip Specimen for Fatigue Crack Growth Rate Testing in the High-Rate Regime

REFERENCE: Mowbray, D. F., "**Use of a Compact-Type Strip Specimen for Fatigue Crack Growth Rate Testing in the High-Rate Regime,**" *Elastic-Plastic Fracture, ASTM STP 668*, J. D. Landes, J. A. Begley, and G. A. Clarke, Eds., American Society for Testing and Materials, 1979, pp. 736–752.

ABSTRACT: Fatigue crack growth in chromium-molybdenum-vanadium steel was studied in the high-rate regime of 2.5×10^{-4} to 2.5×10^{-1} mm/cycle with a compact-type strip specimen. The specimen was found to give rise to constant growth rates over large increments of crack length when cycled under simple load control. Use of the crack opening load ranges and an approximate J-integral analysis showed that the growth was occurring under essentially constant ΔJ.

KEY WORDS: fatigue crack growth, fracture mechanics, J-integral, crack propagation

Nomenclature

a Crack length
a_{eff} Effective crack length, $a + r_y$
B Specimen thickness (gross section)
B_n Strip thickness at minimum section
C_1, γ Constants in Dowling-Begley crack growth relationship
e, p Subscripts indicating elastic and plastic parts
E, ν Elastic constants
G Strain energy release rate
J Path-independent integral
K Stress intensity factor
N Cycles

[1] Manager, Mechanics of Materials Unit, Materials and Processes Laboratory, General Electric Company, Schenectady, New York 12345.

P Load
P_L Limit load
r_y Plastic zone size, $(K/2\sigma_y)^2$
R Ratio of minimum to maximum load in a fatigue cycle
U Potential energy
W Specimen width (gross section)
W_G Strip width
W' Effective specimen width, W^2/W_G
δ Load point deflection
Δ Indicates the range of a variable
σ_y Yield stress

Fatigue crack growth at high rates[2] is of current interest because of its role in the low-cycle fatigue damage process. This process consists mainly of the propagation of small cracks, the growth of which occurs at high rates due to the highly (plastically) strained surrounding material. It is commonly reported that as much as 99 percent of the measured life in low-cycle fatigue specimens involves propagation.

The foregoing observations have prompted a limited amount of work on characterizing growth at high rates [1-5],[3] as well as several models which relate crack growth to low-cycle fatigue [6-9]. The models appear to demonstrate a good correspondence between crack growth and low-cycle fatigue relationships. It is suggested that their validity and usefulness can be furthered by the acquisition and characterization of more crack growth rate data.

The number of investigations of crack growth in the high-rate regime has remained, for two prominent reasons, quite small. One reason is connected with difficulties in testing with small-size test specimens and standard control procedures. Unstable crack advance commonly results as such tests proceed into the high-growth-rate regime, thus precluding the possibility of obtaining any useful data. The second reason centers on the apparent absence of a suitable controlling variable with which to correlate data between tests and specimen types. Most crack growth investigations in recent years have been based on the stress intensity factor as the controlling variable. The stress intensity factor is defined for linear elastic material only, and loses meaning as a controlling variable when the crack-tip plastic zone becomes a significant fraction of the crack length. This is normally the case in the high-growth-rate regime with small-size laboratory specimens. Hence, most investigations are terminated at rates less than 10^{-3} mm/cycle.

---

[2] The range of crack growth rates in the high-rate regime is considered in this paper as 10^{-4} to 10^{-1} mm/cycle.

[3] The italic numbers in brackets refer to the list of references appended to this paper.

A nonlinear fracture mechanics approach was recently explored by Dowling and Begley [3] with apparent success. They obtained crack growth rate data for A533-B steel in the high-growth-rate regime with deep-notched compact specimens. In order to keep the growth rates stable, they utilized an analog computer to decrease the load-point deflection in a prescribed fashion as the crack length increased. The test results were successfully correlated with an operational definition of the J-integral, which considered that only the loading during crack face opening results in damage. The crack closure point during a cycle was detected by noting the point in the load-deflection curve where the unloading slope changed curvature. The load range corresponding to the crack closure point and maximum load was used to compute ΔJ.

Dowling has subsequently added confirmation to the correlation with tests on deep-notched center-cracked specimens [4] and smooth bar low-cycle fatigue specimens in which the growth of small surface cracks was charted [5]. The results from the first two investigations are shown in Fig. 1. In the high-growth-rate regime, they fit a power type relationship of the form

$$\frac{da}{dN} = C_1 \Delta J^\gamma \tag{1}$$

where da/dN is the crack growth rate and C_1 and γ are material constants.

Considering that the $\Delta J - da/dN$ curve is independent of geometry, the approach of Dowling and Begley should have general applicability. This approach also reduces to, and extends, the linear elastic fracture mechanics approach because of the relationship of J to K.[4] Data obtained from ΔJ-testing overlap and extend to higher growth rates data obtained in the nominally elastic range and correlated with ΔK. Figure 1 illustrates this result for the A533B steel.

There is one prominent objection to, and one practical difficulty in, applying the J-integral to fatigue crack growth. The objection centers about the mathematical definition of J. In the strict mathematical sense, it is valid only within the confines of deformation plasticity theory [10], which excludes consideration of unloading. Dowling and Begley [3] approached this objection on the basis that J may have more applicability than the current mathematical definition indicates. Theirs is an operational definition of J implying that the stress and strain fields near the crack tip during the loading half of a fatigue cycle are defined by J, despite the intermittent unloading.

The difficulty with practical applications is in the determination of $J - a$ relationships. There are only a limited number of configurations for which

---

[4] $J = K^2/\alpha E$, where $\alpha = 1$ for plane stress and $1 - \nu^2$ for plane strain.

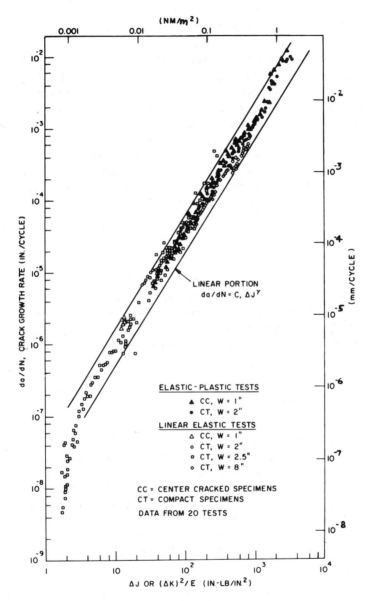

FIG. 1—*Correlation of A533B steel crack growth rate data (from Ref 4).*

J is known or can be directly measured. Most deep-notch configurations allow J to be determined during test by measuring the area enclosed by the load-displacement curve. Most other configurations require, at present, a complex set of experiments or elastic-plastic analysis to determine the $J - a$ relationship. When examining this approach, these difficulties must be balanced against the fact that any approach involving nonlinear material behavior will have similar difficulties.

The present work was undertaken to investigate further the possibility of utilizing the J-integral to correlate crack growth rate data. A somewhat different approach was taken in the testing. A specimen was sought which would allow stable growth rates to be obtained while utilizing simple control procedures. It was found that a compact-type strip specimen first used by McHenry and Irwin [11] possessed these characteristics. A slightly modified version of their specimen design gave rise to constant growth rates over large increments of crack length when cycled under simple load control. No simple J-integral solution could be evolved for this configuration, however, and it was necessary to employ estimation procedures to compute ΔJ.

Test Program

Material

The test material was chromium-molybdenum-vanadium (Cr-Mo-V) steel forging. The chemical composition of the steel is given in Table 1. Table 2 gives the standard tension test properties and cyclic stress-strain curve properties. The latter were determined by means of the incremental step test [12].

Specimen

The modified strip specimen is shown in Fig. 2. It is of the compact type, with deep side grooves part-way across its width. The full-thickness section remaining beyond the grooves provides the stiffness for crack growth rates to remain stable under constant-loading cycling. The specimen differs from that of McHenry and Irwin in that it is 25.4 mm wider and has machined-in knife edges on the load line. The larger width gives the specimen a useful crack length range of 50 mm. The knife edges allow for the measurement of load-line displacement, as required in J-computations.

A stress intensity factor solution for the specimen was determined by the

TABLE 1—*Chemical composition*.

Composition, Weight %				
C	Mn	P, max	S, max	Si
0.25 to 0.35	0.7 to 1.0	0.025	0.025	0.15 to 0.35
Ni, max	Cr	Mo	V	Fe
0.5	0.85 to 1.25	1.0 to 1.5	0.2 to 0.3	balance

TABLE 2—*Mechanical properties*.

	Tension Test
0.2% yield strength, MN/m^2	676
0.02% yield strength, MN/m^2	634
Ultimate tensile strength, MN/m^2	820
Elongation in 50.8 mm, %	22
Reduction in area, %	66
Cyclic Stress-Strain[a]	
Cyclic strain hardening exponent, (n')	0.144
Cyclic strength coefficient (K'), MN/m^2	1207
0.2% offset stress, MN/m^2	503

$$a \frac{\Delta \sigma}{2} = K' \left(\frac{\Delta \varepsilon_p}{2} \right)^{n'}$$

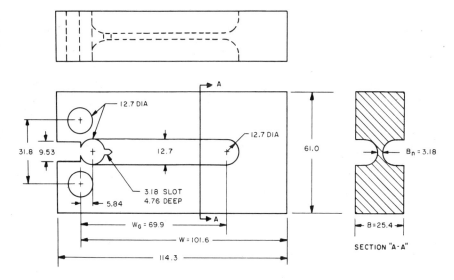

FIG. 2—*Compact-type strip specimen*.

experimental compliance method. A description of the experimental technique utilized is given in Ref *13*. The following polynomial form describes the result obtained (refer to Fig. 2)

$$K = \frac{P}{\sqrt{BB_n W'}} \left[9.48 + 28.98 \left(\frac{a}{W'} \right) + 29.44 \left(\frac{a}{W'} \right)^2 \right] \quad (2)$$

where W' is an effective width defined after McHenry and Irwin [11] as

$$W' = \frac{W^2}{W_G} \quad (3)$$

It was noted by McHenry and Irwin that the K-solution for the strip specimen could be expressed to within a few percent accuracy by

$$K = \frac{CP\sqrt{a}}{\sqrt{BB_n}\,W'} \quad (4)$$

where C is a constant. This is also true for the present design, with $C = 38$. The accuracy is ± 4 percent for a/W' in the range of 0.1 to 0.4.

Test Procedure

The specimens were tested in a servo-hydraulic testing system under load control. All testing was at room temperature. The cyclic frequency was varied in each test from ~10 to 0.01 Hz, depending upon the current crack growth rate. Crack lengths were monitored with a telescopic filar gage having a 0.125-mm division scale in the field of view. Displacement across the knife edges was measured with a clip gage. Loops of load versus displacement were recorded periodically.

The program included crack growth rate tests on four specimens. Each specimen was tested at a series of constant load ranges for $R = 0.1$. Generally, four load ranges per specimen were attemped with each successive range increased above the previous one. The cracks were propagated ~10 mm at each load range.

Test Results

Example crack growth data from two of the tests are plotted in Figs. 3 and 4 on linear coordinates. Each plot represents the results from one specimen in which four successively higher load ranges were imposed. (Note that the data for each load range are defined by different scales on the abscissa.)

Figures 3 and 4 indicate a unique crack growth rate response for the specimen in that the data define linear curves or constant growth rates. This response is apparently insensitive to the absolute load levels and crack length range of ~40 mm. Different starting load levels were used for the four specimens so that in general there were differing load-range/crack-length combinations. The only departure from the linear trend appears in the initial 1/2 mm of crack propagation at a new load level.

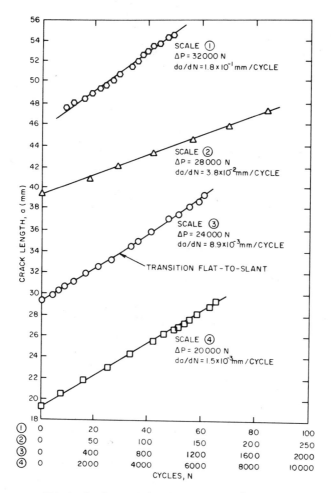

FIG. 3—*Crack growth data for specimen 4 (Cr-Mo-V steel)*.

Note that at one of the load ranges in each test the slope of the linear line defined by the data does change slightly. This appears to always coincide with the transition to fully slant fracture. The growth rate increases approximately 10 percent when this takes place.

Fatigue crack growth rates are summarized in Table 3. They were determined by fitting linear curves to the crack length versus cycles data and differentiating the analytic expression describing the fitted curve. Least-squares regression analysis was used in the fitting. With the initial one or two data points at each load range omitted, deviations from a straight-line relationship were less than 2.0 percent.

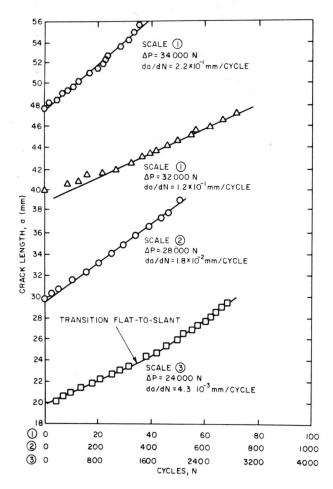

FIG. 4—*Crack growth data for specimen 5 (Cr-Mo-V steel).*

Analysis of Data

An analysis of the data was based on the previously discussed operational definition of ΔJ. In lieu of a simple J-solution for the strip specimen, values of ΔJ were calculated using an estimation procedure developed by Bucci et al [14]. These authors demonstrated the procedure quite accurately by comparing calculated and experimental results for different specimen geometries. Others [15,16] have subsequently shown favorable results with the same approach.

TABLE 3—*Summary of crack growth rate data.*

Specimen No.	Load Range, N	a (Range), mm	da/dN, mm/cycle	ΔJ, MJ/m²	$\Delta K^2/E$, MJ/m²
1[a]	7100	18 to 23	6.1×10^{-5}	0.0039	0.0039
2	10900	18 to 26	1.6×10^{-4}	0.0096	0.0096
	11300	26 to 36	2.1	0.010	0.010
	11800	36 to 46	4.8	0.014	0.014
	12500	46 to 56	1.3×10^{-3}	0.028 to 0.030	0.027 to 0.028
	13100	56 to 66	1.6	0.038 to 0.040	0.033 to 0.035
3	17800	20 to 26	9.9×10^{-4}	0.021	0.021
	20000	26 to 31	2.0×10^{-3}	0.032 to 0.033	0.027 to 0.029
	22000	31 to 36	3.5	0.048 to 0.050	0.040 to 0.042
	24000	36 to 41	8.1	0.080 to 0.086	0.066 to 0.071
	26000	41 to 46	2.2×10^{-2}	0.116 to 0.119	0.094 to 0.097
	30000	46 to 53	1.2×10^{-1}	0.213 to 0.219	0.156 to 0.160
	34000	53 to 58	3.0	0.319 to 0.332	0.221 to 0.230
4	20000	20 to 31	1.5×10^{-3}	0.032 to 0.036	0.027 to 0.030
	24000	31 to 41	8.9	0.078 to 0.084	0.064 to 0.070
	28000	41 to 48	3.8×10^{-2}	0.171 to 0.176	0.129 to 0.133
	32000	48 to 58	1.8×10^{-1}	0.269 to 0.279	0.186 to 0.193
5	24000	20 to 31	4.3×10^{-3}	0.049 to 0.053	0.042 to 0.045
	28000	31 to 41	1.8×10^{-2}	0.108 to 0.114	0.087 to 0.091
	32000	41 to 48	1.2×10^{-1}	0.198 to 0.205	0.147 to 0.153
	34000	48 to 58	2.2	0.289 to 0.310	0.201 to 0.216

[a] Used for experimental compliance analysis.

The estimation procedure is based on the definition of J in terms of potential energy, stated as

$$J = -\frac{1}{B}\left(\frac{\partial U}{\partial a}\right)_\delta \quad (5)$$

where U is the potential energy at load-point displacement δ (or area under the load/load-point displacement curve). J and U are partitioned into elastic and plastic components, such that

$$J = J_e + J_p \quad (6)$$

$$J_e = -\frac{1}{B}\left(\frac{\partial U_e}{\partial a}\right)_\delta \quad (7)$$

$$J_p = -\frac{1}{B}\left(\frac{\partial U_p}{\partial a}\right)_\delta \quad (8)$$

where the subscripts e and p designate elastic and plastic, respectively.

J_e is equivalent to the elastic strain energy release rate, G, and this in turn to (K^2/E) for plane stress. Bucci et al found it most accurate to compute G as a function of the plastic zone corrected crack length; that is

$$J_e = G(a_{\text{eff}}) = \frac{K(a_{\text{eff}})^2}{E} \tag{9}$$

$$a_{\text{eff}} = a + r_y \tag{10}$$

where r_y is the plastic zone length for plane stress

$$r_y = \frac{1}{2\pi}\left(\frac{K}{2\sigma_y}\right)^2 \tag{11}$$

where σ_y is the material yield stress. In this application, the conventional yield stress is multiplied by a factor of 2 to account for cyclic plastic action [17].

Computations of J_e involve two steps. First, K and r_y are computed based on a, and then K is recomputed based on a_{eff}. Because of the cyclic problem, σ_y was taken as the 0.2 percent offset stress on the cyclic stress-strain curve.

U_p is estimated as the product of the limit load, P_L, and the plastic load-line displacement, δ_p, such that

$$J_p = -\frac{1}{B}\left[\frac{\partial(P_L\delta_p)}{\partial a}\right]_\delta = -\frac{\delta_p}{BW}\left(\frac{\partial P_L}{\partial a/W}\right)_\delta \tag{12}$$

Computations of J_p were made utilizing plastic displacements from the recorded load-displacement loops, and a lower-bound limit-load solution for a compact specimen. The latter, developed by Merkle and Corten [18], is expressed as

$$P_L = 1.26\sigma_y BW(1 - a/W)\alpha \tag{13}$$

where the 1.26 multiplier has been added to account for plastic constraint via the Green and Hundy [19] solution for bend specimens, and α is defined by

$$\alpha = \left[\left(\frac{2\,a/W}{1-a/W}\right)^2 + 4\left(\frac{a/W}{1-a/W}\right) + 2\right]^{1/2} - \left(\frac{2\,a/W}{1-a/W} + 1\right) \tag{14}$$

Differentiation of Eq 13 yields the result required for evaluation in Eq 12. The derived expression was applied to the strip specimen assuming it had constant cross-sectional dimensions defined by the equivalent quantities, $\sqrt{BB_n}$ and W^2/W_G.

Values of ΔJ were calculated for each load range in increments of crack length of 1.25 mm. The crack opening load ranges were established by locating the inflection points on the unloading line of the load-displacement records. Figure 5 shows some typical records with the estimated inflection points indicated by arrows. Location of the inflection points was rather subjective with this specimen and material because of the small percentage of plastic deflection which develops. (The heavy ligament at the back face tends to prevent yielding on a gross scale.) The trend in behavior is clear,

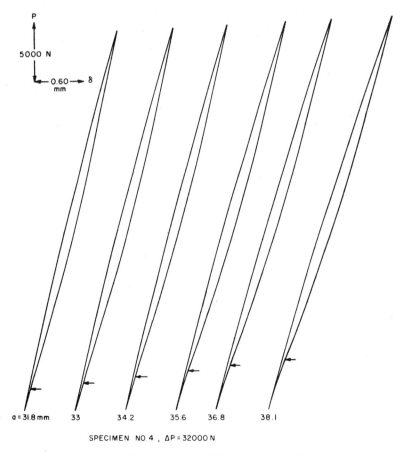

FIG. 5—*Example load-displacement records.*

however, in that the crack closure load increases with increasing crack length.

The values of ΔJ calculated in the fashion described in the foregoing are listed in Table 3. A range of values is given corresponding to each increment of crack length at which the applied load was maintained constant. The range of ΔJ indicates how much variation was calculated for the indicated crack length range. Examination shows that all values were within ±4 percent of the median value. Hence, there appears a near constancy in ΔJ, deriving from a balancing of the increasing crack length with a decreasing crack opening load range. Shown in the adjoining column of Table 3 are the linear elastic fracture mechanics based values, $\Delta K^2/E$. The difference between the ΔJ and $\Delta K^2/E$ values indicates the extent of the nonlinear correction. These differences vary from zero at growth rates of 2.5×10^{-3} mm/cycle to ~40 percent at 2.5×10^{-1} mm/cycle.

Discussion

From the testing point of view, the strip specimen provides an excellent means for generating high growth rate data. It is not limited by any type of instability, and can be utilized to obtain very high growth rates with simple load control. Although rates as high as 3.0×10^{-1} mm/cycle were obtained in this investigation, higher rates could apparently have been achieved without concern for ratchetting.

Of further significance with regard to the specimen performance is the constant growth rate obtained under simple load control. This is a desirable feature in any crack growth rate test because it (1) limits scatter, and (2) means the parameters controlling the crack growth process are being kept constant in the test. In this case, it appears that ΔJ for crack surface opening was being maintained constant. This is stated with caution because ΔJ was not directly measurable and there was an element of subjectivity in selecting the crack closure load points. Further analysis of the specimen and tests on other materials are needed to confirm the potential constant ΔJ feature. In any event, what has been observed lends support for ΔJ as the controlling variable.

A plot of the test results is shown in Fig. 6 on logarithmic coordinates of da/dN versus ΔJ (or $\Delta K^2/E$ in the nominally elastic range). The resulting correlation is very good, showing no dependence on load-range crack-length combination. Also, the scatter is minimal considering the subjective nature of the load range interpretation, and the fact that the customary independent variable (ΔK) is squared in this representation. Further evidence of the generality of the correlation is shown in Fig. 7, where data obtained for the same material with two other specimen types (single-edge notch and standard compact specimens) are plotted together with strip specimen data. Most of the auxiliary data lie in the applicable range for linear elastic

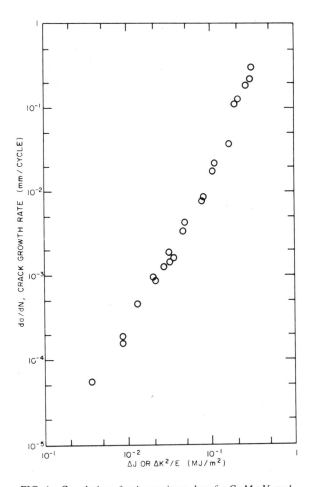

FIG. 6—*Correlation of strip specimen data for Cr-Mo-V steel.*

fracture mechanics, and are correlated on the basis of $\Delta K^2/E$. However, in the growth rate range of 2.5×10^{-3} to 2.5×10^{-2} mm/cycle, the single-edge notch specimen was in the plastic range. Values of ΔJ were determined by the same estimation procedure applied to the strip specimen.[5]

The combined set of data (Fig. 7) covering the growth rate range 5.0×10^{-6} to 5.0×10^{-2} mm/cycle appears to fit in a reasonably narrow scatter band of two parallel lines similar to that observed by Dowling for

---

[5] Use of the SEN specimen to even higher growth rates was not pursued because of buckling problems.

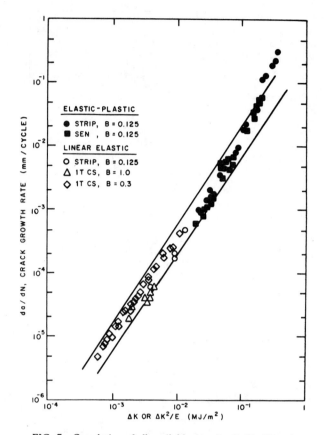

FIG. 7—*Correlation of all available data for Cr-Mo-V steel.*

A533-B steel (see Fig. 1). A least-squares fit of the data in this growth rate range yields the following relationship

$$\frac{da}{dN} = 1.68 \, (\Delta J)^{1.55} \tag{15}$$

For crack growth rates beyond 5.0×10^{-2} mm/cycle the data define an upward swing toward unstable behavior. This is unlike that for A533-B data (Fig. 1), which exhibit a straight-line relationship to rates of at least 2.5×10^{-1} mm/cycle. The difference is reasonable, however, since the Cr-Mo-V steel possesses considerably less toughness: $J_{Ic} \sim 0.01$ MJ/m^2 versus 0.175 for A533-B.

It is noted that the foregoing observations concerning ΔJ as the controlling variable rest in part on the accuracy of the estimation procedure as employed herein. The accuracy has not been checked very well in this investigation. One partial check was obtained by showing that some auxillary data in the elastic range blend fairly well with data in the high-growth-rate regime. A more complete check could be provided by comparing with data obtained on specimens in which J is directly measurable. If the estimation procedure proves an accurate means for determining ΔJ, it will be of considerable practical value in applications involving high strain loadings.

As a final item, it is suggested that the estimation procedure employed for computing ΔJ could be used in a broader sense to correct existing data which are expressed in terms of ΔK, but which are actually beyond the range of validity of linear elastic fracture mechanics. That is, data from an individual test are oftentimes characterized in terms of ΔK even though part of or all of the test was conducted at net-section stress levels too high for the stress intensity factor to remain valid as a crack-tip stress field parameter. For these cases, the proposed estimation procedure could be used to correct for the nonlinear material effect, thus extending the range of validity of the test results to higher growth rates.

Summary

A fatigue crack growth study was carried out in the high-growth-rate regime. A compact-type strip specimen was employed in the testing. This specimen was found to have some excellent qualities for testing in the high-rate regime. It exhibited extremely stable behavior and gave rise to constant crack growth rates under simple load control. The data were analyzed in terms of the J-integral by applying an approximate calculational procedure. The resulting correlation of data tend to support the Dowling and Begley hypothesis that the crack growth rate is controlled by the range of J operative in opening the crack surfaces.

References

[1] Solomon, H. D., *Journal of Materials*, Vol. 7, 1972, p. 299.
[2] Gowda, C. V. B. and Topper, T. H., in *Cyclic Stress-Strain Behavior-Analysis, Experimentation and Fracture Predictions, ASTM STP 519*, 1973, p. 170.
[3] Dowling, N. E. and Begley, J. A. in *Mechanics of Crack Growth, ASTM STP 590*, American Society for Testing and Materials, 1976, p. 83.
[4] Dowling, N. E. in *Cracks and Fracture, ASTM STP 601*, American Society for Testing and Materials, 1976, p. 19.
[5] Dowling, N. E. in *Cyclic Stress-Strain and Plastic Deformation Aspects of Fatigue Crack Growth, ASTM STP 637*, 1977, pp. 97-121.
[6] Mowbray, D. F. in *Cracks and Fracture, ASTM STP 601*, American Society for Testing and Materials, 1976, p. 33.
[7] Tompkins, B., *Philosophical Magazine*, Vol. 18, 1968, p. 1041.
[8] Boettner, R. C., Laird, C., and McEvily, A. J., *Transactions*, Metallurgical Society of the American Institute of Mining Engineers, Vol. 233, 1965, p. 379.
[9] McEvily, A. J., Beukelman, D., and Tanaka, K. in *Proceedings*, International Conference on Mechanical Behavior of Materials, Japan, 1972, p. 269.

[10] Rice, J. R., *Transactions*, American Society of Mechanical Engineers, *Journal of Applied Mechanics*, Vol. 35, June 1968, p. 379.
[11] McHenry, H. I. and Irwin, G. R., *Journal of Materials*, Vol. 7, 1972, p. 455.
[12] Landgraf, R. W., Morrow, J., and Endo, T., *Journal of Materials*, Vol. 4, 1969, p. 176.
[13] LeFort, P. and Mowbray, D. F., *Journal of Testing and Evaluation*, Vol. 6, No. 2, March 1978.
[14] Bucci, R. J., et al in *Fracture Toughness, Proceedings of the 1971 National Symposium on Fracture Mechanics, Part II, ASTM STP 514,* American Society for Testing and Materials, 1972, p. 40.
[15] Shih, C. F. and Hutchinson, J. W., "Fully Plastic Solutions and Large Scale Yielding Estimates for Plane Stress Crack Problems," Harvard University Report DEAP S-14, 1975; to appear in *Transactions,* American Society of Mechanical Engineers, *Journal of Applied Mechanics.*
[16] Sumpter, J. D. G. and Turner, C. E. in *Cracks and Fracture, ASTM STP 601,* 1976, p. 3.
[17] Paris, P. C. in *Fatigue—An Interdisciplinary Approach,* Burke, Reed, and Weiss, Eds., Syracuse University Press, Syracuse N.Y., 1963, p. 107.
[18] Merkle, J. G. and Corten, H. T., *Transactions,* American Society of Mechanical Engineers, *Journal of Pressure Vessel Technology,* Nov. 1974, p. 286.
[19] Green, A. P. and Hundy, B. B., *Journal of the Mechanical and Physics of Solids,* Vol. 4, 1956, p. 128.

Summary

Summary

The papers in this publication can be divided into three major sections: (1) the presentation and analytical evaluation of elastic-plastic fracture criteria; (2) experimental evaluation, including both the toughness evaluation of materials in the elastic-plastic regime and the evaluation of various fracture criteria and characterizing parameters; and (3) application of elastic-plastic methodology to the evaluation of structural components, including the application to fatigue crack growth analysis.

These papers demonstrate that the elastic-plastic fracture field is in a stage of rapid development. New approaches and parameters are emerging and no single approach has been adopted by all of the workers in this field. However, the field has reached a state of development where certain trends can be identified. Ductile fracture characterization, which in the past had largely been based on criteria taken at the point of initiation of stable crack growth, has been extended so that stable crack growth and ductile instability are analyzed. Fracture-characterizing parameters, which are mainly divided into field parameters and local crack-tip parameters, were once viewed as presenting opposing approaches but are now generally regarded as having a common basis. Detailed aspects of developing elastic-plastic fracture techniques, such as establishing limitations on the use of criteria and on test specimen type and size, are now being actively pursued, implying that a level of confidence in the underlying concepts has been reached. Further evidence of this confidence is demonstrated by attempts to apply the methodology to structural analysis and phenomena other than fracture toughness.

Results from individual papers are summarized in the following sections.

Elastic-Plastic Fracture Criteria and Analysis

The papers in this section are concerned with the development of criteria and parameters to characterize elastic-plastic fracture. In many papers, finite-element analysis is applied to determine how well these parameters characterize the crack-tip stress and strain fields under conditions of large-scale plasticity. Much of the emphasis in these papers is on stable crack growth and instability characterizations of fracture.

Paris et al have proposed a method for characterizing fracture at the point of ductile instability. Characterization of stable crack growth is based on the J-integral where the slope of the J versus crack growth resistance curve is given by a nondimensionalized parameter called the tearing modulus, T. The instability condition is formulated in a manner similar to the linear elastic fracture mechanics (LEFM) R-curve approach. When an applied mechanical crack drive of a specimen or structure, labeled T_{applied}, is equal to or greater than the material resistance to crack advance, labeled T_{material} ($T_{\text{applied}} \geq T_{\text{material}}$), tearing instability ensues. The instability condition was formulated for a large number of geometries and loading configurations in this paper. The result is a simple methodology for using laboratory tests to evaluate the tearing instability condition for many types of structures. Some initial experimental verification of this method is given in a second paper by Paris et al in this publication; however, much more remains to be done. This proposed method suggests many areas for future research both in analytical developments and in experimental verification and material property evaluation.

A paper by Hutchinson and Paris provides some rationale for the method proposed in the previous paper by taking a theoretical approach to evaluate the use of the J-integral for characterizing stable crack growth. A criteria for J-controlled crack growth is formulated by determining a region of proportional loading ahead of an advancing crack. The criterion is formulated in terms of a nondimensional size parameter, ω, which is a ratio of the size of the proportional loading region to the uncracked ligament. The condition for J-controlled and growth is $\omega \gg 1$.

Shih et al evaluated five parameters for characterizing stable crack initiation and growth, using nine criteria for the evaluation. The two chosen as the most viable were the J-integral and crack-tip opening displacement, δ. Finite-element investigations show that both parameters characterize the near-tip deformation. For stable crack growth, nondimensional parameters were developed similar to those of Paris et al and labeled T_J for crack growth characterized by J and T_δ for crack growth characterized by a crack opening angle. A J-characterization of the crack growth was determined to be valid up to a crack extension equal to 6 percent of the remaining ligament; however, crack opening angle remained constant over a much larger range of crack extension, suggesting that T_δ would be preferred over T_J. Complete ductile fracture characterization is given by a two-parameter approach, either J_{Ic} and T_J or δ_c and T_δ, which characterizes both the initiation and the growth of a stable crack. The analysis suggested the use of side grooves for experimental evaluations to ensure uniform flat fracture.

Kanninen et al used a finite-element approach to evaluate eight parameters for stable crack growth and instability. They also set nine requirements for choosing the appropriate parameter to characterize stable crack growth. From the parameters evaluated, four were found to vary with crack extension while four did not. The four that did not vary were: (1) crack tip opening angle, (2) work involved in separating crack faces per unit area of crack growth, (3) generalized energy release rate based on a computational process zone, and (4) critical crack-tip force for stable crack growth. These four parameters were judged to be more suitable for stable crack growth and instability characterization. The concept of a J-increasing R-curve was viewed as being fundamentally incorrect because the crack-tip toughness does not increase with an advancing crack.

Sorensen used finite-element techniques to study plane-strain crack advance under small-scale yielding conditions in both elastic-perfectly plastic and power hardening materials. The stress distribution ahead of a growing crack was found to be nearly the same as that ahead of a stationary crack; however, strains are lower for the growing crack. When loads are increased at fixed crack length, the increment in crack-tip opening is uniquely related to the increment in J; when an increment of crack advance is taken at constant load, the incremental crack tip opening is related logarithmically to J. When separation energy rates are calculated for large crack growth steps, the use of J as a correlator is sensitive to strain hardening properties and details of external loading.

McMeeking and Parks used finite-element techniques to study specimen size limitations for J-based dominance of the crack-tip region. They analyzed deeply-cracked center-notched tension and single-edge notched bend specimens using both nonhardening and power loading laws where deformation was taken from small-scale yielding to the fully plastic range. The criterion used to judge the degree of dominance was the agreement between stress and strain for the plastically blunted crack tip with those for small-scale yielding. They found good agreement for the bend specimen when all specimen dimensions were larger than 25 J/σ_o, where σ_o is the tensile yield. This size limitation is equivalent to one proposed for J_{Ic} testing. The center-notched tension specimen, however, would require specimen dimensions about eight times larger (200 J/σ_o), although loss of dominance is gradual and this requirement is somewhat arbitrary.

Nakagaki et al studied stable crack growth in ductile materials using a two-dimensional finite-element analysis. They looked at three parameters: (1) the energy release to the crack tip per unit crack growth, using a global energy balance; (2) the energy release to a finite near-tip "process zone" per unit of crack growth; and (3) crack opening angle. Their work confirmed numerically an earlier observation by Rice that the crack-tip energy release rate approaches zero as the increment of crack advance approaches zero for perfectly plastic material. From these present results, they are not ready to propose an instability criterion. However, they cannot base such a criterion on the magnitudes of an energy release parameter since these depend on the magnitudes of the growth step; therefore, a generalized Griffith's approach cannot be used for ductile instability.

Miller and Kfouri presented results from a finite-element analysis of a center-cracked plate under different biaxial stress states. Comparisons were made of: (1) crack-tip plastic zone size, (2) crack-tip plastic strain intensity and major principal stresses, (3) crack opening displacements, (4) J-integral, and (5) crack separation energy rates. They found that, for biaxial loading, brittle crack propagation can be best correlated with plastic zone size. Crack-tip plastic strain intensity is more relevant to initiation while crack opening displacement is more relevant to crack propagation. Stable crack propagation was not uniquely related to J.

D'Escatha and Devaux used elastic-plastic finite-element computations to evaluate a fracture model based on a three-stage approach—void nucleation, void growth, and coalescence. The purpose of this model is to predict the fracture properties of a material represented as the initiation of cracking, stable crack growth, and maximum load. The problem in a fracture model is to use two-dimensional analysis to predict fracture for a more realistic three-dimensional crack problem. Various parameters used to correlate stable crack growth were evaluated by this model, including crack opening angle, J-integral, and crack-tip nodal force. The next step will be an experimental evaluation of the present results.

The papers in this section were mainly concerned with the presentation and analytical evaluation of ductile fracture criteria. A common theme is that fracture evaluation should include more than simply the initiation of stable crack growth; stable crack growth characterization and ductile instability prediction must also be included. While there is no agreement as to which parameter should be used, the types of parameters are mainly field-type or crack-tip parameters. Field-type parameters such as the J-integral have a lot of appeal and are shown to be useful for correlating stable crack growth under a restricted set of conditions. A crack-tip parameter such as crack opening angle has fewer restrictions and has more general support for correlating stable crack growth. The results presented here suggest many areas for future study. More analysis is needed to determine the best single approach to ductile fracture characterization. The approaches presented must be evaluated with critical experimental studies. The optimum approach must lend itself to relatively simple evaluation of material properties and must be easily applicable to the evaluation of structural components. This approach may include one or a combination of methods suggested here or may be one that is developed in future studies of ductile fracture criteria.

Experimental Test Techniques and Fracture Toughness Data

The papers in this section deal with experimental evaluation of elastic-plastic techniques and fracture toughness determination for several materials. A number of papers deal with various aspects of the analysis used to determine the J-integral from the experimental load versus load-line displacement records for various specimen types. A critical evaluation of the present analysis techniques along with proposed new techniques for elastic-plastic specimen analysis are presented in this section. Also included are a number of papers describing the results of elastic-plastic fracture toughness testing using both J-integral and crack opening displacement (COD) techniques.

The paper by Paris et al outlined the test procedure and results used to verify the tearing instability model described in the previous section. An experimental technique with a variable-stiffness testing system was used by Paris et al to vary the applied tearing modulus, $T_{applied}$, for each test. The value of $T_{applied}$ at the point of ductile instability was determined by continuously increasing the value of $T_{applied}$ until instability occurred. This value of $T_{applied}$ was then compared with the value of the material tearing modulus, $T_{material}$, determined from the slope of the J versus crack extension curve developed for the material of interest. The results of the tests on single-edge notched bend specimens showed extremely good agreement between the predicted value of instability and the actual experimentally determined instability for the material tested. It was emphasized by the authors that, as the applied tearing modulus is a function of the compliance of the overall system, the ductile instability phenomenon is very much dependent on the overall stiffness of the testing system or the structure under consideration. Future research in this area was discussed by the authors and consisted of testing a wider range of specimen types and variable geometries of a given generic specimen type.

Landes et al evaluated the approximation techniques used to calculate the value of J from the area under the load displacement curves for the most commonly used test specimens. This was accomplished by testing compact, three-point bend, and center-crack tension specimens each with blunt notches of various lengths. The values of J determined from the energy rate definition of the J-integral were compared with the various area methods of approximating J to evaluate the accuracy of the various approximation techniques. It was found that a correction factor for the tension component in a compact specimen was necessary. A modified Merkle-Corten correction factor was proposed for both simplicity and accuracy when calculating the value of J for a compact specimen. The three-point bend approximation was found to be accurate if the total energy aplied to the specimen in the approximation formula is used. The value of J calculated from the approximation formula for the center-cracked panel was also found to be quite accurate when compared with the value of J calculated by the energy rate definition.

McCabe and Landes proposed the use of an effective crack length to calculate the resistance to crack growth by the K_R technique. It was found that by using a secant method to calculate the effective crack length, the value of J at any point on the load displacement curve could effectively be calculated by using the relationship between K and J. A comparison of the results from this technique with the values of J calculated from the energy rate definition of J and the value of J calculated from the Ramberg-Osgood approximation of the load displacement curves was presented. The results from the secant method showed that this technique is a very good approximation to the value of J calculated from the energy rate definition.

In the next two papers by Dawes and Royer et al, the effect of specimen thickness on the critical value of J was noted. Dawes presented data showing that the critical values of both COD and J can be affected by section thickness and that therefore care should be taken to match or overmatch the plastic constraint in the test specimen to that of the structure. Dawes also proposed that the crack-tip COD should be defined as the displacement at the original crack-tip position. The data presented by Dawes show that is it possible to overestimate the value of K_{Ic} when using results from a J_{Ic} test on smaller specimens. While Royer et al also note a size effect for the three-point bend specimen, none was found for the compact specimen. It was pointed out that while the compact specimen showed no effect of size on the critical value of J, this result may possibly be fortuitous due to the type of material under investigation.

The paper by Milne and Chell discusses a proposed mechanism which may account for the specimen size effect on J_{Ic} found by them. For ferritic steels, a shift in the transition temperature due to increased triaxiality of larger specimens may well account for a size dependency on J_{Ic}. It is concluded by Milne and Chell that obtaining K_{Ic} from the small-specimen J_{Ic} test can possibly lead to nonconservative values of K_{Ic}. The mechanism attributed to this phenonmenon is one of a loss of through-thickness constraint which may cause crack-tip blunting during the test.

Using an analysis which employed the assumption of elastic-perfectly plastic behavior, Berger et al evaluated the various forms of the Merkle-Corten formula for the correction factors for the tension component in the compact specimen. A comparison was made of the equation which separates the elastic and plastic portions of the load displacement curve with various modified forms of the correction formula. It was found that the simplified form of the Merkle-Corten equation, which utilizes the total displacement as limits of integration, slightly overestimates the previously discussed form. The authors also proposed that a fixed displacement value should be used to determine the critical value of J rather than the intersection of the J versus Δa line and the theoretical blunting line.

Munz suggested in his paper that linear-elastic toughness testing need not be restricted to the size criteria defined in the ASTM Test for Plane-Strain Fracture Toughness of Metallic Materials (E 399-74). It was noted that the size dependence of K_Q as determined by the 5 percent secant offset method is due primarily to the crack-tip plasticity and the existence of a rising plane-strain crack growth resistance curve. A proposed variable secant method is presented which would allow specimens of up to six times smaller than the present size criterion permits to be tested for K_Q values.

Andrews and Shih presented a study on shear lip formation during testing and the effects of side grooving specimens. They noted that the shear lip dimensions found in the specimens were independent of the specimen dimensions. However, side grooving the specimens to a depth of 12½ percent of the thickness completely suppressed shear lip formation. The J versus Δa crack growth resistance curve was shown to be affected both by the thickness of the specimen and side grooving of the specimen. By measuring the crack-tip opening displacement using a linear variable differential transformer (LVDT) near the center of the specimen, Andrews and Shih showed that a crack growth resistance curve could be developed which is independent of specimen geometry and side grooving.

An interactive computerized J-integral test technique was described in the paper by Joyce and Gudas. By using a data acquisition system along with a computer, they showed excellent agreement between the values of J_{Ic} obtained from the unloading compliance technique and those obtained from the heat tinting method. The advantages of an interactive data reduction process occurring while the test is still in progress were discussed. This technique also allows for future reanalysis of the data by storing the data points on a magnetic tape system. The data from a test sequence on the computerized test technique showed that a nonconservative error in J_{Ic} could be obtained when using specimens with subsized remaining ligaments or specimens with insufficient thickness.

In the paper by Wilson an evaluation of a number of toughness testing methods to characterize various plate steels was made. The methods evaluated were Charpy V-notch (CVN), dynamic tear (DT), and J_{Ic}. The materials tested were A516, A533B, and HY130 manufactured by conventional steel-making techniques and also by a calcium-treated technique. A conventionally manufactured A543 material was also evaluated at the centerline and quarter-point positions of a plate. It was found that the J_{Ic} method of testing was far more sensitive to material quality than the other methods. It is postulated that this sensitivity may well be due to the acuity of the crack in the J_{Ic} specimen compared with the machined notch and the pressed notch of the CVN and the DT specimens, respectively. The results of these tests show that the J_{Ic} tests indicate a significant improvement in the toughness of the calcium-treated steels over the conventionally manufactured steels.

Nine pressure vessel materials were evaluated using static and dynamic initiation toughness results in the paper by Server. It was found from these test results that a nine-point average of the crack front gave a higher value of J_{Ic} than a three-point average of the crack front. The dynamic test values always gave greater slopes of the J versus Δa crack growth resistance curves and in many cases the dynamic values of J_{Ic} were higher than the corresponding static values.

Logsdon presented the results of a dynamic fracture toughness test on SA508 C1 2a material using elastic-plastic techniques. A temperature-versus-toughness curve at testing rates up to 4.4×10^4 MPa$\sqrt{m}$/s was developed using the K_{Id} procedure at low temperatures and J_{Id} at higher temperatures. The results of these tests show that this material is suitable for nuclear applications. It was also shown that the necessary deceleration of the J_{Id} multispecimen test, due to the speed of testing to prescribed displacement values, had no effect on the results of the J_{Id} test.

In the paper by Tobler and Reed a presentation of the techniques used to test an electroslag remelt Fe-21Cr material at cryogenic temperatures was made. The toughness values at 4, 76, and 295 K were found by using J_{Ic} techniques. It was noted from the tension test results that, once plastic deformation occurred, a slight martensitic transformation took place at room temperature; at 76 and 4 K, however, an extensive martensitic transformation took place. The toughness of this material was found to be adversely affected as the temperature was reduced from 295 to 4 K while the yield strength increased by a factor of 3.

The problems of testing high-ductility stainless steel were presented in a paper by Bamford and Bush. Tests were conducted on 304 forged and 316 cast stainless steel at both room temperature and 316°C. The authors pointed out that the present recommended size requirements for J_{Ic} may be too restrictive as no change was noted in the slope of the crack growth resistance curve when passing from the proposed valid region to the nonvalid region. An acoustic emission system was also used in order to detect the initiation of crack growth. While the acoustic emission test showed large increases in count rate during the test, there was no obvious means of detecting crack initiation. The extensive plasticity achieved during the test also obscured the crack initiation point as defined by an increase in the electric potential of an electric potential system used. The unloading compliance technique was found to work favorably on the compact specimen; however, difficulty was encountered when using the three-point bend specimen.

The papers in this section were concerned mainly with the evaluation of various elastic-plastic criteria using experimental methods. There were basically two areas of investigation in this section: (1) the evaluation of the actual criteria, and (2) the results of fracture toughness testing when using a particular criterion. While a number of papers show an effect of size on both COD and the J-integral, others do not. Various testing procedures are used to show these size effects, creating a future need for a common method of testing. This section also shows encouraging results in the development of an instability criterion for ductile fracture. Future work in these areas should of course be directed at both size effects on the various elastic-plastic criteria and on the development of a test technique which correctly describes initiation and stable crack growth resistance up to and including ductile instability. The papers presented in this section will aid future studies in elastic-plastic criteria and testing methods.

Applications of Elastic-Plastic Methodology

The use of elastic-plastic fracture methodology to analyze structural components marks its emergence from the status of being mainly a research technique to that of being a useful engineering tool. The papers in this section include generalized methods for applying elastic-plastic fracture methodology, specific applications to structural components, and the application of elastic-plastic parameters to fatigue crack growth-rate correlation.

Chell discussed methods for using a Failure Assessment Curve to make failure predictions for structures subjected to thermal, residual, or other secondary stresses where a failure collapse parameter is not definable. A procedure is introduced which transforms points on a failure diagram from an elastic-plastic fracture analysis into approximate equivalent points on the Failure Assessment Diagram. A method for assessing the severity of a mechanical load superposed on an initial constant load is also presented. The paper concludes that the Failure Assessment Curve will provide a good lower-bound failure criterion for most mechanical loading.

Harrison et al reviewed methods for applying a COD approach to the analysis of welded structures. The COD test is particularly useful in studying fracture toughness of materials in the transition between linear-elastic and fully plastic behavior. Design curves are developed relating a nondimensional COD to applied strain or stress. These curves are useful for (1) selection of materials in design of structures, (2) specification of maximum allowable flaw sizes, and (3) failure analyses. Many examples are cited where the COD design curve has been used for these evaluations on structures designed for real applications. Examples of structures analyzed by COD methods include pipelines, offshore structures, pressure vessels, and nuclear components.

McHenry et al used elastic-plastic fracture mechanics analysis methods to determine size limits for surface flaws in pipeline girthwelds. Four criteria were used: (1) a critical COD method based on a ligament-closure force model, (2) the COD procedure based on the Draft British Standard, (3) a plastic instability method based on critical net ligament strain, and (4) a semi-empirical method based on full-scale pipe rupture tests. The critical flaw sizes determined varied significantly, depending on the fracture criterion chosen, and experimental work will be needed to determine which method most accurately predicts girthweld fracture behavior.

Simpson and Clarke used a crack growth resistance (R-curve) approach based on small fracture mechanics type specimens to determine critical crack lengths in Zr-2.5Nb pressure tubes. The R-curves were based on COD as the mechanical characterizing parameter. Their results showed little specimen size effect on the R-curve shape. R-curves based on a J-integral approach were shown to be consistent with the COD approach. Effects of temperature and hydrogen on the R-curve shape were investigated. Predictions of critical crack lengths in pressure tubes based on an R-curve procedure gave results which were consistent with published burst testing data.

Macdonald used a three-dimensional elastic-plastic fracture model to correlate the fracture strength of two structural steels in the form of beam-column connections. The model was based on the combination of (1) a three-dimensional elastic-plastic finite-element stress analysis, (2) a plastic stress singularity for a crack, and (3) the maximum tensile stress theory of fracture. From these a plastic singularity strength parameter, K_f, was developed. Cracking occurred by mixed mode (crack opening and sliding). Experimental results correlated with K_f showed a relatively small scatterband.

Merkle used approximate elastic-plastic fracture methods to analyze the unstable failure condition for inside nozzle corner cracks in intermediate test vessels. The method was applied to two vessels tested in the heavy section steel technology (HSST) program (Vessels V-9 and V-5). Semi-empirical methods were developed for estimating the nozzle corner pressure-strain curve. Two approximate methods of fracture analysis were used: one used an LEFM approach based on strain which did not consider stable crack growth; the second used a tangent modulus method which incorporated stable crack growth by using a maximum load fracture toughness value. The beneficial effect of transverse contraction was included in the analysis. Calculations of failure strain and fracture toughness agreed well with measured values.

Hammouda and Miller used elastic-plastic finite-element analyses to predict the effect of notch plasticity on the behavior of short cracks under cyclic loading. This analysis was used to predict crack growth behavior in a regime where LEFM methods do not apply. Consideration of the interaction between the crack tip and notch field plasticity can account for fatigue crack growth where a linear elastic analysis would predict that the fatigue threshold stress intensity factor is not exceeded. Crack propagation from a notch initially proceeds at a decreasing rate and in some cases cracks initiate but become nonpropagating.

Brose and Dowling studied the effect of planar specimen size on the fatigue crack growth rate properties of 304 stainless steel on specimen widths of 5.08 and 40.64 cm (2 and 16 in.). The objective was to evaluate size criteria intended to limit crack growth testing to the linear elastic regime and to evaluate the use of a cyclic value of J-integral, ΔJ, for correlation of crack growth rate data on specimens undergoing gross plasticity. The results show that crack growth rates correlated by ΔJ on small specimens having gross plasticity are equivalent to results from large specimens in the linear elastic regime, where the data are correlated by ΔK. No significant size effects were observed.

Mowbray studied fatigue crack growth of chromium-molybdenum-ranadium steel in the high-growth-rate regime where a cyclic J-integral value, ΔJ, was used to correlate growth rate. A compact-type strip specimen was used which gave rise to constant crack growth rate under simple load control at essentially constant ΔJ. These results supported previous results by Dowling and Begley which showed that crack growth rate in the high-growth-rate regime is controlled by ΔJ. An approximate analysis was used to determine ΔJ from cyclic load range for the strip specimen.

The papers in this section consider methods for applying elastic-plastic fracture techniques to the analysis of structures. The prominent technique for using small-specimen results to analyze large structural components is one based on crack opening displacement concepts. The COD was one of the first proposed elastic-plastic fracture parameters and has gained some degree of acceptance as an engineering tool. Other methods for application of elastic-plastic techniques include the Failure Assessment Diagram, R-curve techniques, plastic instability, and the plastic stress singularity. Again, no single method of analysis is generally accepted; many areas for future studies are identified by these papers.

A cyclic value of J-integral, ΔJ, is shown experimentally to correlate fatigue crack growth rate in the high-growth-rate regime. This approach is gaining more acceptance and has promise of becoming a useful tool for analyzing fatigue crack growth under large-scale plasticity.

J. D. Landes
G. A. Clarke
Westinghouse Electric Corp. Research and Development Center, Pittsburgh, Pa.; coeditors.

Index

A

Accoustic emission, 541, 560
Airy's stress function, 201
Antibuckling guides, 131
Area estimation procedure, 271, 276, 286

B

Bauschinger effect, 200
Bend specimens
 Deeply cracked, 38, 45
 Three point, 17, 49, 236, 269, 346, 353
Biaxiality, 215
Blunting line, 393, 489, 544
Body centered cubic, 539
Boundary layer analysis, 186, 216, 219
British Standards Institute, 317, 608, 635
Brittle fracture, 363

C

Center cracked panel, 71, 101, 108, 269, 290
Center cracked strip, 9, 10, 59, 71
Charpy correlation, 490
Charpy energy, 508
Clevage
 Fracture, 365
 Instability, 5, 15, 23, 260, 322
 Rupture, 525

Closure
 Load, 727, 738
 Stress, 12
Compact specimens, 27, 78, 79, 269, 290, 347, 355
Compliance calibration, 252
Complimentary work, 344
Computer interactive testing, 451
Crack driving force, 659
Crack growth
 Initiation, 49
 Simulation, 71
 Stable, 49, 126, 131
 Unstable, 53, 226
Crack opening angle (COA), 71, 88, 98, 115, 116, 124, 203
Crack opening displacement (COD), 88, 118, 195, 316, 328, 386, 608
COD design curve, 309, 623
Crack tip
 Acuity, 370, 465
 Force, 124
 Opening ration, 180
Crack velocity, 715
Creep studies, 305
Criteria
 Failure, 67, 604
 Instability, 13, 27
 Recoverable energy, 128
 Tresca, 20, 691
 Von Mises, 20, 154, 668
Critical
 Crack length, 659
 Crack opening displacement, 634
 Energy release rate, 148
 Thickness, 408

Cyclic J, 725
Cyclic plastic zone, 721

D

Damage function, 231
Deep surface flaw, 13
Deformation theory of plasticity, 43, 61, 80, 94, 112, 115
Double cantilever beams, 14
Double edged cracked strip, 11, 23, 56
Ductile-brittle transition, 332, 373
Ductile fracture, 65, 230
Ductile tearing, 365
Dynamic
  Compact tests, 499
  Fracture toughness, 515, 532, 681
  J-Integral tests, 506
  Resistance curves, 41, 525
  Tear energy, 473
  Yield strength, 530

E

Eddy current, 454
Effective
  Crack size, 289
  Elastic modulas, 291
  Elastic span, 251
Elastic compliance, 427, 444, 562, 741
Elastic-plastic deformation, 6, 40
Elastic shortening, 19
Electrical potential, 336, 415, 559, 648, 661
Elliptical surface flaw, 73, 230
Energy
  Deformation, 380
  Rate definition, 286, 276
  Separation rate, 70, 71
Epoxy model, 694
Equi-biaxial state, 44

Equivalent energy, 379, 386, 403
Equivalent length, 254, 705

F

Face centered cubic, 539
Failure assessment diagram, 582, 597
Failure curve, 586
Fatigue, 704
Fatigue crack growth, 722, 731, 742
Fatigue failure, 716
Finite element
  Constrant strain elements, 76, 125, 155, 165
  Elastic-plastic, 74, 123, 131, 199, 227
  Equations, 153
  Hybrid displacement model, 199
  Isoparametric elements, 76, 165, 216
  Mesh, 80, 97, 157, 165, 246
  Model, 74, 80, 202
  Three dimensional, 664
Finite strain studies, 92
First load drop, 478, 486
Flow theory, 62, 94
  Incremental, 43, 80
  J_2, 68, 70, 80, 113
Fracture parameter, 72, 104, 110

G

G, strain energy release rate, 28, 204, 272, 338
Gaussian integration, 201
Geometry dependance, 359, 654
Girth welds, 626, 633
Grain boundaries, 309

H

Heat tinting, 78, 431, 559
Hydrostatic stress, 166

I

Irradiation damage, 23, 263, 661
Incremental polynomial, 725
Instability, 5, 13, 27, 66, 637
Instrumented Charpy, 495

J

J-controlled crack growth, 38, 42, 43, 113, 186
J-dominance, 177, 186
J-resistance curve, 5, 39, 66, 464, 644

K

K-field, 103
K_{Ic} test, 105
Kirchhoff stress, 178

L

Large-scale yielding, 37
Least squares fit, 504, 750
Limit load, 18, 344
Limit moment, 14, 18, 48
Linear elastic fracture mechanics, 13
Liquified natural gas tanks, 628
Log deviate, 505
Linear variable displacement transducer (LVDT), 131

M

Margin of safety, 66
Martensite transformation, 537, 549
Metallurgical mechanisms, 359
Microstructure, 372
Minimum ligament, 411
Minimum specimen thickness, 421
Mixed mode fracture, 73
Multiple specimen test, 566

N

National Bureau of Standards, 628, 633
Neuber's equation, 689, 697, 705
Nodal force, 171, 197, 218, 248
Nodal release, 155, 168
Nonmetallic inclusions, 309
Nonpropagating cracks, 709
Notch ductility factor, 697
Notch plasticity, 706
Notch round bars, 18, 58
Nozzle corner, 676, 686
Nuclear pressure vessels, 123, 495, 516, 676
Nuclear reactor, 643, 677

O

Offshore oil platform, 623

P

Part through cracks, 610, 695
Path independence, 93
Phase transformation, 546
Photo-elastic analysis, 626
Plane strain, 7, 9, 55, 128
Plane stress, 51, 59, 67
Plastic collapse load, 582, 595
Plastic constraint, 746
Plastic zone
    Cyclic, 721
    Monotonic, 721
    Shape, 168
    Size, 104, 215
Plasticity theory
    Deformation, 43, 61, 80, 94, 112
    Incremental flow, 43, 80
    J_2 flow, 68, 70, 80, 113
    Prandtl-Reuss, 133, 154, 191, 709
Post yield fracture, 582
Prandtl slip line, 167

Prandtl stress, 166
Precracked Charpy specimen, 680
Pressure vessels, 123, 495, 516, 628, 676
Process zone, 70, 103, 118, 127, 138, 210, 435
Proportional loading, 41

Q

Quasi-brittle fracture, 314

R

R-values, 127
Ramberg-Osgood stress strain, 50, 59
Rate sensitive materials, 23
Reactor coolant piping, 553
Reference toughness curve, 553
Residual stress, 592, 619, 629, 640, 653
Resistance curves, 5, 39, 66, 464, 644
Rubber infiltration, 74, 102, 438

S

Secant method, 409, 544
Secant offset, 273
Semi-elliptical surface crack, 236
Separation energy rates, 172, 216, 225
Shear lip, 3, 63, 427, 434
Short crack lengths, 712
Side grooves, 77, 101, 427, 434
Silicone rubber replicas, 438
Single parameter characterization, 67, 358
Single specimen tests, 451
Singularity, 41
Size independence of R-curves, 7
Slip line fields, 10, 22, 106
Slip line solutions, 176
Small-scale yielding, 37
Specimen size effect, 398, 414

Specimen size requirements, 491
Stable crack growth, 49, 126, 131
Steels
  A508, 561
  A516, 471
  A533, 67, 471, 516
  Austenitic stainless, 548, 554
  CrMoV, 740
  Ferritic, 521
  HY103, 471
  NiCrMo, 387
  Rotor forging, 361
Stiffness matrix, 155
Strain energy density, 127
Strain energy release rate, 28, 204, 272, 338
Strain field, 137, 151
Strain hardening, 38, 49, 51, 60, 68, 105, 172, 724
Strain hardening laws,
  Isotropic, 165, 178
  Multilinear, 665
  Power hardening, 176
  Ramberg-Osgood, 207, 294
Stress
  Maximum hoop, 634
  Prandtl field, 107, 166
  Residual, 592, 619, 629, 640, 653
  Secondary, 593
  Thermal, 592
Stress intensity
  Factors, 198, 717
  Magnification factors, 611
Stretch zone, 309, 365, 391
Strip yield model, 31, 309, 609
Submerged arc weld, 512

T

Tangent modulas, 689, 697
Tearing instability, 6, 14, 25, 251
Tearing
  Instability, 6. 14. 25. 251
  Modulas, 8, 24, 118, 255, 574

Resistance, 20
Stable, 6, 365
Temperature dependence of toughness, 373
Tension component in bend specimens, 2, 71
Tension testing, 473
Testing machine stiffness, 28
Through cracks, 236
Triazial
  Stress, 247, 369
  Tension, 176, 372, 375

U

Uncracked body energy, 510
Unloading compliance, 78, 429, 559, 562

V

Variable secant method, 417
Void
  Coalescence, 74, 176, 234
  Growth, 74, 176, 191, 231, 375
  Nucleation, 74, 234

W

Work density, 71

X

X-rays, 627

Y

Yielding
  Large scale, 37
  Small scale 37
  Surface, 191

Z

Zirconium, 644